Transition to College Mathematics, Revised Printing

Transition to College Mathematics, Revised Printing

Franklin D. Demana
Joan R. Leitzel
The Ohio State University

With the assistance of
F. Joe Crosswhite
Alan Osborne
The Ohio State University

ADDISON-WESLEY PUBLISHING COMPANY
Reading, Massachusetts □ Menlo Park, California
New York □ Don Mills, Ontario □ Wokingham, England
Amsterdam □ Bonn □ Sydney □ Singapore □ Tokyo □ Madrid □ San Juan

ISBN 0-201-51523-7

6 7 8 9 10 MA 96959493

Dedicated to the memory of
John W. Riner

Transition to College Mathematics, Revised Printing

Preface

BACKGROUND

This book was developed for students who will need to take college courses in mathematics but who are not adequately prepared for those courses. It grew out of the authors' experience with thousands of such students at The Ohio State University. It was written during a two-year project that developed an alternative course for college-intending high school seniors. This audience consisted of a large group of students, some with limited experience and others with limited success in college preparatory mathematics. The book is equally appropriate for college students who lack the preparation needed for precalculus courses or other beginning mathematics courses. No working skills in algebra are required; however, a commitment by students to do many problems on a daily basis is essential.

APPROACH

The book provides for a highly numerical approach to mathematics. Algebra and geometry are approached through numerical computation in concrete problem settings. The book is particularly intended for students who have been discouraged by, or unsuccessful with, formal, axiomatic approaches to mathematics.

The mathematical ideas of the course are developed through a large collection of problems set in instructional, informative contexts (problem 48, page 88). Problems are frequently approached first arithmetically through numerical computation; later, the same problems are investigated geometrically with graphic representation; and still later, they are represented algebraically by the writing and solving of equations (example 7, page 100; example 1, page 230; example 2, page 266). Students understand the relationship between numerical charts, graphs, and equations and have more than one way to get started on any problem.

Graphing is used to strengthen students' numerical intuition (problem 44, page 95) as well as to provide a problem-solving tool (example 2, page 204). Graphing gives a concrete representation of functional relationships before relationships are dealt with algebraically (example 2, page 308). Throughout the book the approach to algebra remains highly numerical with significant dependence on graphing (example 1, page 526.)

CALCULATORS

Calculators play a key role. They provide insight into arithmetic properties (example 2, page 3) and access to demanding, realistic problems (example 4, page 160). They also enable students to have early experience with the graphing of numerical relationships (example 1, page 190). The calculator used in the development of this course was the TI-30. Other scientific calculators of similar capacity are equally suitable.

NCTM STANDARDS

In 1989 the National Council of Teachers of Mathematics (NCTM) proposed broad changes in the priorities and emphases of school mathematics. This report, entitled "The Curriculum and Evaluation Standards for School Mathematics," was written in response to the same kinds of circumstances that lead to the writing of this text. The text meets all of the standards set by NCTM and fully addresses the areas of emphases described in the 1989 report.

Mathematics as Problem Solving

NCTM points out that problem solving is "much more than applying specific techniques to the solution of classes of word problems." (Standards, page 137). The problems in this text are broader in scope than typical at this level. The availability of calculators permits students to investigate realistic problems and to solve problems in non-algorithmic ways. Frequently problem solutions are represented numerically, geometrically, and algebraically so that students acquire a variety of problem-solving tools.

Mathematics as Communication

The second NCTM recommendation calls for students "to more easily form multiple representations of ideas, express relationships within and among representation systems, and formulate generalizations. In fact, facility with the language of mathematics is an integral part of thinking mathematically, solving problems, and reflecting on one's own mathematical experiences." (Standards, page 140). In this text the relationship of quantities in a problem setting is exhibited through numerical tables, through geometric graphs, and through algebraic expressions. Students understand the relationships between these representation systems. They move from concrete numerical experiences to generalized statements. In addition, the use of calculators requires students to understand precise language, and the explanatory material and the illustrative examples in this text are more extensive than is usual. With guidance, students can learn to read the material with understanding and to acquire the language of mathematics for their own use.

Mathematics as Reasoning

The third NCTM standard states that "the mathematics curriculum should include numerous and varied experiences that reinforce and extend logical reasoning skills." (Standards, page 143). Students using this text make conjectures and test them numerically from the very beginning. Using calculators students gain evidence that suggests generalizations. The numerical reasoning in problem situations that embody functional relationships leads naturally to students' thinking in terms of arguments and exceptions. This experience provides a natural foundation for later, more formal techniques of proof.

Mathematical Connections

The fourth NCTM Standard specifies that the mathematics curriculum should, "include investigations of the connections and interplay among various mathematical topics and their applications. . . ." (Standards, page 146).

The revisitation of problems is a unique element of this text, serving several purposes. Students learn several ways of modeling a problem setting. They learn to apply different representations of the same problem situation and understand the mathematical connections between equivalent representations. The learning of new topics is connected to what is already known and thus learning becomes more stable and permanent. Students consolidate their learning in key application areas, for example, exponential growth.

Algebra and Functions

The fifth and sixth standards for grades 9–12 relate to the study of fundamental algebraic concepts and methods. This text continually provides students with the opportunity to "represent situations that involve variable quantities with expressions, equations, inequalities . . . (and) use tables and graphs as tools to interpret expressions, equations, and inequalities . . ." (Standards, page 150). Throughout the text students investigate problem settings that are based on functional relationships between two quantities. These relationships are demonstrated in numerical tables, later in graphs, and still later in algebraic equations. Within these representations students solve both equalities and inequalities. There is a strong emphasis on algebra as a means of representation.

SCHEDULE

The book has been written for a full-year course. Chapters 1 through 6 can be covered in one semester; Chapters 7 through 11, in a second semester. For schools that are on a quarter system, the material separates naturally into Chapters 1 through 4 for the first quarter, Chapters 5 through 8 for the second quarter, and Chapters 9 through 11 for the third quarter. Teachers with less time may choose to omit later sections of the book. In a rapidly paced version of the course at Ohio State, the text is covered in two quarters, with sections 4.4, 6.1, 7.7, 10.4, 11.2, and 11.5 omitted. More detailed suggestions about pacing and topic selection are given in the Teacher's Guide.

SUPPLEMENTS:

- ANSWER BOOKLET
  Answers to both odd and even exercises for instructors

- TEST BOOKLET
  Diagnostic Test
  Two quizzes per chapter
  Four alternate test forms for each chapter test and final exams with
    answers to all test questions

- INSTRUCTOR'S RESOURCE GUIDE
  Section-by-section objectives and teaching commentaries for each
    section of the text
  Black-line transparency masters for key topics
  Enrichment activities for each chapter

- CALCULATOR SUPPLEMENT WITH GRAPHING CALCULATOR
  ACTIVITIES

- COMPUTERIZED TESTING SYSTEM for MACINTOSH

ACKNOWLEDGMENTS

The authors are indebted to a great many people who participated in the development of this book. Colin B. B. Bull, Dean of the College of Mathematical and Physical Sciences, has provided consistent moral support for the implementation of effective developmental instruction at Ohio State and has worked hard to secure appropriate funding for this instruction and for articulation efforts with the schools of Ohio. Very important contributions to the book were made by Donna Fugate and Robert Mizer, who taught pilot classes in 1981–82 at Whetstone High School and Upper Arlington High School. In 1982–83, teachers in forty Ohio high schools taught the field-test version of this text; the authors wish to acknowledge their significant contributions. Also, we appreciate the constructive criticisms and suggestions of reviewers of our preliminary manuscript: Joseph Elich, Utah State University; Carletta Elich, Logan High School; Bobby Finnell, Portland Community College; and John Spellman, Southwest Texas State University. Special thanks are due to Dodie Shapiro, who skillfully typed the manuscript and coordinated much of the activity surrounding the development of materials, and to Suzanne Mettle, who assisted in the preparation of the index. The staff at Addison-Wesley, especially Patricia Mallion, David Wessel, and Stuart Brewster, have provided enthusiastic support and expert guidance throughout the production process. Finally, the authors wish to express appreciation to their families for their commitment to this project and their patience with its demands.

F.D.D.
J.R.L.

Transition to College Mathematics, Revised Printing

NCTM Objectives Met by this Text

The chart below is designed to provide the teacher with specific examples of each NCTM objective met in the text.

STANDARD 1: MATHEMATICS AS PROBLEM SOLVING

In grades 9–12, the mathematics curriculum should include the refinement and extension of methods of mathematical problem solving so that all students can—

Objective	*Example*
· Use, with increasing confidence, problem-solving approaches to investigate and understand mathematical content	Example 1, page 9 Examples 1 & 2, page 124 Example 1, page 152 Example 4, page 160 Example 2, page 166
· Apply integrated mathematical problem-solving strategies to solve problems from within and outside mathematics	Examples 1 & 2, pages 202–204
· Apply the process of mathematical modeling to real-world problem solutions	Sections 3.7, 3.8, 5.5

STANDARD 2: MATHEMATICS AS COMMUNICATION

In grades 9–12, the mathematics curriculum should include the continued development of language and symbolism to communicate mathematical ideas so that all students can—

Objective	*Example*
· Formulate mathematical definitions and express generalizations discovered through investigations	Section 3.8
· Express mathematical ideas orally and in writing	Example 4, page 254
· Read written presentations of mathematics with understanding	Entire text
· Appreciate the economy, power, and elegance of mathematical notation and its role in the development of mathematical ideas.	Section 3.7

STANDARD 4: MATHEMATICAL CONNECTIONS

In grades 9–12, the mathematics curriculum should include investigation of the connections and interplay among various mathematical topics and their applications so that all students can—

Objective	*Example*
· Recognize equivalent representations of the same concept	Sections 4.4, 5.3
· Relate procedures in one representation to procedures in an equivalent representation	Example 6, page 296

STANDARD 5: ALGEBRA

In grades 9–12, the mathematics curriculum should include the continued study of algebraic concepts and methods so that all students can—

Objective	Example
· Represent situations that involve variable quantities with expressions, equations, inequalities, and matrices	Example 4, page 168 Example 3, page 253
· Use tables and graphs as tools to interpret expressions, equations, and inequalities	Example 1, page 209 Example 2, page 266 Example 6, page 296 Example 2, page 308 Example 3, page 311 Example 1, page 342 Example 1, page 420 Example 3, page 528 Example 3, page 536 Section 10.4
· Operate on expressions and matrices, and solve equations and inequalities	Example 1, page 260 Example 1, page 293 Example 1, page 510 Example 1, page 522 Example 1, page 526 Example 2, page 570
· Demonstrate technical facility with algebraic transformations, including techniques based on the theory of equations	Sections 8.5, 8.6, 8.7

STANDARD 6: FUNCTIONS

In grades 9–12, the mathematics curriculum should include the continued study of functions so that all students can—

Objective	Example
· Model real-world phenomena with a variety of functions	Example 1, page 230 Example 2, page 266

· Represent and analyze relationships   Example 2, page 252
using tables, verbal rules, equations,
and graphs

· Translate among tabular, symbolic, and   Example 1, page 165
graphical representations of functions   Example 3, page 213
                                               Example 3, page 253
                                               Example 1, page 209
                                               Example 3, page 346

· Recognize that a variety of problem   Sections 5.3, 5.4, 5.5
situations can be modeled by the same
type of function

· Analyze the effects of parameter   Example 1, page 353
changes on the graphs of functions

STANDARD 9: TRIGONOMETRY

In grades 9–12, the mathematics curriculum should include the study of
trigonometry so that all students can—

Objective	*Example*
· Apply trigonometry to problem	Example 2, page 650
situations involving triangles	Example 1, page 656

Transition to College Mathematics, Revised Printing

Contents

CHAPTER 3

Using Exponents and Scientific Notation

CHAPTER 4

Using Graphs to Solve Problems

CHAPTER 5

Linear Equations in One Variable 249

CHAPTER 6

Graphing Equations in Two Variables 305

Quadratic Equations and Inequalities 487

CHAPTER 10

Rational Expressions and Fractional Equations 543

CHAPTER 11

Measurement Geometry and Trigonometry   593

Answers to Odd-Numbered Exercises   683

Index   759

Transition to College Mathematics, Revised Printing

Numerical
Mathematics
with a Calculator

1.1 / INSTRUCTING THE CALCULATOR

Calculators permit us to do numerical computation more easily and more accurately than we can do with paper and pencil. However, we must have a good understanding of the way the calculator "thinks" in order to instruct it to do what we want done. Unfortunately, calculators from different manufacturers may not think alike. For example, if we key $\boxed{2}$ $\boxed{\times}$ $\boxed{3}$ $\boxed{+}$ $\boxed{4}$ $\boxed{\times}$ $\boxed{5}$ $\boxed{=}$, some calculators will give the answer 50 and other calculators will give the answer 26. How can this happen?

Calculators that give the answer 50 to the sequence $\boxed{2}$ $\boxed{\times}$ $\boxed{3}$ $\boxed{+}$ $\boxed{4}$ $\boxed{\times}$ $\boxed{5}$ $\boxed{=}$ are performing the operations in the order in which they are received. These calculators first compute 2×3 to get 6, then they add 4 to get 10, then multiply by 5 for the answer 50.

Calculators that give 26 as the answer to $\boxed{2}$ $\boxed{\times}$ $\boxed{3}$ $\boxed{+}$ $\boxed{4}$ $\boxed{\times}$ $\boxed{5}$ $\boxed{=}$ use algebraic logic with **hierarchy**. Going from left to right in a sequence of this type, the calculator computes all pending multiplication and division operations when the $\boxed{+}$ or $\boxed{-}$ key is pressed. Thus the multiplication and division operations are performed before the addition and subtraction operations. In our example, a calculator with hierarchy first computes 2×3 and then 4×5. Finally it computes the sum $6 + 20$ to yield the answer 26.

It is important for you to know how your calculator reacts to the instructions it receives. We will assume in this book that you have a calculator that uses algebraic logic with hierarchy. If you do not, you will need to translate some of our statements for your calculator. Since we are assuming that calculators have hierarchy, an expression like $2 + 3 \times 4$ will have a single meaning, namely, $2 + 12$. We will assume that multiplication and division are done first, left to right, and then addition and subtraction, left to right. We make the same assumption when an expression is given in words rather than symbols. For example, "two more than three times four" means $2 + 12$.

But what if (in the expression $2 + 3 \times 4$) we want to direct the calculator to find the sum $2 + 3$ first and then multiply it by 4? Parentheses are what we need.

Example 1   Find the value of $(2 + 3) \times 4$.

a. The expression $(2 + 3) \times 4$ always means that we should first find the sum of 2 and 3, and then multiply it by 4. We can use this keying sequence for the calculator:

Keying sequence: $(\,\boxed{2}\,\boxed{+}\,\boxed{3}\,)\,\boxed{\times}\,\boxed{4}\,\boxed{=}$

Display: 0   2   2   3   5   5   4   20

Notice that the $\boxed{)}$ key signals the calculator to perform the addition inside the parentheses and the $\boxed{=}$ key signals the calculator to find the product.

b. Another way to key $(2 + 3) \times 4$ is this:

Keying sequence: $\boxed{2}\,\boxed{+}\,\boxed{3}\,\boxed{=}\,\boxed{\times}\,\boxed{4}\,\boxed{=}$

Display: 2   2   3   5   5   4   20

In this sequence the first $\boxed{=}$ key causes the addition to be performed and the second $\boxed{=}$ key causes the multiplication to be performed.   □

Example 2   Find the value of $\dfrac{3 + 5 + 7}{4}$.

a. The line in the fraction $\dfrac{3 + 5 + 7}{4}$ is a division symbol, which is discussed in Chapter 2. We must first add $3 + 5 + 7$ and then divide the sum by 4. If we key $\boxed{3}\,\boxed{+}\,\boxed{5}\,\boxed{+}\,\boxed{7}\,\boxed{\div}\,\boxed{4}\,\boxed{=}$, a calculator with hierarchy will compute $7 \div 4$, or 1.75, and then compute $8 + 1.75$, or 9.75. Given this keying sequence, the calculator does not divide $3 + 5 + 7$ by 4; it only divides 7 by 4. We can correct this by using parentheses and writing $\dfrac{(3 + 5 + 7)}{4}$:

Keying sequence: $(\,\boxed{3}\,\boxed{+}\,\boxed{5}\,\boxed{+}\,\boxed{7}\,)\,\boxed{\div}\,\boxed{4}\,\boxed{=}$

Display: 0   3   3   5   8   7   15   15   4   3.75

Notice that the second $\boxed{+}$ key causes the first addition to be performed in this sequence; the $\boxed{)}$ key completes the sum inside the parentheses.

b. There is another way to evaluate $\dfrac{3 + 5 + 7}{4}$ on a calculator without using parentheses. First we have the calculator compute $\boxed{3}\,\boxed{+}\,\boxed{5}\,\boxed{+}\,\boxed{7}\,\boxed{=}$. Now the number on the display is 15 and we continue with $\boxed{\div}\,\boxed{4}\,\boxed{=}$. The secret is in keying $\boxed{=}$ after keying the symbols in the numerator:

Keying sequence: $\boxed{3}\,\boxed{+}\,\boxed{5}\,\boxed{+}\,\boxed{7}\,\boxed{=}\,\boxed{\div}\,\boxed{4}\,\boxed{=}$

Display: 3   3   5   8   7   15   15   4   3.75   □

Example 3 Find the value of $\dfrac{3 + 5 + 7}{4 + 6 + 8}$.

a. Both of these sequences are correct:

If we were to omit parentheses altogether and key the incorrect sequence

$$\boxed{3}\;\boxed{+}\;\boxed{5}\;\boxed{+}\;\boxed{7}\;\boxed{\div}\;\boxed{4}\;\boxed{+}\;\boxed{6}\;\boxed{+}\;\boxed{8}\;\boxed{=},$$

the calculator would compute $3 + 5 + \frac{7}{4} + 6 + 8$.

b. There is a way to key the computation $\dfrac{3 + 5 + 7}{4 + 6 + 8}$ that enables us
to practice with the STORE and RECALL keys. If there is a number on
the display and the $\boxed{\text{STO}}$ key is pressed, that number is put in memory.
We can then go ahead and do other parts of the computation; when the
number is needed again, keying $\boxed{\text{RCL}}$ will produce it. In the example
$\dfrac{3 + 5 + 7}{4 + 6 + 8}$ we can first compute the denominator and store it, then compute the numerator, and finally recall the denominator to divide into
the numerator. The keying sequence looks like this:

Keying sequence:
$$\boxed{4}\;\boxed{+}\;\boxed{6}\;\boxed{+}\;\boxed{8}\;\boxed{=}\;\boxed{\text{STO}}\;\boxed{3}\;\boxed{+}\;\boxed{5}\;\boxed{+}\;\boxed{7}\;\boxed{=}\;\boxed{\div}\;\boxed{\text{RCL}}\;\boxed{=}$$

Display:

4   4   6   10   8   18   18   3   3   5   8   7   15   15   18   .83333333

□

Sometimes a mathematical phrase is given in words rather than in
symbols. Before attempting to evaluate the phrase using a calculator, it
is helpful to rewrite it using mathematical symbols.

Example 4 Translate these English expressions into mathematical phrases.

Three times four increased by five
Ten divided by the sum of two and three

"Three times four increased by five" can be written as the mathematical
phrase $3 \times 4 + 5$. The value of the expression is $12 + 5$, or 17. "Ten
divided by the sum of two and three" can be written symbolically as

$10 \div (2 + 3)$. This phrase can also be read as "the sum of two and three *divided into* ten."
□

You will be asked to solve two kinds of problems using a calculator. The first is to find a keying sequence that will evaluate a given mathematical phrase. The second is to find a mathematical phrase that is evaluated by a given keying sequence. The following table summarizes the information from the examples in this section and should help you to translate calculator keying sequences into mathematical phrases.

Calculator keying sequence	*Mathematical phrase*
$\boxed{7}\boxed{\times}\boxed{3}\boxed{-}\boxed{2}\boxed{\times}\boxed{5}\boxed{=}$	$(7 \times 3) - (2 \times 5)$
$\boxed{4}\boxed{+}\boxed{6}\boxed{\div}\boxed{3}\boxed{=}$	$4 + \dfrac{6}{3}$
$\boxed{4}\boxed{+}\boxed{6}\boxed{=}\boxed{\div}\boxed{3}\boxed{=}$	$\dfrac{4+6}{3}$
$\boxed{(}\boxed{6}\boxed{+}\boxed{8}\boxed{)}\boxed{\div}\boxed{(}\boxed{5}\boxed{+}\boxed{2}\boxed{)}\boxed{=}$	$\dfrac{6+8}{5+2}$
$\boxed{5}\boxed{+}\boxed{2}\boxed{=}\boxed{\text{STO}}\boxed{6}\boxed{+}\boxed{8}\boxed{=}\boxed{\div}\boxed{\text{RCL}}\boxed{=}$	$\dfrac{6+8}{5+2}$

Exercises 1.1

Use your calculator to evaluate each expression in Exercises 1–10. Try using the STORE and RECALL keys for Problems 5, 6, and 7.

1. $4 + 5 \times 6$

2. $(4 + 5) \times 6$

3. $5 \times (3 \times 4 + 2)$

4. $16 + 4 \times [8 + 3 \times (7 + 11)]$

5. $\dfrac{17.2 + 18.3}{9.5 + 12.3}$

6. $\dfrac{32.3 + 18.9}{13.4 - 4.7}$

7. $\dfrac{12.3 + 25.4 + 17.3}{3.4 + 5.7}$

8. $8 + 4 \times 5 + 6$

9. $9 \times 3 + 7 \times 4$

10. $(12.2 + 15.5) \times (13.3 + 18.3)$

Write a sequence of calculator key strokes that will evaluate each expression in Exercises 11–14. Then evaluate the expression.

11. $\dfrac{16.4}{4.2}$

12. $\dfrac{(4.56) \times (3.45)}{2.31}$

13. $\dfrac{43.2 + 35.7 + 16.8}{3.2}$

14. $4.2 + 5 \times (17.5 + 12.8)$

Write the mathematical phrase evaluated by each sequence of calculator key strokes in Exercises 15–20. Use parentheses to show how the numbers are grouped.

15. $\boxed{3}\boxed{\times}\boxed{5}\boxed{+}\boxed{4}\boxed{\times}\boxed{6}\boxed{=}$

16. $\boxed{3}\boxed{+}\boxed{5}\boxed{\times}\boxed{4}\boxed{+}\boxed{6}\boxed{=}$

17. $\boxed{3}\boxed{\times}\boxed{5}\boxed{\div}\boxed{6}\boxed{=}$

18. $\boxed{3}\boxed{\div}\boxed{6}\boxed{\times}\boxed{5}\boxed{=}$

19. $\boxed{3}\boxed{\div}\boxed{6}\boxed{\div}\boxed{5}\boxed{=}$

20. $\boxed{3}\boxed{\times}\boxed{(}\boxed{4}\boxed{+}\boxed{5}\boxed{)}\boxed{=}$

Translate each of the English expressions in Exercises 21–27 into a mathematical phrase.

21. Five times four increased by three

22. Six times five decreased by two

23. Seven more than three times eight

24. Three less than four times ten

25. Eleven divided into the sum of eight and seven

26. Five times the sum of three and four

27. Twelve divided by the sum of six and eight

28. The second of two numbers is 15 more than the first.

(a) Complete the following table:

First number	Second number
6	21
10	
14	
	27

(b) Explain in words how to find the second number if the first number is known.

(c) Explain in words how to find the first number if the second number is known.

29. The second of two numbers is three more than twice the first number.

    (a) Complete the following table:

First number	Second number
8	19
11	
14	
	29

    (b) Explain in words how to find the second number if the first number is known.

    (c) Explain in words how to find the first number if the second number is known.

30. The length of a rectangle is 2 feet less than three times the width.

    (a) Complete the following table:

Width	Length
10	28
15	
20	
	46

    (b) Explain in words how to find the length if the width is known.

    (c) Explain in words how to find the width if the length is known.

31. The length of a rectangle is 3 feet less than twice the width.

    (a) Complete the following table:

Width	Length
15	27
20	
25	
	45

(b) Explain in words how to find the length if the width is known.

(c) Explain in words how to find the width if the length is known.

1.2 / USING THE K KEY

There is one key on the calculator that is particularly helpful if an operation needs to be performed repeatedly. For example, say we have a list of numbers and we want to add 9 to each one. We can key this sequence:

	(Display)
9 + K 5 =	14
8 =	17
9 =	18

The effect of keying 9 + K is to prepare the calculator to add 9 to any subsequent number given to it. Try the sequence above and extend the list.

The K key is called the **constant** ("konstant") **key** on the calculator. It locks in a number and an operation for repeated use. We can use the K key with each of the four operations—addition, subtraction, multiplication, and division.

Consider multiplication using the K key. (When we want to denote the keying of a number that has more than one digit, say the number 374, we will write 374 rather than 3 7 4 .) If you key the following sequence, you will get the display shown.

	(Display)
9 × K 3 =	27
11 =	99
102 =	918

The effect of keying K after 9 × is to prepare the calculator to multiply each subsequent entry by 9.

Subtraction and division with the K key may be the opposite of what you expect. Study these examples.

	(Display)			(Display)
$\boxed{9}\ \boxed{-}\ \boxed{K}\ \underline{10}\ \boxed{=}$	1		$\boxed{9}\ \boxed{\div}\ \boxed{K}\ \underline{18}\ \boxed{=}$	2
$\underline{12}\ \boxed{=}$	3		$\underline{45}\ \boxed{=}$	5
$\underline{35}\ \boxed{=}$	26		$\underline{81}\ \boxed{=}$	9

The effect of the first sequence is to prepare the calculator to subtract 9 from any number. Notice that by using the $\boxed{K}$ key we compute $10 - 9$ (not $9 - 10$) and $12 - 9$ (not $9 - 12$). Similarly, the effect of the second sequence is to prepare the calculator to divide any number by 9. When we key $\boxed{9}\ \boxed{\div}\ \boxed{K}\ 18\ \boxed{=}$ we compute $18 \div 9$; then when we key $\underline{45}\ \boxed{=}$, we compute $45 \div 9$.

Two examples follow in which the $\boxed{K}$ key is helpful.

Example 1   A car travels at an average speed of 55 mph. Compute the distance it travels in each of the following cases:

Hours spent traveling	0	1	2	3.5	6	8	10	24
Miles traveled								

Since the car goes 55 miles in 1 hour, we must multiply 55 by the number of hours traveled in each case. We prepare the calculator by keying $\underline{55}\ \boxed{\times}\ \boxed{K}$ and then enter

$\boxed{0}\ \boxed{=}$

$\boxed{1}\ \boxed{=}$

$\boxed{2}\ \boxed{=}$

and so on down the row until we have a distance for each time. The completed table looks like this:

Hours spent traveling	0	1	2	3.5	6	8	10	24
Miles traveled	0	55	110	192.5	330	440	550	1320

□

Example 2   A new car is averaging 30 miles per gallon of gasoline. Compute the amount of gasoline needed to travel each of the following distances.

Miles traveled	60	90	135	165	210
Gallons of gas used					

Since the car can travel 30 miles on one gallon of gas, we must determine how many 30s are in each of the numbers in the first row; thus we must divide each of these numbers by 30. The keying sequence 30 $\div$ K prepares the calculator to do this; then we enter 60 $=$, 90 $=$, 135 $=$, and so on. The completed table looks like this:

Miles traveled	60	90	135	165	210
Gallons of gas used	2	3	4.5	5.5	7

□

Repeated Addition

We can use the K key to add a sum when the terms are all the same number. Say we wish to compute $9 + 9 + 9 + 9 + 9$. We can key 9 $+$ K. Since the number 9 is already on the display, we do not need to key it again. Check this keying sequence on your own calculator.

	(Display)
9 $+$ K $=$	18
$=$	27
$=$	36
$=$	45

We can continue to key $=$, each time adding 9 to the previous sum. Thus we get a sum of 9s. To get $9 + 9 + 9 + 9 + 9$ (or 5×9), we will need to key $=$ 4 times. In general, there will be one more 9 in the sum than the number of times we push $=$. Can you see why?

Exercises 1.2

Use the constant key K on your calculator to complete the tables in Exercises 1–4.

1.

+	0	2	3	5	8	12	16
8							

2.

+	0	1	3	5	8	9	11
9							

3.

×	0	3	5	7	10	14	20
5							

4.

×	0	2	3	5	8	11	12
11							

Write the mathematical phrase evaluated by each sequence of calculator key strokes in Exercises 5–9.

5. $\boxed{4}\ \boxed{+}\ \boxed{K}\ \boxed{=}\ \boxed{=}$

6. $\boxed{4}\ \boxed{+}\ \boxed{K}\ \boxed{=}\ \boxed{=}\ \boxed{=}\ \boxed{=}\ \boxed{=}$

7. $\boxed{2}\ \boxed{+}\ \boxed{K}\ \boxed{=}\ \boxed{=}\ \boxed{=}\ \boxed{=}\ \boxed{=}\ \boxed{=}$

8. $\boxed{2}\ \boxed{+}\ \boxed{K}\ \boxed{=}\ \boxed{+}\ \boxed{K}\ \boxed{=}\ \boxed{=}\ \boxed{=}$

9. $\boxed{2}\ \boxed{+}\ \boxed{K}\ \boxed{=}\ \boxed{=}\ \boxed{+}\ \boxed{K}\ \boxed{=}\ \boxed{=}\ \boxed{=}$

Complete the tables in Exercises 10 and 11.

10. A car travels at 65 mph.

Hours traveled	1	2	3	8		12	15	h
Miles traveled					650			

11. A car averages 22 miles per gallon of gasoline.

Gallons used	1	2	5	10		18	25	n
Miles traveled					264			

12. The sales tax is 4 cents on every dollar. Compute the sales tax on the following prices:

(a) $10   (b) $12.50   (c) $37.75   (d) $122.25   (e) $N

13. The bank pays 6 cents on every dollar left on deposit for one year. How much money will you have at the end of one year if today you deposit the following amounts:

(a) $100   (b) $1000   (c) $3500   (d) $4275   (e) $N

14. Each price is reduced by $13. Complete the following table:

Old price in dollars	47	35	29	76		60	121	P
New price in dollars					53			

15. Tennis balls cost $.95 each in the Sporting Goods catalog. You must also pay $.50 handling charge on any order. Find the total cost of ordering

   (a) 6 balls   (b) 10 balls   (c) 25 balls   (d) 100 balls
   (e) n balls

16. If a plane can fly at the rate of 8 miles per minute, how far will it fly in each of these times?

   (a) 20 minutes   (b) 45 minutes   (c) x minutes   (d) 1 hour
   (e) 12 hours

17. If a plane is flying at the rate of 400 miles per hour, how many miles will it fly in each of these times?

   (a) 11 hours   (b) one-half day   (c) 15 hours   (d) 18 hours
   (e) b hours

18. If sound travels 1100 feet per second, how many feet will it travel in

   (a) 30 seconds   (b) 2 minutes   (c) r minutes   (d) 1 hour
   (e) 1 day

19. A 6-ounce tube of toothpaste is priced at 89 cents. What is the price per ounce?

20. A 32-ounce bottle of mouthwash is advertised for $2.00. What is the price per ounce?

21. A car traveled 150.8 miles and used 23.2 gallons of gasoline. How many miles per gallon did the car average on the trip?

22. Auto expenses for driving for a charity may be deducted from income tax at the rate of 7 cents per mile. If you drove 1536 miles last year for charity, how much can you deduct for your driving expenses?

23. Light travels approximately 186,284 miles per second.

   (a) How far will light travel in a year of 365 days? (The answer is a unit of distance called a **light year**.)

(b) The nearest star outside our solar system is Proxima Centauri. It is 4.3 light years away. How many miles is that?

24. You have 30 coins in nickels and dimes.

(a) Complete the following table:

Number of nickels	Number of dimes	Value
5	25	$2.75
15		
25		
	0	
		$1.90

(b) Explain in words how to find the number of dimes if the number of nickels is known.

(c) Explain in words how to find the value of the coins if the number of nickels and dimes is known.

25. You have 40 coins in dimes and quarters.

(a) Complete the following table:

Number of dimes	Number of quarters	Value
0	40	$10.00
10		
20		
	10	
40		
		$ 4.75

(b) Explain in words how to find the number of dimes if the number of quarters is known.

(c) Explain in words how to find the value of the coins if the number of dimes and quarters is known.

26. The second of two numbers is one more than the square of the first.
   (a) Complete the following table:

First number	Second number
2	5
3	
5	
8	
	37

   (b) Explain in words how to find the second number if the first number is known.
   (c) Explain in words how to find the first number if the second number is known.

27. The second of two numbers is one less than twice the square of the first.
   (a) Complete the following table:

First number	Second number
2	7
4	
8	
12	
	71

   (b) Write a sentence that explains how to find the second number if the first number is known.
   (c) Write a sentence that explains how to find the first number if the second number is known.

28. (a) Complete the following table for a rectangle:

Width	Length	Area	Perimeter
3	4	12	14
2	5		
12	26		

Width	Length	Area	Perimeter
25	25		
6		30	
	8		24

(b) Write a sentence that explains how to find the perimeter if the width and length are known.

(c) Write a sentence that explains how to find the length if the width and area are known.

(d) Write a sentence that explains how to find the length if the width and perimeter are known.

1.3 / REPEATED MULTIPLICATION AND EXPONENTS

If we use multiplication and the $\boxed{\text{K}}$ key with a single number entry, we can multiply a number times itself as many times as we wish. Say we want to compute $9 \times 9 \times 9 \times 9$. We could use this keying sequence:

	(Display)
$\boxed{9}\ \boxed{\times}\ \boxed{\text{K}}\ \boxed{=}$	81
$\boxed{=}$	729
$\boxed{=}$	6561

We see a product of 9s each time we key $\boxed{=}$. To compute $9 \times 9 \times 9 \times 9$, we push the $\boxed{=}$ key three times. (Why not four times?)

We often abbreviate $9 \times 9 \times 9 \times 9$ as 9^4, which we read as 9 to the exponent 4, or 9 to the fourth power, or 9 to the fourth. In the expression 9^4, the number 4 is called the **exponent** and the number 9 is called the **base**. The exponent shows how often the base appears in the product. Although we read 9^4 as 9 to the fourth power and 9^5 as 9 to the fifth power, we often read 9^2 as 9 squared because 9×9 is the area of a square with side length 9. We read 9^3 as 9 cubed, because $9 \times 9 \times 9$ is the volume of a cube with edge length 9.

We have seen that an expression like 9^4 can be evaluated on the calculator by using multiplication and the $\boxed{K}$ key. Your calculator also has an exponent key $\boxed{y^x}$. A quick way to compute 9^4 is to key $\boxed{9}\,\boxed{y^x}\,\boxed{4}\,\boxed{=}$. The display should show 6561. Use your own calculator to check the following computations.

Mathematical phrase	Keying sequence	Display
3^5, or $3 \times 3 \times 3 \times 3 \times 3$	$\boxed{3}\,\boxed{y^x}\,\boxed{5}\,\boxed{=}$	243
5^3, or $5 \times 5 \times 5$	$\boxed{5}\,\boxed{y^x}\,\boxed{3}\,\boxed{=}$	125
2^4, or $2 \times 2 \times 2 \times 2$	$\boxed{2}\,\boxed{y^x}\,\boxed{4}\,\boxed{=}$	16
4^2, or 4×4	$\boxed{4}\,\boxed{y^x}\,\boxed{2}\,\boxed{=}$	16

(Can you explain why $2^4 = 4^2$ but $3^5 \neq 5^3$?)

Although 4^2 can be computed using the $\boxed{y^x}$ key, there is also a special key for raising a number to the power of 2. The key for squaring is $\boxed{x^2}$. All we need to do is key $\boxed{4}\,\boxed{x^2}$.

Mathematical phrase	Keying sequence	Display
3^2	$\boxed{3}\,\boxed{x^2}$	9
5^2	$\boxed{5}\,\boxed{x^2}$	25
10^2	$\boxed{10}\,\boxed{x^2}$	100
$(.5)^2$	$\boxed{.5}\,\boxed{x^2}$	.25

In the calculator hierarchy the $\boxed{x^2}$ and $\boxed{y^x}$ keys take precedence over multiplication and division. Thus in the absence of parentheses, raising a number to a power within a phrase is performed before multiplication and division. (Recall that multiplication and division are performed before addition and subtraction.)

Example 1   Give the mathematical phrase evaluated by the keying sequence

$$\boxed{5}\,\boxed{\times}\,\boxed{3}\,\boxed{x^2}\,\boxed{+}\,\boxed{4}\,\boxed{=}.$$

Observe the display on the calculator when this keying sequence is pressed:

Keying sequence: $\boxed{5}\,\boxed{\times}\,\boxed{3}\,\boxed{x^2}\,\boxed{+}\,\boxed{4}\,\boxed{=}$

Display: 5   5   3   9   45   4   49

The $\boxed{x^2}$ key squares the number 3; it does not square 5×3. The $\boxed{+}$ key causes the calculator to multiply 5×3^2. The $\boxed{=}$ key then adds 4 to 5×3^2. Thus the keying sequence evaluates the phrase

$$5 \times 3^2 + 4.$$

□

A Word about Variables

There are times when we want to make a statement about a collection of numbers rather than about a particular number. Here are some examples.

Example 1 *(Exercises 1.2, Problem 10)*

If a car is traveling 65 mph, how many miles will the car travel in h hours?

Computing the distance a car travels, if you know its average speed, is always done the same way. You multiply the speed by the number of hours. The expression $65 \times h$ means 65 times the number of hours. The letter h stands for a whole collection of numbers: 1, 2, $2\frac{1}{2}$, 5, for examples. Presumably h can stand for 0 and any number greater than 0.

□

Example 2 *(Exercises 1.2, Problem 15)*

Find the cost of n tennis balls ordered through the Sporting Goods catalog if each ball costs \$.95 and there is \$.50 handling charge on any order.

In this catalog, each tennis ball costs \$.95. There is always a \$.50 handling charge. Thus to compute the total cost of an order, you need to multiply \$.95 by the number of balls purchased and then add on the service charge, \$.50. That is what the expression $(.95 \times n) + .50$ represents. Here the letter n can stand for any of the whole numbers: 1, 2, 3, 4, ... It does not stand for a fraction such as $\frac{1}{5}$ because, presumably, we must buy whole tennis balls.

□

In Examples 2 and 3 the letters h and n are called **variables**. They represent numbers; in our examples, each represents many numbers.

Sometimes problems state what numbers a variable may represent (or, we might say, what values a variable may assume). If the problem does not state otherwise, you may expect that the variable represents all numbers that are sensible in the expression.

There are a few calculator models that have keys for variables, but we will assume you do not have one of those. We will use the calculator only for numerical computation.

Because we frequently use the letter x for a variable, we will stop using it to denote multiplication. (Surely writing x times x as $x \times x$ is confusing!) Henceforth, either we will denote multiplication with the symbol $\cdot$ or we will use no symbol between the numbers or between the numbers and the variables. Here are some examples:

$$7 \cdot (4 + 8) \quad \text{means} \quad 7 \times (4 + 8),$$

$$7(4 + 8) \quad \text{means} \quad 7 \times (4 + 8),$$

$$8n^2 \quad \text{means} \quad 8 \times n \times n.$$

Exercises 1.3

Identify the base and exponent in each of Exercises 1–5.

1. $(-2)^3$ 2. $(-3)^2$ 3. 4^5 4. 3^{-4} 5. $(-3)^{-2}$

Use your calculator to evaluate each expression in Exercises 6–11.

6. $2 \cdot 3^2 + 5^2 - 3 \cdot 4^2$

7. $4^3 - 3^2$

8. $\dfrac{(4.2)^2 + (3.5)^3}{(2.1)^2 - (1.1)^3}$

9. $\dfrac{(7.1)^4 + (2.4)^3}{(3.1)^2}$

10. $\dfrac{(7.1 + 4.6)^2}{(7.1)^2 + (4.6)^2}$

11. $3[(6.7 - 2.3)^2 + (8.2 + 3.5)^2]$

Write a sequence of calculator key strokes that will evaluate each expression in Exercises 12–17. Then evaluate the expressions.

12. $(3.2)^2 + (4.5)^2$

13. $(3.2 + 4.5)^2$

14. $2 \cdot 3^2 + 12$

15. $(2 \cdot 3)^2 + 12$

16. $4^3 + 3^4$

17. $3 \cdot 2^4 + 4^3$

Translate each of the English expressions in Exercises 18–25 into a mathematical phrase.

18. The sum of three squared and five squared

19. The sum of four squared and six squared

20. The sum of three and five squared

21. The sum of four and six squared

22. The sum of two cubed and three to the fourth power

23. Three cubed minus four squared

24. Five squared minus three squared

25. The sum of two squared, three squared, and five squared

Write the mathematical phrase evaluated by each sequence of calculator key strokes in Exercises 26–35.

26. $\boxed{2}\ \boxed{x^2}\ \boxed{+}\ \boxed{3}\ \boxed{=}$ 27. $\boxed{3}\ \boxed{x^2}\ \boxed{+}\ \boxed{2}\ \boxed{x^2}\ \boxed{=}$

28. $\boxed{2}\ \boxed{+}\ \boxed{3}\ \boxed{=}\ \boxed{x^2}$ 29. $\boxed{3}\ \boxed{x^2}\ \boxed{x^2}$

30. $\boxed{2}\ \boxed{\times}\ \boxed{K}\ \boxed{=}\ \boxed{=}$ 31. $\boxed{2}\ \boxed{\times}\ \boxed{K}\ \boxed{=}\ \boxed{=}\ \boxed{=}\ \boxed{=}\ \boxed{=}$

32. $\boxed{2}\ \boxed{\times}\ \boxed{K}\ \boxed{=}\ \boxed{\times}\ \boxed{K}\ \boxed{=}\ \boxed{=}$ 33. $\boxed{3}\ \boxed{y^x}\ \boxed{2}\ \boxed{=}$

34. $\boxed{3}\ \boxed{y^x}\ \boxed{2}\ \boxed{y^x}\ \boxed{2}\ \boxed{=}$ 35. $\boxed{2}\ \boxed{x^2}\ \boxed{+}\ \boxed{3}\ \boxed{y^x}\ \boxed{4}\ \boxed{=}$

36. Find the area of a square with each side of length 8 inches.

37. Find the area of a square that has a perimeter of 40 inches.

38. Find the length of the side of a square whose area is 400 square inches.

39. Find the volume of a cube that has sides of length 6 inches.

40. Find the length of a side of a cube whose volume is 64 cubic inches.

1.4 / DISTRIBUTIVE PROPERTY

In the first section we discussed the need for parentheses in an expression such as $4(5 + 6)$, which means 4 times the sum of 5 and 6. In this section we want to look more closely at the simple matter of multiplying a number times a sum.

Computation in our number system is guided by several properties of the four operations that are defined on the numbers. Most of these properties become so natural as we learn to compute that we tend to use them without realizing it. One, however, is important enough that we want to emphasize it and give it a name: **the distributive property of multiplication over addition** (or simply, **the distributive property**).

Example 1   Assume that every morning on his way to work, Sam Johnson stops at Krispy Kreme for a doughnut and a cup of coffee. The doughnut always costs $.25 and the coffee, $.20. What does Sam spend on breakfast each week?

There are two different ways to analyze this situation. We might say that each morning Sam spends .25 + .20, so in 5 mornings he spends 5(.25 + .20). Or, we might say that in a week Sam buys 5 doughnuts for (5)(.25) and he also buys 5 coffees for (5)(.20), making the total cost (5)(.25) + (5)(.20). Both solutions are correct. The fact that 5(.25 + .20) = (5)(.25) + (5)(.20) is an example of the distributive property.   □

Example 2   Write expressions that give the area of the rectangle shown in Fig. 1.1.

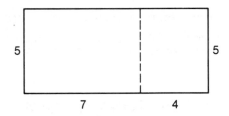

Figure 1.1

When we want to compute the area of a rectangle, we multiply its length times its width. Figure 1.1 has been divided into two smaller rectangles. The large rectangle has area $5 \cdot (7 + 4)$. The smaller rectangles have areas $5 \cdot 7$ and $5 \cdot 4$. The area of the large rectangle equals the sum of the areas of the smaller rectangles. Thus $5 \cdot (7 + 4) = (5 \cdot 7) + (5 \cdot 4)$. This fact is another example of the distributive property.
     □

The distributive property tells us about multiplying a number times a sum. Using the labels in Fig. 1.2, the distributive property says

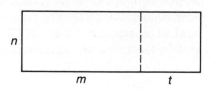

Figure 1.2

that if n, m, and t are any numbers, then $n \cdot (m + t) = (n \cdot m) + (n \cdot t)$. Sometimes we read this statement from the right to the left:

$(n \cdot m) + (n \cdot t) = n \cdot (m + t)$.

Thus we might write $(7 \cdot 6) + (7 \cdot 4) = 7 \cdot (6 + 4)$. Since the order of terms in a product does not change the product, we may also write the distributive property as $(m + t) \cdot n = (m \cdot n) + (t \cdot n)$. Study the following statements to see that they are all examples of the distributive property:

$7 \cdot (100 + 1) = (7 \cdot 100) + (7 \cdot 1)$

$(11 + 19) \cdot 2 = (11 \cdot 2) + (19 \cdot 2)$

$(30 \cdot 5) + (2 \cdot 5) = (30 + 2) \cdot 5$.

Sometimes we may wish to multiply a number times a sum of more than two numbers, for example, $7 \cdot (40 + 10 + 5)$. An analogous statement can be made in this case: $n \cdot (m + t + r) = (n \cdot m) + (n \cdot t) + (n \cdot r)$.

Distributive Property of Multiplication Over Subtraction

In our number system, multiplication distributes over subtraction as well as addition. In other words, for any numbers n, m, and t,

$n \cdot (m - t) = (n \cdot m) - (n \cdot t)$.

To visualize this equation in terms of areas of rectangles, assume that we have a large rectangle that is n units wide and m units long. The rectangle is divided into two smaller rectangles as shown in Fig. 1.3, one of which has length t. Thus the length of the other small shaded rectangle is $m - t$ and its area is our expression $n \cdot (m - t)$. The equation $n \cdot (m - t) = (n \cdot m) - (n \cdot t)$ thus says that the area of the shaded rectangle equals the area of the large rectangle minus the area of the unshaded small rectangle.

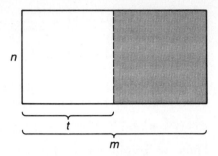

Figure 1.3

Example 3   Compute the cost of 4 bags of 79-cent potato chips by using the distributive property of multiplication over subtraction.

The distributive property can help us in computing products mentally. If a shopper is buying 4 bags of 79-cent potato chips, she may compute the cost in her head by regarding 79 as $80 - 1$. Since $4(.80 - .01) = 4(.80) - 4(.01)$ she can compute $4(.80) = \$3.20$ and subtract .04 to get \$3.16.   □

Computing with Variables

Because variables represent numbers, they are used in computation as though they were numbers. The distributive property is particularly important because it provides statements such as

$$2x + 3x = (2 + 3)x = 5x$$
$$6b^2 + b^2 = (6 + 1)b^2 = 7b^2$$
$$21t^3 - 10t^3 = (21 - 10)t^3 = 11t^3.$$

We are not surprised that

$$2x + 3x = 5x$$
$$6b^2 + b^2 = 7b^2$$
$$21t^3 - 10t^3 = 11t^3.$$

It is the distributive property that makes these statements true.

The terms $2x$ and $3x$ are called **like terms**. They are both multiples of the same variable x and can be combined under addition and subtraction. Similarly, $6b^2$ and b^2 are like terms; so are $21t^3$ and $10t^3$. The terms $2x$ and $3x^2$ are *not* like terms. They cannot be combined using the distributive property. The expressions $2x + 3x^2$ and $2x - 3x^2$ cannot be simplified according to the patterns of our other examples because $2x$ and $3x^2$ are not like terms.

Example 4 Use the distributive property and combine like terms to simplify
$$x + 4(x + 3).$$

The computation can be displayed this way:

$$
\begin{aligned}
x + 4(x + 3) &= 1x + 4x + 4(3) &&\text{(Distributive property)} \\
&= (1 + 4)x + 4(3) \\
&= 5x + 12 &&\text{(Replacing } x + 4x \text{ with } 5x)
\end{aligned}
$$ □

Example 5 Use the distributive property and combine like terms to simplify
$$2x^2 - .05x^2 + x^4 - .06x^4.$$

The distributive property permits us to combine like terms:

$$
\begin{aligned}
2x^2 - .05x^2 + x^4 - .06x^4 &= (2 - .05)x^2 + (1 - .06)x^4 &&\text{(Distributive} \\
&= 1.95x^2 + .94x^4. &&\text{property)}
\end{aligned}
$$ □

Exercises 1.4

Use your calculator to evaluate each expression in Exercises 1–6.

1. $(12.3)(17.1) + (12.3)(18.6)$

2. $(17.6)(18.4) - (17.6)(12.8)$

3. $7(4) + 7(5) + 7(6) + 7(7) + 7(8)$

4. $5(6) + 5(7) - 5(4) - 5(3) + 5(8)$

5. $3[5 + 6(7 - 3) + 7]$

6. $12\{7.5 + 6.2[7 + 4.5(12.1 + 13.3)]\}$

Use the distributive property to write two sequences of calculator key strokes that will evaluate each expression in Exercises 7 and 8.

7. $12(13 + 15)$

8. $16(19) + 16(17) - 16(13)$

Translate each of the English expressions in Exercises 9–12 into a mathematical phrase.

9. Eight times the sum of 3 and 4

10. The sum of 7 times x and 7 times y

11. Six times the difference, 5 minus 3

12. Four times y subtracted from 4 times x

Write the mathematical phrase evaluated by each sequence of calculator key strokes in Exercises 13–16.

13. $\boxed{6}\boxed{\times}\boxed{(}\boxed{4}\boxed{+}\boxed{5}\boxed{)}\boxed{=}$

14. $\boxed{5}\boxed{\times}\boxed{7}\boxed{-}\boxed{5}\boxed{\times}\boxed{2}\boxed{=}$

15. $\boxed{5}\boxed{\times}\boxed{(}\boxed{7}\boxed{+}\boxed{3}\boxed{\times}\boxed{(}\boxed{8}\boxed{-}\boxed{4}\boxed{)}\boxed{)}\boxed{=}$

16. $\boxed{6}\boxed{\times}\boxed{(}\boxed{9}\boxed{+}\boxed{4}\boxed{\times}\boxed{(}\boxed{7}\boxed{-}\boxed{2}\boxed{)}\boxed{+}\boxed{5}\boxed{)}\boxed{=}$

Show how the distributive property can be used to justify the statements in Exercises 17–20.

17. $3a + 2a = 5a$

18. $b^2 + 0.25b^2 = 1.25b^2$

19. $y^3 - 0.35y^3 = 0.65y^3$

20. $13y^4 + 14y^4 - 17y^4 = 10y^4$

Use the distributive property and combine like terms to simplify the expressions in Exercises 21–28.

21. $c + 0.3c + b - 0.7b$

22. $d^2 - 0.45d^2$

23. $0.25b^4 + 0.35b^4 - 0.15b^4 + b^2$

24. $25x + 5(30 - x)$

25. $.06x + .05(18{,}000 - x)$

26. $.05x + .10(2x + 1) + .25(59 - 3x)$

27. $x(2x - 1) + 2x(3x + 1)$

28. $x(3x + 4) + 4(4x - 1) + 5$

29. Use the distributive property to write two different sequences of calculator key strokes to evaluate the expression in Problem 3. Count the number of key strokes in each case.

30. Compute the cost of 7 loaves of 78-cent bread by using the distributive property of multiplication over subtraction.

31. Compute the cost of 40 gallons of 99-cent gasoline by using the distributive property of multiplication over subtraction.

32. Compute the cost of 30 pounds of $2.02 steak by using the distributive property of multiplication over addition.

33. (a) Sally earns $8 per hour. She works 6 hours on Monday and 5 hours on Tuesday. Use the distributive property to compute her wages in two ways.

    (b) Sally receives a 50-cent per hour raise. Compute her new wages in two ways.

34. Bill's car averages 20 miles per gallon of gasoline. He uses 6 gallons of gasoline on Saturday and 8 gallons on Sunday. Use the distributive property to compute the number of miles traveled on Saturday and Sunday in two ways.

35. A car averages 52 mph. The car travels 6 hours on Friday, 7 hours on Saturday, and 8 hours on Sunday. Use the distributive property to compute in two different ways the total number of miles traveled.

36. Candy bars sell for 30 cents each. Jane bought 40 candy bars one day, 65 candy bars another day, and 75 candy bars on a third day. Use the distributive property to compute, in two different ways, the amount of money Jane spent on candy bars.

1.5 / FACTORING WHOLE NUMBERS INTO PRIMES

When we write a number as a product, we say that we have *factored* the number. For example, $12 = 2 \cdot 6$, or $12 = 3 \cdot 4$, or $12 = 1 \cdot 12$. The numbers 1, 2, 3, 4, 6 and 12 are called **factors (or divisors)** of 12. Although 12 can be factored several ways, some numbers can be factored in essentially one way; the number 5 can only be written as $5 = 1 \cdot 5$ (or $5 \cdot 1$).

In this section we consider factoring whole numbers: 0, 1, 2, 3, 4, ... If a whole number greater than 1 can be factored only one way (as the number times 1), then we call the number a **prime**. The number 5 is a prime; so are 2 and 3. There is only one other prime less than 10, and that is 7. The numbers 4, 6, 8, 9, and 10 are not primes because $4 = 2 \cdot 2$, $6 = 2 \cdot 3$, $8 = 2 \cdot 4$, $9 = 3 \cdot 3$, and $10 = 2 \cdot 5$.

It is an important property of whole numbers that every whole number greater than 1 that is not a prime can be written as a product of primes, and, except for rearranging the order of the primes, there is only one way to do this. Here are some examples:

$$9 = 3 \cdot 3$$

$$12 = 2 \cdot 2 \cdot 3$$

$$165 = 3 \cdot 5 \cdot 11$$

An interesting problem is to decide for any whole number whether it is a prime and, if not, how to write it as a product of primes. The general idea is to find one way to factor the number and then, if these factors are not both primes, to factor them again, continuing until all factors are primes.

Example 1   Factor the number 9702 into a product of prime numbers.

$$
\begin{aligned}
9702 &= 2 \cdot 4851 \\
&= 2 \cdot 3 \cdot 1617 \\
&= 2 \cdot 3 \cdot 3 \cdot 539 \\
&= 2 \cdot 3 \cdot 3 \cdot 7 \cdot 77 \\
&= 2 \cdot 3 \cdot 3 \cdot 7 \cdot 7 \cdot 11
\end{aligned}
$$

In this example we first guess that 2 is a factor of 9702. (Why is that a sensible guess?) To see that 2 is a factor, we compute $9702 \div 2$ and get the second factor 4851. Now we attempt to factor 4851. Since 4851 is an odd number, we know that 2 is not a factor. Our next guess is that 3 is a factor of 4851. (We just take the primes in order. It is helpful to have a list of the primes, and we will work on that project near the end of this section.) When we try 3 as a factor of 4851, we key

(Display)

4851 $\boxed{\div}$ $\boxed{3}$ $\boxed{=}$ 1617

Thus we know that 3 is one factor and 1617 is another. Now we have $9702 = 2 \cdot 3 \cdot 1617$ and we factor 1617. Since 1617 is an odd number, 2 cannot be a factor. To try 3, we key

(Display)

<u>1617</u> [÷] [3] [=]   539

Now we know that $1617 = 3 \cdot 539$, so $9702 = 2 \cdot 3 \cdot 3 \cdot 539$. We next attempt to factor 539. We may need to try several primes before we get one that works, so it helps to STORE 539. Then we can RECALL it as often as we need to and not have to continue to key [5] [3] [9]. When we test 3 as a factor of 539, it does not work; neither does 5. The next prime factor is 7. The keying sequence looks like this:

(Display)

<u>539</u> [STO] [÷] [3] [=]   179.66667

[RCL] [÷] [5] [=]   107.8

[RCL] [÷] [7] [=]   77

We replace 539 by $7 \cdot 77$. Now we have $9702 = 2 \cdot 3 \cdot 3 \cdot 7 \cdot 77$ and continue to factor 77 into a product of primes. Since 2, 3, and 5 were not factors of 539, they cannot be factors of 77; the next prime, 7, is the one we need.   □

But what if we have a number that we cannot find factors for? Here is an example.

Example 2   Factor the number 397 into a product of prime numbers.

We start by testing the primes in order:

(Display)

<u>397</u> [STO] [÷] [2] [=]   198.5

[RCL] [÷] [3] [=]   132.33333

[RCL] [÷] [5] [=]   79.4

[RCL] [÷] [7] [=]   56.714286

[RCL] [÷] [11] [=]   36.090909

[RCL] [÷] [13] [=]   30.538462

[RCL] [÷] [17] [=]   23.352941

[RCL] [÷] [19] [=]   20.894737

None of the small primes is a factor of 397. You may ask, "How long do we need to continue this process? Must we check all primes up to 397 before we can conclude that 397 is itself a prime?" Fortunately, no. We are looking for two numbers, a and b, that multiply together to give 397. Now if both a and b were greater than 20, then ab would be greater than 400. Too big! Thus one of the factors must be less than 20. However, we have checked all the primes less than 20, and since none are factors of 397, we can conclude that 397 is itself a prime.   □

Listing the Small Primes

We can use the ideas in Example 2 to list all the prime numbers less than a given number, say 100. If n is a number less than or equal to 100 and is not a prime, then n can be factored into the product of two numbers, ab, where neither a nor b is 1. At least one of the two numbers a or b must be less than or equal to 10 because if both were greater than 10, their product would be greater than 100. Thus a number n that is less than 100 and is not a prime has a prime factor less than 10.

Our project then is to list all the numbers from 2 to 100 and cross out the ones (except for 2, 3, 5, and 7) that have prime factors less than 10. These are the numbers that are not prime; we do not want them. The numbers that are left will be the primes less than 100.

Consider the four primes less than 10: 2, 3, 5, and 7. The numbers that have 2 for a factor are the even numbers. The numbers that have 3 for a factor are 6, 9, 12, 15, 18, 21, and so forth. The numbers that have 5 for a factor are the numbers in the fifth and tenth columns below. (Do you see why?) We cross out all these numbers. Now we look for numbers that have 7 for a factor. Those not already crossed out are 49, 77, and 91. The remaining numbers, which we have circled, are the primes less than 100.

```
 (2) (3)  4  (5)  6  (7)  8   9  10
(11) 12 (13) 14  15  16 (17) 18 (19) 20
 21  22 (23) 24  25  26  27  28 (29) 30
(31) 32  33  34  35  36 (37) 38  39  40
(41) 42 (43) 44  45  46 (47) 48  49  50
 51  52 (53) 54  55  56  57  58 (59) 60
```

(61)	~~62~~	~~63~~	~~64~~	~~65~~	~~66~~	(67)	~~68~~	~~69~~	~~70~~
(71)	~~72~~	(73)	~~74~~	~~75~~	~~76~~	~~77~~	~~78~~	(79)	~~80~~
~~81~~	~~82~~	(83)	~~84~~	~~85~~	~~86~~	~~87~~	~~88~~	(89)	~~90~~
~~91~~	~~92~~	~~93~~	~~94~~	~~95~~	~~96~~	(97)	~~98~~	~~99~~	~~100~~

This procedure for listing primes is a very old one and is called a **Sieve of Eratosthenes.**

Factors and Multiples

In Example 1 we wrote the number 9702 as $2 \cdot 3 \cdot 3 \cdot 7 \cdot 7 \cdot 11$ (or $2 \cdot 3^2 \cdot 7^2 \cdot 11$). Factoring into primes gives information about all the factors of 9702. Since the only prime factors of 9702 are 2, 3, 7, and 11, we know, for example, that 15 cannot be a factor of 9702 because if $9702 = 15 \cdot n$ for some number n, then $9702 = 5 \cdot 3 \cdot n$ and 5 would be a prime factor of 9702. Thus the factors of 9702 are numbers made up only of the primes 2, 3, 7, and 11. In writing the factors of 9702 we can use at most one 2, at most two 3s, at most two 7s, and at most one 11. There is a long list of factors:

$$2 \cdot 3 = 6$$

$$2 \cdot 3 \cdot 3 = 18$$

$$2 \cdot 3 \cdot 3 \cdot 7 = 126$$

$$2 \cdot 3 \cdot 7 = 42$$

$$3 \cdot 3 \cdot 7 \cdot 7 = 441$$

$$2 \cdot 3 \cdot 11 = 66$$

$$3 \cdot 7 = 21$$

$$3 \cdot 7 \cdot 7 \cdot 11 = 1617.$$

There are many more. See if you can write five additional factors of 9702.

We know that if a whole number n is a factor of 9702 that there is another whole number m such that 9702 is the product of n and m. If 9702 and n are both written as products of primes, it is easy to see what the number m is.

Example 3   Find m if $2 \cdot 3 \cdot 3 \cdot 7 \cdot 7 \cdot 11 = 2 \cdot 3 \cdot 3 \cdot 11 \cdot m$.

The number m must consist of those prime factors of the number on the left side that are not contained in $2 \cdot 3 \cdot 3 \cdot 11$.

$$2 \cdot 3 \cdot 3 \cdot 7 \cdot 7 \cdot 11$$
$$2 \cdot 3 \cdot 3 \cdot 11$$

Thus $m = 7 \cdot 7$.   □

Exercises 1.5

Find a in Exercises 1–8.

1. $2^3 \cdot 3^3 = 2 \cdot 3 \cdot a$

2. $2^3 \cdot 3^3 = 2^2 \cdot 3 \cdot a$

3. $2^3 \cdot 3^3 = 2^2 \cdot 3^2 \cdot a$

4. $2^3 \cdot 3^3 = 2 \cdot 3^2 \cdot a$

5. $2^3 \cdot 3^4 \cdot 5^2 \cdot 7^3 \cdot 11^2 = 6a$

6. $2^3 \cdot 3^4 \cdot 5^2 \cdot 7^3 \cdot 11^2 = 18a$

7. $2^3 \cdot 3^4 \cdot 5^2 \cdot 7^3 \cdot 11^2 = 75a$

8. $2^3 \cdot 3^4 \cdot 5^2 \cdot 7^3 \cdot 11^2 = 121a$

List all the factors (not just the primes) of the numbers in Exercises 9–16.

9. 8

10. 11

11. 10

12. 18

13. 24

14. 27

15. p^3, if p is a prime

16. p^2q, if p and q are different primes

Find the prime factorization of each number in Exercises 17–26. Some of the numbers are primes.

17. 5400

18. 65,000

19. 283

20. 2187

21. 373

22. 6137

23. 899

24. 1763

25. 15,625

26. 30,030

27. What is the quotient when you divide

(a) $2^3 \cdot 3^2 \cdot 11$ by $2^2 \cdot 3$ (Remember that $2^3 \cdot 3^2 \cdot 11 = 2 \cdot 2 \cdot 2 \cdot 3 \cdot 3 \cdot 11$ and that $2^2 \cdot 3 = 2 \cdot 2 \cdot 3$.)

(b) $2^3 \cdot 3^2 \cdot 11$ by $2 \cdot 3 \cdot 11$

(c) $2^3 \cdot 3^3 \cdot 5^2 \cdot 7^3$ by $3^2 \cdot 5 \cdot 7$

(d) $2 \cdot 3 \cdot 5^2$ by $2 \cdot 3 \cdot 5^2$

28. Construct a Sieve of Eratosthenes to find all primes less than 200.

1.6 / GREATEST COMMON FACTOR AND LEAST COMMON MULTIPLE

If we list all the factors of two different numbers, both lists will contain the number 1, because 1 is a factor of every whole number. An interesting problem is to find the largest number that is a factor of both of two numbers. Consider, for example, the numbers 9702 and 4950. Both have many factors—more than we want to list. However, if we did list them, we could circle the numbers that are factors of both 9702 and 4950 and then pick out the greatest of these common factors. This number is called the **greatest common factor (GCF)** of 9702 and 4950. Fortunately we do not need to list all the factors to find it.

Example 1   Find the greatest common factor of 9702 and 4950.

Consider the prime factoring of our two numbers:

$$9702 = 2 \cdot 3 \cdot 3 \cdot 7 \cdot 7 \cdot 11,$$
$$4950 = 2 \cdot 3 \cdot 3 \cdot 5 \cdot 5 \cdot 11.$$

If we want a number that is a factor of 9702, we build it from one 2, two 3s, two 7s, and one 11. If we want a number that is a factor of 4950, we build it from one 2, two 3s, two 5s, and one 11. What is the greatest number that will be a factor of both? We can use one 2, two 3s, and one 11. Thus $2 \cdot 3 \cdot 3 \cdot 11$ (or 198) is the greatest common factor of 9702 and 4950.   □

What we see from this example is that when we want to write the greatest common factor of two numbers, we factor the numbers into primes and list their common primes, if there are some. If there are, the greatest common factor is made up of these primes, each used the *smaller* number of times it appears in the original numbers. If the two numbers have no prime factors in common, their greatest common factor is 1.

Least Common Multiples

In addition to talking about factors of a number, we also talk about multiples of a number. The multiples of 9702, for example, are the numbers we get when we multiply 9702 by whole numbers. Here are some multiples of 9702:

$$1 \cdot 9702 = 9{,}702{,}$$

$$2 \cdot 9702 = 19{,}404{,}$$

$$3 \cdot 9702 = 29{,}106{,}$$

$$10 \cdot 9702 = 97{,}020{,}$$

$$18 \cdot 9702 = 174{,}636.$$

If we want to see multiples of 9702 on the calculator, we can key $\underline{9702}$ $\boxed{\times}$ $\boxed{\text{K}}$ and then key any sequence of numbers we wish.

A question we might ask is, "Are there any numbers that are multiples of both 9702 and 4950?" The number $9702 \cdot 4950$ surely is a multiple of both 9702 and 4950. But $9702 \cdot 4950$ is a very large number. Is there a smaller one? The smallest number that is a multiple of both 9702 and 4950 is called their **least common multiple (LCM)**. Factoring into primes helps us find it.

Example 2   Find the least common multiple of 9702 and 4950.

Since $9702 = 2 \cdot 3 \cdot 3 \cdot 7 \cdot 7 \cdot 11$, any number times 9702 will contain at least one 2, at least two 3s, at least two 7s, and at least one 11. Since $4950 = 2 \cdot 3 \cdot 3 \cdot 5 \cdot 5 \cdot 11$, any multiple of 4950 will contain at least one 2, at least two 3s, at least two 5s, and at least one 11. We want the smallest number that is both, so we take one 2, two 3s, two 5s, two 7s, and one 11. The least number that is a multiple of both 9702 and 4950 then is $2 \cdot 3 \cdot 3 \cdot 5 \cdot 5 \cdot 7 \cdot 7 \cdot 11$. You may compute its value if you wish. Notice that we can see quickly what number multiplies 9702 to give $2 \cdot 3 \cdot 3 \cdot 5 \cdot 5 \cdot 7 \cdot 7 \cdot 11$. Since $9702 = 2 \cdot 3 \cdot 3 \cdot 7 \cdot 7 \cdot 11$, it must follow that $(5 \cdot 5) \cdot 9702 = 2 \cdot 3 \cdot 3 \cdot 5 \cdot 5 \cdot 7 \cdot 7 \cdot 11$. Since $4950 = 2 \cdot 3 \cdot 3 \cdot 5 \cdot 5 \cdot 11$, it must follow that $(7 \cdot 7) \cdot 4950 = 2 \cdot 3 \cdot 3 \cdot 5 \cdot 5 \cdot 7 \cdot 7 \cdot 11$.   □

From the example above we see that to build the least common multiple of two numbers, we first factor them into primes and then list all the primes that appear. The least common multiple is made up of all these primes, each used the *larger* number of times it appears in the original numbers.

Example 3   Find the GCF and the LCM of the numbers 85 and 78.

The prime factorizations are $85 = 5 \cdot 17$ and $78 = 2 \cdot 3 \cdot 13$. The two numbers have no primes in common. The GCF thus is 1. The LCM is $5 \cdot 17 \cdot 2 \cdot 3 \cdot 13$.   □

Example 4   Find the GCF and the LCM of the numbers 286 and 605.

The prime factorizations are $286 = 2 \cdot 11 \cdot 13$ and $605 = 5 \cdot 11^2$. The prime 11 is common. The GCF is 11 and the LCM is $2 \cdot 5 \cdot 11^3 \cdot 13$.   □

Exercises 1.6

Find all common factors of each pair of numbers in Exercises 1–4.

1. 15 and 21
2. 12 and 25
3. 18 and 24
4. 10 and 20

Find the GCF and LCM of the numbers given in Exercises 5–14.

5. 11 and 13
6. 9 and 25
7. 60 and 126
8. 12 and 72
9. 45 and 225
10. 28 and 140
11. 63 and 220
12. 363 and 392

13. 105 and 385 and 70   14. 12 and 45 and 50

15. Two race cars go around an oval track. The faster car completes one lap every 8 minutes, while the slower car completes one lap every 12 minutes. When will the two cars first return to the starting line together?

16. Repeat Problem 15 with times 5 minutes and 6 minutes.

17. A large shopping center operates two buses that start from the same point, where riders may change buses. Bus A makes 4 stops; bus B makes 6 stops. Each bus takes 3 minutes to go from one stop to the next. If both buses start their routes at noon, when will they both arrive at station 1 at the same time again?

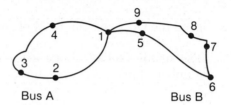

Bus A   Bus B

18. Using the map in Problem 17, suppose that bus A makes one circuit in 15 minutes and bus B makes one circuit in 20 minutes. When will both buses arrive at station 1?

19. A Christmas tree has three blinking lights. One blinks 12 times a minute, another blinks 15 times a minute, and a third blinks 20 times a minute. How often will they blink together?

20. Repeat Problem 19, replacing 12, 15, and 20 by 12, 30, and 10, respectively.

21. Let us suppose that there are 12 small communities distributed more or less evenly around a large lake and that a rapid transit system connects these communities (one station in each community). The stations are numbered from 1 through 12 in order around the lake (clockwise). Starting at station 12, a train has several options for traveling clockwise around the lake. It could move forward one station at a time and thus stop at each station; or it could move forward two stations at a time and stop at stations 2, 4, 6, 8, 10, and 12.

   (a) If the train stops at every third station, which stations will it eventually stop at?

(b) If the train stops at every fourth station, which stations will it eventually stop at?

(c) Complete this table:

Number of stations train moves forward	Stations train stops at	Total number of stations stopped at
1	1, 2, 3, 4, 5, 6, 7, 8, 9, 10, 11, 12	12
2	2, 4, 6, 8, 10, 12	6
3		
4		
5	5, 10, 3, 8, 1, 6, 11, 4, 9, 2, 7, 12	12
6		
7		
8		
9		
10		
11		
12		

1.7 / NEGATIVE NUMBERS AND THE CALCULATOR

There are many problems that require negative numbers in their solutions. Remember that every number n has an opposite $-n$ and that $-0 = 0$. If n is a positive number, $-n$ is a negative number. If n is a negative number, $-n$ is positive.

We can construct a number line by assigning numbers to points on a line. To do this, we first choose a point and label it 0; then, using a fixed distance (called the **unit distance**), we mark equally spaced points

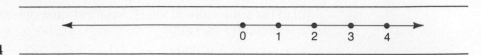

Figure 1.4

to the right of 0, labeling them 1, 2, 3, . . . as shown in Fig. 1.4. Each negative number is located to the left of 0 at the same distance from 0 as its opposite. This can be seen in Fig. 1.5. All the whole numbers and their opposites comprise the set of numbers called the **integers**.

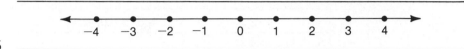

Figure 1.5

Using the Symbols $<$ and $>$

If a number is to the left of a second number on the number line, we say the first is **less than** the second and use the symbol $<$ to express this relationship. For example, -3 is less than -1, or $-3 < -1$. When a number is to the right of a second number, we say it is **greater than** the second number and use the symbol $>$. Thus 0 is greater than -3, or $0 > -3$, because 0 is to the right of -3 on the number line.

Sometimes we want to use inequality (not equal) symbols with a variable. The statement "x is a number such that $x < 10$" means that x is less than 10. Similarly, "$x > 10$" means that x is greater than 10.

Example 1   Rewrite the statement "x is a number between -2 and 3" using mathematical symbols.

The fact that x is between -2 and 3 can be expressed by saying that x is greater than -2 and less than 3. Thus we can write $x > -2$ and $x < 3$. Sometimes these two statements are combined into $-2 < x < 3$. This statement is read, "-2 is less than x and x is less than 3." You should see that this is the same as saying that x is between -2 and 3.   □

Negative Numbers
and the Arithmetic Operations

We must describe the operations of addition, subtraction, multiplication, and division for negative numbers. Fortunately, the calculator can compute with negative as well as positive numbers. To key the number −3, for example, we key $\boxed{3}$ $\boxed{+/-}$. The symbol − in the number −3 is called the sign of the number. It is not the subtraction symbol. Thus we do not use the subtraction key to indicate a negative number but, rather, the key $\boxed{+/-}$, the **sign-change key**. The effect of the sign-change key is to change the sign of the number on display. It changes positive numbers to negative numbers and negative numbers to positive numbers.

Mathematical phrase	Calculator	Display
−3	$\boxed{3}$ $\boxed{+/-}$	−3
−(−3)	$\boxed{3}$ $\boxed{+/-}$ $\boxed{+/-}$	3
−[−(−3)]	$\boxed{3}$ $\boxed{+/-}$ $\boxed{+/-}$ $\boxed{+/-}$	−3
−(3 + 4)	$\boxed{(}$ $\boxed{3}$ $\boxed{+}$ $\boxed{4}$ $\boxed{)}$ $\boxed{+/-}$	−7

It is possible to use the calculator to compute with negative numbers even before knowing how the operations are defined for negative numbers. Here are some examples:

Mathematical phrase	Calculator	Display
(−2) + (−7)	$\boxed{2}$ $\boxed{+/-}$ $\boxed{+}$ $\boxed{7}$ $\boxed{+/-}$ $\boxed{=}$	−9
2 + (−7)	$\boxed{2}$ $\boxed{+}$ $\boxed{7}$ $\boxed{+/-}$ $\boxed{=}$	−5
7 − (−2)	$\boxed{7}$ $\boxed{-}$ $\boxed{2}$ $\boxed{+/-}$ $\boxed{=}$	9
−4 − (−9)	$\boxed{4}$ $\boxed{+/-}$ $\boxed{-}$ $\boxed{9}$ $\boxed{+/-}$ $\boxed{=}$	5
(6)(−3)	$\boxed{6}$ $\boxed{\times}$ $\boxed{3}$ $\boxed{+/-}$ $\boxed{=}$	−18
(−6)(−3)	$\boxed{6}$ $\boxed{+/-}$ $\boxed{\times}$ $\boxed{3}$ $\boxed{+/-}$ $\boxed{=}$	18
6 ÷ (−3)	$\boxed{6}$ $\boxed{\div}$ $\boxed{3}$ $\boxed{+/-}$ $\boxed{=}$	−2
$\dfrac{-6}{-3}$	$\boxed{6}$ $\boxed{+/-}$ $\boxed{\div}$ $\boxed{3}$ $\boxed{+/-}$ $\boxed{=}$	2

As you gain practice computing with negative numbers, the various rules for the operations will become second hand for you. We summarize these rules here so that you will recognize an error if you make one in keying the calculator, and so that you can use them when there are variables in the computation.

Addition

One place where you have undoubtedly had experience with negative numbers is on a thermometer. You can visualize addition with negative numbers in terms of a temperature.

Sum	*Interpretation*	*Result*
$10 + (-10)$	$10°$ followed by a drop of $10°$	0
$0 + (-10)$	$0°$ followed by a drop of $10°$	-10
$10 + (-5)$	$10°$ followed by a drop of $5°$	5
$10 + (-15)$	$10°$ followed by a drop of $15°$	-5
$-10 + 5$	$-10°$ followed by an increase of $5°$	-5
$-10 + (-5)$	$-10°$ followed by a drop of $5°$	-15

Two things are apparent from this interpretation. For any number n, $0 + n = n$ and $n + (-n) = 0$. There are three additional cases in computing sums of signed numbers. Check these statements against the thermometer model and with your calculator.

■   If two numbers are positive, their sum is positive:

   $5 + 2 = 7.$

■   If two numbers are negative, their sum is negative:

   $(-5) + (-2) = -7.$

■   If one number is positive and the other negative, the sign of the sum agrees with the number farther from 0:

   $5 + (-2) = 3,$

   $-5 + (2) = -3.$

Look again at the statement $n + (-n) = 0$. This property helps identify opposites for expressions that contain variables. For example, what is $-(x + y)$? A way to answer this question is to fill the box in the equation $x + y + \boxed{} = 0$. Filling the box with $(-x) + (-y)$ gives $x + y + (-x) + (-y) = x + (-x) + y + (-y) = 0$. We conclude that

   $-(x + y) = (-x) + (-y).$

There will be more examples of this type in the exercises.

Subtraction

Remember that a subtraction statement always corresponds to an addition statement: $a - b =$ ⬚ means the same as ⬚ $+ b = a$. To fill the box in ⬚ $+ b = a$, we can add $-b$ to b to get 0 and then add a. In other words, $(a + (-b)) + b = a$. This means for any number a and b that

$$a - b = a + (-b).$$

This interpretation of subtraction—adding the opposite—is an important one for us. In the two examples below, a positive number is subtracted from another number:

$$7 - 10 = \quad 7 + (-10) = -3$$
$$-7 - 10 = -7 + (-10) = -17.$$

The same rule applies when subtracting a negative number, namely, add the opposite.

$$7 - (-10) = \quad 7 + (10) = 17$$
$$-7 - (-10) = -7 + (10) = 3.$$

The examples above show that subtracting -10 is the same as adding 10.

In computing with expressions that contain variables, we use the same meaning for subtraction because the variables, after all, simply represent numbers. Here is an important consequence of the meaning of subtraction: $z - (x + y) = z + [-(x + y)] = z + (-x) + (-y)$. This is a result of the observation in the section entitled *Addition* that $-(x + y) = (-x) + (-y)$.

Multiplication

When we consider the product $(4)(-6)$ we want it to represent four (-6)s; that is, $(-6) + (-6) + (-6) + (-6)$. Thus $(4)(-6) = -24$. Actually there are three cases to summarize, in addition to stating $0 \cdot n = 0$ for any number n. We do not have a good "real-world" model for multiplication of signed numbers. However, you can check these statements on your calculator.

- If two numbers are positive, their product is positive:

$$(2)(7) = 14.$$

- If two numbers are negative, their product is positive:

$$(-2)(-7) = 14.$$

- If one number is positive and the other negative, their product is negative:

$$(2)(-7) = -14.$$

One application of these rules is to see that multiplying a number by -1 gives the opposite of the number: $(-1)(n) = -n$. Thus whenever convenient for computation, we can view $-n$ as $(-1)n$. For example, $-(x + y) = (-1)x + (-1)y = (-x) + (-y)$.

Division

Division statements always correspond to multiplication statements: $a \div b = \boxed{}$ means the same as $\boxed{} \times b = a$. For this reason the rules determining the signs of quotients are the same as the rules determining the signs of products. Use your calculator to verify the following statements.

- If two numbers are positive, their quotient is positive:

$$\frac{20}{5} = 4.$$

- If two numbers are negative, their quotient is positive:

$$\frac{-20}{-5} = 4.$$

- If one number is positive and another negative, their quotient is negative:

$$\frac{-20}{5} = -4,$$

$$\frac{20}{-5} = -4.$$

Example 2   Find the value of $15.4 - (-8.13)$.

The calculator keying sequence is

(Display)

15.4 $\boxed{-}$ 8.13 $\boxed{+/-}$ $\boxed{=}$ 23.53

Notice the difference in the use of the two keys, $\boxed{-}$ and $\boxed{+/-}$.   □

Example 3 Find the value of $(-2)(-3)(-.5)$.

There are at least two ways to key this product. The first way is to key directly:

(Display)

$\boxed{2}$ $\boxed{+/-}$ $\boxed{\times}$ $\boxed{3}$ $\boxed{+/-}$ $\boxed{\times}$ $\boxed{.5}$ $\boxed{+/-}$ $\boxed{=}$ −3

The second way is to recognize that the product of three negative numbers is negative, so $(-2)(-3)(-.5) = -(2)(3)(.5)$. Thus we can key

(Display)

$\boxed{2}$ $\boxed{\times}$ $\boxed{3}$ $\boxed{\times}$ $\boxed{.5}$ $\boxed{=}$ $\boxed{+/-}$ −3   □

Example 4 Find the value of $\dfrac{(-5.2)(-.006)}{.04}$.

Again there are at least two keying sequences. In the first sequence we key the negative numbers in the numerator:

(Display)

5.2 $\boxed{+/-}$ $\boxed{\times}$ .006 $\boxed{+/-}$ $\boxed{\div}$ .04 $\boxed{=}$ 0.78

In the second sequence we can ignore the negative signs. Since the numerator is the product of two negative numbers and is positive, the quotient is also positive.

(Display)

.52 $\boxed{\times}$ .006 $\boxed{\div}$ .04 $\boxed{=}$ 0.78   □

**Exercises
1.7**

Give the opposite of the numbers in Exercises 1–7.

1. 7   2. −2.4

3. −(−4)   4. 3 − 2

5. $2 - 3$ 6. $2x$

7. $-3x^2$

Find the value of the expressions in Exercises 8–16.

8. $7 + (-5) - 8 - (-4)$

9. $21 - (-11) + (-8) - 13$

10. $(-2)(-2)(-2)(-2)$

11. $(-2)(-2)(-2)(-2)(-2)$

12. $(-3.2)(-4.5) - (-5.6)(-3.1)$

13. $[12 + (-3)](6 - 8)$

14. $(5.1 - 8.5)[17.3 + (-9.3) + (-13.5)]$

15. $\dfrac{(-6.72)(-5.34)}{4.29}$

16. $\dfrac{(-8.6)(-7.3)^2}{-5.2}$

Write a sequence of calculator key strokes that will evaluate each expression in Exercises 17–19. Then evaluate the expressions.

17. $64 - (-32) + (-72) - 49$ 18. $(-7.3)(-9.4) - (8.6)(-3.2)$

19. $\dfrac{(-17.8)(-13.4)}{-10.5}$

Translate each of the English expressions in Exercises 20–22 into a mathematical phrase.

20. Negative eighteen subtracted from three times negative two

21. The sum of negative twelve and negative two times x

22. Negative twelve divided by the sum of six and negative four

Write the mathematical phrase evaluated by each sequence of calculator key strokes in Exercises 23 and 24.

23. $\boxed{2}$ $\boxed{+/-}$ $\boxed{+}$ $\boxed{3}$ $\boxed{+/-}$ $\boxed{x^2}$ $\boxed{=}$

24. $\underline{6.1}$ $\boxed{\times}$ $\underline{7.3}$ $\boxed{+/-}$ $\boxed{+}$ $\underline{8.5}$ $\boxed{+/-}$ $\boxed{\times}$ $\underline{4.2}$ $\boxed{=}$

Insert the proper symbol ($<$, $=$, $>$) between each pair of numbers in Exercises 25–30.

25. $-8, -5$

26. $-1, -8$

27. $-7, 2$

28. $-(-6), -2$

29. $-(-7), 7$

30. $12, -(-12)$

Find all whole numbers x that satisfy the inequalities in Exercises 31–33.

31. $x < 7$

32. $5 < x$

33. $2 < x < 8$

Find all integers x that satisfy the inequalities in Exercises 34–37.

34. $-5 < x < 0$

35. $x > -5$

36. $-2 < x < 4$

37. $-1 < x < 1$

Find the value of each expression in Exercises 38–41, assuming that $x = -3$ and $y = 2$.

38. $2x - 5y$

39. $20 - 3xy$

40. $x^2 - 3y^2$

41. $\dfrac{x + y}{x - y}$

Use the distributive property and combine like terms to simplify the expressions in Exercises 42–45.

42. $5(x - 3) + 2(4 - x)$

43. $(-2)(3x - 4) - 3(x + 2)$

44. $2x^2 + x(1 - x)$

45. $x(4 - x) - 2x(x - 3)$

46. Show that the opposite of $x - y$ is $y - x$.

47. Show that the opposite of $a + b - c$ is $c - b - a$.

48. Each day, Bill's coffee shop buys 10 dozen doughnuts. Here are the daily net "profits" (in dollars) for doughnut sales on 15 days. A plus sign indicates a profit; a minus sign indicates a loss: +12, +9, −3, −14, +7, −8, 0, −13, +17, +10, −2, +6, +15, −3, −9. All in all, did the 15-day total turn out to be a profit or a loss? By how much?

49. Ms. Jones watches the daily price of her stock in Company A. Here are the daily fluctuations in its price for a week. A plus sign indicates that the stock went up; a minus sign indicates that the stock went down. What was the net gain or loss for the week?

Monday: −6   Tuesday: +3   Wednesday: +4
Thursday: +5   Friday: −3

50. George S. Patton died in 1945 at the age of 60. When was he born?

51. The boiling point of water is 212°F. This is 180°F more than the freezing point. What is the freezing point?

52. How much higher is the taller hill than the shorter hill if the heights of two hills are 567 feet and 203 feet?

53. How many more pages of a book remain to be read if the book has 305 pages and 101 pages have been read? N pages read?

54. The second of two numbers is 13 less than twice the first. Complete the following table:

First number	Second number	Sum	Product
−2	$2(-2) - 13 = -17$	−19	34
−1			
0			
1			
2			
x			
	−3		
		−1	

1.8 / CHAPTER 1 PROBLEM COLLECTION

Evaluate each expression in Exercises 1–10.

1. $13^2 - (2.1)(3.4) + 6.7$

2. $(17.5)[12.1 + 4(18.3 - 7.8) + 3(71)]$

3. $(4.2)\{8.1 - 2.3[5 + 6(91 - 13) + 2] + 17\}$

4. $\dfrac{3^2 + 7^2}{5^2 - 2^2}$

5. $\dfrac{12.3 - 29.8 - 13.2}{21.3 - 18.6}$

6. $73(12) + 73(91) + 73(-101) - 73(13) + 73(67) + (-73)(201)$

7. $5 \cdot 4^2 - 2 \cdot 3^3$

8. $3(-2)^2 - 7 \cdot 3^3 + (-5)(8)$

9. $(17.3 + 12.9)^2(37.1 - 21.5)$

10. $(-2)[(7.3 + 4.5)^2 - (8.1 - 7.9)^2]$

Write a sequence of calculator key strokes that will evaluate the expressions in Exercises 11–14. Then evaluate each expression.

11. $\dfrac{22.7 - 49.1 + 86.8}{13.5 + 29.7}$

12. $(5.7)(8.9 - 12.8)$

13. $2 \cdot 3^2 + 4^3$

14. $\dfrac{6(7)}{5^2} - \dfrac{8(-4)^2}{6}$

Write the mathematical phrase evaluated by each sequence of calculator key strokes in Exercises 15–20.

15. $\underline{18}$ $\boxed{\div}$ $\boxed{6}$ $\boxed{+/-}$ $\boxed{\times}$ $\boxed{4}$ $\boxed{=}$

16. $\boxed{5}$ $\boxed{+}$ $\boxed{3}$ $\boxed{+/-}$ $\boxed{=}$ $\boxed{x^2}$

17. $\boxed{5}$ $\boxed{+}$ $\boxed{3}$ $\boxed{+/-}$ $\boxed{x^2}$ $\boxed{=}$

18. $\boxed{5}$ $\boxed{+/-}$ $\boxed{\times}$ $\boxed{K}$ $\boxed{=}$ $\boxed{=}$ $\boxed{=}$ $\boxed{=}$

19. $\boxed{2}$ $\boxed{\times}$ $\boxed{3}$ $\boxed{y^x}$ $\boxed{4}$ $\boxed{+}$ $\boxed{5}$ $\boxed{+/-}$ $\boxed{x^2}$ $\boxed{=}$

20. $\boxed{5}$ $\boxed{\times}$ $\boxed{(}$ $\underline{17}$ $\boxed{+}$ $\boxed{8}$ $\boxed{\times}$ $\boxed{(}$ $\boxed{(}$ $\underline{13}$ $\boxed{-}$ $\boxed{4}$ $\boxed{)}$ $\boxed{x^2}$
$\boxed{+}$ $\boxed{6}$ $\boxed{)}$ $\boxed{)}$ $\boxed{=}$

Translate each of the English expressions in Exercises 21–25 into a mathematical phrase.

21. The sum of four times x and six

22. The sum of five squared and y squared

23. Negative seven times the sum of x and y

24. The sum of negative seven times x and negative seven times y

25. The sum of two times four cubed and three times five to the fourth power

26. A car averages 18 miles per gallon of gasoline. Complete the following table:

Miles traveled	36	63		x
Gallons used			4.5	

27. Complete the following table:

Price per pound	7.20	8.32	14.24		16.16	x	
Price per ounce				.67			y

28. The sales tax is 5 cents on every dollar. Compute the sales tax on the following purchases.

    (a) $3500   (b) $4750   (c) $5260.40   (d) $N

29. Gasoline costs $1.29 per gallon at Unity Station.

    (a) How much does 24 gallons of gasoline cost?

    (b) If you pay $20.64 for gasoline, how many gallons do you purchase?

30. A laborer is paid at the rate of $7.50 per hour.

    (a) How much does he earn in 6 hours and 30 minutes?

    (b) If he earns $76.50, how long did he work?

31. A car averages 17.2 miles per gallon of gasoline.

    (a) How many miles are traveled on 8 gallons?

    (b) How many gallons are needed to drive a distance of 215 miles?

32. Lisa is 99,999,999 seconds old. How old is Lisa to the nearest day? (You may ignore leap years.)

33. List all factors of 50.

Find the prime factorization of the numbers in Exercises 34–37.

34. 5940   35. 899

36. 1212   37. 13,566

38. Prove that 347 is a prime number.

Find the GCF and the LCM of the numbers in Exercises 39–42.

39. 1540; 1755   40. 900; 1176

41. 294; 616; 3276   42. 408; 585; 1925

43. Use the distributive property to write two sequences of calculator key strokes that will evaluate $13(18 + 19)$.

Use the distributive property and combine like terms to simplify the expressions in Exercises 44–49.

44. $7a^3 - a^3 + b^2 - .1b^2$

45. $8z^2 - 2z^2 - z^2 + .2z^2$

46. $x(2.5x + 3) - x(3.2x + 1.2)$

47. $3x(2x - 3) + 2(2x - 3)$

48. $x(x - 3) - 5(1 - x) - 2(x + 2)$

49. $x(x^2 + 2x) + 4x - 2(x^2 + 2x) - 8$

50. Show that the opposite of $1 - x$ is $x - 1$

51. Show that the opposite of $a - b + c$ is $-c + b - a$.

Find all whole numbers x that satisfy the inequalities in Exercises 52–54.

52. $x < 3$ 53. $0 < x < 4$

54. $2 < x < 7$

Find all integers x that satisfy the inequalities in Exercises 55 and 56.

55. $-3 < x < 1$ 56. $-1 < x$

Find the value of $\dfrac{2xy}{x + 2y}$ for the values of x and y given in Exercises 57–59.

57. $x = 2, y = 3$ 58. $x = -2, y = 3$

59. $x = -2, y = -3$

60. Compute the cost of 20 cartons of 98-cent coke by using the distributive property of multiplication over subtraction.

61. Two race cars go around an oval track. The faster car completes one lap every 42 seconds, while the slower car completes one lap every 48 seconds. When will the two cars first return to the starting line together?

62. Find the area of a square with sides of 12 inches.

63. Find the area of a square whose perimeter is 46 inches.

64. Find the length of the side of a square whose area is 289 square inches.

65. The money box contains 40 coins in nickels and quarters.

    (a) Complete the following table:

Number of nickels	Number of quarters	Value of coins
10		
20		
	10	
40		
		$5.00
x		

    (b) Explain in words how to find the number of quarters if the number of nickels is known.

    (c) Explain in words how to find the value of the coins if the number of nickels is known.

66. Bill has 2 more quarters than nickels and 5 times as many nickels as dimes.

    (a) Complete the following table:

Number of nickels	Number of dimes	Number of quarters	Value of coins
20	4	22	$ 6.90
	8		
60			
		77	
			$14.90

(b) Explain in words how to find the number of dimes if the number of nickels is known.

(c) Explain in words how to find the number of nickels if the number of dimes is known.

(d) Explain in words how to find the value of the coins if the number of nickels is known.

67. The second of two numbers is 20 minus the square of the first number. Complete the following table:

First number	Second number	Sum	Product
−2	$20 - (-2)^2 = 16$	14	−32
−3			
3			
5			
−5			

Computing
with Fractions
and Decimals

2

2.1 / COMMON FRACTIONS
AND FINITE DECIMALS

Although whole numbers are very useful when we work with sets of objects that can be counted, there are many numerical problems that do not have whole numbers for their solutions. Here is an example:

> The three Angelino brothers plan to share two large pizzas equally. How much pizza does each brother get?

The solution to this problem is not a whole number. The brothers might cut each pizza into 3 equal slices, and then each brother might take 2 of the 6 slices. In this instance, a slice would be $\frac{1}{3}$ of a pizza and a brother's share would be $\frac{2}{3}$ of a pizza.

To represent the fractions $\frac{1}{3}$ and $\frac{2}{3}$ on a number line, we subdivide the unit interval from 0 to 1 into three equal segments, as shown in Fig. 2.1.

Not only does this procedure locate $\frac{1}{3}$ and $\frac{2}{3}$, it can be extended to locate all fractions that have 3 in the denominator (Fig. 2.2).

An important property of fractions is that each fraction has more than one name. For example, if we locate the fractions with denominator 6 in the unit interval, we see that $\frac{1}{3} = \frac{2}{6}$ and $\frac{2}{3} = \frac{4}{6}$. Also, $\frac{3}{3} = \frac{6}{6} = 1$ (Fig. 2.3). In these cases the effect has been to multiply the numerator and the denominator by 2.

In general, if the numerator and the denominator of a fraction are

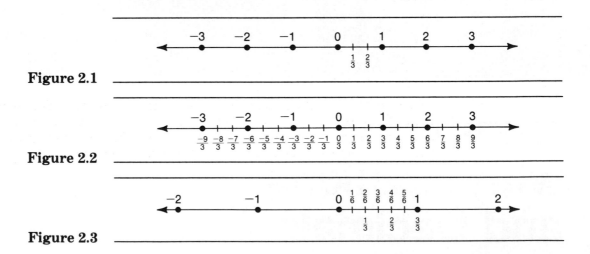

Figure 2.1

Figure 2.2

Figure 2.3

multiplied by the same number, the result is a fraction equal to the original fraction. Study these examples:

$$\frac{2}{3} = \frac{(4)(2)}{(4)(3)} = \frac{8}{12}$$

$$\frac{1}{7} = \frac{(-2)(1)}{(-2)(7)} = \frac{-2}{-14}$$

$$\frac{5}{4} = \frac{3(5)}{3(4)} = \frac{15}{12}.$$

Building and Reducing Fractions

When the numerator and denominator of a fraction are multiplied by the same number, the form of the fraction is changed, but the value is not changed. This fact means that any two fractions can be written with the same denominator. We rename the fractions by choosing a denominator that is a multiple of both of the original denominators.

Example 1   Write $\frac{5}{6}$ and $\frac{3}{8}$ as fractions with the same denominator.

To write the fractions $\frac{5}{6}$ and $\frac{3}{8}$ with the same denominator, we need a number that is a multiple of both 6 and 8. The number $6 \cdot 8$ is one such number. We can write

$$\frac{5}{6} = \frac{8 \cdot 5}{8 \cdot 6} = \frac{40}{48}$$

$$\frac{3}{8} = \frac{6 \cdot 3}{6 \cdot 8} = \frac{18}{48}.$$

Of course, 48 is not the smallest common multiple of 6 and 8. The smallest is the LCM. Since $6 = 2 \cdot 3$ and $8 = 2 \cdot 2 \cdot 2$, the LCM is $2 \cdot 2 \cdot 2 \cdot 3$, or 24. If we want the fractions $\frac{5}{6}$ and $\frac{3}{8}$ to have the same denominator and that denominator to be the smallest possible, we write

$$\frac{5}{6} = \frac{4 \cdot 5}{4 \cdot 6} = \frac{20}{24}$$

$$\frac{3}{8} = \frac{3 \cdot 3}{3 \cdot 8} = \frac{9}{24}.$$

□

We can express the fact that the value of a fraction is not changed if its numerator and denominator are multiplied by the same number by writing the equation $\frac{a}{b} = \frac{na}{nb}$. Reading from right to left, the equation is

$$\frac{na}{nb} = \frac{a}{b}.$$

This means that if a fraction has the same factor n in its numerator and denominator, that common factor can be removed from the numerator and denominator without changing the value of the fraction. How do we decide if the numerator and denominator have a factor in common? We factor both the numerator and denominator. Then we remove all the common factors. This process is called **reducing the fraction to lowest terms**.

Example 2   Reduce the fraction $\frac{156}{455}$.

Factoring the numerator and denominator into a product of primes gives

$$\frac{156}{455} = \frac{2 \cdot 2 \cdot 3 \cdot 13}{5 \cdot 7 \cdot 13}.$$

We can remove the common factor 13 from the numerator and denominator:

$$\frac{156}{455} = \frac{2 \cdot 2 \cdot 3 \cdot 13}{5 \cdot 7 \cdot 13} = \frac{2 \cdot 2 \cdot 3}{5 \cdot 7} = \frac{12}{35}. \qquad \square$$

Example 3   Reduce the fraction $\frac{1596}{5460}$.

Factoring the numerator and denominator into a product of primes gives

$$\frac{1596}{5460} = \frac{2 \cdot 2 \cdot 3 \cdot 7 \cdot 19}{2 \cdot 2 \cdot 3 \cdot 5 \cdot 7 \cdot 13}.$$

There are several common factors that can be removed: 2, 2, 3, and 7. Thus

$$\frac{1596}{5460} = \frac{2 \cdot 2 \cdot 3 \cdot 7 \cdot 19}{2 \cdot 2 \cdot 3 \cdot 5 \cdot 7 \cdot 13} = \frac{19}{5 \cdot 13} = \frac{19}{65}. \qquad \square$$

Notice in the third example that the number $2 \cdot 2 \cdot 3 \cdot 7$ that was removed from the numerator and the denominator is, in fact, the greatest common factor of 1596 and 5460. This, in general, is the goal in reducing fractions. We find the greatest common factor of the numerator and denominator by factoring both into a product of primes; then we remove the greatest common factor to reduce the fraction.

Fractions with Powers of 10 for Denominators

We will see that fractions that have powers of 10 for their denominators are particularly easy to compute with. There is a special notation for these fractions, called **decimal notation**; it is the form used by calculators. We sometimes call numbers in this form **finite decimals**. It is important to realize that these are not new numbers. Finite decimals are merely another way of writing those common fractions that have powers of 10 for their denominators. Study the following examples to remind yourself about decimal notation. (Since common fractions are actually quotients of whole numbers, you can check these examples with your calculator.)

Common fraction	Decimal notation	In words
$\dfrac{1}{10}$	.1	One tenth
$\dfrac{6}{10}$	.6	Six tenths
$-\dfrac{7}{10^2} = -\dfrac{7}{100}$	−.07	Negative seven hundredths
$\dfrac{45}{10^2} = \dfrac{45}{100}$	.45	Forty-five hundredths
$-\dfrac{3}{10^3} = -\dfrac{3}{1000}$	−.003	Negative three thousandths
$\dfrac{26}{10^3} = \dfrac{26}{1000}$	.026	Twenty-six thousandths
$\dfrac{159}{10^3} = \dfrac{159}{1000}$	.159	One hundred fifty-nine thousandths

The examples above illustrate how to write special fractions (those with powers of 10 for denominators) as finite decimals. We also need to be able to write finite decimals as common fractions.

Finite decimal	*Common fraction*
.2	$\dfrac{2}{10}$
.8	$\dfrac{8}{10}$
$-.03$	$-\dfrac{3}{10^2} = -\dfrac{3}{100}$
.85	$\dfrac{85}{10^2} = \dfrac{85}{100}$
.009	$\dfrac{9}{10^3} = \dfrac{9}{1000}$
$-.066$	$-\dfrac{66}{10^3} = -\dfrac{66}{1000}$
.1432	$\dfrac{1432}{10^4} = \dfrac{1432}{10000}$

Notice that the number of decimal places in the finite decimal is the same as the exponent on 10 in the denominator of the common fraction.

Since the fractions $\frac{2}{10}$ and $\frac{20}{100}$ are equal, the two finite decimals .2 and .20 are equal. In fact, $.2 = .20 = .200 = .2000$, and so on. We can add zeros to a finite decimal and not change its value.

Decimals can be located on the number line in the same way as common fractions. To locate .3 (or $\frac{3}{10}$), for example, we subdivide the unit interval into 10 equal segments and label the right endpoint of the third one. In fact, all the decimals that are tenths can be located this way.

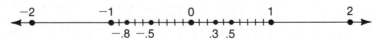

To graph $.35 = \frac{35}{100}$, we theoretically need to subdivide the unit interval into 100 equal subintervals and then label the right-hand endpoint of the thirty-fifth subinterval as .35. Another way to locate .35 is to recognize that $\frac{35}{100}$ is halfway between $\frac{30}{100}$ and $\frac{40}{100}$; thus .35 is halfway between .30 (or .3) and .40 (or .4):

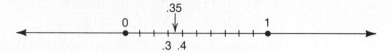

Writing Fractions as Decimals

Even though every finite decimal is a common fraction, it is not true that every common fraction is a finite decimal; only the common fractions that can be written with powers of 10 for denominators are finite decimals. Thus a common fraction is a finite decimal if there is a number that can multiply the denominator to give a power of 10. Here are some examples in which there is such a number. Factoring the denominator into a product of primes helps us to find that number.

$$\frac{1}{2} = \frac{5 \cdot 1}{5 \cdot 2} = \frac{5}{10} = .5$$

$$\frac{3}{8} = \frac{3}{2 \cdot 2 \cdot 2} = \frac{5 \cdot 5 \cdot 5 \cdot 3}{5 \cdot 5 \cdot 5 \cdot 2 \cdot 2 \cdot 2} = \frac{375}{10^3} = .375$$

$$\frac{4}{5} = \frac{2 \cdot 4}{2 \cdot 5} = \frac{8}{10} = .8 .$$

These fractions can be written with powers of 10 in their denominators because the denominators in the original fractions contain only 2 and 5 as prime factors. If the denominator of a reduced fraction contains any primes other than 2 and 5, no multiple of the denominator is a power of 10. We could not, for example, write $\frac{1}{3}$ as $\frac{a}{10^n}$ because there is no whole number that multiplies 3 to give a power of 10.

Thus common fractions like $\frac{1}{3}$, $\frac{2}{7}$, $\frac{11}{15}$, and $\frac{20}{22}$ cannot be written as finite decimals because their denominators contain primes other than 2 and 5. But we have said that $\frac{1}{3} = 1 \div 3$ and if we key $\boxed{1}\ \boxed{\div}\ \boxed{3}\ \boxed{=}$, the calculator gives a finite decimal. It displays .33333333. However, .33333333 means $\frac{33333333}{10^8}$ and this fraction does not equal $\frac{1}{3}$. To see that this is true, we write $\frac{33333333}{10^8}$ and $\frac{1}{3}$ with the same denominator, $3 \cdot 10^8$:

$$\frac{33333333}{10^8} = \frac{3(33333333)}{3(10^8)} = \frac{99999999}{3(10^8)}$$

$$\frac{1}{3} = \frac{10^8(1)}{10^8(3)} = \frac{10^8}{3(10^8)} = \frac{100000000}{3(10^8)}.$$

The numerators, 99999999 and 100000000, are not equal, so these are different fractions. (In fact, $\dfrac{33333333}{10^8}$ is smaller than $\frac{1}{3}$.)

Example 4   Find a finite decimal that approximates $\frac{2}{7}$.

We key $\boxed{2}$ $\boxed{\div}$ $\boxed{7}$ $\boxed{=}$ and read the display .28571429. To see that .28571429 does not equal $\frac{2}{7}$, we can write

$$.28571429 = \frac{28571429}{10^8} = \frac{7(28571429)}{7(10^8)} = \frac{200000003}{7(10^8)}$$

$$\frac{2}{7} = \frac{10^8(2)}{10^8(7)} = \frac{200000000}{7(10^8)}.$$

The denominators of the fractions are the same, but the numerators are different. However, the calculator gives a good finite decimal approximation for $\frac{2}{7}$.   □

Actually, the finite decimal .28571429 in the problem above results from the calculator rounding the number .285714285714285714 . . . . (In Section 2.7 we will explain how we know that the fraction $\frac{2}{7}$ is a decimal that continues and repeats in this pattern.) Since the calculator can only display eight digits, it takes the first eight shown, increasing the last by 1 if the ninth position is filled by 5 or a number greater than 5. We call a number like .285714285714285714 . . . an **infinite repeating decimal**. We sometimes denote it with the symbol $.\overline{285714}$, using the line to indicate how the decimal repeats. For .333333 . . . we would write $.\overline{3}$.

Thus the situation is that some fractions, those that can be written with denominators having powers of 10, are finite decimals. The others turn out to be infinite repeating decimals.

**Exercises
2.1**   Write the reduced name for the fractions indicated by the arrows on the number line in Exercises 1 and 2.

1.

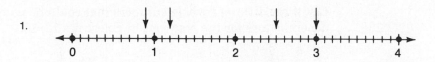

2.

Write the finite decimal indicated by the arrows on the number line in Exercises 3 and 4.

3.

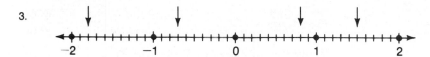

4.

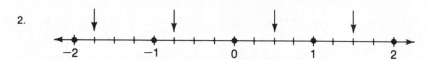

Graph the numbers in Exercises 5–8. (In other words, find the points on the number line associated with the numbers.)

5. $-\dfrac{9}{7}, -\dfrac{2}{7}, \dfrac{6}{7}, \dfrac{11}{7}$

6. $-\dfrac{11}{2}, -\dfrac{3}{2}, \dfrac{1}{2}, \dfrac{9}{2}$

7. $-1.8, -.15, .7, 1.05$

8. $-.19, -.155, -.105$

Write the finite decimal that represents the shaded portion of each of the figures in Exercises 9–12.

9.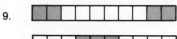

10.

11.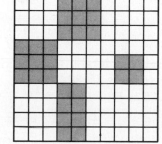

12.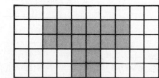

List three different whole numbers that could be used as common denominators for the fractions in Exercises 13–16.

13. $\dfrac{5}{3}, -\dfrac{7}{4}$

14. $\dfrac{4}{5}, \dfrac{7}{10}$

15. $6, \dfrac{4}{3}$

16. $-\dfrac{5}{4}, -\dfrac{7}{6}$

Reduce the fractions in Exercises 17–20 to their lowest terms.

17. $\dfrac{660}{150}$

18. $\dfrac{22{,}869}{480{,}249}$

19. $-\dfrac{899}{713}$

20. $-\dfrac{6500}{8500}$

Write the numbers in Exercises 21–26 as finite decimals.

21. $\dfrac{7}{10^3}$

22. $-\dfrac{12}{10^4}$

23. $-\dfrac{23}{10^2}$

24. $\dfrac{75}{10^3}$

25. $\dfrac{175}{100}$

26. $-\dfrac{38}{10}$

Write the numbers in Exercises 27–30 as common fractions.

27. $.7$

28. $-.45$

29. $-.02$

30. $.00005$

Rewrite each fraction in Exercises 31–34 with denominators having a power of 10, if possible.

31. $\dfrac{3}{40}$

32. $\dfrac{1}{55}$

33. $-\dfrac{1}{12}$

34. $-\dfrac{67}{50}$

Write an eight-place finite decimal that approximates each of the common fractions in Exercises 35–38.

35. $\dfrac{5}{9}$

36. $-\dfrac{2}{11}$

37. $\dfrac{3}{8}$

38. $-\dfrac{7}{16}$

39. Five students share 3 large pizzas equally. How much pizza does each student get? Give three equivalent forms for the answer and explain how you could cut the pizzas to reflect each equivalent form.

40. An inch is what fraction of a foot? a yard? a mile?

41. A second is what part of a minute? an hour? a day?

42. If 2 gallons of salt are added to 20 gallons of water, what fraction of the new mixture is salt? What fraction of the mixture is water?

43. If 4 gallons of alcohol are added to 28 gallons of water, what fraction of the new mixture is alcohol? What fraction of the mixture is water?

44. A 40-gallon salt solution contains 12 gallons of pure salt. If 8 gallons of salt are added, what fraction of the new mixture is salt? What fraction of the mixture is water?

45. A 60-gallon alcohol solution contains 18 gallons of pure alcohol. If 14 gallons of water are added, what fraction of the new solution is alcohol? What fraction of the mixture is water?

2.2 / COMPARING FRACTIONS AND DECIMALS

We have seen that multiplying the numerator and denominator of a fraction by the same number changes the form of the fraction but does not change its value. For example,

$$\frac{4}{5} = \frac{8}{10}, \qquad \frac{3}{7} = \frac{-3}{-7}, \qquad \frac{-2}{3} = \frac{-10}{15}, \qquad \frac{-2}{3} = \frac{2}{-3}.$$

If we have two fractions that we wish to compare to see if they are equal, we can write the fractions with the same denominator and then compare their numerators.

Example 1 Compare $\dfrac{6}{28}$ and $\dfrac{57}{266}$.

One common denominator for these two fractions is (28)(266).

$$\frac{6}{28} = \frac{6(266)}{28(266)}$$

$$\frac{57}{266} = \frac{28(57)}{28(266)}.$$

To see if the fractions are equal, we must compare 6(266) and 28(57). Since 6(266) = 1596 and 28(57) = 1596, we conclude that $\frac{6}{28} = \frac{57}{266}$.

□

Look again at the example above. To decide that $\frac{6}{28} = \frac{57}{266}$, we compared the products 6(266) and 28(57):

$$\frac{6}{28} \times \frac{57}{266}.$$

The same process (called **cross multiplication**) can be used to check any two fractions to see if they are equal. This fact can be expressed this way.

> If $\frac{a}{b}$ and $\frac{c}{d}$ are fractions, then $\frac{a}{b} = \frac{c}{d}$ if $ad = bc$.

Example 2   Find the number x that makes $\frac{x}{22} = \frac{35}{77}$.

In order for the two fractions to be equal, it must be that

$x \cdot 77 = (22)(35)$

$x \cdot 77 = 770.$

Thus x must be 10.

□

Cross multiplication has other consequences. As you know, integers can be viewed as fractions with denominators of 1:

$$-3 = \frac{-3}{1}, \qquad -2 = \frac{-2}{1}, \qquad -1 = \frac{-1}{1}, \qquad 0 = \frac{0}{1}, \qquad 2 = \frac{2}{1}.$$

The number 0 can be written as $\frac{0}{1}$ and also as $\frac{0}{2}$; cross multiplication shows this is true:

$$\frac{0}{1} = \frac{0}{2} \quad \text{because } 0 \cdot 2 = 1 \cdot 0.$$

In fact, $0 = \frac{0}{1} = \frac{0}{2} = \frac{0}{3} = \frac{0}{4}$, and so on. Again you can see this by checking the cross multiplication.

Comparing two finite decimals to see if they are equal or, if not, which one is larger is an even simpler task than comparing fractions. The cross multiplication procedure shows that

$$\frac{7}{10} = \frac{70}{100} = \frac{700}{1000} = \frac{7000}{10000};$$

thus also $.7 = .70 = .700 = .7000$.

In fact, adding zeros to the end of a finite decimal does not change its value. We can compare two finite decimals by making them the same length (that is to say, by giving them the same denominator).

Example 3 Compare .44 and .404.

To make the decimals the same length, we write $.44 = .440$. Now we compare the whole numbers 440 and 404. Since $440 > 404$, we conclude that $\frac{440}{1000} > \frac{404}{1000}$ so $.440 > .404$. □

Example 4 Compare $-.15$ and $-.105$.

Remember that one number is less than another if the first lies to the left of the second on the number line. To compare $-.15$ and $-.105$ we write $-.150$ and $-.105$ and look at the whole numbers -150 and -105. On the number line -150 is farther from 0, hence farther left:

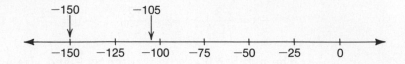

Thus $-.150 < -.105$.   □

Example 5   Order these numbers from smallest to largest:

$$0, \quad \frac{4}{11}, \quad .3636, \quad \frac{-4}{11}, \quad -.3636.$$

Since comparing decimals is somewhat more straightforward than comparing fractions, we can use the calculator to write the decimal approximations for the fractions and use these in the comparisons. The symbol $\doteq$ means **approximately equals**.)

$$\frac{4}{11} \doteq .36363636,$$

$$-\frac{4}{11} \doteq -.36363636.$$

The smallest number in the list is one of the two negative numbers: $-.36363636$, or $-.3636$. Since $-.36363636$ is farther from 0 than $-.36360000$, we start by observing that

$$-.36363636 < -.3636.$$

Each of these numbers is less than 0, so

$$-.36363636 < -.3636 < 0.$$

To compare the two positive numbers in the list, we write .36363636 and .36360000. Since $36360000 < 36363636$, we have

$$.3636 < .36363636.$$

Both are greater than 0. Thus the decimals are put in order this way:

$$-.36363636 < -.3636 < 0 < .3636 < .36363636.$$

If we replace the decimal approximations by the fractions in the problem, we find that

$$-\frac{4}{11} < -.3636 < 0 < .3636 < \frac{4}{11}.$$ □

The procedures we have used to compare two finite decimals can be used to compare any two decimals. We compare the digits, decimal place by decimal place, until we find the first place where they differ. The number with the larger digit in that place is the larger number. For example, the number $.34567\overline{3}$ is larger than the number $.34528\overline{3}$ because the digits agree in the first three places, but 6 is greater than 2 in the fourth place.

Example 6 Determine which is the better buy: a 15-ounce can of ham for $2.90 or a 35-ounce can of ham for $6.77.

Using a calculator to compute the price per ounce, we find that

(Display)

2.90 ÷ 15 = .19333333 dollars per ounce

6.77 ÷ 35 = .19342857 dollars per ounce

We need to go to the fourth place to see that .19333333 is less than .19342857. Thus the first ham is the better buy. (But not much better!)

□

Exercises 2.2 Insert the proper symbol ($<, =, >$) between the pairs of numbers in each of Exercises 1–17.

1. $\frac{3}{5}, \frac{231}{385}$

2. $\frac{17}{13}, \frac{19}{14}$

3. $\frac{37}{80}, \frac{37}{81}$

4. $-\frac{144}{126}, -\frac{8}{7}$

5. $-\frac{32}{21}, -\frac{29}{21}$

6. $-\frac{1}{22}, -\frac{1}{21}$

7. $\dfrac{3}{-2}, \dfrac{-3}{2}$

8. $\dfrac{-2}{3}, -.66$

9. $.333, \dfrac{1}{3}$

10. $-\dfrac{11}{12}, \dfrac{11}{-12}$

11. $-1.44, -1.404$

12. $.3, .30$

13. $.55, .505$

14. $-1.22, -1.220$

15. $.3700, .370$

16. $.\overline{3}, .\overline{32}$

17. $.4\overline{35}, .\overline{435}$

Solve for x in Exercises 18–21.

18. $\dfrac{x}{5} = \dfrac{56}{20}$

19. $\dfrac{3}{x} = \dfrac{15}{-10}$

20. $\dfrac{9}{-24} = \dfrac{6}{x}$

21. $\dfrac{15}{6} = \dfrac{x}{4}$

Order the numbers in Exercises 22–27 from smallest to largest.

22. $0.55, 0.055, 0.505$

23. $.\overline{3}, .33, .3$

24. $-.\overline{4}, -.44, -.4$

25. $1.6, 1.\overline{6}, 1.66, 1.666$

26. $0, 0.22, -.22, -.202, 0.202$

27. $0, \dfrac{6}{7}, -1.1427, -\dfrac{8}{7}, .8571$

Find all the whole numbers x that satisfy the inequalities in Exercises 28 and 29.

28. $\dfrac{x}{3} < 2$

29. $\dfrac{x}{4} < \dfrac{7}{4}$

Find all the integers x that satisfy the inequalities in Exercises 30–33.

30. $-\dfrac{6}{7} < \dfrac{x}{7} < 1$

31. $-\dfrac{11}{2} < \dfrac{x}{2} < \dfrac{3}{2}$

32. $-1 < \dfrac{x}{3} < 1$

33. $-\dfrac{1}{4} < \dfrac{x}{4} < 2$

Determine which is the better buy in Exercises 34–40.

34. 40 ounces of soap powder for $9.36 or 92 ounces for $21.16?

35. 3 bars of soap for $.41 or 7 bars for $.96?

36. 5 pounds of apples for $1.98 or 4 pounds for $1.69?

37. 6.4-ounce tube of toothpaste for $1.29 for 4.7 ounces for $.79?

38. 40 ounces of peanut butter for $3.49 or 18 ounces for $1.49?

In Exercises 39 and 40, determine which solution has the larger fraction of salt.

39. A 40-gallon salt solution that contains 5 gallons of salt or a 60-gallon salt solution that contains 8 gallons of salt?

40. A 45-gallon salt solution that contains 30 gallons of water or a 52-gallon salt solution that contains 39 gallons of water?

2.3 / ADDING AND SUBTRACTING FRACTIONS AND DECIMALS

In order to put fractions and decimals to work in solving problems, we need to describe ways of combining them using addition, subtraction, multiplication, and division. In this section we review addition and subtraction of fractions and decimals.

If two fractions have the same denominator, it is easy to visualize their sum and difference. Consider $\frac{4}{3}$ and $\frac{1}{3}$.

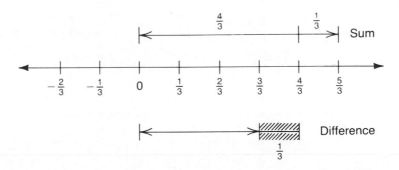

The segment above the number line has been formed by combining segments of lengths $\frac{4}{3}$ and $\frac{1}{3}$. We see that it has length $\frac{5}{3}$; that is to say,

$\frac{4}{3} + \frac{1}{3} = \frac{5}{3}$. To add two fractions with the same denominator, we simply add their numerators. The segment below the number line is the result of removing a segment of length $\frac{1}{3}$ from a segment of length $\frac{4}{3}$. We are left with a segment of length $\frac{3}{3}$; that is to say, $\frac{4}{3} - \frac{1}{3} = \frac{3}{3}$. To subtract one fraction from a larger one with the same denominator, we subtract the numerators.

When we have two fractions with different denominators that we want to add or subtract, we first rename the fractions so that they have the same denominator. Then we combine numerators as in the examples above. Any multiple of the two denominators can be chosen for the new denominator. One apparent multiple is the product of the two denominators; the smallest multiple is the LCM.

Example 1   Compute the sum $\dfrac{5}{6} + \dfrac{3}{4}$ and the difference $\dfrac{5}{6} - \dfrac{3}{4}$.

a. We need to rewrite $\frac{5}{6}$ and $\frac{3}{4}$ as fractions with the same denominator. If we use the product $6 \cdot 4$, then we have $\frac{5}{6} = \frac{5 \cdot 4}{6 \cdot 4} = \frac{20}{24}$ and $\frac{3}{4} = \frac{6 \cdot 3}{6 \cdot 4} = \frac{18}{24}$, so

$$\frac{5}{6} + \frac{3}{4} = \frac{20}{24} + \frac{18}{24} = \frac{38}{24} = \frac{19}{12}$$

$$\frac{5}{6} - \frac{3}{4} = \frac{20}{24} - \frac{18}{24} = \frac{2}{24} = \frac{1}{12}.$$

b. Say we choose the LCM of 6 and 4. Since $6 = 2 \cdot 3$ and $4 = 2 \cdot 2$, the LCM is $2 \cdot 2 \cdot 3$, or 12. Then

$$\frac{5}{6} = \frac{2 \cdot 5}{2 \cdot 6} = \frac{10}{12} \quad \text{and} \quad \frac{3}{4} = \frac{3 \cdot 3}{3 \cdot 4} = \frac{9}{12}.$$

Thus

$$\frac{5}{6} + \frac{3}{4} = \frac{10}{12} + \frac{9}{12} = \frac{19}{12},$$

and

$$\frac{5}{6} - \frac{3}{4} = \frac{10}{12} - \frac{9}{12} = \frac{1}{12}. \qquad \square$$

If we want to say what the sum and difference are for any two fractions $\frac{a}{b}$ and $\frac{c}{d}$, we use the common denominator that is easy to identify, bd. This is one way to say how addition and subtraction are performed for fractions.

$$\text{If } \frac{a}{b} \text{ and } \frac{c}{d} \text{ are two fractions, then}$$

$$\frac{a}{b} + \frac{c}{d} = \frac{ad}{bd} + \frac{bc}{bd} = \frac{ad + bc}{bd}, \text{ and}$$

$$\frac{a}{b} - \frac{c}{d} = \frac{ad}{bd} - \frac{bc}{bd} = \frac{ad - bc}{bd}.$$

When numbers are written in decimal form, the calculator can do the addition and subtraction for us. To see the underlying principle, consider first the example $.543 + .36$. As common fractions, this sum equals $\frac{543}{1000} + \frac{36}{100}$. We can use 1000 for a common denominator and add the numerators: $\frac{543}{1000} + \frac{360}{1000} = \frac{903}{1000} = .903$. Changing $\frac{36}{100}$ to $\frac{360}{1000}$ is the same as changing .36 to .360. Thus to add the two numbers in decimal notation without changing to common fractions, we make the two decimals the same length, add as if they were whole numbers, and set the decimal point to give the same number of decimal places in the sum.

The procedure for subtraction is the same. To compute $.543 - .36$ without a calculator, we write $.543 - .360$. Since $543 - 360 = 183$, the solution is $.543 - .360 = .183$.

Example 2   Compute $\frac{5}{6} + \frac{3}{4}$ and $\frac{5}{6} - \frac{3}{4}$ as decimals.

In Example 1 we computed the sum and difference in fraction form. Since $\frac{5}{6}$ is not a finite decimal, the computation with decimals will give us approximate answers. This keying sequence changes the fractions to decimal form and adds them:

Keying sequence	Display	Comments
[5]	5	
[÷]	5	
[6]	6	

Keying sequence	*Display*	*Comments*
$+$	.83333333	Pressing $+$ causes $\div$ to be performed.
3	3	
$\div$	3	
4	4	
$=$	1.5833333	Pressing $=$ causes the second $\div$ and then $+$ to be performed.

In Example 1 we computed $\frac{5}{6} + \frac{3}{4} = \frac{19}{12}$. You can check that the calculator gives 1.5833333 as the decimal approximation to $\frac{19}{12}$. The keying sequence for the difference $\frac{5}{6} - \frac{3}{4}$ is this:

Keying sequence	*Display*	*Comments*
5	5	
$\div$	5	
6	6	
$-$	.83333333	Pressing $-$ causes $\div$ to be performed.
3	3	
$\div$	3	
4	4	
$=$	.08333333	Pressing $=$ causes the second $\div$ and then $-$ to be performed.

The decimal .08333333 approximates $\frac{1}{12}$, our answer in Example 1. (It is easy to change a fraction to a decimal on the calculator. In the next section we will learn how to change a decimal such as $.08\overline{3}$ to its fraction form.)   □

Example 3 Compute $\dfrac{1}{2} - \left(-\dfrac{2}{5}\right)$.

a. Remember that subtracting a number in fractional form is the same as adding its opposite:

$$\frac{1}{2} - \left(-\frac{2}{5}\right) = \frac{1}{2} + \frac{2}{5}.$$

We can use 10 as the common denominator:

$$\frac{1}{2} - \left(-\frac{2}{5}\right) = \frac{1}{2} + \frac{2}{5}$$

$$= \frac{5}{10} + \frac{4}{10}$$

$$= \frac{9}{10}.$$

b. To compute in decimal form, we can use this keying sequence:

Keying sequence: $\boxed{1}\ \boxed{\div}\ \boxed{2}\ \boxed{-}\ \boxed{2}\ \boxed{+/-}\ \boxed{\div}\ \boxed{5}\ \boxed{=}$

Display: 1   1   2   .5   2   −2   −2   5   0.9

Of course, the decimal .9 and the fraction $\frac{9}{10}$ are the same number.   □

An important consequence of the method for adding fractions is $\dfrac{2}{3} + \dfrac{(-2)}{3} = \dfrac{2 + (-2)}{3} = \dfrac{0}{3} = 0$. Thus $\dfrac{-2}{3}$ is the opposite of $\dfrac{2}{3}$:

$$-\frac{2}{3} = \frac{-2}{3}.$$

Also, $\dfrac{-2}{3} = \dfrac{(-1)(-2)}{(-1)(3)}$, so we have three ways to write this fraction:

$$-\frac{2}{3} = \frac{-2}{3} = \frac{2}{-3}.$$

Fractions and Decimals Greater Than 1 (or Less Than −1)

A positive fraction that has a larger whole number numerator than denominator is a fraction greater than 1. Such fractions can always be written as a whole number plus a fraction less than 1.

Example 4   Write the fraction $\dfrac{12}{5}$ as a whole number plus a positive fraction less than 1.

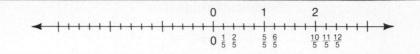

Figure 2.4

The number line in Figure 2.4 suggests that

$$\frac{12}{5} = 2 + \frac{2}{5}.$$

To see that this is true, we write

$$\frac{12}{5} = \frac{10 + 2}{5} = \frac{10}{5} + \frac{2}{5} = 2 + \frac{2}{5}.$$

We wrote 12 as $10 + 2$ because 10 is the largest multiple of 5 in 12.   □

Numbers like $2 + \frac{2}{5}$ that consist of a whole number plus a positive fraction less than 1 are often called **mixed numbers**. Rather than $2 + \frac{2}{5}$, this number is usually written $2\frac{2}{5}$. Recall that when we write a number and a variable with no operation sign between them, the operation is understood to be multiplication: $2x$ means $2 \cdot x$. However, $2\frac{2}{5}$ means $2 + \frac{2}{5}$. We will be careful when we mean multiplication of a whole number and a fraction to use either a $\cdot$ sign or parentheses; that is, we will either write $2 \cdot \frac{2}{5}$ or $(2)(\frac{2}{5})$ when we mean 2 times $\frac{2}{5}$. The symbol $2\frac{2}{5}$ always means $2 + \frac{2}{5}$.

The symbol $-2\frac{2}{5}$ denotes the opposite of $2\frac{2}{5}$. Thus $-2\frac{2}{5} = -(2 + \frac{2}{5})$. See where this number is located on the number line.

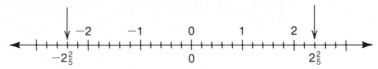

Mixed numbers are also written in decimal form. Since $\frac{2}{5} = .4$, the mixed number $2\frac{2}{5}$ is written as 2.4 and $-2\frac{2}{5}$ as -2.4.

Example 5   Write $7\frac{3}{4}$ in decimal form and as a common fraction.

Since $\frac{3}{4} = .75$, the mixed number $7\frac{3}{4} = 7.75$. The calculator keying sequence that gives $7\frac{3}{4}$ in decimal form is

$$\boxed{7}\ \boxed{+}\ \boxed{3}\ \boxed{\div}\ \boxed{4}\ \boxed{=}.$$

To convert $7\frac{3}{4}$ to a common fraction, remember that $7\frac{3}{4}$ is a sum:

$$7\frac{3}{4} = 7 + \frac{3}{4}$$

$$= \frac{28}{4} + \frac{3}{4}$$

$$= \frac{31}{4}.$$

□

Example 6 Write $-3\frac{5}{7}$ as a common fraction.

Parentheses make this computation clear:

$$-3\frac{5}{7} = -\left(3 + \frac{5}{7}\right)$$

$$= -\left(\frac{21}{7} + \frac{5}{7}\right)$$

$$= -\frac{26}{7}.$$

□

Example 7 Compute $-3\frac{1}{4} + 2\frac{5}{6}$.

a. If we choose to compute with fractions rather than decimals, first we need to write the mixed numbers as fractions and then add:

$$-3\frac{1}{4} + 2\frac{5}{6} = -\left(3 + \frac{1}{4}\right) + \left(2 + \frac{5}{6}\right)$$

$$= -\left(\frac{12}{4} + \frac{1}{4}\right) + \left(\frac{12}{6} + \frac{5}{6}\right)$$

$$= -\frac{13}{4} + \frac{17}{6}$$

$$= -\frac{39}{12} + \frac{34}{12}$$ (Change fractions to common denominator)

$$= \frac{-39 + 34}{12}$$

$$= \frac{-5}{12}.$$

b. If we choose to compute with decimals, we realize that the calculator will give an approximation because $\frac{5}{6}$ is not a finite decimal. The keying sequence below uses the fact that $-3\frac{1}{4}$ means $-(3 + \frac{1}{4})$ and that $2\frac{5}{6}$ means $2 + \frac{5}{6}$.

Mathematical phrase $-3\frac{1}{4} + 2\frac{5}{6}$

Keying sequence	Display	Comments
(	0	
3	3	
+	3	
1	1	
÷	1	
4	4	
)	3.25	Pressing) causes ÷, then + to be performed.
+/−	−3.25	
+	−3.25	
2	2	
+	−1.25	Pressing + causes the prior + to be performed.
5	5	
÷	5	
6	6	

Keying sequence	*Display*	*Comments*
$\boxed{=}$	$-.41666667$	Pressing $\boxed{=}$ causes $\boxed{\div}$, then $\boxed{+}$ to be performed.

You can check that $\frac{-5}{12} \doteq -.41666667$.   □

Exercises 2.3

Write the fractions in Exercises 1 and 2 as mixed numbers and decimals.

1. $\dfrac{111}{13}$ 2. $-\dfrac{212}{7}$

Write the mixed numbers in Exercises 3 and 4 as fractions and decimals.

3. $-17\dfrac{3}{4}$ 4. $21\dfrac{5}{6}$

Write the decimals in Exercises 5–8 as common fractions and mixed numbers.

5. 17.2 6. -8.34

7. -17.567 8. 5.0003

Perform the indicated operations in Exercises 9–24 and write the answers in fractional form.

9. $\dfrac{5}{12} + \dfrac{7}{18}$ 10. $\dfrac{11}{30} + \dfrac{9}{20}$

11. $1 + \dfrac{3}{4}$ 12. $\dfrac{8}{15} - \dfrac{4}{35}$

13. $3\dfrac{2}{3} + 4\dfrac{3}{4}$ 14. $2 - \dfrac{7}{3}$

15. $\dfrac{1}{2} - \dfrac{1}{3} + \dfrac{1}{4}$ 16. $\dfrac{3}{10} + \dfrac{4}{15} - \dfrac{5}{6}$

17. $\dfrac{8}{39} - \dfrac{15}{26}$ 18. $6\dfrac{1}{5} - 2\dfrac{3}{4}$

19. $1\frac{1}{3} - 2\frac{5}{6}$

20. $5\frac{3}{8} + 2\frac{3}{4}$

21. $\frac{2}{3} + \frac{2}{-3}$

22. $\frac{2}{3} + \frac{-2}{3}$

23. $-\frac{7}{6} - \frac{-7}{6}$

24. $-\frac{7}{6} - \frac{7}{-6}$

25. Use decimals to check your computations in Exercises 9, 12, and 19.

26. Convert 23,429 feet to miles and feet.

27. The table below shows the parts of three parks in a city that have been set aside for recreation, picnicking, and parking. In each, find the remaining part.

	Recreation	Picnicking	Parking	Remaining part
Shadeywood	$\frac{3}{8}$	$\frac{1}{4}$	$\frac{3}{16}$	?
10th Avenue	$\frac{1}{3}$	$\frac{2}{9}$	$\frac{1}{18}$	?
Linden	$\frac{2}{5}$	$\frac{1}{3}$	$\frac{1}{6}$	?

28. Find the total number of yards in 3 rolls of cloth if their measurements are $50\frac{1}{2}$ yards, $48\frac{5}{8}$ yards, and $36\frac{3}{4}$ yards.

29. Normal body temperature is $98\frac{3}{5}$ degrees. How many degrees above or below normal is a temperature of 100 degrees? $97\frac{2}{3}$ degrees? $103\frac{3}{10}$ degrees?

30. A share of stock was worth $22\frac{1}{8}$ dollars per share on Monday. On Thursday it was worth $18\frac{3}{4}$ dollars per share. How much did the stock fall per share?

31. A carpenter cut pieces that were $2\frac{1}{2}$ feet long and $5\frac{3}{5}$ feet long from a 20-foot board. How much is left?

Use fractions to write the mathematical phrase evaluated by each sequence of calculator key strokes in Exercises 32–34.

32. $\boxed{7}\boxed{\div}\boxed{3}\boxed{-}\boxed{4}\boxed{\div}\boxed{5.}\boxed{=}$

33. $\boxed{(}\boxed{3}\boxed{+}\boxed{1}\boxed{\div}\boxed{2}\boxed{)}\boxed{+/-}\boxed{+}\boxed{2}\boxed{+}\boxed{1}\boxed{\div}\boxed{5}\boxed{=}$

34. $\boxed{(}\boxed{4}\boxed{+}\boxed{2}\boxed{\div}\boxed{3}\boxed{)}\boxed{+/-}\boxed{-}\boxed{(}\boxed{5}\boxed{+}\boxed{5}\boxed{\div}\boxed{6}\boxed{)}\boxed{=}$

In Exercises 35–39, combine like terms to simplify each expression.

35. $\dfrac{3}{2}x + \dfrac{4}{3}x + 5y$

36. $y^2 + \dfrac{1}{2}y^2 + x + 1$

37. $z^3 - \dfrac{3}{2}z^3 + 4$

38. $x + \dfrac{1}{2}x + \dfrac{1}{3}x$

39. $a^2 - \dfrac{1}{3}a^2 + \dfrac{1}{2}b - \dfrac{2}{3}b$

40. Add the following, giving your answers as fractions:

(a) $\dfrac{1}{2} + \dfrac{1}{4}$

(b) $\dfrac{1}{2} + \dfrac{1}{4} + \dfrac{1}{8}$

(c) $\dfrac{1}{2} + \dfrac{1}{4} + \dfrac{1}{8} + \dfrac{1}{16}$

(d) Write and evaluate the next three sums suggested by (a) through (c).

(e) Write the hundredth sum as suggested by (a) through (c).

(f) Guess the value of the sum in part (e) and explain your guess.

41. Repeat Problem 40, using the fractions $\frac{1}{3}, \frac{1}{9}, \frac{1}{27}, \ldots$.

42. Repeat Problem 40, giving your answers as decimals.

43. Repeat Problem 41, giving your answers as decimals.

2.4 / MULTIPLYING AND DIVIDING FRACTIONS AND DECIMALS

The product of two fractions describes the area of a rectangle, just as the product of two whole numbers describes the area of a rectangle. Say we want to visualize the product $\left(\frac{2}{3}\right) \cdot \left(\frac{1}{5}\right)$.

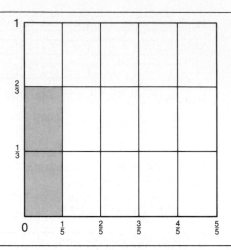

Figure 2.5

The area of the whole square shown in Fig. 2.5 is $1 \cdot 1 = 1$ square unit. The shaded rectangle has area $\left(\frac{2}{3}\right) \cdot \left(\frac{1}{5}\right)$. Its area consists of 2 parts out of the 15 parts that make up the whole square. Thus we say its area is $\frac{2}{15}$ and that $\left(\frac{2}{3}\right) \cdot \left(\frac{1}{5}\right) = \frac{2}{15}$.

Notice that the 15 parts of the square above are a result of 3 rows with 5 rectangles each; that is, $3 \cdot 5$. The 2 parts of the shaded rectangle are a result of 2 rows with 1 rectangle each; that is, $2 \cdot 1$. Thus when we multiply $\left(\frac{2}{3}\right) \cdot \left(\frac{1}{5}\right)$, the numerator is $2 \cdot 1$ and the denominator is $3 \cdot 5$. For any two fractions $\frac{a}{b}$ and $\frac{c}{d}$, the product is

$$\frac{a}{b} \cdot \frac{c}{d} = \frac{a \cdot c}{b \cdot d}.$$

Assume that we have a fraction $\frac{a}{b}$ and look also at $\frac{b}{a}$. (Be sure $a \neq 0$ and $b \neq 0$.) Computing the product, we get $\frac{a}{b} \cdot \frac{b}{a} = \frac{a \cdot b}{b \cdot a} = \frac{a \cdot b}{a \cdot b} = 1$. Always $\frac{a}{b} \cdot \frac{b}{a} = 1$. The fraction $\frac{b}{a}$ is called the **reciprocal** of $\frac{a}{b}$. A fraction times its reciprocal is always 1.

With a little reasoning we can see what the quotient $\frac{a}{b} \div \frac{c}{d}$ equals. Every division statement corresponds to a multiplication statement. Study these corresponding division and multiplication statements for whole numbers:

$$8 \div 4 = 2 \quad \text{because} \quad 2 \cdot 4 = 8$$

$$54 \div 6 = 9 \quad \text{because} \quad 9 \cdot 6 = 54$$

$$45 \div 9 = 5 \quad \text{because} \quad 5 \cdot 9 = 45.$$

If we want to fill in the box so that $\frac{a}{b} \div \frac{c}{d} = \boxed{}$, we need to fill it so that $\boxed{} \cdot \frac{c}{d} = \frac{a}{b}$. What do we multiply times $\frac{c}{d}$ to get $\frac{a}{b}$? First we multiply $\frac{c}{d}$ by its reciprocal $\frac{d}{c}$ to get 1 and then multiply by $\frac{a}{b}$. We fill the box with $\boxed{\frac{a}{b} \cdot \frac{d}{c}}$ because $\boxed{\frac{a}{b} \cdot \frac{d}{c}} \cdot \frac{c}{d} = \frac{a}{b} \cdot 1 = \frac{a}{b}$. Thus the quotient of the two fractions is given as

$$\frac{a}{b} \div \frac{c}{d} = \frac{a}{b} \cdot \frac{d}{c}.$$

Sometimes we say, the quotient of two fractions is the first times the reciprocal of the second. Perhaps you have said, invert the second and multiply. The meaning is the same. This is how we write the definitions for multiplication and division of fractions:

Assume that $\frac{a}{b}$ and $\frac{c}{d}$ are fractions. Then

$$\frac{a}{b} \cdot \frac{c}{d} = \frac{ac}{bd} \text{ and}$$

$$\frac{a}{b} \div \frac{c}{d} = \frac{a}{b} \cdot \frac{d}{c} \text{ if } c \neq 0.$$

Example 1   Compute the product $\frac{-5}{2} \cdot \frac{7}{12}$ and the quotient $\frac{-5}{2} \div \frac{7}{12}$.

We can compute in fraction form using the definitions above:

$$\frac{-5}{2} \cdot \frac{7}{12} = \frac{-5 \cdot 7}{2 \cdot 12} = \frac{-35}{24}$$

$$\frac{-5}{2} \div \frac{7}{12} = \frac{-5}{2} \cdot \frac{12}{7} = \frac{-5 \cdot 12}{2 \cdot 7} = \frac{-5 \cdot 6}{7} = \frac{-30}{7}.$$ □

Using the Reciprocal Key for Fractions

We can use the calculator to get a decimal approximation for the product and quotient in Example 1. Of course, the fraction $\frac{-5}{12}$ can be keyed with the $\boxed{\div}$ key: $\boxed{5}\,\boxed{+/-}\,\boxed{\div}\,\boxed{12}\,\boxed{=}$. Another key that is often useful in computing with fractions is the reciprocal key $\boxed{1/x}$. Pressing $\boxed{1/x}$ gives the reciprocal of a number on display.

Keying sequence	Mathematical phrase	Display
$\boxed{5}$ $\boxed{1/x}$	$\dfrac{1}{5}$	.2
$\underline{10}$ $\boxed{+/-}$ $\boxed{1/x}$	$-\dfrac{1}{10}$	−.1
$\underline{.4}$ $\boxed{1/x}$	$\dfrac{1}{.4}$	2.5

Since $\dfrac{7}{12} = 7 \cdot \dfrac{1}{12}$, the fraction $\dfrac{7}{12}$ can be keyed as

$\boxed{7}$ $\boxed{\times}$ $\underline{12}$ $\boxed{1/x}$ $\boxed{=}$.

Example 2   Give a finite decimal approximation for

$$\frac{-5}{2} \cdot \frac{7}{12} \quad \text{and} \quad \frac{-5}{2} \div \frac{7}{12}.$$

The $\boxed{1/x}$ key permits us to key this product and quotient without using parentheses. Observe first that $\dfrac{-5}{2} \cdot \dfrac{7}{12} = -5 \cdot \dfrac{1}{2} \cdot 7 \cdot \dfrac{1}{12}$.

Mathematical phrase $\dfrac{-5}{2} \cdot \dfrac{7}{12}$

Keying sequence	Display	Comments
$\boxed{5}$	5	
$\boxed{+/-}$	−5	
$\boxed{\times}$	−5	
$\boxed{2}$	2	
$\boxed{1/x}$	.5	Reciprocal of 2.
$\boxed{\times}$	−2.5	Pressing $\boxed{\times}$ causes the first $\boxed{\times}$ to be performed.
$\boxed{7}$	7	

Keying sequence	Display	Comments
$\boxed{\times}$	-17.5	Pressing $\boxed{\times}$ causes the second $\boxed{\times}$ to be performed.
$\underline{12}$	12	
$\boxed{1/x}$	.08333333	Reciprocal of 12.
$\boxed{=}$	-1.4583333	Pressing $\boxed{=}$ causes the third $\boxed{\times}$ to be performed.

To compute the quotient $\dfrac{-5}{2} \div \dfrac{7}{12}$, we are not able merely to imitate what we did above and key $\boxed{5}\boxed{+/-}\boxed{\times}\boxed{2}\boxed{1/x}\boxed{\div}\boxed{7}\boxed{\times}\underline{12}\boxed{1/x}\boxed{=}$ because this sequence evaluates $\dfrac{-5}{2 \cdot 7 \cdot 12}$. One thing we can do is key the sequence with parentheses:

$$\boxed{5}\boxed{+/-}\boxed{\div}\boxed{2}\boxed{\div}\boxed{(}\boxed{7}\boxed{\div}\underline{12}\boxed{)}\boxed{=} \qquad \begin{array}{c}\text{(Display)}\\ -4.2857143\end{array}$$

Another keying sequence can be found by writing $\dfrac{-5}{2} \div \dfrac{7}{12} = \dfrac{-5}{2} \cdot \dfrac{12}{7}$ and then imitating the the procedure for the products shown above:

$$\boxed{5}\boxed{+/-}\boxed{\times}\boxed{2}\boxed{1/x}\boxed{\times}\underline{12}\boxed{\times}\boxed{7}\boxed{1/x}\boxed{=} \qquad \begin{array}{c}\text{(Display)}\\ -4.2857143\end{array} \quad \square$$

Example 3 Find $2\frac{1}{3} \cdot 1\frac{5}{6}$ and $2\frac{1}{3} \div 1\frac{5}{6}$.

a. To compute with the numbers in fraction form, we first change the mixed numbers to common fractions:

$$2\frac{1}{3} \cdot 1\frac{5}{6} = \frac{7}{3} \cdot \frac{11}{6} = \frac{77}{18}$$

$$2\frac{1}{3} \div 1\frac{5}{6} = \frac{7}{3} \div \frac{11}{6} = \frac{7}{3} \cdot \frac{6}{11} = \frac{7 \cdot 6}{3 \cdot 11} = \frac{7 \cdot 2}{11} = \frac{14}{11}.$$

b. We can use the calculator to get a decimal approximation. The two keying sequences are given below. Observe that parentheses must be used for the mixed numbers $2\frac{1}{3}$ and $1\frac{5}{6}$.

$$2\frac{1}{3} \cdot 1\frac{5}{6} = \left(2 + \frac{1}{3}\right) \cdot \left(1 + \frac{5}{6}\right)$$

Keying sequence:

(2 + 3 1/x) × (1 + 5 × 6 1/x) =

(Display)
4.2777778

$$2\frac{1}{3} \div 1\frac{5}{6} = \left(2 + \frac{1}{3}\right) \div \left(1 + \frac{5}{6}\right)$$

Keying sequence:

(2 + 3 1/x) ÷ (1 + 5 × 6 1/x) =

(Display)
1.2727273   □

Multiplying and Dividing Decimals by Powers of 10

Example 4   Compute (3.41)(10).

The calculator gives (3.41)(10) = 34.1. The explanation is clear when the computation is written in fraction form:

$$(3.41)(10) = \frac{341}{10^2} \cdot \frac{10}{1} = \frac{341 \cdot 10}{10 \cdot 10} = \frac{341}{10} = 34.1.$$ □

When a finite decimal is multiplied by 10, we see that one digit of the decimal is moved from the right of the decimal point to the left to give a larger number. If a number is multiplied by 10^2, then it is multiplied twice by 10 and two digits are moved to the left of the decimal point. Multiplying by 10^3 moves three digits across the decimal point. However, what if we divide a decimal by 10?

Example 5   Compute 3.41 ÷ 10.

Try this on the calculator. You should get .341. The reason is that in order to fill in the box for 3.41 ÷ 10 = ☐, we must have ☐ · 10 = 3.41. What number times 10 gives 3.41? The answer is .341. □

The example suggests that dividing by 10 moves one digit from the left side of the decimal point to the right side to give a smaller number. Dividing by 10^2 will move two digits from the left side to the right. Thus multiplying and dividing decimals by powers of 10 are particularly simple operations. We can write these products and quotients without using a calculator.

Using Fractions in Markdown and Markup Problems

It is not uncommon to read that a business has marked up the wholesale price of an item by a certain fraction to get its retail price or that a store is marking down its price to convince us that we should buy immediately.

Example 6 Sinks Department Store is having a $\frac{1}{4}$-off sale. Complete the following table:

Old price ($)	8	12	24.40		x
New price ($)				21	

Each old price is reduced by $\frac{1}{4}$. We must multiply $\frac{1}{4}$ times each old price and subtract this amount from the old price:

$$8 - \left(\frac{1}{4} \cdot 8\right) = 8 - 2 = 6$$

$$12 - \left(\frac{1}{4} \cdot 12\right) = 12 - 3 = 9$$

$$24.40 - \frac{1}{4}(24.40) = 24.40 - 6.10 = 18.30.$$

If the old price is x dollars, then the same reasoning gives $x - \frac{1}{4}x$ as the new price.

One advantage of working through this example for a variable amount, namely x dollars, is that we obtain a short cut to the arithmetic involved. The distributive property tells us that $x - \frac{1}{4}x = (1 - \frac{1}{4})x = \frac{3}{4}x$. Thus to get the new price we need only multiply the old price by $\frac{3}{4}$. This observation helps us find the old price if \$21 is the new price. We ask, "\$21 is $\frac{3}{4}$ times what number?" The answer is \$28. The completed table is this:

Old price (\$)	8	12	24.40	28	x
New Price (\$)	6	9	18.30	21	$x - \frac{1}{4}x = \frac{3}{4}x$

□

Exercises 2.4

Perform the indicated operations in Exercises 1–12. Write your answers in fractional form.

1. $\dfrac{20}{63} \cdot \dfrac{21}{40}$

2. $\dfrac{-92}{87} \cdot \dfrac{58}{23}$

3. $\dfrac{7}{-55} \div \dfrac{1}{15}$

4. $\dfrac{247}{187} \div \dfrac{39}{34}$

5. $\left(16\dfrac{1}{4}\right)\left(-2\dfrac{14}{15}\right)$

6. $\left(2\dfrac{19}{22}\right)\left(5\dfrac{16}{21}\right)$

7. $21 \cdot \dfrac{15}{28}$

8. $22 \cdot \dfrac{6}{33}$

9. $\dfrac{-100}{63} \cdot \dfrac{21}{110} \cdot \dfrac{-33}{70}$

10. $3\dfrac{3}{7} \div 1\dfrac{11}{14}$

11. $-7\dfrac{7}{10} \div 6\dfrac{1}{15}$

12. $\dfrac{1}{2} \cdot \dfrac{2}{3} \cdot \dfrac{3}{4} \cdot \dfrac{4}{5}$

13. Use decimals to check your computations in Exercises 1, 4, 5, and 10.

Write the numbers in Exercises 14–17 in decimal form.

14. $(2.123)(10^5)$

15. $(-3413.5) \div 10^3$

16. $(-75.624)(10^2) \div 10^4$

17. $(4567.9 \div 10^3) \div 10^2$

Solve for n in Exercises 18–21.

18. $8643 = (86.43)(10^n)$

19. $-76{,}540 = (-7.654)(10^n)$

20. $97.345 = \dfrac{9734.5}{10^n}$

21. $8.765 = \dfrac{8765}{10^n}$

Make each statement in Exercises 22–25 true by replacing ☐ by an appropriate fraction.

22. $6 \cdot$ ☐ $= 9$ 23. $28 \cdot$ ☐ $= -315$

24. $(-12) \cdot$ ☐ $= 9$ 25. $252 \cdot$ ☐ $= 30$

Use the distributive property and combine like terms to simplify the expressions in Exercises 26–29.

26. $a + \dfrac{1}{2}(b - a)$ 27. $\dfrac{7}{6}(x + 1) + \dfrac{5}{4}(2x - 3)$

28. $\dfrac{x}{2}(3x - 1) + \dfrac{x}{4}(2x + 3) + \dfrac{1}{3}(x - 2)$ 29. $\dfrac{1}{2}a(a - b) - \dfrac{1}{2}b(b - a)$

Write the mathematical phrase evaluated by each sequence of calculator key strokes in Exercises 30–34.

30. ⟨1⟩⟨÷⟩⟨2⟩⟨+⟩⟨1⟩⟨÷⟩⟨4⟩⟨−⟩⟨1⟩⟨÷⟩⟨8⟩⟨=⟩

31. ⟨2⟩⟨1/x⟩⟨+⟩⟨4⟩⟨1/x⟩⟨−⟩⟨8⟩⟨1/x⟩⟨=⟩

32. ⟨(⟩⟨2⟩⟨÷⟩⟨5⟩⟨+/−⟩⟨)⟩⟨×⟩⟨(⟩⟨7⟩⟨÷⟩⟨3⟩⟨)⟩⟨=⟩

33. ⟨2⟩⟨×⟩⟨5⟩⟨+/−⟩⟨1/x⟩⟨×⟩⟨7⟩⟨×⟩⟨3⟩⟨1/x⟩⟨=⟩

34. ⟨(⟩⟨11⟩⟨×⟩⟨7⟩⟨1/x⟩⟨)⟩⟨÷⟩⟨(⟩⟨4⟩⟨×⟩⟨5⟩⟨1/x⟩⟨)⟩⟨=⟩

35. A carpenter is paid at the rate of $4\frac{1}{2}$ dollars per hour.

(a) If he works for $13\frac{1}{5}$ hours, how much does he earn?

(b) If he earned \$24.75, how many hours did he work?

36. A car averages 21.5 miles per gallon of gasoline.

(a) How far will the car travel on 24.5 gallons of gasoline?

(b) If the car travels 387 miles, how many gallons of gasoline does it use?

37. How much money is earned in 40 hours at the hourly rate of

(a) \$3.50?   (b) \$5.75?

(c) \$8.92?   (d) \$y?

38. At \$3.75 per hour, how many hours will it take to earn

(a) \$22.50?   (b) \$41.25?

(c) \$31.50?   (d) \$y?

39. The telephone rate between two zones is 45 cents for the first three minutes and 12 cents for each additional minute. How many minutes was a call that cost

    (a) $1.41?　　　　　　　　　(b) $1.89?

    (c) $2.85?　　　　　　　　　(d) x?

40. You had 120 silver dollars and you gave your brother Joe $\frac{1}{3}$ of them, your brother Paul $\frac{1}{4}$ of them, and your sister Kate $\frac{1}{5}$ of them. How many did each receive? What fraction of the total did you give away? What fraction of the total do you have left?

41. A gambler spent 3 days at Las Vegas. He started with $726. Each day his gambling losses were $2 more than $\frac{2}{3}$ of his initial capital for that day. How much of his original capital was left at the end of his stay?

42. Shears Department Store is having a $\frac{1}{3}$-off sale. Complete the following table:

Old price ($)	12	21	24	30		45.30	x
New price ($)					22		

43. The Three D Discount Store marks the wholesale price up $\frac{2}{5}$. Complete the following table:

Wholesale price ($)	10	15	20.50		40	x
Retail price ($)				35		

44. A rectangle is $\frac{4}{3}$ feet longer than it is wide. Complete the following table:

Width	Length	Area
2	$2 + \dfrac{4}{3} = \dfrac{10}{3}$	$\dfrac{20}{3}$
$2\dfrac{1}{2}$		
	4	

Width	Length	Area
3		
	$2\frac{3}{5}$	
x		

45. The second of two numbers is one more than half of the first. Complete the following table:

First number	Second number	Sum
4	3	7
10		
−16		
	12	
	−11	
		28
x		

46. The second of three numbers is one-half of the first, and the third number is one-third of the first. Complete the following table:

First number	Second number	Third number	Sum
6	3	2	11
18			
	15		
		14	
			66
x			

47. A chain is cut into three parts. The largest piece is four times the smallest, and the other piece is one-half the difference between the largest and the smallest. Complete the following table:

Smallest piece	Largest piece	Other piece	Total length
4	16	6	26
10			
	64		
		33	
			117
x			

48. Hank Aaron hits a fly ball deep to center field. The center fielder stands 350 feet from home plate.

   (a) If sound travels 1087 feet per second, how soon does the center fielder hear the crack of the bat?

   (b) How soon does the TV sound microphone, standing 50 feet away from home plate, hear (receive) the crack of the bat?

   (c) The electronic signal of that sound now travels at the speed of light 20 miles to a home where you are watching the game. You are 10 feet from your set. How soon after the crack of Aaron's bat do you hear the sound? (The speed of light is approximately 186,284 miles per second.)

   (d) Who hears it first, you or the center fielder?

2.5 / CONSEQUENCES OF MULTIPLICATION AND DIVISION

Two important properties of fractions can be explained in terms of the definition of division of fractions.

Important
Property 1 Fractions are quotients of whole numbers.

Since whole numbers are fractions, we can write

$$2 \div 5 = \frac{2}{1} \div \frac{5}{1} = \frac{2}{1} \cdot \frac{1}{5} = \frac{2}{5}$$

$$7 \div 9 = \frac{7}{1} \cdot \frac{1}{9} = \frac{7}{9}.$$

Thus the fraction $\frac{2}{5}$ is the quotient $2 \div 5$; the fraction $\frac{7}{9}$ is the quotient $7 \div 9$. This gives us an important way of viewing fractions; namely, a fraction is a quotient of whole numbers. Of course, we have been using this fact in the early exercises. We have written $\frac{2+3+4}{7}$ as $(2 + 3 + 4) \div 7$, but an explanation of why fractions are quotients of whole numbers had to be delayed until we talked about division of fractions.

Important
Property 2 No fraction has a denominator 0.

We can explain now why there are no fractions with denominators 0. If n is a whole number different from 0, the symbol $\frac{n}{0}$, if it had meaning, would mean $n \div 0$; now $n \div 0 = \boxed{}$ only if $\boxed{} \cdot 0 = n$. But there is no whole number to put in this box. Any whole number times 0 is 0, not n. Thus there is no number $n \div 0$ or $\frac{n}{0}$. This is why textbooks talk about fractions $\frac{a}{b}$, where $b \neq 0$.

Complex Fractions

Since $\frac{4}{7} = 4 \div 7$ and $\frac{1}{3} = 1 \div 3$, it has become a convention to read the line between the whole numbers in a fraction as $\div$. Sometimes you will see a line used even when the numbers above and below it are not whole numbers. For example,

$$\frac{\frac{1}{6}}{\frac{2}{5}} \quad \text{means} \quad \frac{1}{6} \div \frac{2}{5}$$

$$\frac{\dfrac{-3}{7}}{\dfrac{5}{4}} \quad \text{means} \quad \frac{-3}{7} \div \frac{5}{4}.$$

Notice what happens when we evaluate these expressions.

Example 1   Evaluate $\dfrac{\dfrac{1}{6}}{\dfrac{2}{5}}$ means $\dfrac{\dfrac{-3}{7}}{\dfrac{5}{4}}.$

$$\frac{\dfrac{1}{6}}{\dfrac{2}{5}} = \frac{1}{6} \div \frac{2}{5} = \frac{1}{6} \cdot \frac{5}{2} = \frac{1 \cdot 5}{6 \cdot 2} = \frac{5}{12}$$

$$\frac{\dfrac{-3}{7}}{\dfrac{5}{4}} = \frac{-3}{7} \div \frac{5}{4} = \frac{-3}{7} \cdot \frac{4}{5} = \frac{-3 \cdot 4}{7 \cdot 5} = \frac{-12}{35}.$$

$\square$

Notice that the quotients above give a fraction made up of a product of whole numbers in the numerator and a product of whole numbers in the denominator. We can mark the expressions to indicate how to compute the values of the numerator and denominator:

$$\text{Numerator} \left[\frac{\dfrac{1}{6}}{\dfrac{2}{5}} \right] \text{Denominator}$$

$$\text{Numerator} \left[\frac{\dfrac{-3}{7}}{\dfrac{5}{4}} \right] \text{Denominator}.$$

In general, the expression $\dfrac{\dfrac{a}{b}}{\dfrac{c}{d}}$ has the value of $\dfrac{ad}{bc}$.

Writing Infinite Repeating Decimals as Fractions

We have seen that finite decimals are fractions. For example,

$$.13 \ = \frac{13}{100}$$

$$.205 = \frac{205}{1000}$$

$$7.4 \ \ = \frac{74}{10}.$$

Infinite repeating decimals are also fractions. To understand the method for writing an infinite repeating decimal as a fraction, you need to know that multiplying an infinite decimal by 10 causes one digit to move from the right side to the left side of the decimal point.

Example 2   Write the decimal $.\overline{6}$ as a common fraction.

We are looking for a fraction $F = .66666. \ldots$ We reason this way: $10F = 6.66666. \ldots$, so $10F - F = (6.66666. . .) - (.66666. . .) = 6$. Thus $9F = 6$. The fraction F that is multiplied by 9 to give 6 is $\frac{6}{9}$, or $\frac{2}{3}$. Does your calculator agree that $\frac{2}{3} = .\overline{6}$?   □

Example 3   Write the decimal $.1\overline{85}$ as a common fraction.

We want a fraction $F = .185858585. \ldots$ We can get two numbers with the same decimal part by multiplying F by 10 and multiplying F by 1000:

$$10F = 1.85858585. . .$$

$$1000F = 185.858585. \ldots$$

Subtracting $10F$ from $1000F$ gives a whole number:

$$1000F - 10F = 185.\overline{85} - 1.\overline{85} = 184$$

so

$$990F = 184.$$

The fraction F that is multiplied by 990 to give 184 is $\frac{184}{990}$. Use your calculator to check that $\frac{184}{990} = .1\overline{85}$.   □

In general, to change an infinite repeating decimal to a fraction, we start by multiplying the decimal by two powers of 10 to get two different numbers with the same decimal part. Then we subtract these numbers to get a whole number, which is the numerator of the fraction.

Finding Midpoints

If two points on the number line are given, we can find the point halfway between the two. For example, if the two points are 3 and 5, the point halfway between them is 4. The problem can be analyzed this way:

Find the distance between the points (that is to say, subtract the smaller from the larger). $5 - 3$

Compute half the distance between the points. $\frac{1}{2}(5 - 3)$

Add half the distance to the smaller number. $3 + \frac{1}{2}(5 - 3) = 4$

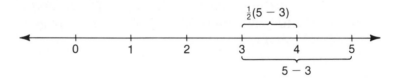

The distributive law permits us to write this computation another way:

$$3 + \frac{1}{2}(5 - 3) = 3 + \frac{1}{2} \cdot 5 - \frac{1}{2} \cdot 3$$

$$= 3 - \frac{1}{2} \cdot 3 + \frac{1}{2} \cdot 5$$

$$= \left(1 - \frac{1}{2}\right) \cdot 3 + \frac{1}{2} \cdot 5$$

$$= \frac{1}{2} \cdot 3 + \frac{1}{2} \cdot 5$$

$$= \frac{1}{2}(3 + 5).$$

Thus a second way to find the point midway between two numbers is this:

Compute half the sum of the two numbers. $\frac{1}{2}(3 + 5) = 4$

Sometimes $\frac{1}{2}$ times the sum of two numbers (or the sum divided by 2) is called the **average** of the two numbers. It represents the point halfway between the two numbers.

Example 4   Find the point midway between $-.12$ and 3.3.

a. We can compute the distance between 3.3 and $-.12$, find half the distance, and add it to $-.12$:

$$-.12 + \frac{1}{2}[3.3 - (-.12)] = 1.59.$$

b. We can compute the average of $-.12$ and 3.3:

$$\frac{1}{2}(-.12 + 3.3) = 1.59. \qquad \square$$

Exercises 2.5

Write the numbers in Exercises 1–6 as common fractions.

1. 6.7
2. -3.45
3. -37.4
4. 371.79
5. 23.4567
6. $.003$

Perform the indicated operations in Exercises 7–14 and write your answers in fractional form.

7. $\dfrac{\dfrac{6}{77}}{\dfrac{35}{8}}$

8. $\dfrac{-\dfrac{15}{28}}{\dfrac{45}{88}}$

9. $\dfrac{-12}{\dfrac{10}{13}}$

10. $\dfrac{\dfrac{143}{42}}{13}$

11. $\dfrac{2\frac{1}{6}}{2\frac{1}{3}}$

12. $\dfrac{-1\frac{13}{15}}{12\frac{5}{6}}$

13. $1 + \dfrac{1}{1 + \dfrac{1}{1 + \frac{1}{3}}}$

14. $1 - \dfrac{1}{1 - \dfrac{1}{1 - \frac{1}{2}}}$

Write the first four blocks of the repeating decimals in Exercises 15–18.

15. $2.0\overline{56}$

16. $0.\overline{347}$

17. $1.24\overline{36}$

18. $0.\overline{6739}$

Find the point midway between each pair of numbers in Exercises 19–27.

19. $0.7, 3.4$

20. $-1.2, 2.7$

21. $-3.4, -1.8$

22. $\dfrac{11}{3}, \dfrac{23}{5}$

23. $-\dfrac{7}{4}, \dfrac{13}{6}$

24. $\dfrac{-24}{7}, \dfrac{-3}{7}$

25. $1\frac{2}{3}, 7\frac{4}{5}$

26. $-4\frac{1}{2}, 2\frac{1}{6}$

27. $-3\frac{2}{3}, -1\frac{2}{5}$

Use the distributive property and combine like terms to simplify the expressions in Exercises 28–31.

28. $.03x + .08(2000 - x)$

29. $.05x + .10(2x + 1) + .25(60 - 3x)$

30. $3.5x + 2.5(3x - 1) + 1.5(3000 - 4x)$

31. $1.2x(2.4x - 3.4) - 3.4x(1.7x + 1.4)$

Write the repeating decimals in Exercises 32–37 as common fractions.

32. $1.\overline{2}$ 33. $0.\overline{38}$

34. $0.8\overline{7}$ 35. $0.0\overline{72}$

36. $0.0\overline{72}$ 37. $0.\overline{9}$

Evaluate $\dfrac{x + y}{x^2 - y^2}$ for the values of x and y given in Exercises 38–40.

38. $x = 2, y = 2$ 39. $x = \dfrac{1}{2}, y = -\dfrac{1}{3}$

40. $x = 2\dfrac{1}{5}, y = -3\dfrac{1}{4}$

41. Write the 3 numbers that divide the interval from 2 to 4 on the number line into 4 pieces of equal length.

42. Write the 7 numbers that divide the interval from 2 to 4 on the number line into 8 pieces of equal length.

43. Write the 5 numbers that divide the interval from −1 to 2 on the number line into 6 pieces of equal length.

44. Write the 11 numbers that divide the interval from −1 to 2 on the number line into 12 pieces of equal length.

2.6 / PERCENT

A common application of decimals is to problems that involve percent. The word percent means **per hundred**. Thus 15 percent means 15 per 100, or $\frac{15}{100}$, or .15. When we introduce percent, we introduce a *third* way of writing certain fractions. No wonder students often get confused about percent problems!

Our first task is to practice writing fractions and decimals as percents, and then practice writing percents as fractions and decimals. The abbreviation for percent is %; instead of writing 15 percent, we usually write 15%. Study the three forms of these numbers:

Percent form	Decimal form	Fraction form
25%	.25	$\dfrac{25}{100}$
3%	.03	$\dfrac{3}{100}$
120%	1.20	$\dfrac{120}{100}$
6.2%	.062	$\dfrac{6.2}{100}$, or $\dfrac{62}{1000}$
$8\dfrac{1}{2}$%	.085	$\dfrac{8.5}{100}$, or $\dfrac{85}{1000}$
$\dfrac{1}{5}$%	.002	$\dfrac{.2}{100}$, or $\dfrac{2}{1000}$

Example 1   Complete the following table:

Percent form	Decimal form	Fraction form
4.5%		
	.518	
		$\dfrac{8}{5}$

In the first line, 4.5% means $\dfrac{4.5}{100}$, or $\dfrac{45}{1000}$. As a decimal this is .045. In the second line, the decimal .518 is $\dfrac{51.8}{100}$. As a fraction this is $\dfrac{518}{1000}$; as a percent it is 51.8%. In the third line, we can change $\dfrac{8}{5}$ to a decimal by keying ⑧ ➗ ⑤ ＝. We see that $\dfrac{8}{5} = 1.6$ as a decimal. To read the percent we write either $\dfrac{8}{5} = \dfrac{160}{100}$, or 1.6 = 1.60 to see that the correct percent is 160%. The completed table looks like this:

Percent form	Decimal form	Fraction form
4.5%	.045	$\dfrac{45}{1000}$
51.8%	.518	$\dfrac{518}{1000}$
160%	1.6	$\dfrac{8}{5}$

□

Example 2   Find 45% of 110.

The phrase "45% of" means ".45 times." Thus the problem is asking for the value of $(.45)(110)$. On the calculator, the product is 49.5.   □

In Example 2, the problem was translated to the mathematical equation $(.45)(110) = \bigcirc$. The solution is $(.45)(110) = 49.5$. Commonly, questions concerning percent can be translated to mathematical equations of this type. Since there are three numbers involved in such an equation, there are three questions that can be asked. In our example we might have asked for a solution to any one of these equations:

$(.45)(110) = \bigcirc$ (Solution: 49.5)

$(.45) \bigcirc = 49.5$ (Solution: 110)

$\bigcirc \cdot (110) = 49.5$ (Solution: .45)

Each of these equations corresponds to a commonly phrased percent question. The first one we translated in Example 1. Study the translations below.

Question	Mathematical equation
What is 45% of 110?	$(.45)(110) = \bigcirc$
45% of what number gives 49.5?	$(.45) \bigcirc = 49.5$
What percent of 110 is 49.5?	$\bigcirc (110) = 49.5$

We have already talked about answering the first type of question. To solve the other two types, we remember the correspondence between division statements and multiplication statements:

$(.45)(110) = 49.5$ corresponds to $49.5 \div 49.5 = 110$

$(.45)(110) = 49.5$ corresponds to $49.5 \div 110 = .45.$

Example 3   45% of what number gives 49.5? (Or, 49.5 is 45% of what number?)

The mathematical translation of the question is

$.45\ \square = 49.5.$

We can rewrite this as a division statement:

$49.5 \div .45 = \square.$

The calculator gives $49.5 \div .45 = 110$.   □

Example 4   What percent of 110 is 49.5?

The mathematical translation of this question is

$\square\ (110) = 49.5.$

Rewriting this multiplication statement as a division statement, we find that

$49.5 \div 110 = \square.$

The calculator shows $49.5 \div 110 = .45$. Thus the answer is 45%.   □

The examples above demonstrate that an important step in solving a percent problem is translating the problem statement into a mathematical statement. The exercises will give you practice with this translation. It is important that you *know what the words mean* before you begin to compute.

As a consumer you frequently hear about sales in which items are reduced by a certain percent, or you may be told that a product is marked up only a small percent over its wholesale price. Here are two examples.

Example 5   Carter Cars advertises that for the month of October every car will be priced 5% above wholesale price. What will be the selling price of a car that has a wholesale price of $6500?

According to the problem, the selling price is

$6500 + .05(6500)$.

We can key this as

6500 $\boxed{+}$ .05 $\boxed{\times}$ 6500 $\boxed{=}$

or, if the calculator has a percent key, as

6500 $\boxed{+}$ $\boxed{5}$ $\boxed{\%}$ $\boxed{=}$.

A third possibility is to apply the distributive property, writing $6500 + .05(6500) = (1 + .05)(6500) = (1.05)(6500)$. In any case, the selling price is $6825.   □

Example 6   A blouse on a 20%-off table has sale price of $10. What was its price before the sale?

Here we are asking, "What price reduced by 20% is $10?" One way to solve this problem is to guess and then check the guess. For example, we might guess that if a $12 blouse were reduced by 20%, it would have a sale price of $10. To check our guess we compute

$12 - (.20)(12)$.

If the calculator has a percent key, then this computation can be keyed as 12 $\boxed{-}$ 20 $\boxed{\%}$ $\boxed{=}$. The value is $9.60. A $12 blouse reduced by 20% would sell for $9.60, not $10. Thus the blouse in the problem must have been priced at more than $12. We might try $13. If a $13 blouse were reduced by 20%, the sale price would be

$13 - (.20)(13)$

or $10.40. Too high. The price must be between $12 and $13. If you persevere, you will find the correct answer. Our experimentation shows that we need a number to fill this box:

$\boxed{} - (.20) \boxed{} = 10$.

Using the distributive property, we can write

$\boxed{}(1 - .20) = 10$

or

$\boxed{}(.80) = 10.$

As a division statement, this is $10 \div .80 = \boxed{}$. Using the calculator, we find that $10 \div .80 = 12.50$. □

Example 7 A mixture is formed by adding 15 gallons of salt to 25 gallons of water. What percent of the new mixture is salt? What percent is water?

There are $15 + 25 = 40$ gallons of the mixture. Since 15 out of 40 gallons are salt, the fraction of salt in the mixture is $\frac{15}{40}$, or 37.5%. The fraction of water in the mixture is $\frac{25}{40}$, or 62.5%. Notice that the percent of salt plus the percent of water is 100%. □

Example 8 A container holds 50 gallons of salt water that is 28% salt. How many gallons of salt are in the container? How many gallons of water?

The amount of salt is 28% of 50 gallons:

$.28(50) = 14$ gallons.

Thus there are 14 gallons of salt in the container. The remainder is water; the amount of water is

$50 - 14 = 36$ gallons. □

Exercises 2.6

1. Complete the following table:

Percent form	Decimal form	Fraction form
12.6%		
132%		
.5%		

Percent form	Decimal form	Fraction form
$\frac{1}{2}\%$		
	.38	
	1.67	
	.004	
	.0005	
		$\frac{72}{100}$
		$\frac{246}{100}$
		$\frac{35}{1000}$
		$\frac{3}{8}$

2. What is 12.7% of 80? What is 12.7% of x?

3. 11.4 is 15.2% of what number?

4. What percent of 90 is 15.3?

5. What percent of 90 is 100.8?

6. If the sales tax is 5%, how much tax is due on a $450 television set? How much tax is due on a N set?

7. If the sales tax is 5% and you pay $13.75 in sales tax on an item, what is the purchase price of the item?

8. A worker's salary is raised from $150 to $200 a week. What percent of the original salary is the increase that the worker received? (This percent is called the **percent increase**.)

9. If your bill at a restaurant is $25 and you wish to leave a 15% tip, how much should you leave? How much should you leave if the bill is N?

10. If Jane leaves a $4.05 tip and that is 15% of the check, how much is the check?

11. If a coat has been reduced from $62 to $38, what percent of the original price is the reduction?

12. State income tax is $1\frac{1}{2}$%. If Bill earns $370, how much is deducted for state income tax? How much is deducted if he earns x?

13. If the state deducts $6.90 from Jim's check for an income tax that is $1\frac{1}{2}$%, how much did Jim earn?

14. A baseball player has 89 hits of which 15 are home runs. What percent of his hits are home runs?

15. If a television set costs $580 before sales tax and the sales tax is $5\frac{1}{2}$%, what is the total price?

16. Senior citizens receive a 14% discount at Hall's Music Store. What does a senior citizen pay for a stereo priced at $190?

17. A worker earning $175 a week receives a 12% salary increase. What is the worker's new salary?

18. Jim's weekly salary was $180 before an 8.5% cut in pay. What is Jim's new salary?

19. A department store marks wholesale prices up 40%. What is the retail price of a dress that was $85 wholesale?

20. If a suit originally priced at $120 has been marked down 18%, what does it now cost?

21. The Three D Department Store marks up wholesale prices by 40%. Complete the following table:

Wholesale price ($)	Mark-up ($)	Retail price ($)
10	4	14
20		
30		

Wholesale price ($)	Mark-up ($)	Retail price ($)
40		
50		
	22	
x		
		126

22. Shears Department Store is having a 12%-off sale. Complete the following table:

Old price ($)	Discount ($)	Sale price ($)
10	1.20	8.80
20		
30		
40		
50		
	5.40	
x		
		30.80

23. In a 16-gallon salt solution, 6 gallons are salt. What percent of the solution is salt? What percent is water?

24. A container of 20 gallons of salt water is 35% salt. How many gallons are salt? How many gallons are water?

25. A vat containing 40 liters of an alcohol and water solution is 38% pure alcohol. How many liters are pure alcohol? How many liters are water?

26. A container of 60 gallons of salt water is 35% salt. If 4 gallons of salt are added, what percent of the new mixture is salt? What percent is water?

27. A $30 dress was marked up 20% and then later marked down 20%. What is the final price?

Use a guess-and-check approach to solving Exercises 28–30.

28. A suit costs $158.08, including 4% sales tax. What is the price of the suit without the tax?

29. Jane's new salary after a 9.2% pay increase is $15,834. What was her salary before the raise?

30. Bill's new salary after an 8.7% cut in pay is $12,234.20. What was his salary before the cut in pay?

2.7 / THE DIVISION ALGORITHM

Remember the procedure for writing a fraction greater than 1 as a mixed number. For example, we can write

$$\frac{17}{6} = \frac{12 + 5}{6} = \frac{12}{6} + \frac{5}{6} = 2 + \frac{5}{6}.$$

In the first step of this procedure, we write 17 as $12 + 5$ because 12 is the largest multiple of 6 in 17. The number 5 that is left is necessarily smaller than the divisor, 6.

The fact that $\frac{17}{6} = 2 + \frac{5}{6}$ is sometimes expressed by saying that "17 divided by 6 equals 2 plus a remainder of 5," or, more simply, "17 equals 2 times 6 plus 5." A similar statement can be made for any two positive whole numbers: the first number equals a multiple of the second plus a remainder that is less than the second. This property of whole numbers is referred to as the **division algorithm**.

DIVISION ALGORITHM FOR WHOLE NUMBERS

If a and b are positive whole numbers, then we can always find whole numbers q and r such that

$$a = q \cdot b + r$$

and $r < b$.

In the statement of the division algorithm, the number q is called the **quotient** of a divided by b, and r is called the **remainder**. The remainder is always less than the number b, the **divisor**.

Example 1 Find the quotient and remainder if 132 is divided by 17.

The quotient comes from the largest multiple of 17 in 132. Consider the multiples of 17:

$$
\begin{aligned}
1 \cdot 17 &= 17 \\
2 \cdot 17 &= 34 \\
3 \cdot 17 &= 51 \\
4 \cdot 17 &= 68 \\
5 \cdot 17 &= 85 \\
6 \cdot 17 &= 102 \\
7 \cdot 17 &= 119 \\
8 \cdot 17 &= 136.
\end{aligned}
$$

Since $7 \cdot 17 < 132$ and $8 \cdot 17 > 132$, we have the quotient 7. The remainder is what is left when $7 \cdot 17$ (or 119) is subtracted from 132. The remainder is 13. Check to see that $132 = 7 \cdot 17 + 13$. □

Example 2 Find the quotient and remainder if 1506 is divided by 13.

Rather than look at the multiples of 13, we can divide 13 into 1506. If we have no calculator, we use the elementary division procedure for this computation:

```
         1 1 5
   1 3 ⟌1 5 0 6
       1 3
       ②0
       1 3
        ⑦6
        6 5
        ⑪
```

The result is that the quotient is 115 and the remainder is 11. Thus $1506 = (115)(13) + 11$. The circled numbers are remainders that occur at the various steps in the division procedure. Each is necessarily less than 13. In the next paragraph we use calculators to find quotients and remainders.   □

The Division Algorithm on the Calculator

Even though a calculator cannot always write the full decimal form of a fraction, it can replace paper and pencil procedures for finding quotients and remainders. Consider again the example $\frac{17}{6}$, which we wrote as the mixed number $2 + \frac{5}{6}$. If we key $\underline{17}$ $\boxed{\div}$ $\boxed{6}$ $\boxed{=}$, the calculator displays 2.8333333. The whole number 2 in the display is the quotient. Where is the remainder 5? The decimal .8333333 approximates the fraction $\frac{5}{6}$. We can recover the remainder 5 from the decimal .8333333 using this reasoning. Since .8333333 is approximately $5 \div 6$, then 5 is approximately 6(.8333333):

$$.8333333 \doteq 5 \div 6 \quad \text{corresponds to} \quad 6(.8333333) \doteq 5.$$

Thus to get the remainder, we divide 17 by 6 and then take the decimal part of the display times 6. In order to display the decimal part of $17 \div 6$, we subtract the whole number part (the quotient) 2.

Keying sequence	*Display*	*Comments*
$\underline{17}$	17	
$\boxed{\div}$	17	
$\boxed{6}$	6	
$\boxed{-}$	2.8333333	Pressing $\boxed{-}$ causes $\boxed{\div}$ to be performed.
$\boxed{2}$	2	Subtracting the quotient 2.
$\boxed{=}$	.8333333	Remainder displayed in decimal form.
$\boxed{\times}$	.8333333	
$\boxed{6}$	6	
$\boxed{=}$	5	Remainder.

Example 3   Use the calculator to find the quotient and remainder when 132 is divided by 17.

When we key $\underline{132}$ $\boxed{\div}$ $\underline{17}$, the calculator displays 7.7647059. Thus the quotient is 7. To get the remainder, we compute $(.7647059)(17) = 13$. A keying sequence that does the computation quickly is

$$\underline{132} \;\boxed{\div}\; \underline{17} \;\boxed{-}\; \boxed{7} \;\boxed{=}\; \boxed{\times}\; \underline{17} \;\boxed{=}\,.$$ □

Example 4   Write $\dfrac{13470}{78}$ as a mixed number.

The display after keying $\underline{13470}$ $\boxed{\div}$ $\underline{78}$ $\boxed{=}$ is 172.69231. Thus the quotient is 172. When we key the sequence

$$\underline{13470} \;\boxed{\div}\; \underline{78} \;\boxed{-}\; \underline{172} \;\boxed{=}\; \boxed{\times}\; \underline{78} \;\boxed{=}$$

we get 53.999993 for a remainder. Yet we know the remainder is a whole number. An error has occurred because the calculator rounded the decimal $13470 \div 78$. We conclude that the remainder is 54 and write $\frac{13470}{78}$ as $172\frac{54}{78}$. □

Writing Fractions as Decimals (Again)

In Section 2.1 we made an assertion about fractions that cannot be written with a power of 10 in the denominator, that is, fractions that are not finite decimals: such fractions are infinite repeating decimals. The property of whole numbers that makes this a true statement is the division algorithm. Consider the fraction $\frac{2}{7}$. Since $\frac{2}{7} = 2 \div 7$, we can compute the decimal form of $\frac{2}{7}$ by dividing 2 by 7. We regard 2 and 7 as decimals and compute:

```
        . 2 8 5 7 1
    7 ) 2 . 0 0 0 0 0
        1 4
          ⑥0
          5 6
            ④0
            3 5
              ⑤0
              4 9
                ①0
                  7
                  ③
```

The numbers circled are those that occur as remainders after division by 7. Each is necessarily less than 7. So far in the computation they are all different. However, what if the division procedure is continued? Because the only possible remainders that can occur in division by 7 are 0, 1, 2, 3, 4, 5, and 6, after a time we will get a remainder that has occurred before. Therefore, we continue the procedure:

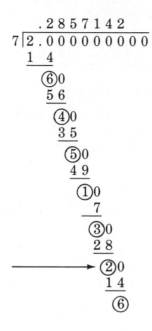

Now it has happened: we are getting remainders that have occurred before and the numbers in the quotient must also repeat. The pattern becomes .285714. A remainder of 0 never occurs and the decimal repeats indefinitely. The same thing happens for any fraction that cannot be written with a denominator that has a power of 10: it is an infinite repeating decimal. The division algorithm guarantees this fact.

Since we are using calculators for much of our computation, it is natural to ask if we can use calculators to find infinite repeating decimals that represent certain fractions. The answer is sometimes.

Example 5   Write the fraction $\dfrac{1}{11}$ as a repeating infinite decimal.

When we key ①　÷　**11**　=, the calculator display is .09090909, strongly suggesting that $\frac{1}{11} = .\overline{09}$. To be absolutely sure, we can inspect the hand computation:

$$
\begin{array}{r}
.0\,9\,0\,9 \\
1\,1\,\overline{)\,1\,.\,0\,0\,0\,0} \\
9\,9 \\
\hline
①0\ 0 \\
9\ 9 \\
\hline
①
\end{array}
$$

We see that the remainders repeat and the pattern .090909 . . . will continue to repeat.   □

Example 6   Write the fraction $\dfrac{1}{17}$ as a repeating infinite decimal.

When we key ①　÷　**17**　=, the calculator displays .05882353. There is a possibility that the repeating decimal is .0588235$\overline{3}$, or perhaps .058823$\overline{53}$, or perhaps .05882$\overline{353}$. However, when you inspect the hand computation of $1 \div 17$ (and you should write this out), you see that every remainder from 1 to 16 occurs before a remainder is repeated. Thus the repeating block has 16 digits in it. In fact, $\frac{1}{17} = .\overline{0588235294117647}$.   □

Exercises 2.7

Write the first four blocks of the repeating decimals in Exercises 1–4.

1. $1.0\overline{35}$ 2. $0.\overline{675}$

3. $2.35\overline{42}$ 4. $0.\overline{3567}$

Find the quotient (q) and the remainder (r) if a is divided by b with a and b as given in Exercises 5–8. Also check that $a = bq + r$.

5. $a = 456, b = 72$ 6. $a = 2016, b = 112$

7. $a = 75, b = 83$ 8. $a = 1112, b = 11$

9. Use the equation $1132 = 5(212) + 72$ to find the quotient and remainder when 1132 is divided by 212.

10. Use the equation $140 = 1(117) + 23$ to find the quotient and remainder when 140 is divided by 117.

Use the division algorithm to rewrite each fraction in Exercises 11–14 as a mixed number.

11. $\dfrac{835}{18}$

12. $-\dfrac{356}{213}$

13. $-\dfrac{13469}{78}$

14. $\dfrac{1111}{111}$

Rewrite the mixed numbers in Exercises 15–18 as common fractions.

15. $17\dfrac{11}{12}$

16. $-459\dfrac{87}{94}$

17. $-7\dfrac{35}{99}$

18. $14\dfrac{101}{179}$

Give the decimal form of the fractions in Exercises 19–30. Place a bar over the block that repeats if the fraction is a repeating decimal.

19. $\dfrac{1}{8}$

20. $\dfrac{1}{9}$

21. $\dfrac{1}{12}$

22. $\dfrac{1}{16}$

23. $\dfrac{1}{22}$

24. $\dfrac{1}{24}$

25. $\dfrac{41}{333}$

26. $\dfrac{323}{12920}$

27. $\dfrac{611}{4950}$

28. $\dfrac{1}{14}$

29. $\dfrac{1}{19}$

30. $\dfrac{1}{21}$

2.8 / THE EUCLIDEAN ALGORITHM

Reducing a fraction can be difficult if the numerator and denominator are whole numbers that are hard to factor into products of primes. For example, to reduce $\frac{1147}{1333}$ we need to factor both 1147 and 1333. Since 31 is the smallest prime factor in each of these numbers, the process of checking for prime factors could be time consuming. The reason we factor the numerator and denominator into primes is to help find the GCF so that the fraction can be reduced to its lowest terms. The division algorithm provides a different way of finding the GCF of two numbers—a way that does not require factoring them into primes. This method is called the **Euclidean algorithm**.

The Euclidean algorithm works by applying the division algorithm more than once. Consider the two numbers in our example: 1147 and 1333. Applying the division algorithm to 1333 and 1147 (dividing the larger number by the smaller one), we find a quotient of 1 and a remainder of 186 as follows:

Keying sequence	*Display*	*Division statement*
1333 ÷ 1147 − 1 = × 1147 =	186	$1333 = 1 \cdot (1147) + 186$

Now we apply the division algorithm to the numbers 1147 and 186 (the divisor and remainder in the previous step):

Keying sequence	*Display*	*Division statement*
1147 ÷ 186 − 6 = × 186 =	31	$1147 = 6 \cdot (186) + 31$

Again we use the division algorithm, this time with the divisor and remainder 186 and 31.

Keying sequence	*Display*	*Division statement*
186 ÷ 31 − 6 = × 31 =	0	$186 = 6 \cdot (31) + 0$

This process stops when we get 0 for a remainder. Summarizing the three steps, we have:

$$1333 = 1 \cdot (1147) + 186$$
$$1147 = 6 \cdot (186) + 31$$
$$186 = 6 \cdot (31) + 0.$$

Analyze what happens in this sequence of division statements. In the first statement we divide by 1147 and get a remainder of 186 that is necessarily less than 1147. In the second step we divide by 186 so the remainder in the second step is less than 186, the remainder of the first step. In the third step, we divide by the second remainder so the third remainder must be less than the second remainder, which must be less than the first remainder. We have set up a sequence of remainders that get smaller at each stage. Eventually one will have to be 0, and then the process will stop because we cannot divide by 0. In our example this happens in the third step.

The important part of the above procedure is that the remainder appearing in the step before the 0 remainder is the GCF of the two original numbers. The GCF of 1147 and 1333 is 31. Knowing this we can remove the factor 31 from the numerator and denominator and write $\frac{1147}{1333} = \frac{(31)(37)}{(31)(43)} = \frac{37}{43}$.

But how do we know that 31 is the GCF of 1147 and 1333? We must analyze the Euclidean algorithm carefully to see that the last nonzero remainder is the GCF. We are looking at the particular example using 1147 and 1333. However, the conclusions are the same for any other example.

The analysis requires two observations. First, if 31 is a factor of a number, it is a factor of any multiple of that number. Second, if 31 is a factor of each of two numbers, it is a factor of their sum and their difference.

Now look again at the three equations in the example:

$$1333 = 1 \cdot (1147) + 186$$

$$1147 = 6 \cdot (186) + 31$$

$$186 = 6 \cdot (31) + 0.$$

The third equation demonstrates that 31 is a factor of 186. Looking then at the second equation, since 31 is a factor of 186, it is a factor of $6 \cdot (186)$ and hence of $6 \cdot (186) + 31$, which equals 1147. Finally, looking at the first equation, we know that 31 is a factor of 186 and of 1147, so it is a factor of $1 \cdot (1147) + 186$, or 1333. Thus we know that 31 is a factor of both 1147 and 1333. But can we conclude it is the greatest common factor? We must inspect the equations again.

Assume that d is any common factor of 1147 and 1333. Then d is a factor of $1333 - 1 \cdot (1147)$, which, according to the first equation, equals 186. Since d is a factor of 1147 and 186, d will be a factor of $1147 - 6(186)$, which, according to the second equation, is 31. Therefore, since any common factor d must be a factor of 31, we know what the greatest common factor is: 31 is the GCF. Thus the Euclidean algorithm process enables us to find the GCF of two numbers.

Example 1 Find the GCF of 151 and 5745 and reduce the fraction $\dfrac{151}{5745}$ to its lowest terms.

We start the Euclidean algorithm by using the division algorithm with the numbers 151 and 5745. Our keying sequence of the last section can be improved by recognizing that since we both divide by 151 and multiply by 151 it helps to STORE 151 the first time it appears. The first sequence therefore is

$$\underline{5745}\ \boxed{\div}\ \underline{151}\ \boxed{\text{STO}}\ \boxed{-}\ \underline{38}\ \boxed{=}\ \boxed{\times}\ \boxed{\text{RCL}}\ \boxed{=}$$

(Display)
6.9999999

The number on the display is 6.9999999; since it is the remainder when 5745 is divided by 151, it must represent the whole number 7. The quotient is 38 and the division statement is

$$5745 = 38 \cdot (151) + 7.$$

Now we apply the division algorithm to 38 and 7 using an analogous keying sequence and continue until the Euclidean algorithm gives a 0 remainder. The division statements are

$$5745 = 38 \cdot (151) + 7$$
$$151 = 21 \cdot (7) + 4$$
$$7 = 1 \cdot (4) + 3$$
$$4 = 1 \cdot (3) + 1$$
$$3 = 3 \cdot (1) + 0$$

Thus the GCF of 5745 and 151 is 1. We conclude that the fraction $\dfrac{151}{5745}$ is already in its lowest terms.

□

Finding the LCM

Since the Euclidean algorithm provides a way of finding the GCF of two numbers without factoring them into primes, it is natural to ask if there is a way to find the LCM without factoring them into primes. This can be done by first finding the GCF. We must recognize that the product of two numbers contains the GCF twice because both numbers contribute the GCF. Other prime powers that occur in the product are contributed by one of the two numbers, but not both. Thus the product of two numbers divided by the GCF equals the LCM.

Example 2   Find the LCM of 1147 and 1333.

In the early part of the section we used the Euclidean algorithm to determine that the GCF of 1147 and 1333 is 31. Thus the LCM is $\frac{(1147) \cdot (1333)}{31}$, or 49,321.   □

Example 3   Find the LCM of 151 and 5745.

In Example 1 we determined that the GCF of 151 and 5745 is 1. Thus the LCM is $\frac{(151)(5745)}{1}$; that is to say, the LCM is the product of the two numbers, or 867,495.   □

Exercises 2.8

Write the division statement that corresponds to each sequence of calculator key strokes in Exercises 1 and 2.

1. 787 [÷] 56 [−] 14 [=] [×] 56 [=]   (Display) 3

2. 17203 [÷] 112 [STO] [−] 153 [=] [×] [RCL] [=]   (Display) 66.999999

Use the Euclidean algorithm to find the GCF and LCM of the pairs of numbers in Exercises 3–8.

3. 2892, 2340   4. 5712, 4550

5. 14911, 1147   6. 58212, 5292

7. 4312, 2925   8. 2993, 938

Perform the indicated operations in Exercises 9–11 using the LCMs from Exercises 4, 6, and 8 as the common denominator. Leave your answers in fractional form.

9. $\frac{111}{5712} + \frac{33}{4550}$ 10. $\frac{3401}{58212} - \frac{353}{5292}$

11. $\frac{1}{2993} + \frac{37}{938}$

Reduce the fractions in Exercises 12–15 to their lowest terms.

12. $\dfrac{2232}{5425}$

13. $\dfrac{2970}{3822}$

14. $\dfrac{945}{308}$

15. $\dfrac{1186}{1575}$

Write the mathematical phrase evaluated by each sequence of calculator key strokes in Exercises 16 and 17.

16. $\underline{117}$ $\boxed{\div}$ $\underline{23}$ $\boxed{-}$ $\boxed{5}$ $\boxed{=}$ $\boxed{\times}$ $\underline{23}$ $\boxed{=}$

17. $\underline{1853}$ $\boxed{\div}$ $\underline{72}$ $\boxed{\text{STO}}$ $\boxed{-}$ $\underline{25}$ $\boxed{=}$ $\boxed{\times}$ $\boxed{\text{RCL}}$ $\boxed{=}$

18. When we apply the Euclidean algorithm to 702 and 246, the division statements are

$$702 = 2 \cdot (246) + 210$$
$$246 = 1 \cdot (210) + 36$$
$$210 = 5 \cdot (36) + 30$$
$$36 = 1 \cdot (30) + 6$$
$$30 = 5 \cdot (6) + 0.$$

Verify that 6 is a factor of 30, 36, 210, 246, and 702.

19. Determine the GCF of 2411 and 1097 from the following division statements:

$$2411 = 2 \cdot (1097) + 217$$
$$1097 = 5 \cdot (217) + 12$$
$$217 = 12 \cdot (18) + 1$$
$$18 = 18 \cdot (1) + 0.$$

20. Complete the following table:

ab	GCF of a and b	LCM of a and b
8,100	6	
784,784		28,028
	18	9,828

2.9 / CHAPTER 2 PROBLEM COLLECTION

Evaluate each expression in Exercises 1–11.

1. $\dfrac{5}{6} - \dfrac{14}{15}$

2. $\dfrac{3}{14} - \dfrac{4}{21} - \dfrac{5}{12}$

3. $3\dfrac{5}{12} + 2\dfrac{1}{18} - 5\dfrac{1}{6}$

4. $\dfrac{3\frac{1}{4}}{-6\frac{1}{2}}$

5. $\dfrac{44}{30} \div \dfrac{11}{5}$

6. $-\dfrac{27}{14} \div -\dfrac{3}{7}$

7. $\left(\dfrac{-21}{20}\right)\left(-\dfrac{10}{63}\right)$

8. $\left(-\dfrac{11}{2}\right)\left(\dfrac{2}{3}\right)\left(\dfrac{3}{4}\right)\left(\dfrac{4}{5}\right)$

9. $\dfrac{(-3.25)^2(-2.18)}{(-1.49)(-7.32)}$

10. $\dfrac{(12.3 + 15.4)^2}{(12.3)^2 + (15.4)^2}$

11. $-\dfrac{3}{4} - \dfrac{3}{-4}$

Solve for x in Exercises 12 and 13.

12. $\dfrac{4}{11} = \dfrac{10}{x}$

13. $\dfrac{27}{x} = \dfrac{-6}{55}$

Find all whole numbers x that satisfy the inequalities in Exercises 14 and 15.

14. $\dfrac{x}{5} < 2$

15. $1 < \dfrac{x}{6} < 2$

Find all integers x that satisfy the inequalities in Exercises 16 and 17.

16. $\dfrac{x}{3} < 1$

17. $-3 < \dfrac{x}{4} < \dfrac{1}{2}$

Write the mathematical phrase evaluated by each sequence of calculator key strokes in Exercises 18–23.

18. $\boxed{2}\ \boxed{\div}\ \boxed{3}\ \boxed{+}\ \boxed{3}\ \boxed{\div}\ \boxed{4}\ \boxed{=}$

19. $\boxed{6}\ \boxed{\times}\ \boxed{5}\ \boxed{1/x}\ \boxed{-}\ \boxed{3}\ \boxed{\times}\ \boxed{8}\ \boxed{1/x}\ \boxed{=}$

20. $\boxed{(}\ \boxed{2}\ \boxed{+}\ \boxed{3}\ \boxed{1/x}\ \boxed{)}\ \boxed{+/-}\ \boxed{-}\ \boxed{(}\ \boxed{2}\ \boxed{+}\ \boxed{4}\ \boxed{\times}\ \boxed{5}\ \boxed{1/x}\ \boxed{)}\ \boxed{=}$

21. $\underline{235}$ ÷ $\underline{34}$ STO − 6 = × RCL =

22. $\underline{50}$ + $\underline{12.3}$ % =

23. $\underline{50}$ − $\underline{12.3}$ % =

Order the numbers in Exercises 24–26 from smallest to largest.

24. $0, \dfrac{3}{11}, 0.2728, -\dfrac{3}{11}, -0.2727$

25. $0, \dfrac{11}{45}, -\dfrac{11}{45}, \dfrac{17}{60}, -\dfrac{17}{60}$

26. $0, .\overline{5}, -.\overline{5}, .5, -.5, .55, -.55, .555, -.555$

Determine which is the better buy in each of Exercises 27 and 28.

27. A 10-pound, 11-ounce box of detergent for $4.39, or a 5-pound, 4-ounce box for $2.09?

28. A half-gallon of milk for $1.39, or a quart of milk for 73 cents?

29. A worker is paid at the rate of $8\frac{3}{4}$ dollars per hour.

    (a) If he works 6 hours, how much does he earn?

    (b) If he earns $63, how many hours has he worked?

30. Gold sells for $500 an ounce. Complete the following table:

Ounces of gold	$1\dfrac{3}{4}$	$2\dfrac{1}{2}$	3.2		x	
Price ($)				200		y

Find the point midway between each pair of numbers in Exercises 31–33.

31. $\dfrac{2}{3}, \dfrac{7}{4}$

32. $-2.7, 1.4$

33. $-5\dfrac{1}{3}, -2\dfrac{1}{2}$

34. Write the 3 numbers that divide the interval from −1 to 1 on the number line into 4 pieces of equal length.

35. Write the 8 numbers that divide the interval from 0 to 3 on the number line into 9 pieces of equal length.

Solve for n in Exercises 36 and 37.

36. $641.2 = (6.412)(10^n)$ 37. $0.07621 = \dfrac{762.1}{10^n}$

38. Find the quotient and remainder when 1327 is divided by 31.

Sketch a number line and graph the numbers in Exercises 39–42.

39. $0, 1.4, (1.4)^2, (1.4)^3, (1.4)^4, (1.4)^5, (1.4)^6$

40. $0, 0.8, (0.8)^2, (0.8)^3, (0.8)^4, (0.8)^5, (0.8)^6$

41. $0, -1.2, (-1.2)^2, (-1.2)^3, (-1.2)^4, (-1.2)^5, (-1.2)^6$

42. $0, -0.6, (-0.6)^2, (-0.6)^3, (-0.6)^4, (-0.6)^5, (-0.6)^6$

Write the numbers in Exercises 43–46 as common fractions.

43. $1.\overline{23}$ 44. $0.\overline{79}$

45. $0.65\overline{9}$ 46. $0.23\overline{45}$

47. Complete the following table:

Decimal form	Percent form	Fraction form
0.67		
1.23		
0.452		
	37%	
	112%	
	75.6%	
		$\dfrac{13}{100}$

Decimal form	Percent form	Fraction form
		$\dfrac{142}{100}$
		$\dfrac{17}{1000}$

48. What is 72.3% of 110?

49. What is 124% of 80?

50. What percent of 80 is 45?

51. What percent of 80 is 110?

52. 42 is 13% of what number?

53. 42 is 110% of what number?

54. A container of 30 gallons of salt water is 42% salt. How many gallons are salt? How many gallons are water?

55. A vat containing 60 gallons of an alcohol and water solution is 52% pure alcohol. How many gallons are pure alcohol? How many gallons are water?

56. If a $32 coat is marked up 22% and then this new price is marked down 22%, what is the final price?

57. LaSalle Department Store marks up wholesale prices by 45%. Complete the following table:

Wholesale ($)	30	40		50	x
Retail ($)			66.70		

58. Berry's Department Store is having a 27%-off sale. Complete the following table:

Original price ($)	35	50		70	x
Sale price ($)			45.26		

59. Bob and Patty each receive a 12% salary increase.

(a) If Patty's salary was $13,200 before the raise, what is her new salary?

(b) If Bob's salary after the raise is \$16,576, what was his salary before the raise?

Use the distributive property and combine like terms to simplify the expressions in Exercises 60–63.

60. $\frac{1}{3}c^3 - c^3 + \frac{1}{2}d - \frac{3}{4}d$

61. $1.32x^2 - 2x^2 + y - .71y$

62. $.10x + .25(2x - 3) + .50(x + 1)$

63. $\frac{x}{3}(2x + 1) - \frac{x}{2}(x + 1)$

64. (a) Use the Euclidean algorithm to find the GCF and LCM of 990 and 2100.

   (b) Reduce $\dfrac{990}{2100}$ to its lowest terms.

   (c) Add $\dfrac{7}{990} + \dfrac{11}{2100}$.

65. (a) Use the Euclidean algorithm to find the GCF and LCM of 2436 and 2604.

   (b) Reduce $\dfrac{2604}{2436}$ to its lowest terms.

   (c) Subtract $\dfrac{11}{2604} - \dfrac{13}{2436}$.

Use a guess-and-check approach to solve Exercises 66–68.

66. The cost of a TV set, including 4.5% sales tax, is \$537.08. What was the price of the TV before the tax was added?

67. Jim's salary after a 12.5% pay increase is \$18,697.95. What was his salary before the raise?

68. Jane's bill, including 15% tip, was \$71.76. What was the cost of the meal before the tip was added?

Compute the value of $\dfrac{2xy}{x + 2y}$ for the given values of x and y in Exercises 69–72.

69. $x = 2, y = 3$

70. $x = -2, y = 3$

71. $x = 4, y = -1$

72. $x = -2, y = -3$

73. Jane has twice as many quarters as nickels and 3 more dimes than nickels. Complete the following table:

Number of nickels	Number of dimes	Number of quarters	Value of coins
4			
	9		
		20	
x			

74. Mr. Smith raises horses and ducks. Complete the following table:

Number of horses	Number of ducks	Number of feet	Number of heads
15	45	150	60
20	40		
25	35		
30	30		
35			60
	20		60
		188	60

75. A safety device beeps every 54 seconds. If it beeps at 2:00 P.M., when will it beep on the minute again?

Using Exponents
and Scientific
Notation

3.1 / POSITIVE WHOLE NUMBER EXPONENTS ON THE CALCULATOR

Recall that when a positive whole number is used as an exponent, it indicates the number of times the base appears in a product. For example, in the expression 3^4, the number 3 is the base, the number 4 is the exponent, and 3^4 means $3 \cdot 3 \cdot 3 \cdot 3$. Exponents give us a way of abbreviating long factored expressions.

You have had some practice evaluating expressions with exponents using the $\boxed{K}$ key.

Example 1   Find a number n so that $3^n = 81$.

One way to find the exponent on 3 that gives 81 is to key $\boxed{3}\,\boxed{\times}\,\boxed{K}$ and then count the number of times the $\boxed{=}$ key must be pressed to get 81 on the display:

	(Display)
$\boxed{3}\,\boxed{\times}\,\boxed{K}\,\boxed{=}$	9
$\boxed{=}$	27
$\boxed{=}$	81

The exponent, you recall, is one more than the number of $\boxed{=}$s. Thus $n = 4$.   □

Example 2   Find the smallest whole number n for which $(-2)^n < -1000$.

Using the $\boxed{K}$ key, we can display the powers of -2:

	(Display)
$\boxed{2}\,\boxed{+/-}\,\boxed{\times}\,\boxed{K}\,\boxed{=}$	4
$\boxed{=}$	-8
$\boxed{=}$	16
$\boxed{=}$	-32
$\boxed{=}$	64

$$\boxed{=} \qquad -128$$
$$\boxed{=} \qquad 256$$
$$\boxed{=} \qquad -512$$
$$\boxed{=} \qquad 1024$$
$$\boxed{=} \qquad -2048$$

We see that the $\boxed{=}$ key must be pressed 10 times to get a power of -2 less than -1000. Thus $n = 10 + 1 = 11$. ◻

Two other calculator keys are specifically used with exponents. The $\boxed{x^2}$ key squares any number that precedes it. Here are two examples:

Mathematical phrase	Keying sequence	Display
5^2	$\boxed{5}\ \boxed{x^2}$	25
$(-2)^2$	$\boxed{2}\ \boxed{+/-}\ \boxed{x^2}$	4

For whole number exponents greater than 2, we use the $\boxed{y^x}$ key. Actually the $\boxed{y^x}$ key works for the exponent 2, but it is slower than the $\boxed{x^2}$ key. You should compare these two ways of evaluating 5^2 by observing the display on your calculator:

$$\boxed{5}\ \boxed{x^2} \qquad \text{and} \qquad \boxed{5}\ \boxed{y^x}\ \boxed{2}\ \boxed{=}.$$

The calculator uses a different procedure when it evaluates expressions with the $\boxed{y^x}$ key than when it uses $\boxed{x^2}$. With the $\boxed{y^x}$ key the calculator uses logarithms, and this means the calculator cannot raise a negative number to a power with the $\boxed{y^x}$ key. Key these examples on your calculator:

Mathematical phrase	Keying sequence	Display
5^7	$\boxed{5}\ \boxed{y^x}\ \boxed{7}\ \boxed{=}$	78125
$(-5)^7$	$\boxed{5}\ \boxed{+/-}\ \boxed{y^x}\ \boxed{7}\ \boxed{=}$	Error

Example 3 Find the values of $(-5)^7$ and $(-5)^{10}$.

We see that the calculator cannot compute directly a negative number to a power when we use the $\boxed{y^x}$ key. Using the $\boxed{K}$ key, we observe a pattern similar to that in Example 2:

Mathematical phrase	Keying sequence	Display
$(-5)^2$	$\boxed{5}\ \boxed{+/-}\ \boxed{\times}\ \boxed{K}\ \boxed{=}$	$25 = \quad 5^2$
$(-5)^3$	$\boxed{=}$	$-125 = -5^3$
$(-5)^4$	$\boxed{=}$	$625 = \quad 5^4$
$(-5)^5$	$\boxed{=}$	$-3125 = -5^5$

The pattern is that

$(-5)^n = 5^n$ if n is even,

$(-5)^n = -5^n$ if n is odd.

This is no surprise because we know that the product of two negative numbers is a positive number. Thus the product of an even number of negative numbers is positive, but the product of an odd number of negative numbers is negative. Since $(-5)^7 = -5^7$ and $(-5)^{10} = 5^{10}$, we can use these keying sequences:

Mathematical phrase	Keying sequence	Display
$(-5)^7$	$\boxed{5}\ \boxed{y^x}\ \boxed{7}\ \boxed{=}\ \boxed{+/-}$	-78125
$(-5)^{10}$	$\boxed{5}\ \boxed{y^x}\ \underline{10}\ \boxed{=}$	9765625 □

It is important to remember that the symbols $(-5)^7$ and -5^7 denote different numbers:

$(-5)^7$ means $(-5) \cdot (-5) \cdot (-5) \cdot (-5) \cdot (-5) \cdot (-5) \cdot (-5)$;

-5^7 means the opposite of 5^7, or $-(5 \cdot 5 \cdot 5 \cdot 5 \cdot 5 \cdot 5 \cdot 5)$.

You can think of -5^7 as $(-1)(5^7)$. In computation, we can replace an expression like $-b^3$ by $(-1)b^3$.

Exponents and Calculator Hierarchy

Often we key an exponential expression in a sequence that also includes other operations. On our calculators, raising to a power has priority over multiplication and division as well as over addition and subtraction. Pending $\boxed{y^x}$ operations are performed when the $\boxed{+}$, $\boxed{-}$, $\boxed{\times}$, and $\boxed{\div}$ keys are pressed. When $\boxed{x^2}$ is pressed, the operation is performed immediately. Study the following examples:

Mathematical phrase

$4^3 \cdot 6^2$

Keying sequence: $\boxed{4}\ \boxed{y^x}\ \boxed{3}\ \boxed{\times}\ \boxed{6}\ \boxed{x^2}\ \boxed{=}$

Display: $\quad$ 4 $\quad$ 4 $\quad$ 3 $\quad$ 64 $\quad$ 6 $\quad$ 36 $\quad$ 2304

$(-4)^3 + (-5)^4$

Keying sequence $\quad \boxed{4}\ \boxed{y^x}\ \boxed{3}\ \boxed{=}\ \boxed{+/-}\ \boxed{+}\ \boxed{5}\ \boxed{y^x}\ \boxed{4}\ \boxed{=}$

Display: $\quad$ 4 $\quad$ 4 $\quad$ 3 $\quad$ 64 $\quad$ −64 $\quad$ −64 $\quad$ 5 $\quad$ 5 $\quad$ 4 $\quad$ 561

The keying sequence that evaluates $(-4)^3 + (-5)^4$ uses the fact that

$\quad (-4)^3 + (-5)^4 = -4^3 + 5^4.$

To evaluate $(-4)^3 + (-5)^3$, we use the fact that

$\quad (-4)^3 + (-5)^3 = -4^3 - 5^3.$

Mathematical phrase

$(-4)^3 + (-5)^3$

Keying sequence: $\boxed{4}\ \boxed{y^x}\ \boxed{3}\ \boxed{=}\ \boxed{+/-}\ \boxed{-}\ \boxed{5}\ \boxed{y^x}\ \boxed{3}\ \boxed{=}$

Display : $\quad$ 4 $\quad$ 4 $\quad$ 3 $\quad$ 64 $\quad$ −64 $\quad$ −64 $\quad$ 5 $\quad$ 5 $\quad$ 3 $\quad$ −189

Exercises 3.1

Insert the proper symbol $(<, =, >)$ between the numbers in Exercises 1–4.

1. $(-1.7)^{53}, 0$

2. $(2.3)^{212}, 0$

3. $(-5)^8, 5^8$

4. $(-4)^{17}, 4^{17}$

Solve for x in Exercises 5–8.

5. $2^x = 32$

6. $3^x = 6561$

7. $\dfrac{4}{5^x} = \dfrac{4}{3125}$

8. $(-2)^x = -2048$

Find the smallest whole number n that makes the statements in Exercises 9–12 true.

9. $2^n > 1000$

10. $2^n > 1{,}000{,}000$

11. $(-3)^n < -1000$

12. $(-3)^n < -1{,}000{,}000$

Use your calculator to evaluate each expression in Exercises 13–25.

13. 2^{10}

14. -5^8

15. -3^9

16. $3(2)^4$

17. $4(-3)^3$

18. $\dfrac{3^8}{3^3}$

19. $(7^5)(7^3)$

20. $(12.7 + 13.4)^2 - (12.7)^2 - (13.4)^2$

21. $-3^3 + (-2)^5$

22. $2(5)^4 - 3(5)^3 + 5^2 - 3(5) + 1$

23. $(-2)^4 - (-4)^3$

24. $2(-3)^2 - 5(-3)^2 + 4(-3) + 1$

25. $(-2.1)^3 + (-3.2)^5 - (-1.3)^3$

Write a sequence of calculator key strokes that will evaluate each expression in Exercises 26–28. Then evaluate the expression.

26. $3^4 + 7(4)^5 - 2^2$

27. $3(-2)^5 + 4(-3)^6$

28. $\dfrac{(-5)^3 - 2(-3)^4}{6^5 - 5^6}$

Write the mathematical phrase evaluated by each sequence of calculator key strokes in Exercises 29–31.

29. $\boxed{4}\ \boxed{\times}\ \boxed{5}\ \boxed{x^2}\ \boxed{+}\ \boxed{3}\ \boxed{y^x}\ \boxed{4}\ \boxed{=}$

30. $\boxed{2}\ \boxed{\times}\ \boxed{3}\ \boxed{y^x}\ \boxed{3}\ \boxed{-}\ \boxed{3}\ \boxed{\times}\ \boxed{2}\ \boxed{y^x}\ \boxed{4}\ \boxed{=}$

31. $\boxed{3}\ \boxed{\times}\ \boxed{2}\ \boxed{y^x}\ \boxed{4}\ \boxed{=}\ \boxed{+/-}\ \boxed{-}\ \boxed{2}\ \boxed{x^2}\ \boxed{\times}\ \boxed{3}\ \boxed{x^2}\ \boxed{=}$

Sketch a number line and graph each of the numbers in Exercises 32 and 33.

32. $1.5, (1.5)^2, (1.5)^3, (1.5)^4, (1.5)^5$ (on the same number line)

33. $.5, (.5)^2, (.5)^3, (.5)^4, (.5)^5$ (on the same number line)

34. Arrange the following numbers in increasing order:

   $0, -2, (-2)^2, (-2)^3, (-2)^4, (-2)^5, (-2)^6.$

35. Compute $(-1)^n$ for every whole number n.

3.2 / RULES FOR COMPUTING WITH EXPONENTS

By remembering what positive whole number exponents mean, we can find ways to simplify numerical computations and mathematical expressions that contain exponents. First consider the product of two exponential expressions that contain the same base, for example, $4^2 \cdot 4^3$. Writing down what the expressions mean suggests a way to shorten the computation:

$$4^2 \cdot 4^3 = (4 \cdot 4) \cdot (4 \cdot 4 \cdot 4) = 4^5$$

$$x \cdot x^5 = x \cdot (x \cdot x \cdot x \cdot x \cdot x) = x^6$$

$$(2y^3)^2(2y^3)^2 = (2y^3 \cdot 2y^3) \cdot (2y^3 \cdot 2y^3) \cdot = (2y^3)^4.$$

Since each exponent counts the number of times the base occurs as a factor, the sum of the exponents counts the number of times the base occurs as a factor in the product. We call this observation our first rule for computing with exponents.

RULE 1 FOR EXPONENTS

$$b^n \cdot b^m = b^{n+m}$$

Notice that Rule 1 applies only when two expressions have the same base. It lets us write directly $7^3 \cdot 7^6 = 7^9$ and $(4y)^2(4y)^6 = (4y)^8$. It says nothing at all about simplifying $2^3 \cdot 5^2$ or $x^3 \cdot y^4$ or $(5n)^2(3m)^3$ because the bases of the two factors are different.

A second observation also simplifies computations with exponents. Look at these examples:

$$(3x)^2 = (3x)(3x) = 3 \cdot x \cdot 3 \cdot x = 3 \cdot 3 \cdot x \cdot x = 3^2 \cdot x^2$$

$$(ab)^4 = ab \cdot ab \cdot ab \cdot ab = aaaabbbb = a^4 b^4.$$

You should observe that raising a product to a power is the same as raising each factor separately to the power and then multiplying. This is our second rule for computing with exponents.

RULE 2 FOR EXPONENTS

$$(a \cdot b)^n = a^n \cdot b^n$$

Example 1 Simplify the expression $(-2x)^3 \cdot x^4$.

We use both Rule 2 and Rule 1 in this computation:

$$
\begin{aligned}
(-2x)^3 \cdot x^4 &= (-2)^3 \cdot x^3 \cdot x^4 \qquad \text{(Rule 2)} \\
&= (-2)^3 \cdot x^7 \qquad\qquad \text{(Rule 1)} \\
&= -8x^7.
\end{aligned}
$$

Consider now expressions in which the base itself is an exponential expression, for example, $(4^2)^3$ or $(t^3)^5$.

$$
\begin{aligned}
(4^2)^3 &= 4^2 \cdot 4^2 \cdot 4^2 \\
&= 4^{2+2+2} \qquad\qquad \text{(Rule 1)} \\
&= 4^{3 \cdot 2} \\
&= 4^6,
\end{aligned}
$$

$$
\begin{aligned}
(t^3)^5 &= t^3 \cdot t^3 \cdot t^3 \cdot t^3 \cdot t^3 \\
&= t^{3+3+3+3+3} \qquad\qquad \text{(Rule 2)} \\
&= t^{5 \cdot 3} \\
&= t^{15}.
\end{aligned}
$$

There is a pattern in raising an exponential expression to a power. This pattern is our third rule for computing with exponents.

RULE 3 FOR EXPONENTS

$$(b^n)^m = b^{n \cdot m}$$

Rule 3 permits us to write directly $(x^6)^8 = x^{48}$ and $(6^2)^3 = 6^6$.

Using all three rules appropriately permits us to rewrite an expression so that each variable appears only once in the expression. This is what we mean by **simplifying** an expression.

Example 2   Simplify the expression $(2x^2)^3 \cdot (-7x^4)^2$.

Follow each step in this process:

$$
\begin{aligned}
(2x^2)^3 \cdot (-7x^4)^2 &= 2^3 \cdot (x^2)^3 \cdot (-7)^2 \cdot (x^4)^2 && \text{(Rule 2 twice)} \\
&= 2^3 \cdot x^6 \cdot (-7)^2 \cdot x^8 && \text{(Rule 3 twice)} \\
&= 2^3 \cdot (-7)^2 \cdot x^6 \cdot x^8 && \text{(Rearrange terms)} \\
&= 8 \cdot 49 \cdot x^{14} && \text{(Rule 1)} \\
&= 392x^{14}. && \text{(Computation)} \quad \square
\end{aligned}
$$

Example 3   Simplify the expression $(-ab^3)^2(-b^2)^5$.

Observe that all three rules are used in this simplification:

$$
\begin{aligned}
(-ab^3)^2(-b^2)^5 &= [(-1)ab^3]^2[(-1)b^2]^5 \\
&= (-1)^2a^2(b^3)^2(-1)^5(b^2)^5 && \text{(Rule 2)} \\
&= (1)a^2b^6(-1)b^{10} && \text{(Rule 3 and powers of } -1) \\
&= -a^2b^6b^{10} \\
&= -a^2b^{16}. && \text{(Rule 1)} \quad \square
\end{aligned}
$$

Our rules for computing with exponents help us compute efficiently. However, a person who forgets all the rules is still able to simplify an expression such as the one in Example 2 just by remembering what exponents mean. This is what the simplification might look like:

$$(2x^2)^3 \cdot (-7x^4)^2 = (2x^2)(2x^2)(2x^2)(-7x^4)(-7x^4)$$

$$= 2 \cdot x \cdot x \cdot 2 \cdot x \cdot x \cdot 2 \cdot x \cdot x \cdot (-7) \cdot x \cdot x \cdot x \cdot x \cdot (-7) \cdot x \cdot x \cdot x \cdot x$$
$$= 2 \cdot 2 \cdot 2 \cdot (-7) \cdot (-7) \cdot x \cdot x \cdot x \cdot x \cdot x \cdot x \cdot x \cdot x \cdot x \cdot x \cdot x \cdot x \cdot x \cdot x$$
$$= 8 \cdot 49 \cdot x^{14}$$
$$= 392x^{14}.$$

You can see that the rules for computing with exponents make simplification more efficient.

Remember that exponents abbreviate products of numbers and products of expressions. The rules for computing with exponents apply to products and not to sums. They do not help us simplify expressions such as $2x^3 + x^5$ or $x + x^2$. Indeed, these terms cannot be combined further.

Exercises 3.2

Insert the proper symbol ($<, =, >$) between the numbers in Exercises 1–4.

1. $(2.1)^3(2.1)^4$, $(2.1)^7$
2. $(5^3)^4$, 5^7
3. $(4.3)^5$, $[(4.3)^2]^3$
4. $(3.4)^4(4.5)^4$, $(15.3)^4$

Simplify the expressions in Exercises 5–22.

5. $(x^3)^5$
6. x^4x^7
7. x^{3^2}
8. $(x^3)^2$
9. $(x^3y^2)^4$
10. $(a^3b)^5$
11. $(-2x^5y^2)(3xy^3)$
12. $[(a^2)^3]^4$
13. $(-y)^4(-y)^3(-y)^5$
14. $(-a^2b^3)(ab^4)(a^3b)$
15. $(2x^3y^2)^3(xy^3)^2$
16. $(-a^2)^3$
17. $(-x^4)^2$
18. $[(-a)^3]^7$
19. $(-x^2y)^5$
20. $(-a^2b^2)^3(-a^4b^2)^2$

21. $(x + 2)^3(x + 2)^2$

22. $[(a + 1)^2(b + 1)^3]^2[(a + 1)(b + 1)^4]^3$

Solve for x in Exercises 23–30.

23. $3^2 3^x = 3^6$

24. $5^3 5^8 = 5^x$

25. $(-2)^x(-2)^4 = (-2)^7$

26. $(a^2)^x = a^8$

27. $(-a^2 b^3)^x = -a^6 b^9$

28. $a^4 + a^2 = a^2(a^x + 1)$

29. $b^3 + b^6 = b^x(1 + b^3)$

30. $c^4 + c^6 = c^x(c^2 + c^4)$

Find the value of the expressions in Exercises 31–36 if $a = -2$ and $b = 3$.

31. $(ab)^3$

32. $(-a^2)^3$

33. $(a + b)^2$

34. $a^2 + b^2$

35. $a^2 a^3$

36. $(-a^2 b)^3$

Use the distributive property and combine like terms to simply the expressions in Exercises 37–40.

37. $x^2(x^2 - y^2) + y(x^2 y - y^3)$

38. $x(x^3 - x) + 4(x^2 - 1)$

39. $y^2(xy + x^2 y^2) - xy^4(1 + x)$

40. $x(x^2 + y^2) - y(x^2 + xy) + y(x^2 - y^2)$

Write the mathematical phrase evaluated by each sequence of calculator key strokes in Exercises 41 and 42.

3.3 / USING 0 AND NEGATIVE INTEGERS AS EXPONENTS

Whole numbers are used as exponents to indicate the number of times the base appears in the product. Negative numbers and zero can also be used as exponents. However, negative numbers are not counting numbers. Their meaning as exponents is different than the meaning of positive numbers as exponents.

You can understand how zero and negative exponents are defined by looking at the rules for computing with exponents from the last section. Zero and negative exponents are defined so that these same rules can be used in computing with them. For example, Rule 1 requires that $b^0 \cdot b^n = b^{0+n} = b^n$ for any positive exponent n. But what value can b^0 have then? The only number than can be multiplied by b^n to give b^n is the number 1. Thus b^0 should be 1 for any nonzero base b. This fact leads to the definition for b^0.

DEFINITION

If $b \neq 0$, then $b^0 = 1$.

For example, $7^0 = 1$, $32.5^0 = 1$, $(xy)^0 = 1$, $(2x^2 - 3)^0 = 1$.

Rule 1 also suggests how to define b^{-n}, where n is a whole number. Since $b^n \cdot b^{-n} = b^{n+(-n)} = b^0 = 1$, it must follow that b^{-n} times b^n equals 1. Thus if b is not zero, b^{-n} should be the reciprocal of b^n. This leads to the definition for b^{-n}.

DEFINITION

If $b \neq 0$, then $b^{-n} = \dfrac{1}{b^n}$.

For example,

$$2^{-1} = \frac{1}{2},\ 5^{-2} = \frac{1}{5^2},\ \left(\frac{1}{x}\right)^{-3} = \frac{1}{\left(\frac{1}{x}\right)^3},\ (2y)^{-5} = \frac{1}{(2y)^5}.$$

The definitions we have given for zero and negative exponents permit Rules 1, 2, and 3 to be valid for all the integer exponents, . . . -3, -2, -1, 0, 1, 2, 3 . . . . This can be checked by replacing b^0 by 1 and b^n by $\dfrac{1}{b^n}$ in the statement of the rules. Thus we compute with zero and negative exponents much the same as we compute with positive exponents.

Example 1   Rewrite $x^{-4}y^5z^{-2}$ so that it has no negative exponents.

We need only to remember the definitions:

$$x^{-4}y^5z^{-2} = \frac{1}{x^4} \cdot y^5 \cdot \frac{1}{z^2}$$

$$= \frac{y^5}{x^4z^2}.$$

□

Example 2   Rewrite $\dfrac{a^{-2}b^{-3}}{c^{-1}}$ so that it has no negative exponents.

Again using the definitions for negative exponents, we can write

$$\frac{a^{-2}b^{-3}}{c^{-1}} = \frac{\dfrac{1}{a^2} \cdot \dfrac{1}{b^3}}{\dfrac{1}{c}} = \frac{\dfrac{1}{a^2b^3}}{\dfrac{1}{c}} = \frac{c}{a^2b^3}.$$

Notice that the effect of changing negative exponents to positive exponents is that factors in the numerator are moved down to the denominator and factors in the denominator are moved up to the numerator.

□

Example 3   Rewrite $(-a^{-2})^3(ab)^{-1}$ so that it has no negative exponents.

The rules for exponents work the same way for negative exponents as for positive ones. Remember that $-a^{-2}$ can be replaced by $(-1)(a^{-2})$.

$$(-a^{-2})^3(ab)^{-1} = [(-1)a^{-2}]^3(ab)^{-1}$$

$$= (-1)^3(a^{-2})^3a^{-1}b^{-1} \qquad \text{(Rule 2)}$$

$$= (-1)a^{-6}a^{-1}b^{-1} \qquad \text{(Rule 3 and powers of } -1)$$

$$= -a^{-7}b^{-1} \qquad \text{(Rule 1)}$$

$$= -\frac{1}{a^7} \cdot \frac{1}{b}$$

$$= -\frac{1}{a^7 b}.$$

□

Example 4   Rewrite $(3x^4)x^{-2} + (3x)^{-3}(3x^4)$ so that it has no negative exponents.

$$(3x^4)x^{-2} + (3x)^{-3}(3x^4) = 3 \cdot x^4 \cdot x^{-2} + 3^{-3}x^{-3}3x^4 \qquad \text{(Rule 2)}$$

$$= 3 \cdot x^4 \cdot x^{-2} + 3^{-3} \cdot 3 \cdot x^{-3} \cdot x^4 \qquad \text{(Rearrange terms)}$$

$$= 3x^{4 + (-2)} + 3^{-3 + 1}x^{-3 + 4} \qquad \text{(Rule 1)}$$

$$= 3x^2 + 3^{-2}x$$

$$= 3x^2 + \frac{1}{3^2}x$$

$$= 3x^2 + \frac{x}{3^2}$$

□

The Calculator and Negative Exponents

Fortunately the calculator can compute a positive number to any integer power: positive, negative, or zero. Check these examples on your calculator.

Mathematical phrase	Keying sequence	Display
2^0	$\boxed{2}\ \boxed{y^x}\ \boxed{0}\ \boxed{=}$	1
$2^{-1} \left(\text{or } \dfrac{1}{2} \right)$	$\boxed{2}\ \boxed{y^x}\ \boxed{1}\ \boxed{+/-}\ \boxed{=}$	0.5
$2^{-2} \left(\text{or } \dfrac{1}{2^2} \right)$	$\boxed{2}\ \boxed{y^x}\ \boxed{2}\ \boxed{+/-}\ \boxed{=}$	0.25
$2^{-3} \left(\text{or } \dfrac{1}{2^3} \right)$	$\boxed{2}\ \boxed{y^x}\ \boxed{3}\ \boxed{+/-}\ \boxed{=}$	0.125

Since the calculator cannot compute a negative number to a power, such expressions must be rewritten in terms of a positive base. Consider the powers of -2:

$$(-2)^4 = (-2)(-2)(-2)(-2) = 16 = 2^4,$$

$$(-2)^3 = (-2)(-2)(-2) = -8 = -2^3,$$

$$(-2)^2 = (-2)(-2) = 4 = 2^2,$$

$$(-2)^1 = -2,$$

$$(-2)^0 = 1,$$

$$(-2)^{-1} = \frac{1}{-2} = \frac{-1}{2} = -2^{-1},$$

$$(-2)^{-2} = \frac{1}{(-2)^2} = \frac{1}{(2)^2} = 2^{-2},$$

$$(-2)^{-3} = \frac{1}{(-2)^3} = \frac{1}{-2^3} = -2^{-3}.$$

From the above computation we conclude that $(-2)^0 = 1$ (of course!) and

$$(-2)^{-4} = 2^{-4} \qquad (-2)^{-3} = -2^{-3}$$
$$(-2)^{-2} = 2^{-2} \qquad (-2)^{-1} = -2^{-1}$$
$$(-2)^2 \ = 2^2 \qquad (-2)^3 \ = -2^3$$
$$(-2)^4 \ = 2^4 \qquad (-2)^5 \ = -2^5$$
$$(-2)^6 \ = 2^6$$

In general,

if n is even, $(-2)^n = 2^n$

if n is odd, $(-2)^n = -2^n$.

This is true whether n is positive, negative, or zero. Remembering this we can use the calculator to compute all the integer powers of -2.

Mathematical phrase	Keying sequence	Display
$(-2)^0$	[2] [yˣ] [0] [=]	1
$(-2)^{-1}$	[2] [yˣ] [1] [+/−] [=] [+/−]	−0.5
$(-2)^{-2}$	[2] [yˣ] [2] [+/−] [=]	0.25
$(-2)^{-3}$	[2] [yˣ] [3] [+/−] [=] [+/−]	−0.125

The keying sequences replace

$(-2)^0$ by 2^0

$(-2)^{-1}$ by -2^{-1}

$(-2)^{-2}$ by 2^{-2}

$(-2)^{-3}$ by -2^{-3}.

Exercises 3.3

Evaluate each expression in Exercises 1–18.

1. 2^{-3}

2. -3^2

3. $(-2)^3$

4. $(-3)^{-2}$

5. -2^{-3}

6. 3^2

7. $-(-2)^3$

8. $-(-3)^{-2}$

9. $-(-7)^0$

10. $(3^{-1})^{-1}$

11. $(2^{-3})^{-2}$

12. $(2^{-2})^3$

13. $(-5^{-1})^{-1}$

14. $(5^2)^{-3}$

15. $(2^2 3)^0 (5^3 7^{-1})^0$

16. $(5^{-2} 7^3)^4 (5^4 7^{-6})^2$

17. $2^{-1} + 3^{-1} - 2^2 3^{-2}$

18. $3(2^{-2}) + 5(2^{-1} 3^{-1}) - 2^{-1} 5^{-1}$

In Exercises 19 and 20, find the smallest whole number n that makes the statement true.

19. $2^{-n} < 0.0001$

20. $3^{-n} < 0.0001$

Solve for x in Exercises 21–26.

21. $2^x = \dfrac{1}{16}$

22. $3^{-x} = 27$

23. $4^5 4^x = 4$

24. $(-3)^x = -\dfrac{1}{27}$

25. $(b^x)^2 = b^{10}$

26. $5^x = \dfrac{1}{25}$

Arrange the numbers in Exercises 27 and 28 in order, from smallest to largest.

27. $0, 3^0, 3^{-1}, 3^{-2}, 3^{-3}, 3^{-4}$

28. $0, (-3)^0, (-3)^{-1}, (-3)^{-2}, (-3)^{-3}, (-3)^{-4}$

Simplify the expressions in Exercises 29–40. There should be no negative or zero exponents in your answers.

29. $(a^3)^{-2}$

30. $(ab^{-2})^{-4}$

31. $a^{-2}b^{-3}$

32. $\dfrac{x^{-3}y^2}{z^{-3}}$

33. $\dfrac{r^{-4}}{2^{-5}}$

34. $(-xy^{-2})^{-3}$

35. $(-a^2b^{-1})^3(a^{-1}b^{-1})^{-1}$

36. $(-x^3y^{-1})(x^{-2}y)$

37. $(3x^{-1}y^2)^{-3}(2^{-3}x^{-2}y^{-1})^{-2}$

38. $(-a^3b^5)^{-1}(-a^3b)^2(a^{-1}b)^3$

39. $(2y^5)y^{-3} + (2y)^{-2}y^3$

40. $(-2x^{-1}y)^{-4}(2^{-3}xy^{-1})^{-2} + (3xy)^{-3}(3^2x^2y)^2$

Write the mathematical phrase evaluated by each sequence of calculator key strokes given in Exercises 41 and 42.

41. $\boxed{5}\,\boxed{y^x}\,\boxed{3}\,\boxed{+/-}\,\boxed{y^x}\,\boxed{4}\,\boxed{+/-}\,\boxed{=}$

42. $\boxed{3}\,\boxed{+/-}\,\boxed{x^2}\,\boxed{y^x}\,\boxed{3}\,\boxed{+/-}\,\boxed{=}$

Use the distributive property and combine like terms to simplify the expressions in Exercises 43 and 44.

43. $x^{-1}(1 + x + x^2) + x^{-2}(1 - x - x^2)$

44. $x^{-2}(2x^3y + xy^2) + y^{-1}(3xy^2 - x^{-1}y^3)$

45. Compute $(-1)^n$ for every integer n.

3.4 / EXPONENTS AND QUOTIENTS

There are two additional rules for computing with exponents that we need to add to the three already on our list. The examples suggest the first of these:

$$\left(\frac{4}{5}\right)^2 = \frac{4}{5} \cdot \frac{4}{5} = \frac{4^2}{5^2}$$

$$\left(\frac{x}{y}\right)^4 = \frac{x}{y} \cdot \frac{x}{y} \cdot \frac{x}{y} \cdot \frac{x}{y} = \frac{x^4}{y^4}.$$

Rule 4 is sometimes summarized by saying, a power of a quotient is the quotient of the powers.

RULE 4 FOR EXPONENTS

If $b \neq 0$, then $\left(\dfrac{a}{b}\right)^n = \dfrac{a^n}{b^n}$.

Rule 4 is true for any exponent n.

Example 1   Write $\left(\dfrac{a}{b}\right)^{-2}$ without negative exponents.

a. We can start with Rule 4 and then use the definition of negative exponents:

$$\left(\frac{a}{b}\right)^{-2} = \frac{a^{-2}}{b^{-2}} \qquad \text{(Rule 4)}$$

$$= \frac{b^2}{a^2} \qquad \text{(Definition of negative exponents)}$$

b. We can first use the definition of negative exponents and then use Rule 4:

$$\left(\frac{a}{b}\right)^{-2} = \frac{1}{\left(\dfrac{a}{b}\right)^2} \qquad \text{(Definition of negative exponents)}$$

$$= \frac{1}{\dfrac{a^2}{b^2}} \qquad \text{(Rule 4)}$$

$$= \frac{b^2}{a^2}.$$

□

The last pattern we need to observe for exponents is suggested by these examples:

$$\frac{5^4}{5^2} = \frac{5 \cdot 5 \cdot 5 \cdot 5}{5 \cdot 5} = \frac{5 \cdot 5}{1} = 5^2$$

$$\frac{x^3}{x^5} = \frac{x \cdot x \cdot x}{x \cdot x \cdot x \cdot x \cdot x} = \frac{1}{x \cdot x} = \frac{1}{x^2} = x^{-2}.$$

Each fraction above was reduced by removing common factors from the numerator and denominator. The result is summarized in Rule 5.

RULE 5 FOR EXPONENTS

If $b \neq 0$, then $\dfrac{b^n}{b^m} = b^{n-m}$.

This rule is true for all integer exponents: positive, negative, and zero.

With the definitions of exponents and the five rules for computing with exponents, we can rewrite an exponential expression so that each variable appears only once and no negative exponents are used. This is what we mean when we say to **simplify** such an expression.

Example 2   Simplify $\dfrac{a^2 b^4}{a^3 b^2}$.

Observe how Rule 5 is applied in this computation:

$$\frac{a^2 b^4}{a^3 b^2} = \frac{a^2}{a^3} \cdot \frac{b^4}{b^2}$$

$$= a^{2-3} \cdot b^{4-2} \qquad \text{(Rule 5)}$$

$$= a^{-1} \cdot b^2$$

$$= \frac{b^2}{a}.$$

$\square$

Example 3   Simplify $\dfrac{(a^4 b^{-3})^4}{(a^{-8} b^2)^{-2}}$.

$$\frac{(a^4 b^{-3})^4}{(a^{-8} b^2)^{-2}} = \frac{a^{16} b^{-12}}{a^{16} b^{-4}} \qquad \text{(Rule 3)}$$

$$= a^{16-16} b^{-12-(-4)} \qquad \text{(Rule 5)}$$

$$= a^0 b^{-8}$$

$$= 1 \cdot \frac{1}{b^8}$$

$$= \frac{1}{b^8}. \qquad \qquad \square$$

Example 4   Simplify $\left(\dfrac{a^2 b^{-3}}{a^{-2} b}\right)^3$.

a. $\left(\dfrac{a^2 b^{-3}}{a^{-2} b}\right)^3 = \dfrac{(a^2 b^{-3})^3}{(a^{-2} b)^3} \qquad \text{(Rule 4)}$

$$= \frac{(a^2)^3 (b^{-3})^3}{(a^{-2})^3 (b)^3} \qquad \text{(Rule 2)}$$

$$= \frac{a^6 b^{-9}}{a^{-6} b^3} \qquad \text{(Rule 3)}$$

$$= a^{6-(-6)} b^{-9-3} \qquad \text{(Rule 5)}$$

$$= a^{12} b^{-12}$$

$$= a^{12} \cdot \frac{1}{b^{12}} = \frac{a^{12}}{b^{12}}.$$

b. $\left(\dfrac{a^2 b^{-3}}{a^{-2} b}\right)^3 = (a^{2-(-2)} b^{-3-1})^3 \qquad \text{(Rule 5)}$

$$= (a^4 b^{-4})^3$$

$$= a^{12} b^{-12}$$

$$= \frac{a^{12}}{b^{12}}. \qquad \qquad \text{(Rule 3)}$$

$\square$

Exercises
3.4

Evaluate each expression in Exercises 1–6.

1. $\left(\dfrac{1}{3}\right)^{-1}$

2. $\left(\dfrac{1}{3}\right)^{-2}$

3. $\left(\dfrac{3}{4}\right)^{-2}$

4. $\left(\dfrac{5}{7}\right)^{-2}$

5. $\left(\dfrac{5}{4}\right)^{-3}$

6. $\left(\dfrac{2^{-2}3^{-1}}{2^{-3}3^{-2}}\right)^{-3}$

Solve for x in Exercises 7 and 8.

7. $\left(\dfrac{1}{2}\right)^{x} = 16$

8. $\left(\dfrac{3}{4}\right)^{x} = \dfrac{16}{9}$

Simplify the expressions in Exercises 9–36. Each answer should be a single fraction with positive exponents.

9. $\dfrac{x^3}{x^{-3}}$

10. $\dfrac{a^8}{a^{11}}$

11. $\dfrac{a^7 a^5}{a^4}$

12. $\dfrac{x^3 y^4}{x^5 y}$

13. $\dfrac{a^4 a^3}{a^5 a^6}$

14. $\left(\dfrac{a^2}{b^3}\right)^4$

15. $\left(\dfrac{a^{-3}}{b^4}\right)^2$

16. $\left(\dfrac{-x^2 y}{z^3}\right)^3$

17. $\left(\dfrac{x^{-3}}{y^{-2}}\right)^3$

18. $\left(\dfrac{x^{-2}}{y^{-4}}\right)^{-3}$

19. $\left(\dfrac{1}{x}\right)^{-2}$

20. $\left(\dfrac{a}{b}\right)^{-1}$

21. $\left(\dfrac{a^4 b^2}{a^2 b^5}\right)^3$

22. $\left(\dfrac{-4x^{-2}y}{2^3 xy^{-1}}\right)^3$

23. $(2a^{-1})^{-2}(4a^2)$

24. $\dfrac{(3x^{-2})(-3x^3)}{18x^2}$

25. $\dfrac{(x+3)^5}{(x+3)^2}$

26. $\dfrac{(2x+1)^{-3}}{(2x+1)^{-6}}$

27. $\left[\left(\dfrac{x^{-1}y^3}{z^2}\right)^{-2}\right]^2$

28. $\dfrac{(x^2y)^3(xy^3)^2}{(x^3y^2)^4}$

29. $\left(\dfrac{a^2b}{c^4}\right)^3\left(\dfrac{c^2}{ab^2}\right)^6$

30. $\left(\dfrac{x^3y^{-2}}{z^{-2}}\right)^3\left(\dfrac{x^{-4}y^4}{z^4}\right)^2$

31. $\left(\dfrac{-a^2b^{-1}}{c^{-2}}\right)^3\left(-\dfrac{a^{-3}b^{-2}}{c^2}\right)^4$

32. $\left(\dfrac{r^{-3}s}{t^2}\right)^{-2}\left(\dfrac{r^4s^{-1}}{t^{-2}}\right)^{-3}$

33. $\dfrac{(x^2y^{-3})^{-2}(x^{-1}y^2)^3}{(xy^2)^{-3}(x^{-1}y^{-3})^{-1}}$

34. $\dfrac{(a^2b)^3(a^{-2}b^{-1})^2}{(a^{-2}b^{-2})^{-1}}$

35. $\dfrac{(a^3b^2)^{-2}(a^{-1}b^{-2})^{-4}}{(a^2b^{-4})^{-1}}$

36. $(a+b)^{-1}$

3.5 / SCIENTIFIC NOTATION

Because our calculator can display only eight digits, it is not possible for this calculator to display numbers such as 364,000,000 or .0000000581 in their decimal form. However, very large numbers and very small numbers are often part of numerical computations. For example, astronomers work with very large numbers in describing distances in space; biologists work with very small numbers in describing the size of microorganisms. Fortunately, the calculator is able to compute with these numbers by writing them in another form—one that uses what you have learned about multiplying and dividing by powers of 10.

In Section 2.4 we observed that multiplying a number by a power of 10 has the effect of moving digits across the decimal point from the right side to the left and that dividing by a power of 10 has the effect of moving digits across the decimal point from the left side to the right. These examples are illustrations:

$4211.064 \times 10^2 = 421106.4$

$4211.064 \div 10^2 = 42.11064.$

Observe that the second example above could be rewritten using a negative exponent on 10:

$4211.064 \div 10^2 = 4211.064 \times \dfrac{1}{10^2} = 4211.064 \times 10^{-2}.$

Thus dividing by a positive power of 10 is the same as multiplying by a negative power of 10. This means that multiplying a number by a negative power of 10 has the effect of moving digits from the left side of the decimal point to the right. Multiplying a number by a positive power of 10 has the effect of moving digits from the right side of the decimal point to the left. We could write the two examples above this way:

$$4211.064 \times 10^2 = 421106.4$$

$$4211.064 \times 10^{-2} = 42.11064.$$

We use our observations about multiplying by powers of 10 to write numbers in a special form that makes computation on a calculator possible with very large and very small numbers. Any positive number can be written as a number between 1 and 10, multiplied by a power of 10. A number in this form is said to be in **scientific notation**. To see how to write a number this way, study these products:

$$2.113 \times 10^3 = 2113$$

$$8.755601 \times 10^4 = 87556.01$$

$$2.175 \times 10^{-2} = .02175$$

$$3.4 \times 10^{-4} = .00034$$

$$4.16 \times 10^0 = 4.16.$$

Each of these products illustrates the effect of multiplying by a power of 10. In each case the product on the left is the scientific notation for the number on the right.

Number	Scientific notation
2113	2.113×10^3
87556.01	8.755601×10^4
.02175	2.175×10^{-2}
.00034	3.4×10^{-4}
4.16	4.16×10^0

In scientific notation, a number is written as a number between 1 and 10, multiplied times a power of 10. We will call the number between 1 and 10 the **stem** and the power on 10 the **exponent** for a number in scientific notation. For example, in scientific notation, 2113 has a stem of 2.113 and an exponent of 3. The stem always has one digit on the left of the decimal and it is one of the numbers 1, 2, 3, 4, 5, 6, 7, 8, or 9. Notice that numbers greater than 10 have positive exponents; numbers between 1 and 10 have zero exponent; and numbers between 0 and 1 have negative exponents.

Example 1   Write 364,000,000 in scientific notation.

The stem is 3.64. To find the exponent, we ask what number n makes

$$3.64 \times 10^n = 364,000,000.$$

To change 3.64 to 364,000,000 we must move 8 digits across the decimal point from the right side to the left; thus $n = 8$, so 3.64×10^8 is the scientific notation for 364,000,000.   □

Example 2   Write .0000000581 in scientific notation.

The exponent is the number n for which

$$5.81 \times 10^n = .0000000581.$$

To change 5.81 to .0000000581 we must move 8 digits from the left side of the decimal point to the right. Thus $n = -8$ and 5.81×10^{-8} is the scientific notation for .0000000581.   □

We have limited this discussion to scientific notation for positive numbers. Negative numbers can also be written in scientific notation. The stem carries the sign of the number. Using Examples 1 and 2, we conclude that $-364,000,000$ has scientific notation -3.64×10^8; and $-.0000000581$ has scientific notation -5.81×10^{-8}.

Scientific Notation and the Calculator

The calculator cannot display all the digits of the number 364,000,000, and we cannot enter that number directly into the calculator. If, however, we write 364,000,000 as 364(1,000,000) and key <u>364</u> $\boxed{\times}$ 1,000,000 , the calculator display is 3.64 08. This display is the way the calculator indicates the number 3.64×10^8. Thus the calculator displays 364,000,000 in scientific notation. (Notice that it does not display the base 10.) Similarly, we cannot enter directly the number .0000000581 because it requires more than eight digits. However, if we view .0000000581 as $581 \div 10^{10}$ and key <u>581</u> $\boxed{\div}$ 10 $\boxed{y^x}$ 10 $\boxed{=}$, the display will be 5.81 -08, indicating that 5.81×10^{-8} is the scientific notation for .0000000581.

The calculator reserves the first five digits of the display for the stem and the last three for the exponent. If the stem has more than five digits, the machine displays the first five, raising the fifth digit by one if the sixth digit is 5 or more; for example, the calculator displays 123,456,000 as 1.2346 08. Since three positions of the display are reserved for the exponent, the exponent can be any number between -99 and 99.

A number can be given to the calculator in scientific notation. The key that makes this possible is $\boxed{\text{EE⬇}}$. To enter directly the number 1.2346×10^8, we key

<table>
<tr><td></td><td>(Display)</td></tr>
<tr><td>1.2346 $\boxed{\text{EE⬇}}$ $\boxed{8}$</td><td>1.2346 08</td></tr>
</table>

The calculator is constructed to display very large and very small numbers in scientific notation. You might enter the number 12.346×10^7 by keying 12.346 $\boxed{\text{EE⬇}}$ $\boxed{7}$. This is the same number as the one above, but it is not in scientific notation (because 12 is too large). It is interesting to observe that if the calculator uses this number in computation (that is, if you press a key such as $\boxed{y^x}$, $\boxed{)}$, $\boxed{+}$, $\boxed{-}$, $\boxed{\times}$, $\boxed{\div}$, $\boxed{=}$, or $\boxed{\text{STO}}$ after the number), the calculator puts the number in its scientific form: 1.2346×10^8. To see that this is true, key the sequence <u>12.346</u> $\boxed{\text{EE⬇}}$ $\boxed{7}$ $\boxed{=}$.

Multiplying and Dividing in Scientific Notation

The exponents in scientific notation simplify multiplying and dividing of very large and very small numbers. To see how, study the following examples:

Example 3   Find the product $(3 \times 10^8)(6 \times 10^{10})$.

a. Without a calculator, we might write

$$(3 \times 10^8)(6 \times 10^{10}) = 3 \times 10^8 \times 6 \times 10^{10}$$
$$= 3 \times 6 \times 10^8 \times 10^{10}$$
$$= 18 \times 10^{18}. \qquad \text{(Rule 1)}$$

The number 18×10^{18} is not in scientific notation because 18 is greater than 10. If we want the answer in scientific notation, we continue:

$$18 \times 10^{18} = 1.8 \times 10 \times 10^{18}$$
$$= 1.8 \times 10^{19}.$$

b. To key this product on the calculator, we use the following keying sequence:

$$\boxed{3}\ \boxed{\text{EE↓}}\ \boxed{8}\ \boxed{\times}\ \boxed{6}\ \boxed{\text{EE↓}}\ 10\ \boxed{=} \qquad \begin{array}{c}\text{(Display)}\\ 1.8\ \ 19\end{array}$$

Notice that the calculator displays the product in scientific notation. No further computation is needed. □

Example 4   Evaluate $\dfrac{(-3.64 \times 10^{-8})(57.6 \times 10^{18})}{(3.45 \times 10^5)}$.

We observe that the numerator of this fraction is a negative number and the denominator is a positive number; thus the quotient is negative. We can either key the negative sign with the number -3.64, or we can key all factors as positive numbers and then use the sign-change key at the end. Here is one keying sequence that evaluates the quotient. You can find others.

Keying sequence:

$$\underline{3.64}\ \boxed{+/-}\ \boxed{\text{EE↓}}\ \boxed{8}\ \boxed{+/-}\ \boxed{\times}\ \underline{57.6}\ \boxed{\text{EE↓}}\ \underline{18}\ \boxed{÷}\ \underline{3.45}\ \boxed{\text{EE↓}}\ \boxed{5}\ \boxed{=}$$

Display:

$$-6.0772\ \ 06$$

The answer is -6.0772×10^6. □

Exercises 3.5

Write the numbers in Exercises 1–6 in decimal form.

1. 3.67×10^6
2. 2.75×10^{-5}
3. 37.4×10^7
4. 0.023×10^{-6}
5. 7×10^{-5}
6. -8×10^7

Write the numbers in Exercises 7–10 in scientific notation.

7. $5,670,000$
8. $9,000,000,000$
9. -0.0000432
10. 0.000000003

Solve for y in Exercises 11–16.

11. $6,540,000 = 6.54 \times 10^y$
12. $732,000 = 73.2 \times 10^y$
13. $0.00000213 = 2.13 \times 10^y$
14. $0.00054 = 54 \times 10^y$
15. $7,000,000 = 7 \times 10^y$
16. $0.000006 = 6 \times 10^y$

Insert the proper symbol ($<$, $=$, $>$) between the numbers in Exercises 17–22.

17. 7.6×10^4, 7.6×10^5
18. 8.3×10^5, 8.2×10^5
19. 6.42×10^4, 64.2×10^3
20. 6.7×10^{-5}, 6.7×10^{-6}
21. -4.35×10^{-8}, -4.29×10^{-8}
22. 2.345×10^{-7}, 23.45×10^{-8}

Use your calculator to evaluate the numbers in Exercises 23–32.

23. $(7.67 \times 10^8)(3.43 \times 10^9)$
24. $(34.5 \times 10^9)(8.67 \times 10^{-17})$
25. $(6.73 \times 10^{-11})(5.24 \times 10^{-9})$
26. $\dfrac{8.3 \times 10^{12}}{6.1 \times 10^4}$
27. $\dfrac{27 \times 10^6}{3 \times 10^{-4}}$
28. $\dfrac{9 \times 10^{-8}}{2 \times 10^4}$
29. $\dfrac{1.34 \times 10^{-11}}{3.67 \times 10^{-19}}$
30. $\dfrac{(8.32 \times 10^{-8})(-42.3 \times 10^{22})}{6.7 \times 10^7}$
31. $\dfrac{(48 \times 10^{-18})(3.5 \times 10^7)}{1.2 \times 10^{-6}}$
32. $\dfrac{(7 \times 10^{-8})(6 \times 10^{25})}{(1.8 \times 10^9)(3.3 \times 10^8)}$

33. Light travels approximately 186,284 miles per second. Find how far light will travel in

   (a) 1 day
   (b) 1 week
   (c) 1 year
   (d) 5 years
   (e) x days
   (f) y years

34. Assume that the distance from the earth to the sun is 93,000,000 miles. How long will it take light from the sun to reach the earth?

35. The approximate mass of the sun is 1.99×10^{33} grams, and the mass of the earth is 5.98×10^{27} grams. How many times heavier than the earth is the sun?

36. The mass of a hydrogen atom is 1.7×10^{-24} grams, and the mass of an electron is 9.1×10^{-28} grams. How many times heavier than an electron is a hydrogen atom?

3.6 / ROUND-OFF AND APPROXIMATE SOLUTIONS

It is likely that the calculator you are using retains more digits for computation than it can display. For example, the powers of .5 all end in the digit 5. However, if you use the $\boxed{K}$ key to display the powers of .5, this is what you will see:

	(Display)
.5 $\boxed{\times}$ $\boxed{K}$ $\boxed{=}$	0.25
$\boxed{=}$	0.125
$\boxed{=}$	0.0625
$\boxed{=}$	0.03125
$\boxed{=}$	0.015625
$\boxed{=}$	0.0078125
$\boxed{=}$	.00390625
$\boxed{=}$	.00195313

The last display represents $(.5)^9$. The exact value of $(.5)^9$ requires more than eight digits, so the calculator has displayed an approximation. However, the calculator in this case retains all nine digits of the exact value .001953125 for $(.5)^9$ and can use it in further computation. To see that this is true on your calculator, you can subtract the number the calculator is retaining for $(.5)^9$ from the displayed number .00195313. If the difference is not 0, then you know the value retained is different from the value displayed. The following keying sequence stores the computed value of $(.5)^9$ and then subtracts it from the number displayed:

$$.5 \; \boxed{\times} \; \boxed{K} \; \boxed{-} \; \boxed{=} \; \boxed{=} \; \boxed{=} \; \boxed{=} \; \boxed{=} \; \boxed{=} \; \boxed{=} \; \boxed{\text{STO}} \; .00195313 \; \boxed{-} \; \boxed{\text{RCL}} \; \boxed{=} \; .$$

The display is 5. −09, meaning the difference between the displayed number and the retained value is $5. \times 10^{-9}$, or .000000005. Thus the calculator is displaying the number .00195313 but is retaining the value .001953125 for later computation. The display is obtained by *rounding* the exact decimal to 8 places; this procedure means we take the first 8 places of the exact value, increasing the last by 1 if the ninth place is a digit 5 or more. (If instead of using the $\boxed{K}$ key you use the $\boxed{y^x}$ key for $(.5)^9$, you will discover that the calculator retains a number slightly different than .001953125. This approximation is the result of the calculator using logarithms to compute powers when the $\boxed{y^x}$ key is pressed.)

Numbers can be rounded to any number of decimal places. The examples below show the effect of rounding the number .001953125 to

8 decimal places: .00195313

7 decimal places: .0019531

6 decimal places: .001953

5 decimal places: .00195

4 decimal places: .0020

3 decimal places: .002

2 decimal places: .00

1 decimal place: .0

Rounding is a way of approximating a number with another number that requires fewer digits. When you compute with decimals that have been rounded, you need to remember that small errors are involved.

Accuracy

Sometimes you will be asked to find the solution to a problem accurate to a certain number of decimal places. In these cases you are not required to find the exact solution to the problem but, rather, the solution rounded to the given number of decimal places. Commonly we look for solutions rounded to a certain number of decimal places in situations where it is difficult or impossible to write exact solutions, or in situations where approximations are as good for our purposes as exact solutions. Measurement is one type of computation in which we must be content with approximation.

Example 1   A number plus its square equals 1. Find the number accurate to 2 decimal places.

The number N that we seek is not a whole number. If we try 0 and 1, we get these results:

Number	Number plus number squared
0	0
1	2

Thus N must be a decimal between 0 and 1. Our task is to use the calculator to get a good estimate. Trying .5, .6, and .7 shows that .5 and .6 are both too small and .7 is too large.

Number	Number plus number squared
0	0
.5	.75
.6	.96
.7	1.19
1	2

Now we see that the number N is between .6 and .7. The table suggests that N is closer to .6 than to .7. Perhaps we should try .61 and .62.

Number	Number plus number squared
0	0
.5	.75
.6	.96
.61	.9821
.62	1.0044
.7	1.19
1	2

We are able to conclude from this information that N is between .61 and .62. We want to find N rounded to two decimal places. Thus we must determine whether the third digit of N is less than 5 or greater than or equal to 5. The last number we need to try is .615.

Number	Number plus number squared
0	0
.5	.75
.6	.96
.61	.9821
.615	.993225
.62	1.0044
.7	1.19
1	2

There is a positive solution N between .615 and .620. We conclude that if N is rounded to two decimal places, it will have the value .62.

Our computation has shown that any number greater than .62 will be too large to be a solution to the problem. Thus we have an approximation to the single positive solution. Could there be a negative solution? Try -1 and -2.

Number	Number plus number squared
−2	2
−1	0

Since −2 gives a value greater than 1 and −1 gives a value less than 1, we expect that there is a number between −2 and −1 that gives a value equal to 1. Call this solution M; we want to write M rounded to two decimal places. Next we might try −1.5, −1.6, −1.7:

Number	Number plus number squared
−2	2
−1.7	1.19
−1.6	.96
−1.5	.75
−1	0

From these computations we conclude that M is between −1.7 and −1.6, and we suspect that M is closer to −1.6. Thus we try −1.61 and −1.62.

Number	Number plus number squared
−2	2
−1.7	1.19
−1.62	1.0044
−1.61	.9821
−1.6	.96
−1.5	.75
−1	0

Now we conclude that M is between −1.62 and −1.61. We must decide whether the digit in the third decimal place is less than 5 or greater than or equal to 5. We take the number −1.615.

Number	Number plus number squared
−2	2
−1.7	1.19
−1.62	1.0044
−1.615	.993225
−1.61	.9821
−1.6	.96
−1.5	.75
−1	0

Since M is between −1.615 and −1.620, we know that rounding M to two decimal places will give −1.62. Now we have a second solution to the problem.

Our computation suggests that if we take numbers less than −2 and compute the number plus the number squared, we will get a sum greater than 2. Thus we do not need to search further for additional solutions to the problem. □

In the example, we found good approximations to two numbers that add to their squares to give 1. In fact, we have found the best approximations that are possible using two decimal places. We will see how to find the exact value of these numbers in a later chapter.

Exercises 3.6

Round 1.2345678 to the number of decimal places specified in Exercises 1–7.

1. 6 places
2. 5 places
3. 4 places
4. 3 places
5. 2 places
6. 1 place
7. To the nearest integer

8. Round each of the following numbers to two decimal places.

(a) 2.344   (b) −5.678

(c) 0.225

9. Write all possible 4-decimal-place numbers between 1.3445 and 1.3454.

10. Write all possible 3-decimal-place numbers between 2.455 and 2.464.

11. Write all 4-decimal-place numbers that round to 2.567.

12. Write all 3-decimal-place numbers that round to 0.45.

Write the mathematical phrase evaluated by each sequence of calculator key strokes in Exercises 13–16.

13. $\boxed{3}\ \boxed{x^2}\ \boxed{-}\ \boxed{3}\ \boxed{=}$ 14. $\boxed{4}\ \boxed{x^2}\ \boxed{-}\ \boxed{4}\ \boxed{=}$

15. $\boxed{3}\ \boxed{x^2}\ \boxed{+}\ \boxed{x^2}\ \boxed{=}$ 16. $\boxed{4}\ \boxed{x^2}\ \boxed{+}\ \boxed{x^2}\ \boxed{=}$

Find the numbers, accurate to 3 decimal places, that satisfy the conditions in Exercises 17–19.

17. The square of the number is 2.

18. The square of the number is 7.

19. The fourth power of the number is 17.

Find the numbers, accurate to 2 decimal places, that satisfy the conditions in Exercises 20–22.

20. The number subtracted from the square of the number equals 1.

21. The square of the number plus the square of the square of the number equals 1.

22. The number minus the number divided by one more than the number equals 2.

3.7 / COMPOUND INTEREST

Interest is an important component in the economics of modern living. Banks pay interest on savings accounts and sometimes even on checking accounts; homeowners pay interest on mortgage loans; consumers pay interest on charge accounts. To investigate the mathematics used in computing interest, we must understand the vocabulary that describes different ways of computing interest.

Simple Interest

Some investments, such as those made in bonds and certain certificates of deposit, pay an investor the same amount of money each year. The amount depends on the rate of interest and the amount of money invested. For example, if an investor buys a $10,000 bond that pays 15% interest, he will receive .15(10,000) or $1500 each year that he holds the bond. The bank may send the investor one-fourth of his earnings each quarter (that is, a check for $375 every three months); the important thing is that the total of the interest checks each year is .15(10,000). Interest paid this way is said to be **simple**.

In the example above, where $10,000 is invested at 15%, the original investment of $10,000 is called the **principal**, 15% is called the **rate of interest**, and the earnings of .15(10,000) are referred to as **interest**.

Example 1   A $5000 certificate of deposit pays 12% simple interest. How much interest will it pay over a 4-year period?

Each year the certificate pays .12(5000) =$600. Over a 4-year period the investor will receive 4(600) = $2400.   □

Example 2   Mr. Henderson holds a $500 Public Power and Light bond that pays $32.50 each year. What is the simple rate of interest on the bond?

The rate of interest is the number r, such that

$$r(500) = 32.50.$$

Thus r is the fraction $\dfrac{32.50}{500}$ or .065. As a percent, the rate of interest is 6.5%.   □

Compound Interest

Savings and loan associations and banks have accounts where interest, instead of being mailed to investors, is added to their accounts. Sometimes you see the phrase 8% *compounded annually*. This phrase indicates that after 1 year the bank pays 8% of the investment as interest (just like with simple interest) but adds the interest to the investor's account; after another year the 8% interest is paid on the sum of the original investment plus the first year's interest. Thus the account is permitted to build, with interest being added to the account each year and new interest being paid on the total amount of money in the account. Here is an example.

Example 3   If $1000 is invested at 8% compounded annually for a 3-year period, how much money is in the account at the end of the period?

We compute the amount of money in the account at the end of each of the 3 years.

	Amount of money ($) interest is paid on	Amount of interest ($)	Total in account ($)
After 1 year	1000	.08(1000) = 80	1000 + 80 = 1080
After 2 years	1080	.08(1080) = 86.40	1080 + 86.40 = 1166.40
After 3 years	1166.40	.08(1166.40) =93.31	1166.60 + 93.31 = 1259.71

At the end of the third year, the amount of money in the account is $1259.71. (Of this amount, $1000 is principal and $259.71 is interest.)

□

Analyzing the computation in this example shows a pattern that simplifies the procedure for computing interest compounded annually. To see the pattern, we write out fully how the total amount of money is computed each year and then use the distributive property.

	Total in account ($)
After 1 year	$1000 + .08(1000) = 1000(1 + .08)$
After 2 years	$1000(1 + .08) + .08(1000)(1 + .08) =$ $[1000(1 + .08)](1 + .08) = 1000(1 + .08)^2$

	Total in account (\$)
After 3 years	$1000(1 + .08)^2 + .08(1000)(1 + .08)^2 =$ $[1000(1 + .08)^2](1 + .08) = 1000(1 + .08)^3$

Thus the pattern for computing interest on \$1000 at 8% compounded annually is the following:

After 1 year: $1000(1 + .08)$

After 2 years: $1000(1 + .08)^2$

After 3 years: $1000(1 + .08)^3$

After n years: $1000(1 + .08)^n$

You should see that an analogous result is obtained if we replace \$1000 by any principal P and .08 by any interest rate r written as a decimal. The total amount T in an account in which P dollars have been invested at interest rate r compounded annually for n years is $T = P(1 + r)^n$.

Actually it is not common for banks to compound interest on an annual basis. Usually interest is compounded more often. Indeed, frequent compounding favors the investor. When interest is paid more than once a year, a fraction of the rate is paid each time. Generally, you divide the rate by the number of payments in a year. However, when interest is advertised as being paid daily, the interest rate is usually divided by 360 rather than by 365. This procedure continues that of an earlier day when there were no sophisticated computing machines, and dividing by 360 was easier than dividing by 365. The summary below should clarify common procedures.

Type of account	*When interest is paid*	*Percent paid each time*
8% compounded annually	Once a year	.08
8% compounded semiannually	Twice a year	$\dfrac{.08}{2} = .04$
8% compounded quarterly	Four times a year	$\dfrac{.08}{4} = .02$
8% compounded monthly	Each month	$\dfrac{.08}{12}$
8% compounded daily	Each day	$\dfrac{.08}{360}$

For each such account there is a pattern that simplifies computing the total amount in the account after a given number of payments. These patterns are analogous to the one we found for compounding interest annually. In general, the expression is $P(1 + \boxed{})^\Delta$, where P is the principal, $\boxed{}$ is filled with the percent paid each time, and the exponent Δ is the number of payments made in the period. If $1000 is invested for 3 years, the following chart indicates the total amount of money accumulated in the various types of accounts.

$1000 Investment for 3 Years

Type of account	*Total money accumulated*
8% compounded annually	$1000(1 + .08)^3$
8% compounded semiannually	$1000\left(1 + \dfrac{.08}{2}\right)^6$
8% compounded quarterly	$1000\left(1 + \dfrac{.08}{4}\right)^{12}$
8% compounded monthly	$1000\left(1 + \dfrac{.08}{12}\right)^{36}$
8% compounded daily	$1000\left(1 + \dfrac{.08}{360}\right)^{3(365)} = 1000\left(1 + \dfrac{.08}{360}\right)^{1095}$

Example 4 Compare the total amounts of money after 3 years in one account where $1500 is invested at 8% compounded annually, and in another account where $1500 is invested at 8% compounded daily.

In the account where interest is compounded annually, the total after 3 years is

$$1500(1 + .08)^3.$$

$\boxed{1}\ \boxed{+}\ \underline{.08}\ \boxed{=}\ \boxed{y^x}\ \boxed{3}\ \boxed{\times}\ \underline{1500}\ \boxed{=}$ (Display)
1889.568

If the interest is compounded daily for 3 years, the amount accumulated is

$$1500\left(1 + \frac{.08}{360}\right)^{3(365)}.$$

(Display)

$\boxed{1}$ $\boxed{+}$ .08 $\boxed{\div}$ 360 $\boxed{=}$ $\boxed{y^x}$ 1095 $\boxed{\times}$ 1500 $\boxed{=}$ 1913.186

Thus when the interest is compounded annually, the amount in the account is $1889.57; when the interest is compounded daily, the amount is $1913.19. You can see that the investor prefers the account where interest is compounded daily.   □

The calculator rounds off some numbers involved in the computation in Example 4. The display at each step is shown below. You should mark those numbers that have been rounded off.

Keying sequence	Display
$\boxed{1}$	1.
$\boxed{+}$	1.
.08	.08
$\boxed{\div}$	.08
360	360
$\boxed{=}$	1.0002222
$\boxed{y^x}$	1.0002222
1095	1095
$\boxed{\times}$	1.2754573
1500	1500
$\boxed{=}$	1913.186

There are several ways that lending institutions can handle rounding-off. The exercises will show you that real money differences can result. The federal Truth in Lending law requires that investors be told not only the rate of interest and how often it is compounded, but also what the interest rate would be if the returns had accumulated in an account where interest is compounded annually. This rate is called the **effective annual yield**. For a one-year period, interest compounded annually is the same as simple interest. Thus effective annual yield can be viewed as the rate of simple interest that would give the same amount of interest in one year as does the advertised scheme for computing interest on an account.

Example 5 $1000 is invested at 8% compounded daily. What is the effective annual yield for that investment?

To find effective annual yield, we compute three steps:

    (a) total amount in account after one year,
    (b) interest paid in one year,
    (c) rate of simple interest that gives the same return.

The total amount in this account after one year is

$$T = 1000\left(1 + \frac{.08}{360}\right)^{365} = 1084.48.$$

Of this total amount, $1000 is principal and $84.48 is interest. If the rate r of simple interest pays $84.48 on $1000 in one year, then

$$r(1000) = 84.48$$

so $r = \dfrac{84.48}{1000} = .08448$. As a percent, $r = 8.448\%$.   □

The example shows that investing $1000 at 8% compounded daily gives the same return in one year as investing $1000 at 8.448% simple interest. Indeed, an analogous computation would show that investing any amount of money at 8% compounded daily for a year would have the same result as investing the same amount of money at 8.448% simple interest. Over a period of any number of years, 8% compounded daily gives the same return as 8.448% compounded annually. You should repeat the example for investments of $100, $10, and $1.

Exercises
3.7

1. $500 is deposited in an account that pays simple interest. Find the amount of interest earned in one year if the rate of interest is

    (a) 4%   (b) $5\frac{1}{2}$%   (c) 8.75%

2. $1500 is deposited in an account. Find the rate of simple interest paid if the amount in the account one year later is

    (a) $1584   (b) $1601.25

If $1000 is invested at $5\frac{1}{4}$% interest compounded daily for 10 years, the amount accumulated is given by $1000(1.0001458)^{3650}$. Compute this amount if 1.0001458 is first rounded to the number of decimal places specified in Exercises 3–5.

3. 6 places   4. 5 places

5. 4 places

6. How much money is accumulated in 300 years if $20 is invested at 4% interest compounded annually?

7. $5000 is invested at 7.25% interest compounded quarterly. Find how much money is accumulated in

    (a) 1 year   (b) 5 years   (c) 69 months

8. $1500 is invested at 6.75% interest compounded monthly. Find how much is accumulated in

    (a) 1 year   (b) 5 years   (c) 71 months

9. $6000 is invested at 6.85% interest compounded daily. Find how much money is accumulated in

    (a) 1 year   (b) 5 years   (c) 1857 days

10. $2500 is invested at 5.75% interest compounded semiannually. Find how much money is accumulated in

    (a) 1 year   (b) $2\frac{1}{2}$ years   (c) 5 years

11. $3500 is invested at $6\frac{1}{4}$% interest compounded annually. Find how much money is accumulated in

    (a) 1 year   (b) 5 years   (c) 10 years

12. $2000 was deposited in a bank on January 1, 1980, at 6.25% interest. Compute how much *interest* would have been earned by January 1, 1990, if interest is

(a) simple, paid yearly   (b) compounded annually

(c) compounded semiannually   (d) compounded quarterly

(e) compounded monthly   (f) compounded daily

13. $2000 was deposited on January 1, 1980. Compute the effective annual yield if the amount in the account on January 1, 1981 was

(a) $2109.80   (b) $2123

14.   Find the effective annual yield accurate to 3 decimal places if the rate of interest is 6.25% compounded

(a) quarterly   (b) daily

15.   Compute how much money would need to be deposited now at 5.5% interest to produce $10,000 in the next ten years (this amount is called the **present value** of $10,000) if the interest is compounded

(a) annually   (b) monthly

16.   Determine how long it will take $1000 to double at 7.25% if the interest is compounded

(a) monthly   (b) daily

17.   Determine how long it will take $1000 to triple at 7.25% if the interest is compounded

(a) annually   (b) quarterly

3.8 / INFLATION, POPULATION GROWTH, CARBON DATING, AND OTHER SUCH MATTERS

There are many phenomena that, like compound interest, give rise to exponents in their mathematical formulation. We will analyze three such situations in the examples that follow. The exercises give others.

Inflation

A high rate of inflation is a particular worry to individuals whose incomes cannot be expected to increase each year. In simple terms, the statement that the "annual rate of inflation is 8%" means that the average cost of purchases made at the end of a year is 8% above the cost of the same purchases made at the beginning of that year. Thus a consumer can expect that maintaining the same standard of living will cost 8% more money than it did the year before. Over several years, inflation has an alarming effect on prices.

Example 1 An average loaf of bread cost $.29 in 1975. If the annual rate of inflation were 8% from 1975 to 1985, what would you predict the average price of bread to be in 1985?

Displaying the situation in the form of a chart will show the pattern of growth. We compute the cost of bread each year at 8% more than the previous year. Notice the repeated use of the distributive property.

Year	Average cost of bread ($)
1975	.29
1976	$.29 + .08(.29) = .29(1 + .08) = .29(1.08)$
1977	$.29(1.08) + .08(.29)(1.08) = .29(1.08)(1 + 08) = .29(1.08)^2$
1978	$.29(1.08)^3$
1979	$.29(1.08)^4$
1980	$.29(1.08)^5$

Seeing the pattern, we conclude that in 1985, 10 years after 1975, the average cost of a loaf of bread will be $.29(1.08)^{10}$, or $.63, more than double the cost in 1975. □

Population Growth

We can look at population increases the same way that we look at price increases. In both cases, if an annual rate is known, the computation is equivalent to the increases in money resulting from compound interest.

Example 2   The country of Lumidia had a population of 10,000,000 people in 1975. If the population is growing at an annual rate of 3%, when will the population of Lumidia be 20,000,000?

Once again, a table will show the growth pattern. To see the pattern, let P represent the population in 1975 and assume that each year the population will be 3% more than the year before. Notice the use of the distributive property.

Year	Population
1975	P
1976	$P + .03P = P(1 + .03) = P(1.03)$
1977	$P(1.03) + .03P(1.03) = P(1.03)(1 + .03) = P(1.03)^2$
1978	$P(1.03)^2 + .03P(1.03)^2 = P(1.03)^2(1 + .03) = P(1.03)^3$
1979	$P(1.03)^4$
1980	$P(1.03)^5$

After n years the population will be $P(1.03)^n$. In our problem $P = 10,000,000$, so the population after n years will be

$$10,000,000(1.03)^n.$$

If $10,000,000(1.03)^n$ is to equal 20,000,000, n must be chosen so that the factor $(1.03)^n$ is 2. We can use the calculator to experiment.

n	$(1.03)^n$
5	1.1592741
10	1.3439164
15	1.5579674
20	1.8061112
25	2.0937779

We see that by the end of the twenty-fifth year the population will have more than doubled. If $n = 24$, the calculator gives $(1.03)^n =$

2.0327941, and if $n = 23$, then $(1.03)^n = 1.9735865$. Thus our best whole number answer is that 24 years after 1975, that is to say by the end of 1999, the population of Lumidia will have doubled.

It is important to observe that 24 years after 1999, if the population continues to grow at a rate of 3%, the number of people in Lumidia will again have at least doubled. The reason is that the population after 24 years will be $(1.03)^{24}$ times the 1999 population and $(1.03)^{24} = 2.0327941$. Thus every 24 years the population will double. We say that the population of Lumidia has a *doubling period* of approximately 24 years.   □

Example 3 The Centerville *Journal* had a circulation of 13,440 in 1980 and a circulation of 14,112 in 1981. If the *Journal* experiences the same annual rate of growth until 1985, what will its circulation be at that time?

To compute the annual rate r of growth, we observe that the number of new subscriptions in 1981 was $14{,}112 - 13{,}440 = 672$. Thus

$$r(13{,}440) = 672$$

so

$$r = \frac{672}{13{,}440} = .05.$$

The number of subscriptions after 5 years (in 1985) is given by

$$13{,}440(1 + r)^5 = 13{,}440(1 + .05)^5$$

$$\doteq 17{,}153. \qquad\qquad □$$

Carbon Dating

Some quantities, instead of increasing like population, decrease in a way that uses exponents. An example is a radioactive element that emits particles and changes into another substance at a predictable rate. In these instances we talk about the **half-life** of a substance, meaning the length of time that it takes a substance to decrease to half its original weight.

The isotope carbon-14 has a half-life of about 5600 years. This means that if a substance contained 20,000 grams of carbon-14 in the year 0, then in the year 5600 it would contain 10,000 grams; in the year 5600 + 5600 (or 11,200) it would contain 5000 grams, and in the year 16,800 it would contain 2,500 grams, and so on. This information permits anthropologists and geologists to estimate the age of fossils. When one atom of carbon-14 disintegrates, it emits one beta ray; beta rays can be counted with a Geiger counter. A living body radiates approximately 918 rays per gram from carbon-14 each hour. By measuring the number of rays per gram emitted by a fossil each hour, we can estimate how old it is (that is, how long it has been dead).

Example 4   A fossil found in a cave is giving off about 7 rays per gram from carbon-14 each hour. How old is the fossil?

The decrease in carbon-14 rays emitted is the result of the decrease in amount of carbon-14 in the fossil; every 5600 years the number of rays per gram emitted each hour decreases by a fraction $\frac{1}{2}$. We need to see how many years are required to decrease 918 rays to 7 rays.

Rays emitted per gram per hour	Age of fossil
918	0
$\frac{1}{2}(918)$	5,600
$\frac{1}{2}\left(\frac{1}{2}\right)(918) = \left(\frac{1}{2}\right)^2(918)$	11,200
$\frac{1}{2}\left(\frac{1}{2}\right)^2(918) = \left(\frac{1}{2}\right)^3(918)$	16,800
$\frac{1}{2}\left(\frac{1}{2}\right)^3(918) = \left(\frac{1}{2}\right)^4(918)$	22,400
$\left(\frac{1}{2}\right)^n(918)$	$n \cdot 5,600$

We need to find the number n that makes $\left(\frac{1}{2}\right)^n(918)$ equal to 7. This is the same as finding how many times we must multiply 918 by .5 to get 7. By experimenting with the calculator, we find that

n	$(.5)^n (918)$
10	.89648438
8	3.5859375
7	7.171875

Thus the best whole number estimate for n is 7, and the number of years that have passed since the fossil was a living creature must be approximately $7 \cdot 5600$, or 39,200 years.   □

**Exercises
3.8**

1. The population of Fairfield was 200,000 in 1980 and 207,000 in 1981. Find the annual rate of growth.

2. The number of bacteria present at 9:00 A.M. was 3000; at 10:00 A.M., 3240. Find the hourly rate of growth.

3. Assume there are 50 grams of a radioactive isotope present on the first day and that there are 47.75 grams present on the second day. Find the daily rate of decay.

Complete the tables in Exercises 4–9.

4. Assume that the population of Bakersfield is increasing at the rate of 4% per year.

5. Assume that a bacterial culture is growing at the rate of 20% per hour.

Year	Population
1970	100,000
1971	
1972	
1975	
1980	

Time	Number of bacteria
9:00 A.M.	2000
10:00 A.M.	
11:00 A.M.	
1:00 P.M.	
8:00 P.M.	

6. Assume that a radioactive isotope is decaying at the rate of 5% per day.

Time	Number of grams
$t = 0$	40
1 day	
2 days	
3 days	
8 days	
14 days	

7. Assume that a car is depreciating at the rate of 20% per year.

Year	Value of car ($)
1975	6000
1976	
1977	
1978	
1981	
1985	

8. Assume that the population of Jamestown is decreasing at the rate of 2.5% per year.

Year	Population
1970	150,000
1971	
1972	
1981	
1990	

9. Assume that the consumption of electricity in Edwardsville is increasing at the rate of 12.5% per year.

Year	Number of kilowatt hours
1980	400,000
1981	
1982	
1992	
2000	

10. During 1980 the price of food increased 12.5%. Assuming that this rate will continue and that bread cost $.52 a loaf in 1980, find the price of a loaf of bread in

(a) 1981

(b) 1996

(c) t years after 1980

(d) In what year will the price of a loaf of bread have at least doubled the 1980 price?

11. During 1978 the price of food increased 10.4%. Assuming that this rate were to continue and that hamburger cost $1.09 a pound in 1978, find the price of a pound of hamburger in

    (a) 1979   (b) 1995

    (c) t years after 1978

    (d) In what year will the price of a pound of hamburger have at least doubled the 1978 price?

12. A family-sized car that costs $10,000 new is known to depreciate at the rate of 25% per year. Find the value of the car when it is

    (a) 1-year old   (b) 5-years old   (c) t-years old

13. A compact car that costs $6000 new is known to depreciate at the rate of 15% per year. Find the value of the car when it is

    (a) 1-year old   (b) 6-years old   (c) t-years old

14. Assume that the cars in Problems 12 and 13 were purchased the same year. In how many years will the compact car be worth more than the family-sized car?

15. The population of Jacksonville was 200,000 in 1979 and 205,000 in 1980.

    (a) Find the annual rate of growth.

    Assuming that this rate of increase will continue, determine what the population in Jacksonville will be in

    (b) 1981   (c) 1991   (d) 1979 + t years

16. In what year will the population of Jacksonville (Problem 15) be 400,000?

17. The population of Huntersville was 100,000 in 1970 and 95,000 in 1971.

    (a) Find the annual rate of decrease.

    Assuming that this rate of decrease will continue, determine what the population of Huntersville will be in

    (b) 1972   (c) 1982   (d) 1970 + t years

18. What is the first year after 1970 that the population of Huntersville (Problem 17) will be 50,000 or less?

19. There were 10 grams of a radioactive substance in a laboratory and one day later there were only 9.5 grams of the substance remaining.

(a) Find the daily rate of decay.

(b) Complete the following table:

Time elapsed	Number of grams remaining
0 days	10
1 day	9.5
2 days	
5 days	
10 days	
t days	

(c) Find the half-life of this substance.

20. Assume that radioactive carbon-14 has a half-life of about 5600 years and that living bodies radiate 918 rays per gram per hour of carbon-14. Find the approximate age of the fossils emitting

(a) 57 rays of carbon-14 per gram per hour

(b) 1.8 rays of carbon-14 per gram per hour

21. The radioactive isotope carbon-11 decays at the rate of 3.5% per minute. If you start with 100 grams of carbon-11, find an expression for the number of grams remaining t minutes later. Find the half-life of carbon-11 accurate to the nearest minute.

22. What constant rate of growth is necessary for the population of a city to double every 10 years? Give the answer accurate to 3 decimal places.

23. What constant rate of decrease is necessary for the population of the city to be cut in half every 10 year?

3.9 / FRACTIONAL EXPONENTS AND SQUARE ROOT

Assume that a bacterial culture has a doubling period of 10 hours; that is to say, every 10 hours the number of bacteria in the culture doubles. If the original culture contained 1000 bacteria, we could make this chart:

Age of culture (hours)	Number of bacteria in culture
0	1000
10	$2(1000) = 2,000$
20	$2^2(1000) = 4,000$
30	$2^3(1000) = 8,000$
40	$2^4(1000) = 16,000$
50	$2^5(1000) = 32,000$

The chart gives information about the number of bacteria in the culture every 10 hours. But there are many ages of the culture about which the chart gives no information. For example, how many bacteria are in the culture after 15 hours? The pattern suggests the number should be $2^{3/2}(1000)$, or $2^{1.5}(1000)$. After 5 hours, the pattern suggests that the number of bacteria should be $2^{1/2}(1000)$. In fact, if we were to fill in the chart for 0 to 10 hours, we would expect the following numbers in the second column:

Age of culture (hours)	Number of bacteria in culture
0	1000
1	$2^{1/10}(1000)$
2	$2^{2/10}(1000)$
3	$2^{3/10}(1000)$
4	$2^{4/10}(1000)$
5	$2^{5/10}(1000)$

Age of culture (hours)	Number of bacteria in culture
6	$2^{6/10}(1000)$
7	$2^{7/10}(1000)$
8	$2^{8/10}(1000)$
9	$2^{9/10}(1000)$
10	$2(1000)$

We have described carefully what whole number exponents mean and what negative integer exponents mean, but we have not yet described the meaning of exponents that are fractions. Actually, the calculator can give a value for these numbers even before we say what fractional exponents mean. For example, to find the value of $2^{1/10}(1000)$, we key

$$\boxed{2}\ \boxed{y^x}\ \underline{10}\ \boxed{1/x}\ \boxed{\times}\ \underline{1000}\ \boxed{=}$$ (Display)
1071.7735

or

$$\boxed{2}\ \boxed{y^x}\ \boxed{(}\ \boxed{1}\ \boxed{\div}\ \underline{10}\ \boxed{)}\ \boxed{\times}\ \underline{1000}\ \boxed{=}$$ (Display)
1071.7735

To find a value for $2^{3/10}(1000)$, we key

$$\boxed{2}\ \boxed{y^x}\ \boxed{(}\ \boxed{3}\ \boxed{\div}\ \underline{10}\ \boxed{)}\ \boxed{\times}\ \underline{1000}\ \boxed{=}$$ (Display)
1231.1444

Since the numbers in the chart refer to bacteria in a culture, they must be whole numbers. Use your calculator to evaluate each of the numbers and, after rounding to whole numbers, be sure you get these values:

Age of culture (hours)	Number of bacteria in culture
0	1000
1	$2^{1/10}(1000) = 1072$
2	$2^{2/10}(1000) = 1149$
3	$2^{3/10}(1000) = 1231$

Age of culture (hours)	Number of bacteria in culture
4	$2^{4/10}(1000) = 1320$
5	$2^{5/10}(1000) = 1414$
6	$2^{6/10}(1000) = 1516$
7	$2^{7/10}(1000) = 1625$
8	$2^{8/10}(1000) = 1741$
9	$2^{9/10}(1000) = 1866$
10	$2(1000) = 2000$

Example 1 Assume that a bacterial culture has a doubling period of 10 hours and that 1000 bacteria are present initially. How many are present after 23 hours?

Continuing the pattern in the chart above, after 23 hours, the number of bacteria should be $2^{23/10}(1000)$, or 4925. □

The Exponent 1/2

The rules for computing with exponents that we summarized in Sections 3.2 and 3.4 suggest what fractional exponents should mean. We require that these rules also apply when the exponents are fractions. Thus it must follow that $(3^{1/2})^2 = 3^{(1/2)\cdot 2} = 3^1 = 3$, so $3^{1/2}$ is defined to be a number whose square is 3. Similarly, $2^{1/2}$ is a number whose square is 2, and $4^{1/2}$ is a number whose square is 4. Although -2 is a number whose square is 4, the exponent 1/2 always designates a *positive* number: $4^{1/2} = 2$.

The numbers $2^{1/2}$ and $3^{1/2}$ are positive numbers, but their values are not as apparent as $4^{1/2}$. No whole number squared equals 2 or 3. The calculator gives these approximations for $2^{1/2}$ and $3^{1/2}$:

(Display)

[2] [y^x] .5 [=]   1.4142136

[3] [y^x] .5 [=]   1.7320508

Indeed, if you key these sequences and then press the [x^2] key, you will get the numbers 2 and 3. However, you should see that $(1.4142136)^2$ does not exactly equal 2, and $(1.7320508)^2$ does not exactly equal 3. If you used paper and pencil to multiply $(1.4142136)(1.4142136)$, the last digit in the product would be 6. In fact, no finite decimal starting with 1.414 . . . can square to give 2 because there will be a nonzero digit in the decimal part. We have seen that a fraction is either a finite decimal or an infinite repeating decimal. The numbers that square to give 2 and 3 are not finite decimals and, in fact, they are not infinite repeating decimals, so they are not fractions at all. The numbers $2^{1/2}$ and $3^{1/2}$ are said to be **irrational** (that is, not rational). Their decimal representations are infinite and nonrepeating. We cannot write down an infinite nonrepeating decimal. Thus we cannot write irrational numbers in their full decimal forms. We can, however, write good approximations.

Other Fractions as Exponents

The exponent 1/3 has a meaning analogous to the exponent 1/2. Since $(2^{1/3})^3 = 2^{(1/3)\cdot 3} = 2^1 = 2$, the number $2^{1/3}$ is the number whose cube is 2. This is another irrational number, but the calculator gives a good approximation. We can evaluate $8^{1/3}$ without a calculator: $8^{1/3} = (2^3)^{1/3} = 2$.

The exponents 1/4, 1/5, 1/6, . . . have meanings analogous to the exponents 1/2 and 1/3. What about fractions like 3/2 or 4/5? We agree that $2^{3/2} = (2^{1/2})^3$ and $2^{4/5} = (2^{1/5})^4$. Thus every positive fraction has meaning as an exponent. Negative exponents have the same meaning when they are fractions as they did for whole numbers:

$$10^{-3/2} = \frac{1}{10^{3/2}}, \qquad 2^{-5/4} = \frac{1}{2^{5/4}}.$$

At this time in our work, we will use fractional exponents only with positive bases.

Example 2   Find the value of $9^{-4/5}$ accurate to three decimal places.

There are several keying sequences that will give a value for $9^{-4/5}$. Study these and see if you can supply others.

(Display)
.17242729

$$\boxed{9}\;\boxed{y^x}\;\boxed{5}\;\boxed{1/x}\;\boxed{=}\;\boxed{y^x}\;\boxed{4}\;\boxed{+/-}\;\boxed{=} \qquad .17242729$$

$$\boxed{4}\;\boxed{\div}\;\boxed{5}\;\boxed{+/-}\;\boxed{=}\;\boxed{STO}\;\boxed{9}\;\boxed{y^x}\;\boxed{RCL}\;\boxed{=} \qquad .17242729$$

Thus, accurate to 3 decimal places, the value of $9^{-4/5}$ is .172. □

Example 3   Find a number x, accurate to 3 decimal places, such that $x^6 = 10$.

We need to fill in this box: $\boxed{}^{\,6} = 10$. Because 6 is an even exponent, we know there is both a positive and a negative solution. Since $(10^{1/6})^6 = 10$, the positive solution is $10^{1/6} = 1.468$. The negative solution is -1.468. □

Example 4   Find a number x, accurate to 3 decimal places, such that $x^{1/3} = 12$.

This time we must fill in this box: $\boxed{}^{\,1/3} = 12$. Since $(12^3)^{1/3} = 12$, we know that $x = 12^3 = 1728$ is the value that works. □

Problems that Use Fractions as Exponents

Fractions occur as exponents in problems of growth and decay. We have already looked at a problem involving the doubling period of a bacterial culture. Here is an example concerning population growth.

Example 5   The 1970 census gave the population of Oilstown as 8300. The 1980 census reported a population of 18,500. Find the population each year from 1970 to 1980, assuming that the annual rate of growth was the same each year during that period.

We could start with a chart similar to the ones used in the last section. Our job is to fill in the populations between 1970 and 1980.

Year	Population
1970	8,300
1971	
1972	
1973	

Year	Population
1974	
1975	
1976	
1977	
1978	
1979	
1980	18,500

If r is the annual rate of growth, then we have seen that the population can be written each year in terms of r:

Year	Population
1970	8300
1971	$8300(1 + r)$
1972	$8300(1 + r)^2$
1973	$8300(1 + r)^3$
1974	$8300(1 + r)^4$
1975	$8300(1 + r)^5$
1976	$8300(1 + r)^6$
1977	$8300(1 + r)^7$
1978	$8300(1 + r)^8$
1979	$8300(1 + r)^9$
1980	$8300(1 + r)^{10} = 18,500$

We can find the value of $1 + r$ from the 1980 entry and use it to find the population in every other year:

$$8300(1 + r)^{10} = 18,500$$

so

$$(1 + r)^{10} = \frac{18500}{8300} \doteq 2.2289157.$$

To find $1 + r$ we must fill the box: $\boxed{}^{10} = 2.2289157$. Since $(2.2289157^{1/10})^{10} = 2.2289157$, we know that $1 + r = 2.2289157^{1/10}$, or $1 + r = 1.0834512$. Now we can complete the chart.

Year	Population
1970	8300
1971	$8300(1.0834512) = 8{,}993$
1972	$8300(1.0834512)^2 = 9{,}743$
1973	$8300(1.0834512)^3 = 10{,}556$
1974	$8300(1.0834512)^4 = 11{,}437$
1975	$8300(1.0834512)^5 = 12{,}392$
1976	$8300(1.0834512)^6 = 13{,}426$
1977	$8300(1.0834512)^7 = 14{,}546$
1978	$8300(1.0834512)^8 = 15{,}760$
1979	$8300(1.0834512)^9 = 17{,}075$
1980	$8300(1.0834512)^{10} = 18{,}500$

Notice that we have rounded to whole numbers because these results describe population. □

Square Roots

We have said that $2^{1/2}$ designates the positive number whose square is 2: that is, $(2^{1/2})^2 = 2$. Similarly, $3^{1/2}$ is the positive number whose square is 3. Geometrically, $3^{1/2}$ is the length of the side of a square that has area 3. Numbers with exponent 1/2 occur often. There is another symbol that is frequently used instead of the exponent 1/2, and that is the symbol $\sqrt{}$:

$$\sqrt{2} = 2^{1/2}$$
$$\sqrt{3} = 3^{1/2}$$

$$\sqrt{100} = 100^{1/2}.$$

We read the symbol $\sqrt{100}$ as "the square root of 100," or as "radical 100," or sometimes as "root 100." It denotes a positive number whose square is 100; thus $\sqrt{100} = 10$. We also agree that $\sqrt{0} = 0$. Notice that negative numbers do not have square roots because no number squared is negative.

We stated earlier that $\sqrt{2}$ and $\sqrt{3}$ are irrational numbers. We cannot write their decimal form. We can write decimal approximations, but if we want to denote exactly the number whose square is 2, we write $\sqrt{2}$. The calculator gives approximations with the $\boxed{\sqrt{x}}$ key. To get an approximation for $\sqrt{2}$, key $\boxed{2}$ $\boxed{\sqrt{x}}$.

Exercises 3.9

Find the value of each expression in Exercises 1–16 accurate to 3 decimal places.

1. $\sqrt{2}$

2. $\sqrt{3}$

3. $\sqrt{9}$

4. $\sqrt{.81}$

5. $2^{1/2}$

6. $3^{1/2}$

7. $7^{-5/3}$

8. $2^{-1/2}$

9. $10^{3/7}$

10. $\left(\dfrac{1}{2}\right)^{5/4}$

11. $(5^{2/3})^{3/2}$

12. $(2^{-1/5})^{-5}$

13. $1200(2^{7/8})$

14. $1200(2^{11/8})$

15. $60(.5)^{3/10}$

16. $60(.5)^{13/10}$

Insert the proper symbol ($<, =, >$) between the numbers in Exercises 17–21.

17. $3^{-2/3}$, $3^{2/3}$

18. $(10^5)^{1/2}$, $(10^{1/2})^5$

19. $2^{3/2}$, $2^{1/2}$

20. $6^{4/3}$, $(6^{1/3})^4$

21. $(5^3)^{-1/2}$, $(5^{1/2})^{-3}$

Use the properties of exponents to solve for x in Exercises 22–28.

22. $2^{1/3}2^x = 2^{5/3}$

23. $(5^{2/3})^{4/5} = 5^x$

24. $(3^4)^x = 9$

25. $3^{1/4}3^x = 3$

26. $\dfrac{2^x}{2^{1/5}} = 2^{2/5}$

27. $\sqrt{x} = 2$

28. $\sqrt{x} = 3$

In Exercises 29–39, find a solution accurate to 3 decimal places.

29. $x^{10} = 2$

30. $x^{20} = \dfrac{1}{2}$

31. $20x^4 = 50$

32. $40x^5 = 15$

33. $(1 + x)^6 = \dfrac{1}{2}$

34. $(1 + x)^7 = 2$

35. $200(1 + x)^8 = 600$

36. $800(1 + x)^{10} = 200$

37. $x^{1/3} = 14$

38. $x^{1/4} = 10$

39. $x^{2/3} = 15$

Complete the tables in Exercises 40 and 41.

40. The amount of electricity used in Kansas City doubles every 8 years.

Year	Number of kilowatt hours
1970	200,000
1971	
1972	
1973	
1974	
1975	
1976	
1977	
1978	400,000

41. The half-life of a radioactive isotope is 4 hours.

Time	Number of grams
8:00 A.M.	60
9:00 A.M.	
10:00 A.M.	
11:00 A.M.	
12 noon	30
1:00 P.M.	
2:00 P.M.	
3:00 P.M.	
4:00 P.M.	15

Assuming a constant rate of change, complete the tables in Exercises 42–47.

42.

Time	Number of bacteria
8:00 A.M.	1000
9:00 A.M.	1260
10:00 A.M.	
11:00 A.M.	
12 noon	
1:00 P.M.	
2:00 P.M.	
3:00 P.M.	
4:00 P.M.	

43.

Date	Savings account ($)
Jan. 1, 1970	2000.00
Jan. 1. 1971	
Jan. 1, 1972	2519.84
Jan. 1, 1973	
Jan. 1, 1974	
Jan. 1, 1975	
Jan. 1, 1976	
Jan. 1, 1977	
Jan. 1, 1978	
Jan. 1, 1979	
Jan. 1, 1980	
Jan. 1, 1981	

44.

Year	Population
1975	500,000
1976	
1977	
1978	615,572
1979	
1980	
1981	
1982	
1983	
1984	
1985	

45.

Time	Number of grams
8:00 A.M.	100.00
9:00 A.M.	79.37
10:00 A.M.	
11:00 A.M.	
12 noon	
1:00 P.M.	
2:00 P.M.	
3:00 P.M.	
4:00 P.M.	

46.

Year	Value of car ($)
1972	4000.00
1973	
1974	3031.43
1975	
1976	
1977	
1978	
1979	
1980	
1981	
1982	

47.

Year	Population
1968	300,000
1969	
1970	
1971	252,269
1972	
1973	
1974	
1975	
1976	
1977	
1978	
1979	
1980	

48. Find the constant rate of decay necessary for a given radioactive isotope to have a half-life of

(a) 10 hours   (b) 10 days   (c) 10 weeks

49. Find the constant rate of growth necessary for a given bacteria to double every

(a) 5 hours   (b) 10 hours   (c) 15 days

50. Write three different sequences of calculator key strokes that evaluate $5^{3/4}$.

3.10 / CHAPTER 3 PROBLEM COLLECTION

Evaluate each expression in Exercises 1–23.

1. $2 \cdot 3^4$

2. -3^{17}

3. -3^{-2}

4. $(-5)^{-3}$

5. $(3^{-2})^{-1}$

6. $(5^{-1})^2$

7. $\left(\dfrac{1}{5}\right)^{-2}$

8. $(-4)^3$

9. $3^5 - 3 \cdot 3^4 + 3^3 - 4 \cdot 3^2 + 5 \cdot 3 - 1$

10. $(-2)^4 + 5(-2)^3 - 2(-2)^2 - 5(-2) + 1$

11. $\dfrac{(67.2 \times 10^{23})(3.45 \times 10^{-8})}{2.45 \times 10^7}$

12. $\dfrac{(-34.5 \times 10^{-21})(1.11 \times 10^6)}{3.42 \times 10^{-7}}$

13. $(\sqrt{11})^2$

14. $\sqrt{155^2}$

15. $\sqrt{\sqrt{10}}$

16. $25^{-3/2}$

17. $12^{4/5}$

18. $12^{-3/5}$

19. $(3^{-1/4})^{-4}$

20. $(5^{3/2})^{2/3}$

21. $(7^{-3})^{1/3}$

22. $\left(\dfrac{4}{9}\right)^{-1/2}$

23. $\left(\dfrac{8}{27}\right)^{1/3}$

Insert the proper symbols ($<, =, >$) between the numbers in Exercises 24–26.

24. $(-3.5)^{73}, \quad 0$

25. $(-1.8)^{44}, \quad 0$

26. $(7^3)^{1/2}, \quad (7^{1/2})^3$

Sketch a number line and graph the numbers in Exercises 27 and 28.

27. $0, -1.5, (-1.5)^2, (-1.5)^3, (-1.5)^4, (-1.5)^5, (-1.5)^6$

28. $0, -0.8, (-0.8)^2, (-0.8)^3, (-0.8)^4, (-0.8)^5, (-0.8)^6$

Write the numbers in Exercises 29 and 30 in decimal form.

29. 3.67×10^7

30. 2.13×10^{-6}

Write the numbers in Exercises 31 and 32 in scientific notation.

31. $2{,}670{,}000$

32. 0.0000023

Solve for x in Exercises 33–42.

33. $4^2 4^x = 64$

34. $(-3)^2(-3)^x = (-3)^5$

35. $3^{2/3} \cdot 3^x = 3^2$

36. $(-2)^x = -\dfrac{1}{8}$

37. $3^{-x} = \dfrac{1}{27}$

38. $\left(\dfrac{1}{3}\right)^x = 27$

39. $0.000025 = 2.5 \times 10^x$

40. $76,200 = 7.62 \times 10^x$

41. $2\sqrt{x} = 3$

42. $c^6 + c^8 = c^x(c^3 + c^5)$

Simplify the expressions in Exercises 43–52. Your answers should not contain negative or zero exponents.

43. $(4x^2 y^3)(2x^3 y^2)$

44. $(-2x^2 y^3)^3(-3xy)^{-1}$

45. $\dfrac{(a^{-2})^{-5}}{(a^3)^{-4}}$

46. $\left(\dfrac{3x^2 y}{2x^{-3} y^2}\right)^0$

47. $(x^{-1})^{-1}$

48. $\left(\dfrac{a^4 b^{-3} c^2}{ab^{-4} c^{-1}}\right)^2$

49. $(-x)^5$

50. $\left(\dfrac{a^{-2}}{b^{-2}}\right)^{-2}$

51. $(x^3 y^{-1})^2(xy)^{-2}(xy^{-1})^{-4}$

52. $\dfrac{(-a^4 b^{-1})^{-1}(-a^2 b^{-2})^2}{(a^{-1}b^{-1})^2}$

Use the distributive property and combine like terms to simplify the expressions in Exercises 53–55.

53. $x^3(x+1) - x^2(x - x^2)$

54. $xy(3xy + 4y) - y^2(2x^2 - 3x)$

55. $x^{-1}y^{-1}(3x^3 y^2 + x^{-1}y^3) - y^{-2}(4x^2 y^3 - x^{-2}y^4)$

56. Find the value of $\dfrac{a^2 - b^2}{a + b}$ if $a = -3$ and $b = 2$.

57. \$1500 is invested at 7.6% interest compounded monthly. Find how much money is accumulated in

    (a) 1 year   (b) 10 years

58. \$2500 is invested at 6.2% interest compounded daily. Find how much money is accumulated in

    (a) 1 year   (b) 6 years and 113 days

59. $5000 is invested for 5 years at 6.75% interest. Find how much interest is earned if interest is

    (a) simple, paid yearly   (b) compounded annually

    (c) compounded quarterly

60. Compute the effective annual yield if money is invested at 5.5% interest compounded

    (a) semiannually   (b) daily

61. Determine how much must be invested for 5 years at 6.4% interest to accumulate $5000 if the interest is compounded

    (a) semiannually   (b) daily

62. Find how long it will take for a 7.25% investment to triple if the interest is compounded

    (a) annually   (b) quarterly

63. Assume that there are 60 grams of a radioactive isotope present on the first day and that there are 58.56 grams present on the second day. Find the daily rate of decay.

Complete the tables in Exercises 64–67.

64. Assume that bacteria are growing at the rate of 9% per hour.

65. Assume that a radioactive isotope is decaying at the rate of 7% per day.

Time (hours)	Number of bacteria
$t = 0$	1200
1	
2	
5	
10	
t	

Time (days)	Number of grams
$t = 0$	60
1	
2	
5	
10	
t	

66. Assume that population is growing at a constant rate.

Year	Population
1980	180,000
1981	
1982	
1983	
1984	
1985	264,552
1986	
$1980 + t$	

67. Assume that gasoline consumption is decreasing at a constant rate.

Year	Gallons used
1980	1,000,000
1981	
1982	
1983	
1984	793,700
1985	
$1980 + t$	

68. Assume that the price of food increased by 12.4% this year and that this rate will continue. If a pound of bacon costs $2.10 this year, determine how much it will cost

(a) 1 year later

(b) $3\frac{1}{2}$ years later

69. Assume that a new car which costs $7500 this year depreciates at the rate of 18% per year. Find what the car will be worth

(a) t years later

(b) 3 years, 7 months later

70. A radioactive isotope has a half-life of 8 years. If there are 80 grams present now, determine how many grams there will be

(a) 1 year later

(b) t years later

(c) $4\frac{1}{2}$ years later

71. If the population of a city doubles every 18 years and the population is 250,000 now, determine what it will be

(a) 1 year later

(b) t years later

(c) 3.25 years later

72. Assume that radioactive carbon-14 has a half-life of 5600 years and that living bodies radiate 918 rays per gram per hour. Find the approximate age of a fossil emitting 3.58 rays of carbon-14 per gram per hour.

In Exercises 73–80, find a solution accurate to 3 decimal places.

73. $x^3 = 5$

74. $6x^8 = 12$

75. $100(1 + x)^5 = 40$

76. $x^3 + x = 1$

77. $x - \sqrt{x} = 1$

78. $5\sqrt{x} - x = 6$

79. $x^{1/5} = 10$

80. $x^{3/5} = 12$

Using Graphs
to Solve Problems

4.1 / USING GRAPHS TO PICTURE INFORMATION

We often draw pictures to communicate information. Indeed, it is said that "a picture is worth a thousand words." In mathematics we frequently can summarize a great deal of numerical information by drawing a picture called a **graph**.

Example 1 The Fullers plan a 300-mile car trip for next Saturday. Draw a graph that shows all the combinations of average speed and time they might use on their trip.

We know, for example, that if the Fullers drive at an average speed of 60 mph, the trip will take 5 hours. However, if they stop for a long lunch and do some sightseeing along the way, so that their average speed is only 30 mph, then the time for the trip will be 10 hours. Here are some possibilities for the 300-mile trip:

Average speed (mph)	10	20	30	40	50	60	70
Time (hours) accurate to tenths	30	15	10	7.5	6	5	4.3

We can picture each of these possibilities as a point on a graph. The point representing an average speed of 10 and time of 30 is located above 10 on the horizontal scale and across from 30 on the vertical scale. You should see that each point in Fig. 4.1 represents one of the possibilities we listed for the 300-mile trip. However, we listed only seven of the many, many possibilities. In fact, any positive number could be the speed and there would be a corresponding time; for example, 100 mph theoretically could be the average speed with a corresponding time of 3 hours. We cannot use 0 as an average speed because, in that case, there would be no trip at all. However, every other number on the horizontal scale shown has a point on the graph associated with it. We cannot draw (usually we say *plot* or *graph*) all these points individually because there are too many. Nevertheless, we must plot enough points to be sure what the completed graph is. Let us limit our attention to average speeds between 0 and 70 mph. This time, we consider average speeds midway between the values we have already used and compute corresponding times as shown on p. 192.

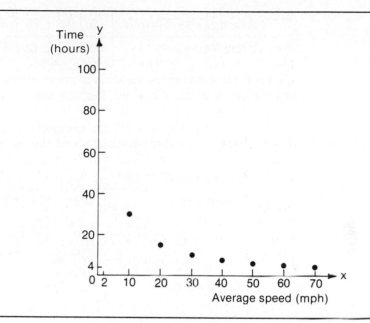

Figure 4.1

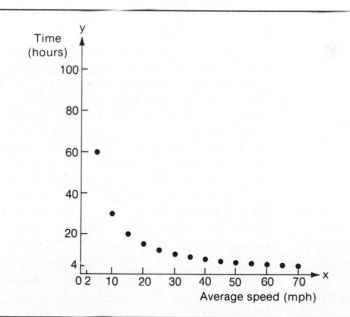

Figure 4.2

Average speed (mph)	5	15	25	35	45	55	65
Time (hours)	60	20	12	8.6	6.7	5.5	4.6

We need to estimate the location of some of the points representing the above information. When we include these points with the seven we already have, we get Fig. 4.2.

To get a better idea of the complete graph, we add even more points. Again we choose points between the points already on the graph.

Average speed (mph)	3	8	13	18	23	28
Time (hours)	100	37.5	23.1	16.7	13.0	10.7

Average speed (mph)	33	38	43	48	53	58	63	68
Time (hours)	9.1	7.9	7.0	6.3	5.7	5.2	4.8	4.4

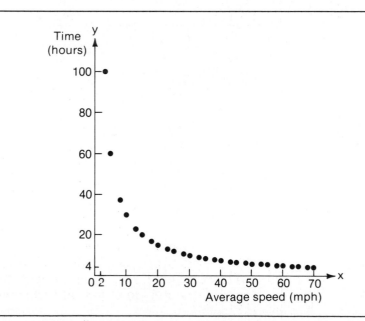

Figure 4.3

When these points are plotted together with the ones that were already on the graph, the picture looks like Fig. 4.3.

Since every possible speed corresponds to a time, there is a point on the graph for every possible speed. If we were able to fill in all the points of the graph, they would form a curve. The points shown on Fig. 4.3 give us a good idea of the shape of the curve between average speeds of 3 and 70. If we tried a few more points for speeds greater than 70, we would see that they get very close to the horizontal line. If we tried additional speeds between 0 and 3, we would see that the corresponding times are very large. We speculate that the completed graph looks like Fig. 4.4. (The arrows suggest that the curve continues beyond the part shown.)

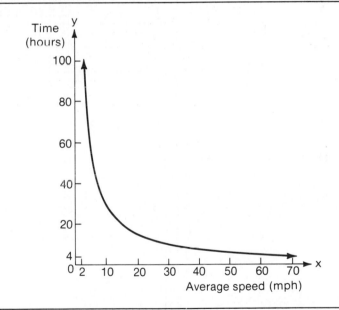

Figure 4.4

Although plotting many points usually gives a good idea of the shape of a graph, we must be sure we do not jump to conclusions. There are always many ways to plot a curve that contains the points we are able to graph.

Example 2   The telephone book gives the charges for a weekday call from Columbus to Washington, D.C., as $2.10 for the first 3 minutes and $.35 for each additional minute or part of a minute. Draw a graph that shows the possible combinations of the length of calls and charges.

The phone book gives us enough information to compute the charges for a call of any length. Here are some of the possibilities.

Length of call (minutes)	Charge ($)
1	2.10
2	2.10
3	2.10
4	2.10 + .35 = 2.45
5	2.10 + 2(.35) = 2.80
6	3.15
7	3.50
8	3.85

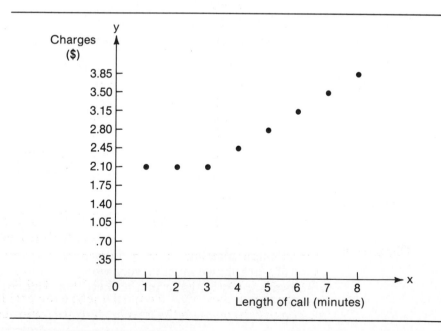

Figure 4.5

When we plot the points that represent this information, we get the points shown in Fig. 4.5. These few points do not give a very good idea of the shape of this graph. A natural next step is to add the points that correspond to the length of calls midway between the ones we have already used.

Length of call (minutes)	$\frac{1}{2}$	$1\frac{1}{2}$	$2\frac{1}{2}$	$3\frac{1}{2}$
Charge ($)	2.10	2.10	2.10	2.45

Length of call (minutes)	$4\frac{1}{2}$	$5\frac{1}{2}$	$6\frac{1}{2}$	$7\frac{1}{2}$
Charge ($)	2.80	3.15	3.50	3.85

These points are shown in Fig. 4.6.

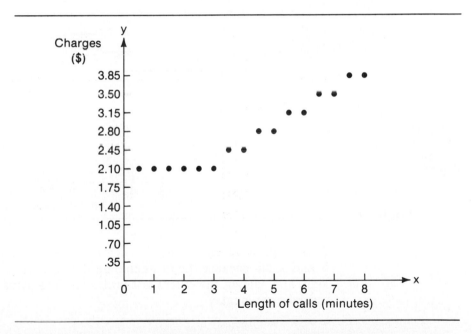

Figure 4.6

There are many curves that contain these points. What do you guess this graph will look like? We must analyze the problem carefully before we attempt to complete the graph. Here are some observations.

- The cost of a call of 0 minutes is $0.
- For a call longer than 0 minutes and up to 3 minutes (including 3), the charge is $2.10.
- For a call longer than 3 minutes and up to 4 minutes (including 4), the charge is $2.10 + $.35 = $2.45.
- For a call longer than 4 minutes and up to and including 5 minutes, the charge is $2.45 + $.35 = $2.80.

When we consider the observations just made, we see that the graph takes the shape of Fig. 4.7.

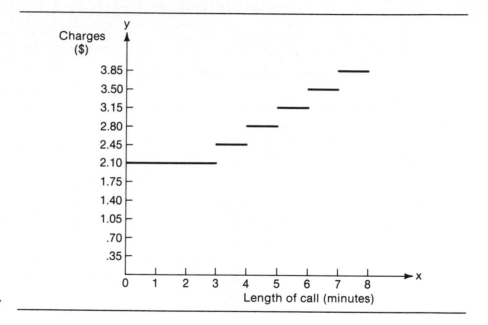

Figure 4.7

This graph is not connected. It has jumps above the whole numbers 3, 4, 5, and so on, because the rates change that way. The first step is 3 units long, and each of the others is 1 unit long. The steps continue to rise as the calls get longer.   □

In Examples 1 and 2 there is a point on the graph above each point on the horizontal scale. However, there are situations in which this does not happen.

Example 3   The Super Sports catalog lists tennis balls at $2.00 for a can of three balls. Also there is a $.50 handling charge on each order. Draw a graph that shows the cost for all possible orders of tennis balls.

First we list several possibilities and plot those points on the graph in Fig. 4.8.

Number of cans of balls	Cost ($)
1	2.00 + .50 = 2.50
2	2(2.00) + .50 = 4.50
3	3(2.00) + .50 = 6.50
4	4(2.00) + .50 = 8.50
5	5(2.00) + .50 = 10.50

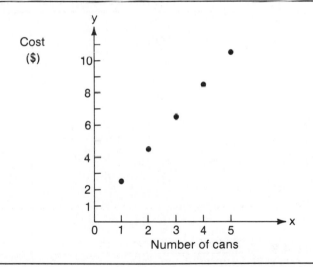

Figure 4.8

It is now possible to guess, without computing, where the points for larger numbers of cans will be located. What does the graph describing this situation look like? There are no points between the points that we have graphed because, presumably, it is not possible to buy a part of a can of tennis balls. The graph is not a connected curve; it is a collection of points—one for each whole number on the horizontal scale.   □

The Vocabulary of Graphing

Any time we have a situation in which one number is associated with another number, we can use a point to represent the pair of numbers and begin to make a graph. What is required is that we have a horizontal scale to locate the first number and a vertical scale to locate the second number. These scales are actually number lines. Usually they are positioned at right angles, with their point of intersection labeled 0 on each scale. The two number lines form the **coordinate system** of the graph; each is called an **axis**. If the axes (*axes* is the plural of *axis*) intersect at 0 on each, the point of intersection is called the **origin**.

If we have a coordinate system, then any point can be identified by a pair of numbers, the first indicating the position of the point with respect to the horizontal axis and the second indicating the position of the point with respect to the vertical axis, as shown in Fig. 4.9.

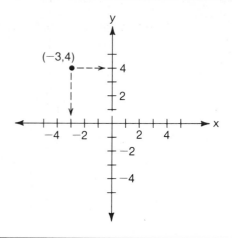

Figure 4.9

These two numbers on the axes are called the **coordinates** of the point. There are some additional examples of points identified by pairs of numbers in Fig. 4.10.

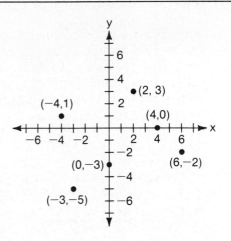

Figure 4.10

Exercises 4.1

1. Write the coordinates of the points in the graph below.

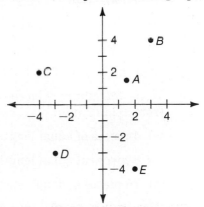

2. Draw a coordinate system and graph the following points:

$(2, -3)$ $(-2, 3)$ $(-2, -3)$ $(2, 3)$ $\left(-\dfrac{3}{2}, -\dfrac{5}{2}\right)$

Draw three different curves containing the points given in each of Exercises 3–6.

3.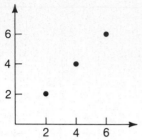

4.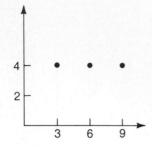

5.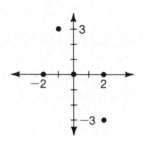

6.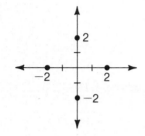

7. Write the numbers on the number line that divide the interval from 0 to 3 into

   (a) 3 pieces of equal length

   (b) 9 pieces of equal length

   Give the length of each piece in (a) and (b).

8. Write the numbers on the number line that divide the interval from −2 to 2 into

   (a) 4 pieces of equal length

   (b) 8 pieces of equal length

   (c) 16 pieces of equal length

   Give the length of each piece in (a), (b), and (c).

9. Jim plans to take a 600-mile trip. Let the numbers on the horizontal axis represent average speed, and let the numbers on the vertical axis represent the time required to make the trip. Draw a graph that shows all the combinations of average speed and time he might use on the trip.

10. Jane plans to take a 200-mile trip. Let the numbers on the horizontal axis represent the time required to make the trip, and let the numbers on the vertical axis represent the average speed on the trip. Draw a graph that shows all the combinations of average speed and time she might use on the trip.

11. A car is traveling at a constant speed of 50 mph. Let the numbers on the horizontal axis represent the time the car travels, and let the numbers on the vertical axis represent the distance the car travels. Draw a graph that shows all the combinations of distance and time the car might travel for a 24-hour period.

12. Tom leaves Columbus at 10:00 A.M. and drives north at a constant speed of 50 mph. Bill also leaves Columbus at 10:00 A.M. and drives north on the same highway at 60 mph. Let the numbers on the horizontal axis represent the time each travels, and let the numbers on the vertical scale represent the distance between them. Draw a graph that shows the distance between Tom and Bill up until 8:00 P.M. the same day.

13. Bill rides his bicycle 10 miles at one speed. He then rides an additional 5 miles at that speed reduced by 2 mph. Let the numbers on the horizontal axis represent the original speed, and let the numbers on the vertical axis represent the time it takes to make the 15-mile trip. Draw a graph that shows all the combinations of original speed and time Bill might use for the trip for original speeds between 2 mph and 20 mph.

14. The charges for a telephone call between two cities are $1.75 for the first 3 minutes and $.25 for each additional minute or part of a minute. Let the horizontal axis represent the length of the call and the vertical axis, the charges. Draw a graph that shows the charges for all calls up to 10 minutes in length.

15. There are 30 coins, all nickels and dimes, in the money box. Let the numbers on the horizontal axis represent the number of nickels and the numbers on the vertical axis, the value of the 30 coins. Draw a graph that shows the value of the coins for all possible numbers of nickels.

4.2 / USING GRAPHS TO SOLVE PROBLEMS ABOUT PERIMETER AND AREA

We have solved many problems using a trial-and-error method. Often a graph can be used to picture all possibilities that arise in a problem situation. Once the graph is drawn, questions can be answered directly by reading information from the graph.

In Example 1 below we first draw a graph that shows how the area of a rectangle with a perimeter of 100 feet depends on its width. Then, in Example 2, the graph is used to answer a question about the areas of rectangles with perimeters of 100 feet.

Example 1　Draw a graph showing for all rectangles with a perimeter of 100 feet the relationship between the width of a rectangle and its area.

In a rectangle with a perimeter of 100 feet, the length plus the width equals 50 feet. In order to start the graph, we compute the areas of several rectangles with perimeters of 100, but with different widths. Observe in Fig. 4.11 that the width can be any number between 0 and 50 but cannot be greater than 50, because the two sides labeled w would then be more than the perimeter of 100.

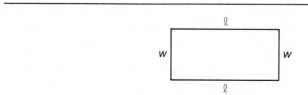

Figure 4.11

Width	0	5	10	15	20	25	30	35	40	45	50
Length	$50 - 0 = 50$	$50 - 5 = 45$	40	35	30	25	20	15	10	5	0
Area	0	225	400	525	600	625	600	525	400	225	0

Notice that the areas for widths 0, 5, 10, 15, and 20 are the same as the areas for widths 50, 45, 40, 35, and 30. Can you see why? If we plot each point determined by a width and the corresponding area, we get the points on the graph in Fig. 4.12.

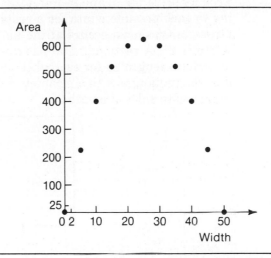

Figure 4.12

If we could graph all points determined by a width and the corresponding area, the completed graph would look like Fig. 4.13.

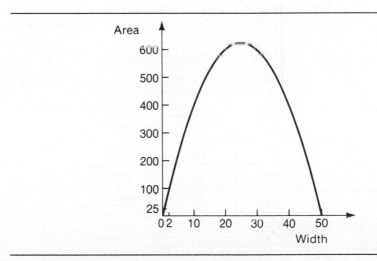

Figure 4.13

□

Example 2   Of all the possible rectangles with perimeters of 100, which has the largest area?

Look again at the graph in Fig. 4.13, which shows the area corresponding to each possible width for a rectangle with a perimeter of 100. The largest area shown occurs at the highest point on the graph when the width is 25. A rectangle that has a width of 25 and a perimeter of 100 must have length 25 for each of the four sides. Our investigation shows that the rectangle with a perimeter of 100 that has the largest area is a square with sides of length 25.

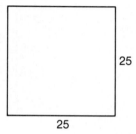

25

25   □

Example 3   Draw a graph that shows the relationship between the width and the area of a rectangle if the rectangle is twice as long as it is wide.

Again we compute the area of several rectangles that are twice as long as they are wide. The width can be any number that is not negative.

Width	Length	Area
0	$2 \cdot 0 = 0$	0
5	$2 \cdot 5 = 10$	50
10	20	200
15	30	450
20	40	800
25	50	1250
30	60	1800

Graphing the points given by the widths and areas in the chart above shows the general shape of the graph in Fig. 4.14.

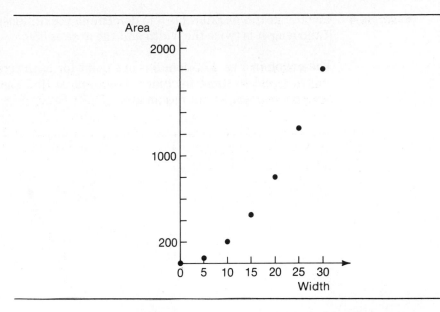

Figure 4.14

If we were able to graph all the points for every possible width, the points would form the curve shown in Fig. 4.15.

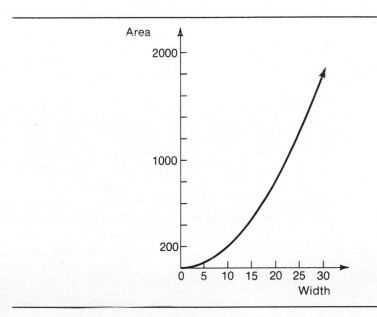

Figure 4.15

□

Example 4   Use the graph in Example 3 to determine the dimensions of a rectangle if the length is twice the width and the area is 650.

The graph in Fig. 4.15 contains one point for each rectangle that has its length equal to twice its width. We want to find the point that represents the rectangle that has an area of 650. Study Fig. 4.16.

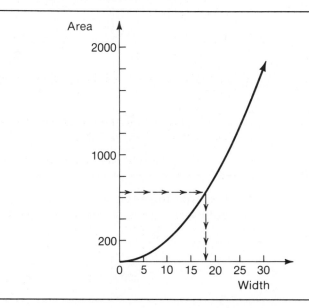

Figure 4.16

The graph suggests that an area of 650 corresponds to a width of 18. We check this approximation to the solution: if the width is 18, the length would be 36, and the area would be $18 \cdot 36 = 648$. Thus 18 is a little bit too small, but it gives us a good first approximation.

Since graphing always involves approximating, we expect that problem solutions obtained from graphs will be approximations. A guess-and-check method can be used to improve the approximation. The answer accurate to three decimal places is 18.028.   □

Exercises 4.2

1. The length of a rectangle is 2 feet more than twice its width.

    (a) Find the length of the rectangle if its width is 5. Find the length if its width is 10.

(b) Find the area of the rectangle if its width is 5. Find the area if its width is 10.

(c) Find the perimeter of the rectangle if its width is 5. Find the perimeter if its width is 10.

2. The perimeter of a rectangle is 200 feet.

(a) Find the length of the rectangle if its width is 20. Find the length if its width is 25.

(b) Find the area of the rectangle if its width is 20. Find the area if its width is 25.

3. The graph below shows the area of a rectangle as a function of its width.

(a) Use the graph to find the area of the rectangle if its width is 5, 30, or 42.5.

(b) Use the graph to find the width of the rectangle if its area is 0, 500, or 675.

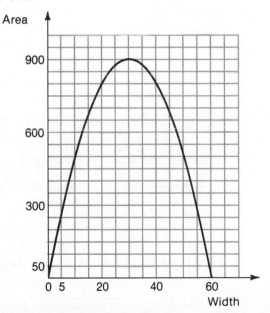

4. The perimeter of a rectangle is 200 feet.

(a) Compute the lengths of rectangles with the following widths: 10, 30, 50, 70, and 90 feet.

(b) Draw a graph that shows the relationship between the width and the length of a rectangle.

5. The area of a rectangle is 600 square feet.

(a) Compute the lengths of rectangles with the following widths: 50, 100, 150, 200, and 300 feet.

(b) Draw a graph that shows the relationship between the width and the length of a rectangle.

6. The length of a rectangle is 1 foot more than twice its width.

(a) Compute the perimeters of rectangles with the following widths: 5, 10, 15, 20, 25, and 30 feet.

(b) Draw a graph that shows how the perimeter of a rectangle depends on the width.

(c) Use the graph in (b) to find the dimensions of a rectangle if the perimeter is 110 feet.

7. A rectangular garden is enclosed by 100 feet of fence and one side of a barn.

(a) Draw a graph that shows the relationship between the width of the rectangular garden and its area.

(b) Use the graph in (a) to find the width of a rectangular garden with an area of 912 square feet. What is its length?

(c) Use the graph in (a) to find the width of the rectangular garden of maximum area. What is its length?

8. A rectangle is 3 feet longer than it is wide.

(a) Compute the areas of rectangles with the following widths: 5, 10, 15, 20, 25, and 30 feet.

(b) Draw a graph that shows the relationship between the width of a rectangle and its area.

(c) Find the dimensions of a rectangle if its area is 180 square feet. Find the dimensions if its area is 300 square feet.

9. The length of a rectangle is 11 feet less than twice its width.

(a) Draw a graph that shows how the area of a rectangle depends on its width.

   (b) Find the dimensions of a rectangle if its area is 450 square feet. Find the dimensions if its area is 250 square feet.

10. The area of a rectangle is 400 square feet.

   (a) Compute the lengths of rectangles with the following widths: 5, 10, 20, 25, and 40 feet.

   (b) Compute the perimeters of rectangles with the following widths: 5, 10, 20, 25, and 40 feet.

   (c) Draw a graph that shows how the perimeter of a rectangle is a function of its width.

   (d) Use the graph in (c) to find the dimensions of the rectangle that has an area of 400 square feet and the smallest perimeter.

4.3 / USING GRAPHS TO SOLVE PROBLEMS INVOLVING EXPONENTS

Powers of expressions arise in many problems, including problems concerning compound interest, inflation, population growth, and carbon dating. In this section we will study graphs of this type and use them to answer questions about the problems described by the graphs.

Consider first a compound-interest problem. If $1000 is put in an account that compounds interest quarterly at 18%, then the amount of money M in the account is computed by

$$M = 1000\left(1 + \frac{.18}{4}\right)^t$$

where t indicates how many times interest has been paid on the account. For example, after one year t would be 4 because, in this account, interest is paid quarterly. We want to draw a graph that shows how the amount of money increases over time.

Example 1 Nancy deposits $1000 in an account that pays 18% interest compounded quarterly. Draw a graph that shows the amount of money in the account over a 20-year period.

For each time t that interest is paid, we want to compute

$$M = 1000\left(1 + \frac{.18}{4}\right)^t$$

Since we need to cover a 20-year period, t will need to represent the numbers from 1 to 80. Here are the values of M for certain values of t.

t	M	t	M
8 (2 years)	1,422.10	48	8,271.46
16 (4 years)	2,022.37	56	11,762.84
24	2,876.01	64	16,727.95
32	4,089.98	72	23,788.82
40	5,816.36	80	33,830.10

By plotting these points in Fig. 4.17, we see how the account grows.

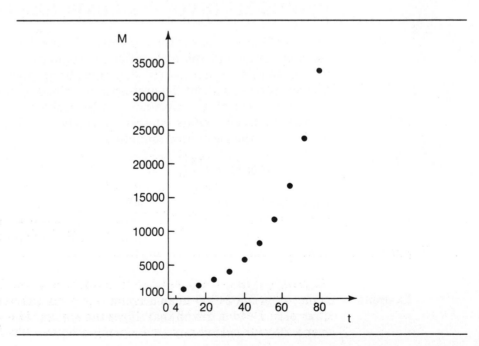

Figure 4.17

We have computed the value of M for the even-numbered years in this 20-year period. Computing the value of M for the odd-numbered years provides us with these additional points.

t	M
4	1,192.52
12	1,695.88
20	2,411.71
28	3,429.70
36	4,877.38

t	M
44	6,936.12
52	9,863.87
60	14,027.41
68	19,948.39
76	28,368.61

These points have been added to the graph in Fig. 4.18.

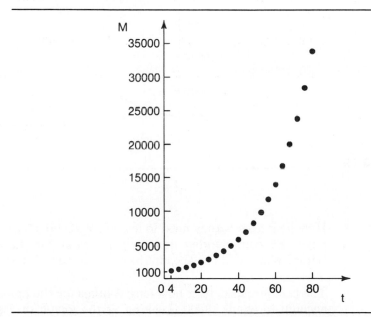

Figure 4.18

We have not made any assumptions about how the bank pays interest on days that are between the last days of each quarter. In this example, let us assume that the bank will pay an appropriate amount of interest on money withdrawn any time during the year. Then we can complete the curve to show the interest paid at any time during the 20-year period (Fig. 4.19).

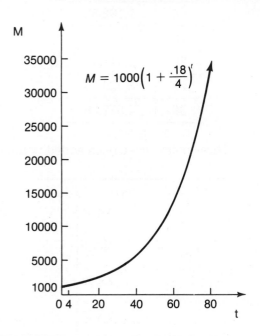

$$M = 1000\left(1 + \frac{.18}{4}\right)^t$$

Figure 4.19

Example 2   How long does Nancy need to leave her $1000 deposit in an account that pays 18% compounded quarterly in order for the original deposit to triple? When will the account have a balance of $20,000?

The problem asks first how long it takes for the amount of money in the account to equal $3000. Looking at the graph in Fig. 4.20, we can see that Nancy must leave her money in the account for more than 24 quarters but fewer than 28 quarters. The graph suggests that the value of t is closer to 24 than to 28.

If $t = 25$, then $M = 1000\left(1 + \frac{.18}{4}\right)^{25} = 3005.43$. Thus Nancy must

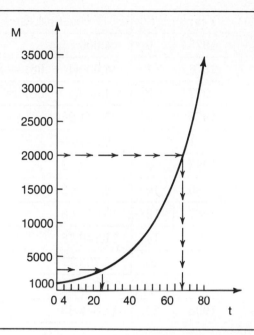

Figure 4.20

have her money on deposit through 25 interest payments to accumulate $3000. This means she must leave it in the account for $6\frac{1}{4}$ years. To reach $20,000, the money must be left in the account for 68 quarters, or 17 years. □

A second problem situation that uses exponents is one involving inflation. Consider, for example, the effect of an 8% annual rate of inflation on the price of an automobile.

Example 3   Assume that the average cost of a new car in 1975 was $4000. Draw a graph that shows the effect of an 8% annual rate of inflation on the price of an automobile.

The chart below gives approximate increases in cost for some of the years during the period from 1975 through 1995. For convenience, we let t denote the number of years after 1975.

Year	t	Cost ($)
1975	0	4,000
1976	1	$4,000(1 + .08) = 4320$
1977	2	$4,000(1 + .08)^2 = 4665.60$
1979	4	5,441.96
1981	6	6,347.50
1983	8	7,403.72
1985	10	8,635.70
1987	12	10,072.68
1989	14	11,748.78
1991	16	13,703.77
1993	18	15,984.08
1995	20	18,643.83

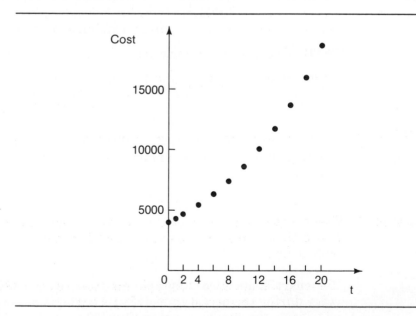

Figure 4.21

The chart gives twelve points to add to the graph we are constructing (Fig. 4.21) to show the effect of 8% inflation over the 20-year period. If we assume that price increases occur continually during this time (and not just on January 1, for example), we can complete the graph in Fig. 4.22.

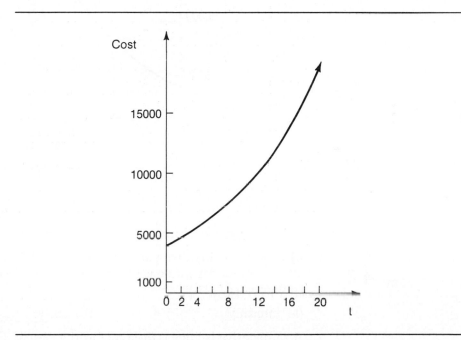

Figure 4.22

Example 4   Use the graph in Example 3 to predict when 8% annual inflation will push the price of an average car to $8000 or more.

It appears, inspecting the graph, that the cost hits $8000 after 9 years, that is to say, very early in 1984 (Fig. 4.23).

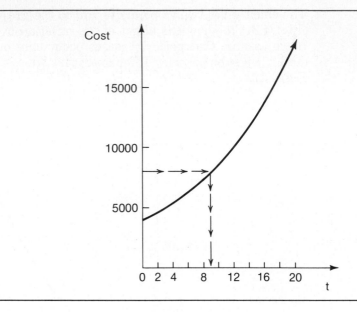

Figure 4.23

□

Exercises
4.3

1. $1500 is deposited in an account that pays 9.5% interest. Find the amount of money in the account 1 year later if the interest is compounded

   (a) monthly   (b) daily

2. A radioactive isotope is decaying at the rate of 3% per week. If there are 50 grams of the isotope present initially, find the amount present

   (a) 3 weeks later   (b) 12 weeks later

3. The half-life of a radioactive isotope is 4 years. If there are 40 grams of the isotope present initially, find the amount present

   (a) 1 year later   (b) 8 years later

4. The graph below shows how the amount of money in a certain savings account depends on the number of years the amount is left on deposit.

(a) Use the graph to find the amount of money in the account initially, 2 years later, and 16 years later.

(b) Use the graph to find the number of years the initial amount of money must be left on deposit to accumulate $900 and $4500.

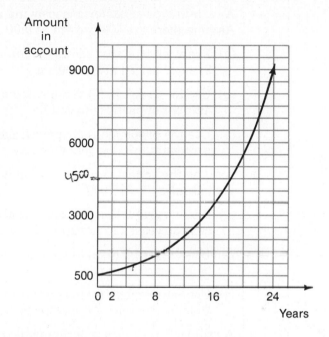

5. $2000 is deposited in an account that pays 14% interest compounded daily.

(a) Draw a graph that shows the amount of money in the account for a 25-year period.

(b) Use the graph in (a) to approximate the time required for the amount of money in the account to double.

(c) Use the graph in (a) to approximate the time required for the amount of money in the account to reach $24,000.

6. $2000 is deposited in an account that pays 14% interest compounded annually.

(a) Draw a graph that shows the amount of money in the account over a 25-year period.

(b) Use the graph in (a) to approximate the time required for the amount of money in the account to double.

(c) Use the graph in (a) to approximate the time required for the amount of money in the account to triple.

7. A certain type of bacteria is growing at the rate of 12% per hour. Assume there are 500 present at 9:00 A.M.

(a) Draw a graph that shows the number of bacteria present for a 24-hour period after 9:00 A.M.

(b) Use the graph in (a) to approximate the time required for the number of bacteria to double.

(c) Use the graph in (a) to approximate the number of bacteria present at 6:30 P.M. the same day.

8. Assume the annual rate of inflation is 6% and that Mike's current annual salary is $12,000.

(a) Draw a graph that shows what Mike's salary should be for the next 30 years to keep up with inflation.

(b) Use the graph in (a) to approximate what Mike's salary should be 18 years from now.

(c) Use the graph in (a) to approximate when Mike's salary will be $60,000 per year if it just keeps up with inflation.

9. A radioactive isotope is decaying at the rate of 10% per year. Assume that there are 40 grams present initially.

(a) Draw a graph that shows the amount of the isotope present over the next 30 years.

(b) Use the graph in (a) to approximate the half-life of the isotope.

(c) Use the graph in (a) to approximate when there will be 10 grams of the isotope present.

10. The half-life of a radioactive isotope is 6 hours. Assume that initially there are 60 grams of the isotope.

(a) Draw a graph that shows the amount of isotope present over a 48-hour period.

(b) Use the graph in (a) to approximate when there will be 12 grams of the isotope present.

(c) Use the graph in (a) to approximate when there will be 10.6 grams of the isotope present.

4.4 / USING GRAPHS TO SOLVE PERCENT PROBLEMS

Whenever one quantity can be computed in terms of another quantity, the relationship between the quantities can be shown in the form of a graph. This is often the situation in problems involving percent. We have had some experience solving these problems by constructing charts and using a guess-and-check procedure. Graphing gives a second way to analyze these problems.

Example 1   Draw a graph that shows the relationship between a number and 22% of that number. Then answer this question: 16 is 22% of what number?

First we compute the coordinates of several points on the graph.

Number	−20	−10	0	10	20	30	40
22% of the number	−4.4	−2.2	0	2.2	4.4	6.6	8.8

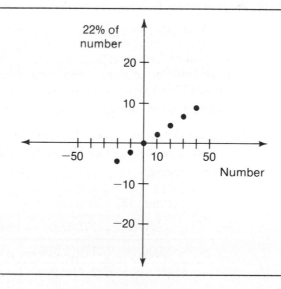

Figure 4.24

The horizontal axis shows the number and the vertical axis shows 22% of the number (Fig. 4.24). The points appear to lie on a straight line and, in this example, do lie on a line. The completed graph is shown in Fig. 4.25. We can see that 16 is 22% of a number approximately equal to 72.5. A guess-and-check procedure would show that, accurate to two decimal places, 16 is 22% of 72.73.

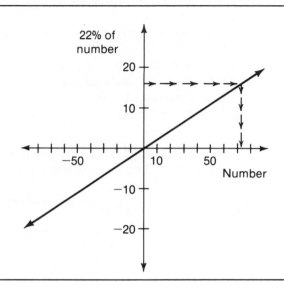

Figure 4.25

□

Example 2 An account has an effective annual yield of 17.2%. Draw a graph that shows how the total amount accumulated in 1 year is a function of the original investment. If an investment yields $9610.40 after 1 year, what was the original investment?

Remember that having an effective annual yield of 17.2% means that the interest paid at the end of the first year is the same as 17.2% simple interest.

Investment ($)	Interest ($)	Total After 1 Year ($)		
500	.172(500)	500	+ .172(500)	= 586.00
1,000	.172(1,000)	1000	+ .172(1000)	= 1,172.00

Investment ($)	Interest ($)	Total After 1 Year ($)
5,000	.172(5,000)	5000 + .172(5000) = 5,860.00
10,000	.172(10,000)	10000 + .172(10000) = 11,720.00

In Fig. 4.26 we use the horizontal axis for the investment and the vertical axis for the total in the account after 1 year.

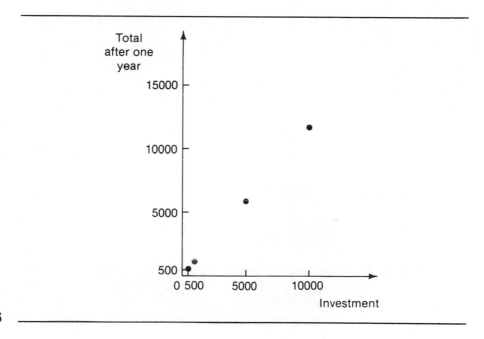

Figure 4.26

Again, the graph is a straight line. We can use Fig. 4.27 to estimate the investment that yields $9610.40. It appears that the investment is between $8000 and $8500. If we guess $8200 and key

8200 [+] 17.2 [%] [=]

we see that the yield is, in fact, $9610.40.

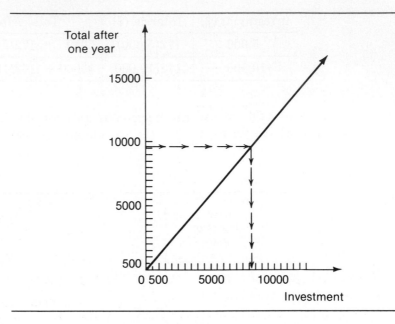

Figure 4.27

□

Example 3   A hardware store is marking all items down 15% for its autumn sale. Draw a graph that shows the relationship between the original price and the sale price. If the sale price of an item is $76.50, what was its original price?

To compute the sale price we must find

     original price − .15(original price).

Here are some points to get the graph in Fig. 4.28 started:

Original price ($)	5	10	40	70	100	130
Sale price ($)	4.25	8.50	34.00	59.50	85.00	110.50

These points lie on a line. From the completed graph in Fig. 4.29 we can find the original price that gives a sale price of $76.50. To check that an original price of $90 does, in fact, give a sale price of $76.50, we can key

$$\underline{90} \boxed{-} \underline{15} \boxed{\%} \boxed{=} \qquad \begin{array}{c}\text{(Display)}\\ 76.5\end{array}$$

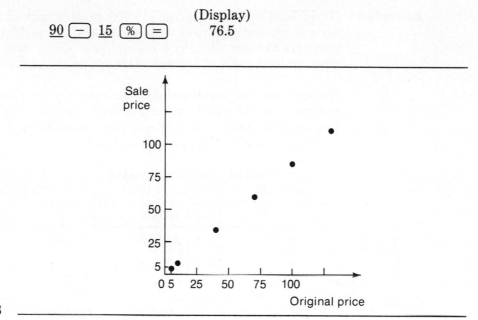

Figure 4.28

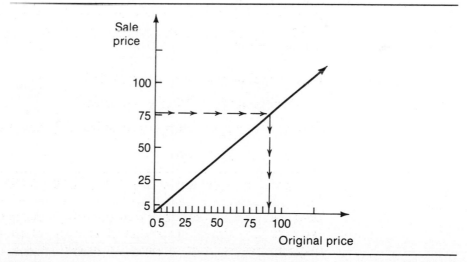

Figure 4.29

Example 4   David Scott intends to invest $18,000, putting part of the money in one savings account that pays 5% annually and the rest in another account that pays 8% annually. Draw a graph that shows how the total interest received at the end of one year depends on the amount invested at 8%.

We will use the horizontal axis to show the amount invested at 8% and the vertical axis to show the total interest received in one year. The amount invested at 8% can vary from $0 to $18,000. Here are some examples:

Amount invested at 8% ($)	Amount invested at 5% ($)	Total interest ($)
0	18,000	.08(0) + .05(18000) = 900
3,000	15,000	.08(3000) + .05(15000) = 990
6,000	12,000	.08(6000) + .05(12000) = 1,080
9,000	9,000	.08(9000) + .05(9000) = 1,170
12,000	6,000	.08(12000) + .05(6000) = 1,260
15,000	3,000	.08(15000) + .05(3000) = 1,350
18,000	0	.08(18000) + .05(0) = 1,440

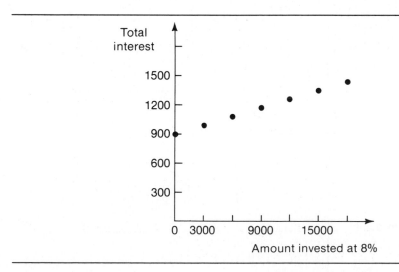

Figure 4.30

The points that have for their coordinates the above amounts invested at 8% and the corresponding total interest are shown in Fig. 4.30. If we were to compute the total amount of interest for all possible amounts between $0 and $18,000 that could be invested at 8%, we would find that the points representing these values all lie on the line shown in Fig. 4.31.

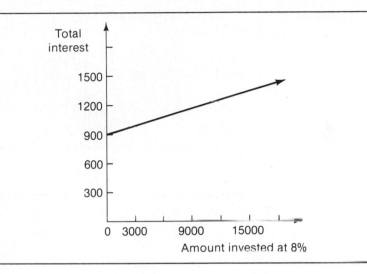

Figure 4.31

□

Now that we have the graph in Example 4, we can answer questions such as the following.

Question: If David's annual interest is $1020, how much of his $18,000 did he invest at 8%?
Answer: Study Fig. 4.32.

The amount invested at 8% that corresponds to $1020 interest is $4000. You should check to see that investing $4000 at 8% and $14,000 at 5% gives $1020 interest.

Question: If David's annual interest is $1230, how much of his $18,000 did he invest at 8%?
Answer: Study Fig. 4.33.

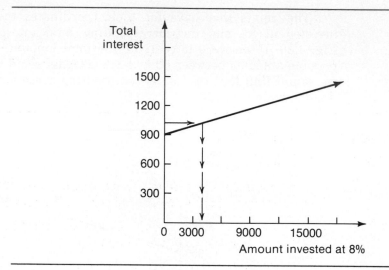

Figure 4.32

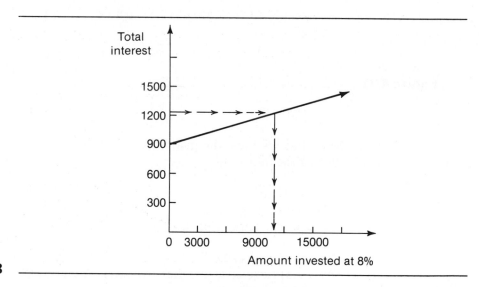

Figure 4.33

Reading this information from the graph requires some estimations. The graph suggests that $11,000 is the amount invested at 8%. We must check the estimate:

Amount invested at 8% ($)	Amount invested at 5% ($)	Total interest ($)
11,000	7,000	$.08(11000) + .05(7000) = 1230$

Exercises 4.4

1. What is 32% of 70?

2. 11.2 is what percent of 70?

3. A department store marks up wholesale prices 28%. Find the retail prices if the wholesale prices are $40 and $62.

4. A department store is having a 30%-off sale. Find the sale prices if the original prices are $52 and $64.

5. The graph below shows how 25% of a number depends on the number.

   (a) Use the graph to find 25% of 60, 110, and −70.

   (b) Use the graph to answer the question, 25% of what numbers equal 30, 27.5, and −15?

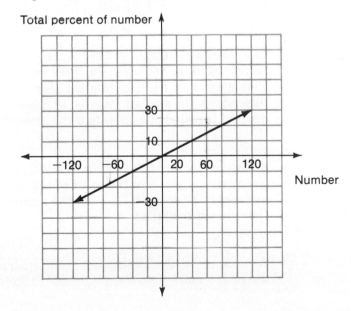

6. Draw a graph that shows the relationship between a number and 12% of a number. Use the graph to answer the following questions:

    (a) 21 is 12% of what number?

    (b) 35 is 12% of what number?

7. An account has an effective annual yield of 14.8%. Draw a graph that shows the total amount accumulated in one year for all possible investments. Use the graph to find the amount of the original investment if the amount of money in the account at the end of one year is

    (a) $11,939.20

    (b) $8000

8. A department store marks up wholesale prices 40%. Draw a graph that shows the relationship between the wholesale price and the retail price. Use the graph to find

    (a) the retail price if the wholesale price is $45.00

    (b) the wholesale price if the retail price is $87.50

9. A department store is having a 28%-off sale. Draw a graph that shows the relationship between the original price and the sale price. Use the graph to find

    (a) the sale price if the original price is $90.00

    (b) the original price if the sale price is $99.00

10. Sue invests $15,000 in two accounts—one part at 5.5% annual interest and the remainder at 7.5% annual interest. Draw a graph that shows how the total interest received at the end of one year depends on the amount invested at 5.5%. Use the graph to find the amount invested at each rate if the annual interest is

    (a) $895

    (b) $1000

11. A money box contains 60 bills in 20- and 50-dollar bills. Draw a graph that shows how the value of the money depends on the number of 20-dollar bills. Find the number of each type of bill if the value of the bills is

    (a) $2940

    (b) $2020

12. A purse contains nickels, dimes, and quarters. The number of dimes is 3 more than the number of nickels and the number of

quarters is 2 less than the number of nickels. Draw a graph that shows how the value of the coins is determined by the number of nickels. Find the number of each type of coin if the value of the coins is

(a) $5.20   (b) $8.60

4.5 / USING GRAPHS TO SOLVE MIXTURE PROBLEMS

If salt and water are mixed to give a salt solution, the **concentration** of salt in the solution is the percent of salt in the solution. For example, say we mix 2 gallons of salt and 8 gallons of water:

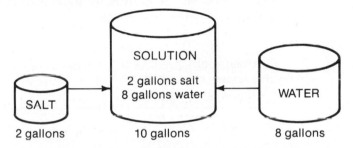

In the mixture, 2 gallons out of 10 are salt. Thus the fraction of salt is $\frac{2}{10} = .2 - 20\%$. We say that the mixture is a 20%-salt solution. To find the concentration of a salt solution, we divide the number of gallons of salt in the solution by the number of gallons of solution.

On the other hand, say we have 50 gallons of a 10%-salt solution. What is in the mixture? 10% of the 50 gallons is salt; the rest is water.

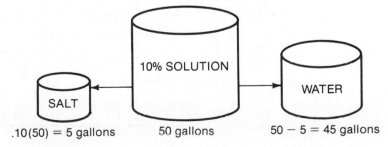

If we want to summarize the effect of adding different amounts of one substance to another substance, we can start with a chart that shows what happens in a few cases and then draw a graph that illustrates what happens in all possible cases. For example, if we start with 20 gallons of water and add increasing amounts of alcohol to the water, we know that the concentration of alcohol in the solution will increase. We can draw a graph to show the result of adding any amount of alcohol to 20 gallons of water.

Example 1   Draw a graph that shows the effect on the concentration of the solution when varying quantities of alcohol are added to 20 gallons of water.

For convenience we can assume that alcohol is also measured in gallons. We first make a chart that shows, in several cases, the number of gallons of alcohol, the number of gallons of new solution, and the percent of alcohol in the new solution.

Amount of water (gallons)	Amount of alcohol (gallons)	Amount of new solution (gallons)	Percent alcohol (%)
20	0	$20 + 0$	$\dfrac{0}{20 + 0} = 0$
20	20	$20 + 20$	$\dfrac{20}{20 + 20} = 50$
20	40	$20 + 40$	$\dfrac{40}{20 + 40} \doteq 66.7$
20	60	$20 + 60$	$\dfrac{60}{20 + 60} = 75$
20	80	$20 + 80$	$\dfrac{80}{20 + 80} = 80$
20	100	$20 + 100$	$\dfrac{100}{20 + 100} \doteq 83.3$

Amount of water (gallons)	Amount of alcohol (gallons)	Amount of new solution (gallons)	Percent alcohol (%)
20	120	20 + 120	$\dfrac{120}{20 + 120} \doteq 85.7$
20	140	20 + 140	$\dfrac{140}{20 + 140} \doteq 87.5$

To show how the percent of alcohol in the solution depends on the amount of alcohol added, we display the amount of alcohol added on the horizontal axis and the percent of alcohol on the vertical axis. The points described by the chart above then have the following coordinates: $(0, 0), (20, 50), (40, 66.7), (60, 75), (80, 80), (100, 83.3), (120, 85.7), (140, 87.5)$. These have been graphed in Fig. 4.34.

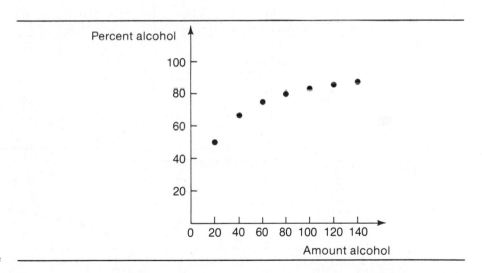

Figure 4.34

We can see that the points in Fig. 4.34 do not lie on a straight line. To get a better idea of the complete graph, we compute 8 additional points and add them to get Fig. 4.35.

Amount of water (gallons)	Amount of alcohol (gallons)	Amount of new solution (gallons)	Percent alcohol (%)
20	10	20 + 10	$\dfrac{10}{20 + 10} \doteq 33.3$
20	30	20 + 30	$\dfrac{30}{20 + 30} = 60$
20	50	20 + 50	$\dfrac{50}{20 + 50} \doteq 71.4$
20	70	20 + 70	$\dfrac{70}{20 + 70} \doteq 77.8$
20	90	20 + 90	$\dfrac{90}{20 + 90} \doteq 81.8$
20	110	20 + 110	$\dfrac{110}{20 + 110} \doteq 84.6$
20	130	20 + 130	$\dfrac{130}{20 + 130} \doteq 86.7$
20	150	20 + 150	$\dfrac{150}{20 + 150} \doteq 88.2$

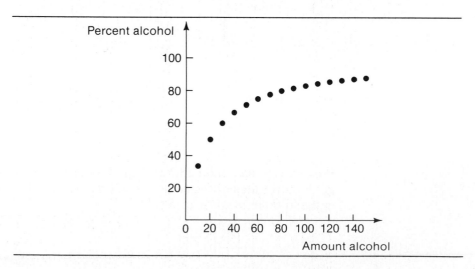

Figure 4.35

The complete graph is shown in Fig. 4.36.

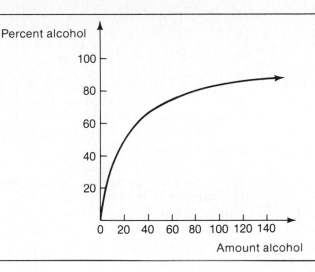

Figure 4.36

☐

Example 2 How much alcohol must be added to 20 gallons of water to give a 68%-alcohol solution?

The graph in Fig. 4.37 suggests that adding 42.5 gallons of alcohol to 20 gallons of water produces a 68% solution.

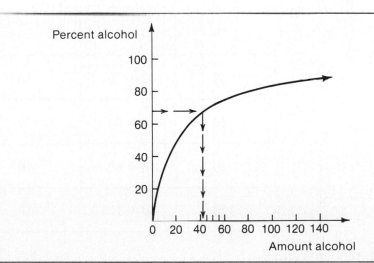

Figure 4.37

Checking this estimate, we get

$$\frac{42.5}{20 + 42.5} = 68\%.$$ □

Example 3 The lab has received 40 gallons of a 25%-salt solution. Draw a graph that shows what happens to the concentration of this solution when different amounts of water are added.

Use the graph to determine how much water should be added to get a 7% solution.

To compute the percent of salt in a new solution, we will need to know the number of gallons of salt in the new solution and the number of gallons of new solution. These are important headings in the chart. The amount of salt in the original solution is .25(40) = 10 gallons. Since water only is being added, the amount of salt in the other cases remains at 10 gallons.

Amount of water added (gallons)	Amount of new solution (gallons)	Amount of salt in new solution (gallons)	Percent salt (%)
0	40 + 0	10	$\frac{10}{40} = 25$
20	40 + 20	10	$\frac{10}{60} \doteq 16.7$
40	40 + 40	10	$\frac{10}{80} = 12.5$
60	40 + 60	10	$\frac{10}{100} = 10$
80	40 + 80	10	$\frac{10}{120} \doteq 8.3$
100	40 + 100	10	$\frac{10}{140} \doteq 7.1$

Amount of water added (gallons)	Amount of new solution (gallons)	Amount of salt in new solution (gallons)	Percent salt (%)
120	40 + 120	10	$\frac{10}{160} \doteq 6.3$
140	40 + 140	10	$\frac{10}{180} \doteq 5.6$

When we graph these eight points, measuring the amount of water added on the horizontal axis and the percent of salt on the vertical axis, we can see the shape of the graph (Fig. 4.38).

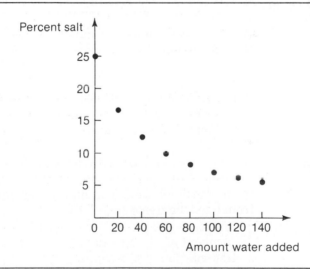

Figure 4.38

To answer the question, "How much water should be added to get a 7% solution?" we complete the graph and find the amount of water that corresponds to 7% (Fig. 4.39). The graph suggests that adding 102.5 gallons of water will produce a 7% solution. We can check that estimate as follows:

$$\frac{10}{40 + 102.5} \doteq 7.02\%.$$

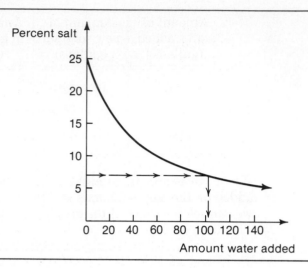

Figure 4.39

Actually, the number of gallons of water that produces a 7% solution, accurate to two decimal places, is 102.86.   □

Example 4   A 10%-salt solution is mixed with a 25%-salt solution to make 20 gallons of solution. Draw a graph that shows how the percent of salt in the final 20 gallons of solution depends on the number of gallons of 10% solution.

We will use the horizontal axis to show the number of gallons of 10% solution in the mixture and the vertical axis to show the percent of salt in the final solution. The number of gallons of 10% solution can vary from 0 to 20 gallons.

Amount of 10% solution (gallons)	Amount of 25% solution (gallons)	Amount of salt in mixture (gallons)	Percent salt in mixture (%)
0	$20 - 0 = 20$	$.10(0) + .25(20) = 5$	$\dfrac{5}{20} = 25$
4	$20 - 4 = 16$	$.10(4) + .25(16) = 4.4$	$\dfrac{4.4}{20} = 22$
8	$20 - 8 = 12$	$.10(8) + .25(12) = 3.8$	$\dfrac{3.8}{20} = 19$

Amount of 10% solution (gallons)	Amount of 25% solution (gallons)	Amount of salt in mixture (gallons)	Percent salt in mixture (%)
12	$20 - 12 = 8$	$.10(12) + .25(8) = 3.2$	$\dfrac{3.2}{20} = 16$
16	$20 - 16 = 4$	$.10(16) + .25(4) = 2.6$	$\dfrac{2.6}{20} = 13$
20	$20 - 20 = 0$	$.10(20) + .25(0) = 2$	$\dfrac{2}{20} = 10$

The points whose coordinates are the amounts of 10% solution shown above, with the corresponding percent of salt in the mixture, are $(0, 25)$, $(4, 22)$, $(8, 19)$, $(12, 16)$, $(16, 13)$, and $(20, 10)$. They have been graphed in Fig. 4.40.

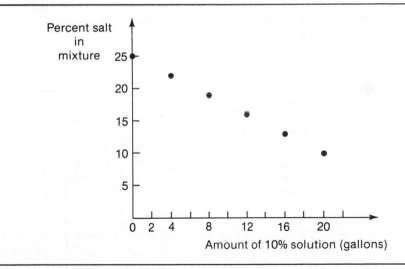

Figure 4.40

If we were able to compute the percent of salt in the mixture for every possible amount of 10% solution between 0 and 20 gallons, we would have the coordinates of all the points on the line segment in Fig. 4.41.

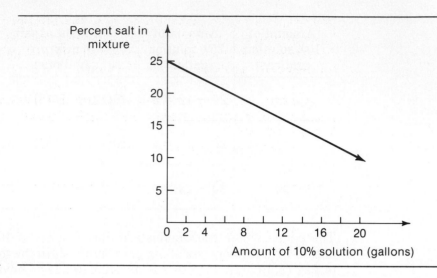

Figure 4.41

From this graph we can answer questions such as these:

Question: How much each of a 10%-salt solution and a 25%-salt solution should be mixed together to make 20 gallons of a solution that is 17.5% salt?

Answer: Study Fig. 4.42.

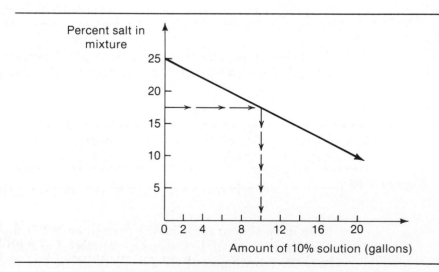

Figure 4.42

The graph indicates that the amount of 10% solution is 10 gallons. Thus there must also be 10 gallons of a 25% solution. You should check that 10 gallons of a 10%-salt solution mixed with 10 gallons of a 25%-salt solution gives a solution that is 17.5% salt.

Question: How much each of a 10%-salt solution and a 25%-salt solution should be mixed together to make 20 gallons of a solution that is 12% salt?

Answer: Study Fig. 4.43.

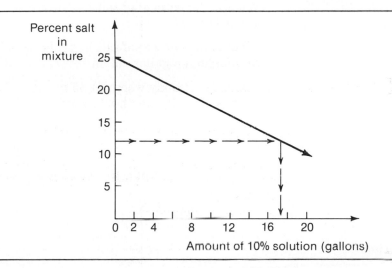

Figure 4.43

The exact value here is difficult to read. We might estimate that the amount of 10% solution is 17.5 gallons and check that guess:

Amount of 10% solution (gallons)	Amount of 25% solution (gallons)	Amount of salt in mixture (gallons)	Percent salt in mixture (%)
17.5	$20 - 17.5 = 2.5$	$.10(17.5) + .25(2.5) = 2.375$	$\dfrac{2.375}{20} = .11875$

This guess gives a percent of salt that is too small. There must, therefore, be slightly more of the 25% solution and slightly less of the 10% solution. For a second guess, we might try 17.3.

Amount of 10% solution (gallons)	Amount of 25% solution (gallons)	Amount of salt in mixture (gallons)	Percent salt in mixture (%)
17.3	$20 - 17.3 = 2.7$	$.10(17.3) + .25(2.7) = 2.405$	$\dfrac{2.405}{20} = .12025$

The second guess is very close. The actual value is $17\frac{1}{3}$ gallons of 10% solution and $2\frac{2}{3}$ gallons of 25% solution.

Exercises 4.5

1. What is the percent of salt in a mixture of 3 gallons of salt and 12 gallons of water?

2. What is the percent of salt in a mixture of 3 gallons of a 12%-salt solution and 5 gallons of an 18%-salt solution?

3. 8 pounds of candy worth $1.50 per pound are mixed with 12 pounds of candy worth $2.25 per pound.

   (a) What is the total value of the 20-pound mixture?

   (b) What is the value per pound of the 20-pound mixture?

4. Varying amounts of a 10%-alcohol solution and an 80%-alcohol

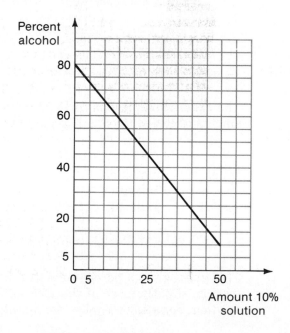

solution are mixed to form 50 gallons of solution. The graph above shows how the percent of alcohol in the 50 gallons depends on the amount of 10%-alcohol solution used.

(a) Use the graph to find the percent of alcohol in the 50-gallon mixture if the number of gallons of 10% solution used is 25; if the number of gallons of 10% solution used is 30.

(b) Use the graph to find the amount of 10% solution used if the percent of alcohol in the 50-gallon mixture is 52%; if the percent of alcohol in the 50-gallon mixture is 35%.

5. Varying amounts of alcohol are added to 30 gallons of water. Draw a graph that shows how the percent of alcohol in the new mixture depends on the amount of alcohol added. Use the graph to find the amount of alcohol added if the percent of alcohol in the new mixture is

    (a) 76%   (b) 28%

6. Varying amounts of water are added to 40 gallons of a 30%-salt solution. Draw a graph that shows how the percent of salt in the new mixture depends on the amount of water added. Use the graph to find the amount of water added if the percent of salt in the new mixture is

    (a) 25%   (b) 9%

7. An 8%-salt solution is mixed with a 28%-salt solution to produce 40 gallons of a new solution. Draw a graph that shows how the percent of salt in the new mixture is a function of the amount of 8%-salt solution used. Use the graph to find how much of each solution must be mixed to make 40 gallons of a new solution containing

    (a) 13% salt   (b) 25.5% salt

8. A 12%-alcohol solution is mixed with a 36%-alcohol solution to produce 30 gallons of a new solution. Draw a graph that shows how the percent of alcohol in the new solution is a function of the amount of 36%-alcohol solution used. Use the graph to find how much of each solution must be mixed to make 30 gallons of a new solution containing

    (a) 22% alcohol   (b) 17% alcohol

9. Varying amounts of candy worth $1.00 per pound are added to 20 pounds of candy worth $3.50 per pound. Draw a graph that shows the relationship between the amount of $1.00-per-pound candy added and the value per pound of the new mixture. Use the graph to find how many pounds of candy worth $1.00 per pound must be added to the 20 pounds of candy worth $3.50 per pound to produce a new mixture worth

(a) $1.80 per pound　　　　　(b) $1.40 per pound

10. Candy worth $1.25 per pound is mixed with candy worth $2.85 per pound to produce a 50-pound mixture. Draw a graph that shows the relationship between the amount of $1.25-per-pound candy used and the value per pound of the 50-pound mixture. Use the graph to find how much of each type of candy must be mixed to make 50 pounds of candy worth

(a) $1.97 per pound　　　　　(b) $2.27 per pound

4.6 / CHAPTER 4 PROBLEM COLLECTION

1. Write the coordinates of the points in the graph below.

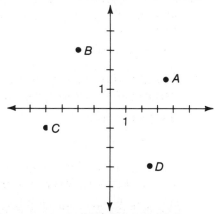

2. Draw a coordinate system and graph the following points:

$$(3, 4), \left(\frac{5}{2}, -1\right), (-3, 1), \left(-\frac{3}{2}, -\frac{5}{2}\right).$$

3. The area of a rectangle is 800 square feet.

   (a) Find the length of the rectangle if its width is 20. Find the length if its width is 50.

   (b) Find the perimeter of the rectangle if its width is 20. Find the perimeter if its width is 50.

4. A rectangle is 4 feet longer than it is wide.

   (a) Find the length of the rectangle if its width is 12. Find the length if its width is 21.

   (b) Find the area of the rectangle if its width is 12. Find the area if its width is 21.

   (c) Find the perimeter of the rectangle if its width is 12. Find the perimeter if its width is 21.

5. What is 38% of 80?

6. 25.2 is what percent of 60?

7. A certain type of bacteria is growing at the rate of 8% per hour and there are 800 present initially. Find the number present

   (a) 1 hour later   (b) 6 hours later

8. $3000 is deposited in an account that pays 10.5% interest. Find the amount of money in the account 2 years later if the interest is compounded

   (a) quarterly   (b) daily

9. The half-life of a radioactive isotope is 12 years. If there are 60 grams of the isotope present initially, find the amount present

   (a) 1 year later   (b) 6 years later

   (c) 24 years later

10. All federal employees are receiving a 6.5% raise. Find the new salary of an employee who made $12,500 before the raise.

11. What is the percent of alcohol in a mixture of 5 gallons of water and 20 gallons of a 35%-alcohol solution?

12. What is the percent of alcohol in a mixture of 30 gallons of a 42%-alcohol solution and 20 gallons of a 96%-alcohol solution?

13. Part of $12,000 is invested at 5% and the remainder is invested at 9%. The graph below shows how the interest received at the end of one year depends on the amount invested at 5%.

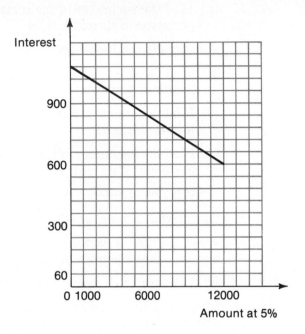

(a) Use the graph to find the interest received if the amount invested at 5% is $3000. Find the interest if the amount invested at 5% is $10,500.

(b) Use the graph to find the amount invested at 5% if the interest received is $840. Find the amount invested at 5% if the interest received is $885.

14. The graph below shows the average cost of a TV set starting with January 1, 1975 ($t = 0$) and continuing for a 24-year period.

(a) Use the graph to find the average cost of a TV set on January 1, 1975, on January 1, 1976, and in mid-1982.

(b) Use the graph to find when the average cost of a TV set will be $1600, and when it will be $3600.

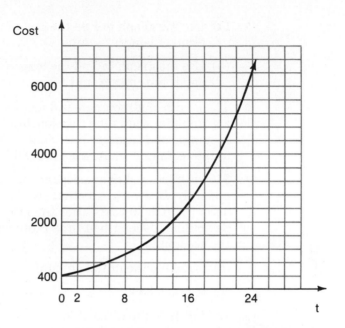

15. The graph below gives the charge for a telephone call between two cities as a function of the length of the call in minutes.

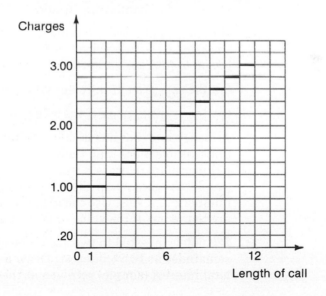

CHAPTER 4   USING GRAPHS TO SOLVE PROBLEMS

(a) Use the graph to find the cost of a call of length 1 minute and a call of length $7\frac{1}{2}$ minutes.

(b) Use the graph to find the length of a call if the charge is $1.00 and if the charge is $1.60.

16. The length of a rectangle is 3 feet more than twice its width. Draw a graph that shows the relationship between the width of the rectangle and its area. Use the graph to find the width and length of the rectangle whose area is 1080.

17. $100 is deposited in an account that pays 10.4% compounded annually. Draw a graph that shows the amount of money in the account up to 30 years after the initial deposit. Use the graph to determine the length of time required for the amount in the account

(a) to double the initial amount

(b) to reach $800

18. During 1970 the price of food increased 8%. Assume that this rate of increase continued and that the cost of a pound of hamburger was $.89 in 1970. Draw a graph that shows the cost of a pound of hamburger from 1970 to 2000. Use the graph to find when the cost of a pound of hamburger would be $2.00 or more.

19. A department store is having a 30%-off sale. Draw a graph that shows how the sale price is a function of the original price. Use the graph to find

(a) the sale price if the original price is $55.00

(b) the original price if the sale price is $87.50

20. A money box contains nickels, dimes, and quarters. The number of dimes is one more than twice the number of nickels, and the number of quarters is 3 less than the number of nickels. Draw a graph that shows how the value of the coins is determined by the number of nickels. Use the graph to find how many of each coin are in the box if the value of the coins is $10.35.

21. $20,000 is invested in two accounts—part at 17.5% interest and the remainder at 12.5% interest. Draw a graph that shows how the total interest (simple) received at the end of one year depends on

the amount invested at 17.5%. Use the graph to find the amount invested at each rate if the total interest received is $3,110.

22. A 22%-alcohol solution is mixed with a 64%-alcohol solution to produce 40 gallons of a new solution. Draw a graph that shows how the percent of alcohol in the new solution depends on the number of gallons of 22%-alcohol solution used. Use the graph to find the amount of each solution needed to make the final solution

    (a) 43% alcohol   (b) 52% alcohol

23. Tom leaves Columbus at 8:00 A.M. and drives north at an average speed of 50 mph. Bill leaves Columbus at 10:30 A.M. and drives north on the same highway at 65 mph. Draw a graph that shows the distance between Tom and Bill any time after 10:30 A.M. Use the graph to find when Bill catches Tom.

24. Sue rides her bicycle 18 miles at one speed. She then rides an additional 5 miles at that speed reduced by 4 mph. Draw a graph that shows how the total time for the 23-mile trip depends on the original speed. Use the graph to find the original speed that gives a total travel time of one hour.

5

Linear Equations
in One Variable

5.1 / WRITING EQUATIONS
TO DESCRIBE PROBLEMS

Applications of mathematics usually require reasoning with quantities that can vary over sets of numbers. We have solved such problems by building numerical charts and by drawing graphs. A third way to describe these problems (and often the most efficient) is to use variables to translate the problems into mathematical equations. In this section you will gain practice with that kind of translation.

Algebraic Expressions and Equations

The following familiar example illustrates how variables can be used in describing problem situations.

Example 1   A car averages 22 miles per gallon of gasoline. How many gallons of gasoline are used for a 264-mile trip?

One valid way to investigate this problem is to guess and check, organizing information in the form of a chart until a correct answer is apparent.

Number of gallons of gasoline	Number of miles
1	22
2	2(22) = 44
10	10(22) = 220
12	12(22) = 264

To find the number of miles, we multiply the number of gallons of gasoline used by 22. Thus if N represents the number of gallons of gasoline used, $22N$ represents the number of miles traveled. We want to know the value of N that makes $22N = 264$.   □

In the example above, the symbol N is a variable. It represents all possible numbers of gallons of gasoline. Presumably N could be 0 or any positive number. A **variable** is a symbol, usually a letter, that stands for the numbers in some set. The numbers that a variable represents are called its **values**. Generally the problem will suggest what numbers are represented by the variable. Using a variable permits us to describe a general situation in a concise way.

When variables and numbers are combined using the usual operations of arithmetic (addition, subtraction, multiplication, division, and raising to a power), the result is called an **algebraic expression**. For example, $22N$ is an algebraic expression. Here are others:

$$3x + 5$$

$$\frac{x}{4} + y - 4$$

$$N^2 - \sqrt{2N}$$

$$uv - v^2$$

$$27.$$

The numerical values that an algebraic expression represents are determined by the numbers that the variables represent. For example, if $N = 1$, then $22N = 22$; if $x = \frac{1}{3}$, then $3x + 5 = 3(\frac{1}{3}) + 5 = 6$; if $x = 12$ and $y = 5$, then $\frac{x}{4} + y - 4 = \frac{12}{4} + 5 - 4 = 4$.

A statement that two algebraic expressions are equal is called an **equation**. Here are some examples of equations:

$$22N = 264$$

$$3x + 5 = x - 7$$

$$\frac{x}{4} + y - 4 = x^2 - 5.$$

An equation always has an equals (=) sign in it; an algebraic expression never contains an equals sign. An equation often is written to summarize a problem mathematically. In these cases we are interested in the solutions to the equation. **Solutions** are the values of the variable or variables that make the equation a true statement. For example,

- 12 is a solution to the equation $22N = 264$ because $22(12) = 264$;
- 10 is not a solution to the equation $22N = 264$ because $22(10) \neq 264$;
- -6 is a solution to the equation $3x + 5 = x - 7$ because $3(-6) + 5 = -6 - 7$;
- 6 is not a solution to the equation $3x + 5 = x - 7$ because $3(6) + 5 \neq 6 - 7$.

A solution to the equation $\frac{x}{4} + y - 4 = x^2 - 5$, which has two variables in it, is a pair of numbers: a value for x and a value for y. One solution is $x = 0$ and $y = -1$. Another solution is $x = 1$ and $y = -\frac{1}{4}$. This equation has many solutions.

Not every equation has solutions. The equation $x + 1 = x + 2$ has no solutions because no number plus 1 equals that same number plus 2. Other equations have every number as a solution; for example, $2(x + 3) = 2x + 6$ is a true statement when x is replaced by any number. If we have found all solutions to an equation (or determined there are none), we say we have **solved** the equation.

Equations such as $22N = 264$ are easily solved. Other equations are difficult to solve. By using the calculator with a guess-and-check procedure, we can usually get good approximations to solutions. There are a few types of equations that can be solved exactly with special techniques, and we will save time by learning those techniques. However, before we focus on solving special types of equations, we want to get more practice describing problem situations with equations.

Writing Equations

We have used charts and graphs to summarize information; now we want to use equations. At this time our interest is in writing an equation that describes a problem situation. Later we will learn to solve equations in order to answer specific questions.

Example 2   Judy has twice as many dimes as nickels and two more quarters than nickels. Let x denote the number of nickels; write an algebraic expression for the value of Judy's coins and then write an equation which states that their value is $5.50.

Earlier when we investigated this problem situation, we built a chart that looked like the following.

Number of nickels	Number of dimes	Number of quarters	Value ($)
5	$2 \cdot 5 = 10$	$5 + 2 = 7$	$.05(5) + .10(10) + .25(7) = 3.00$
8	$2 \cdot 8 = 16$	$8 + 2 = 10$	$.05(8) + .10(16) + .25(10) = 4.50$

Number of nickels	Number of dimes	Number of quarters	Value ($)
12	$2 \cdot 12 = 24$	$12 + 2 = 14$	$.05(12) + .10(24) + .25(14) = 6.50$
x	$2x$	$x + 2$	$.05x + .10(2x) + .25(x + 2)$

In the last line of the chart we have an algebraic expression that gives the value of the coins when there are x nickels; the expression is $.05x + .10(2x) + .25(x + 2)$. When we translate the statement "the value of the coins is \$5.50," we get the equation

$$.05x + .10(2x) + .25(x + 2) = 5.50.$$

Solving this equation will tell us how many nickels Judy has if the total value of her coins is \$5.50. Notice that the equation can be simplified by using the distributive property and then combining terms:

$$.05x + .20x + .25x + .50 = 5.50$$

$$.50x + .50 = 5.50.$$

This equation is one of a special type we will practice solving in the next section. □

Example 3 Berry's Department Store is having a 27%-off sale. Let p denote the original price of an item. Write an algebraic expression for the sale price, and then write an equation which states that the sale price of an item is \$45.26.

Remember that the sale price equals the original price minus .27 times the original price. Thus if we denote the original price by the variable p, we have this information in the chart:

Original price	Sale price
p	$p - .27p$

(This may be the bottom line of your chart. If you do not see immediately how to write the sale price using the variable, first fill in the chart for some numerical values of the original price, such as \$60 or \$50.) The expression $p - .27p$ gives the sale price when the original price is p.

The equation that expresses the sale price as being equal to $45.26 is

$$p - .27p = 45.26.$$

We could find the value of p that makes this true by guessing and checking. However, this equation is also one of those that we will solve by special techniques in the next section. Once we have solved the equation, we will have answered the question, "What was the original price of an item that has sale price $45.26?"   □

Example 4 Part of $20,000 is invested at 12% effective annual yield and the remainder is invested at 10.5% effective annual yield. Let x denote the part invested at 12%. Write an expression for the total interest paid by the two accounts in one year, and then write an equation showing that the interest equals $2287.50.

We can make a chart for this situation using x for the amount invested at 12%.

Amount invested at 12%	Interest paid by 12% account	Amount invested at 10.5%	Interest paid by 10.5% account	Total interest
x	$.12x$	$20,000 - x$	$.105(20,000 - x)$	$.12x + .105(20,000 - x)$

The expression $.12x + .105(20,000 - x)$ gives the total interest paid by the two accounts after one year if x dollars are invested at 12% effective annual yield. The equation that says the total interest is $2287.50 is

$$.12x + .105(20,000 - x) = 2287.50.$$

In the next section you will learn how to solve this equation. The solution answers the question, "How much money is invested at 12% if the total interest paid by the two accounts after one year is $2287.50?"   □

Exercises 5.1   In Exercises 1–6, find the value of each algebraic expression for the values of the variables specified.

1. $\frac{1}{2}bh$; $b = 8, h = 10$

2. $p\left(1 + \frac{r}{4}\right)^{4t}$; $p = 1500, r = .06, t = 3$

3. $\frac{a}{1 - r}$; $a = 3, r = -\frac{1}{2}$

4. $3x + 2y$; $x = -2, y = 3$

5. $2x - \frac{3}{2}y + 1$; $x = -3, y = 2$

6. $a + \dfrac{1}{a + \dfrac{1}{a + \dfrac{1}{a}}}$; $a = 1$

In Exercises 7–10, determine if the specified values of the variables give solutions of the equation.

7. $x^2 + y^2 = 4$

  (a) $x = -2, y = 0$ (b) $x = -\sqrt{2}, y = -\sqrt{2}$

  (c) $x = 1.4, y = 1.4$

8. $x^2 - x = 0$

  (a) $x = 0$ (b) $x = 1$ (c) $x = -1$

9. $x^3 + 2x^2 - x - 2 = 0$

  (a) $x = 1$ (b) $x = -1$ (c) $x = -2.5$

10. $2x - 3y = -6$

  (a) $x = 0, y = -2$ (b) $x = -3, y = 0$ (c) $x = -\frac{3}{2}, y = 1$

11. A money box contains 30 coins in nickels and dimes. Write the number of dimes if the number of nickels is

  (a) 12   (b) x

12. $18,000 is invested, part at 8% and the remainder at 5%. Write the amount invested at 5% if the amount invested at 8% is

(a) $13,200   (b) $x

13. Jane has 60 coins in nickels, dimes, and quarters. The number of dimes is one more than twice the number of nickels. Write the number of dimes and the number of quarters if the number of nickels is

(a) 8   (b) 12   (c) x

14. Write an algebraic expression that represents the amount of money accumulated when $1200 is invested at 11.8% interest compounded daily for x years.

15. During 1980 the price of food increased 12.3%. Assume that a gallon of milk sold for $1.92 in 1980. Write an algebraic expression for the cost of a gallon of milk after x years if the increase in cost of food continues at an annual rate of 12.3%. Evaluate the expression for $x = 5$.

16. A money box has 30 coins in nickels and dimes. Let x be the number of nickels.

    (a) Write an algebraic expression that represents the value of the 30 coins.

    (b) Write an equation that expresses the fact that the value of the 30 coins is $2.65.

17. Bill has two more quarters than nickels and three fewer dimes than nickels. Let x be the number of nickels.

    (a) Write an algebraic expression that represents the value of the coins.

    (b) Write an equation that expresses the fact that the value of the coins is $2.60.

18. Shears Department Store is having a $\frac{1}{3}$-off sale. Let x be the original price of an item.

    (a) Write an algebraic expression that represents the sale price of an item.

    (b) Write an equation that expresses the fact that the sale price of an item is $18.50.

19. The Three D Department Store marks up wholesale prices 40%. Let w be the wholesale price of an item.

    (a) Write an algebraic expression that represents the retail price of an item.

    (b) Write an equation that expresses the fact that the retail price is $27.75.

20. Part of $18,000 is invested at 8% simple interest and the remainder is invested at 5% simple interest. Let x be the amount invested at 8%.

    (a) Write an algebraic expression that represents the total interest paid by the two accounts one year later.

    (b) Write an equation that expresses the fact that the total interest paid one year later is $1260.

21. Tickets to a football game cost $3.75 for adults and $2.50 for students; 2000 tickets were sold. Let x denote the number of students tickets sold.

    (a) Write an algebraic expression that represents the total revenue from the sale of the 2000 tickets.

    (b) Write an equation that expresses the fact that the total revenue from the sale of tickets was $5812.50.

22. Two cars start from the same point and travel in opposite directions on a straight highway at constant speeds of 50 mph and 55 mph.

    (a) Write an algebraic expression that represents the distance between the two cars t hours later.

    (b) Write an equation that expresses the fact that the distance between the two cars t hours later is 100 miles.

23. Two cars start from the same point and travel in the same direction on a straight highway at constant speeds of 48 mph and 61 mph.

    (a) Write an algebraic expression that represents the distance between the two cars t hours later.

    (b) Write an equation that expresses the fact that the distance between the two cars t hours later is 100 miles.

In Exercises 24 and 25, let w represent the width of a rectangle and l, its length.

24. Write an equation that expresses the fact that the perimeter of the rectangle is

    (a) 40 feet   (b) p feet

25. Write an equation that expresses the fact that the area of the rectangle is

    (a) 80 square feet   (b) A square feet

5.2 / SOLVING LINEAR EQUATIONS IN ONE VARIABLE

Sometimes a problem situation can be described by an equation that has only one variable; $.05x + .10(2x) + .25(x + 2) = \$5.50$ is an equation in the one variable x. Other problems require more than one variable in an equation to describe the problem; $lw = 100$ is an equation in two variables, l and w.

In the last section we wrote these four equations, each in one variable:

$$22N = 264$$

$$.50x + .50 = 5.50$$

$$p - .27p = 45.26$$

$$.12x + .105(20,000 - x) = 2287.50.$$

In each of these equations the variable appears always with exponent 1 and no variable occurs in the denominator of a fraction. Such equations are said to be **linear**. Here are some other equations in one variable that are linear:

$$10 - 2x = 3x + 15$$

$$\frac{1}{2}(x + 2) = -130$$

$$100 - \sqrt{2}x = 7(x + 1).$$

The following equations have one variable but are not linear. See if you can explain why.

$$x^2 - 3x = 14$$

$$\frac{1}{x} = \frac{x-2}{6}$$

$$3\sqrt{x} = x + 4.$$

It is also possible to talk about equations that are linear in more than one variable. Here are some examples:

$$3x - y = 12$$

$$x + \frac{y}{3} + 10z = 136.$$

In this section we will restrict our attention to linear equations in one variable and find special techniques for solving this kind of equation.

Equivalent Equations

Equations that have exactly the same solutions are said to be **equivalent**. Check to see that the following equations all have one solution, the number 4, and hence are equivalent:

$$3x - 4 = 8 \qquad\qquad 6 - x = 2$$

$$\frac{1}{2}x = 2 \qquad\qquad x = 4$$

Spotting the solution, 4, for the equations above is not difficult. For other equations equivalent to these, such as $3 - 4(2 - x) = x + 7$, guessing a solution might take a long time. A general procedure for finding a solution to a linear equation is to replace it with other equations that are equivalent but simpler, working until we get an equivalent equation, such as $x = 4$, whose solution is obvious. However, we need a method for doing this.

An equation is a statement that two algebraic expressions are equal. The values of the variable that make an equation true are not changed if the same expression is added to both sides of the equation. For example, consider the equation $10 - 2x = 3x$. If we add $2x$ to both sides of this equation, we obtain $10 - 2x + 2x = 3x + 2x$. Any value of x that makes the first equation true will also make the second true; in fact, the first equation and the second equation will have exactly the same solutions. If we combine like terms in the second equation, we get

$10 = 5x$, an easier equation to solve. By adding $2x$ to both sides of the original equation, we have obtained a simpler, equivalent equation.

The same is true for subtraction and for multiplication and division by a nonzero number. We get an equivalent equation if we

- add the same algebraic expression to, or subtract the same algebraic expression from, both sides of an equation;
- multiply or divide both sides of an equation by the same nonzero number.

We can use these procedures to replace an equation with simpler equivalent equations. Usually there is more than one order in which the procedures can be performed, and there may not always be a "best" way. Generally, the strategy is to get the terms containing the variable on one side (it does not matter which) and the terms without the variable on the other.

Example 1   Solve $10 - 2x = 3x + 15$.

$$10 - 2x = 3x + 15$$
$$10 - 2x + 2x = 3x + 15 + 2x \qquad \text{(Add } 2x \text{ to both sides)}$$
$$10 = 5x + 15 \qquad \text{(Combine like terms)}$$
$$10 - 15 = 5x + 15 - 15 \qquad \text{(Subtract 15 from both sides)}$$
$$-5 = 5x \qquad \text{(Combine like terms)}$$
$$\frac{-5}{5} = \frac{5x}{5} \qquad \text{(Divide both sides by 5)}$$
$$-1 = x.$$

All of these equivalent equations, including the original, have the solution -1.   □

There are times when we need to use the distributive property or to combine like terms before replacing the given equation with simpler equivalent equations. Study the steps in this example carefully.

Example 2   Solve $4(x + 2) - (3 - x) = -130$.

$$4(x + 2) - (3 - x) = -130$$

$4x + 8 - 3 + x = -130$	(Distributive property)
$5x + 5 = -130$	(Combine like terms)
$5x + 5 - 5 = -130 - 5$	(Subtract 5 from both sides)
$5x = -135$	(Combine like terms)
$\dfrac{5x}{5} = \dfrac{-135}{5}$	(Divide both sides by 5)
$x = -27.$	

Use your calculator to check that -27 is a solution to $4(x + 2) - (3 - x) = -130$.   □

Example 3   Solve $\dfrac{1}{2}x - \dfrac{1}{3}x + 1 = 5$.

Although we could start the computation by combining like terms on the left side, we can choose first to multiply both sides of the equation by 6, the LCM of the denominators. This produces an equivalent equation that has integers rather than fractions for coefficents. We sometimes say that we *clear the fractions* from the equation:

$\dfrac{1}{2}x - \dfrac{1}{3}x + 1 = 5$	
$6\left(\dfrac{1}{2}x - \dfrac{1}{3}x + 1\right) = 6(5)$	(Multiply both sides by 6)
$6\left(\dfrac{1}{2}x\right) - 6\left(\dfrac{1}{3}x\right) + 6(1) = 6(5)$	(Distributive property)
$3x - 2x + 6 = 30$	
$x + 6 = 30$	(Combine like terms)
$x + 6 - 6 = 30 - 6$	(Subtract 6 from both sides)
$x = 24.$	(Combine like terms) □

Example 4   Solve $\dfrac{2x + 1}{5} = \dfrac{4x - 3}{2}$.

a. We can multiply both sides of the equation by 10 to clear the fractions:

$$\frac{2x + 1}{5} = \frac{4x - 3}{2}$$

$$\frac{10}{1}\left(\frac{2x + 1}{5}\right) = \frac{10}{1}\left(\frac{4x - 3}{2}\right) \qquad \text{(Multiply both sides by 10)}$$

$$2(2x + 1) = 5(4x - 3)$$

$$4x + 2 = 20x - 15 \qquad \text{(Distributive property)}$$

$$4x + 2 - 4x = 20x - 15 - 4x \qquad \text{(Subtract } 4x \text{ from both sides)}$$

$$2 = 16x - 15 \qquad \text{(Combine like terms)}$$

$$2 + 15 = 16x - 15 + 15 \qquad \text{(Add 15 to both sides)}$$

$$17 = 16x \qquad \text{(Combine like terms)}$$

$$\frac{17}{16} = \frac{16x}{16} \qquad \text{(Divide both sides by 16)}$$

$$\frac{17}{16} = x. \qquad\qquad\qquad \square$$

b. A second and somewhat more efficient way to begin this equation is to remember what it means for two fractions to be equal:

$$\frac{2x + 1}{5} \diagdown\diagup \frac{4x - 3}{2}$$

$$2(2x + 1) = 5(4x - 3) \qquad \text{(Cross multiply)}$$

$$4x + 2 = 20x - 15 \qquad \text{(Distributive property)}$$

$$4x + 2 - 4x = 20x - 15 - 4x \qquad \text{(Subtract } 4x \text{ from both sides)}$$

$$2 = 16x - 15 \qquad \text{(Combine like terms)}$$

$$2 + 15 = 16x - 15 + 15 \qquad \text{(Add 15 to both sides)}$$

$$17 = 16x \qquad \text{(Combine like terms)}$$

$$\frac{17}{16} = \frac{16x}{16}$$ (Divide both sides by 16)

$$\frac{17}{16} = x.$$ □

**Exercises
5.2**

State whether the equations in Exercises 1–6 are linear in one variable, linear in two or more variables, or not linear.

1. $\frac{3}{2}x + 13 = x + 6$ 2. $3x + 4y + 5z = 10$

3. $2\sqrt{x} + \sqrt{y} = 3$ 4. $\frac{1}{2}x - \frac{1}{3}y = 6$

5. $2x^2 - y^2 + y = 1$ 6. $x^{4/3} + y^{1/3} = 1$

Determine whether the two equations in each of Exercises 7–10 are equivalent.

7. $10x + 5 = 2x + 3$ 8. $7x + 6 = 5x + 2$

$10x + 5 - 2x = 2x + 3 - 2x$ $7x + 6 - 6 = 5x + 2 - 2$

9. $\frac{2y + 1}{3} = y + 3$ 10. $\frac{x + 1}{2} = \frac{5}{2} + x$

$2y + 1 = y + 3$ $x + 1 = 5 + 2x$

Solve the equations in Exercises 11–46.

11. $22N = 264$ 12. $.50N + .50 = 5.50$

13. $p - .27p = 45.26$ 14. $2x + 4 = 6x + 20$

15. $5(x + 2) = 45$ 16. $x - \frac{1}{2} = 2$

17. $2x + 12 + x = 7x + 8$ 18. $\frac{x}{3} = 4$

19. $4 + \frac{1}{2}y = 1$ 20. $5x - 17 - 2x = 6x - 1 - x$

21. $2.2x - 5 + 4.5x = 1.7x - 20$

22. $\dfrac{x}{2} - \dfrac{x}{3} = -1$

23. $\dfrac{1}{3}z + \dfrac{2}{3} = -\dfrac{1}{3}z + \dfrac{1}{6}$

24. $\dfrac{1}{3}z + \dfrac{1}{3} = \dfrac{1}{6}z + \dfrac{1}{6}$

25. $(0.35)(x + 2) = 0.70$

26. $3x + 48 = 4x - 12$

27. $\dfrac{1}{2}x + \dfrac{1}{3}x + 2 = 17$

28. $\dfrac{x + 1}{2} - \dfrac{x}{6} = \dfrac{1}{3}$

29. $\dfrac{3w + 2}{3} = \dfrac{4w - 3}{2}$

30. $\dfrac{1}{3}w + \dfrac{1}{9} = 1$

31. $\dfrac{x - 1}{5} + \dfrac{x - 4}{2} + 5 = 0$

32. $\dfrac{x + 7}{2} - \dfrac{1}{3} = \dfrac{1}{2} - \dfrac{x + 9}{9}$

33. $(x - 6) - 15 = (3x + 5) - 6$

34. $(2x + 1) - x = -3 - (4x + 10)$

35. $3(x + 5) + 4(x + 5) = 21$

36. $3(x + 2) = x + 4 + 2(x + 1)$

37. $7(3x + 6) = 11 - (x + 2)$

38. $4(5x + 3) = 3(2x - 5)$

39. $8 - 4(x - 1) = 2 + 3(4 - x)$

40. $3(x - 3) + 2(x - 2) = 15 - 5x$

41. $x - (9x - 8) = 5 - 2x - 3(2x - 3) + 29$

42. $2(x + 1) - 6x = 3(x - 1) + 5(1 - x)$

43. $32x + 15 - 17x + 12 + x = 15 - 3 + 6x - 2x$

44. $7x - 12 - 8x + 2x - 10 = 18 - 3x - 31 - 5x$

45. $x(x + 2) + 1 = x(x + 1) + 2$

46. $x(2x - 1) = 2x(x + 3) + 1$

In Exercises 47–50, use a variable to represent the desired number. Then write an equation that represents the problem and solve the equation.

47. Forty-eight is two-thirds of what number?

48. Seventy-two is 25% of what number?

49. A number increased by 22% is 244. What is the number?

50. A number decreased by 18% is 902. What is the number?

5.3 / SOLVING PERCENT PROBLEMS WITH LINEAR EQUATIONS

Whenever we are able to describe a problem with a linear equation in one variable, we can find a solution using the techniques given in the last section. Often, writing an equation that describes a problem is more difficult than solving the equation once we have it. Remember that building a chart to summarize the information in a problem is a helpful way of developing an equation to describe the problem.

In this section we will study some problem situations whose equations may not be immediately clear. We will construct charts to help organize the information.

Example 1   An account has an effective annual yield of 17.2%. If an investment yields $1406.40 after one year, what was the investment?

Having an effective annual yield of 17.2% means that in one year the account pays as though it were 17.2% compounded annually. To see the result of investing different amounts of money, we can make a chart showing several examples.

Investment ($)	Interest ($)	Total after one year ($)
200	.172(200)	200 + .172(200) = 234.40
500	.172(500)	500 + .172(500) = 586.00
1000	.172(1000)	1000 + .172(1000) = 1172.00
5000	.172(5000)	5000 + .172(5000) = 5860.00

The computation in the chart shows that the total amount of money in the account after one year is the investment plus .172 times the investment. If we let x be a variable representing any investment, we can add one more line to the chart.

Investment ($)	Interest ($)	Total after one year ($)
x	$.172x$	$x + .172x$

To answer the question, "How much must be invested to yield \$1406.40?" we must find x so that $x + .172x = 1406.40$:

$$x + .172x = 1406.40$$

$$1.172x = 1406.40 \qquad \text{(Combine like terms)}$$

$$x = 1200. \qquad \text{(Divide both sides by 1.172)}$$

An investment of \$1200 yields \$1406.40. □

The investment problem of Example 1 is an application of percent. Another type of problem that uses percent is one in which two substances are mixed together and we prescribe the percent of one substance in the mixture.

Example 2   How much salt must be added to 20 gallons of water to make a 22%-salt solution?

We first make a chart that shows in several cases the amount of water, the amount of salt, the amount of solution, and the percent of salt. We can add a last line for x gallons of salt to assist in writing an equation to solve the problem:

Amount of water (gallons)	Amount of salt (gallons)	Amount of solution (gallons)	Percent salt (%)
20	2	20 + 2	$\dfrac{2}{20+2} \doteq 9.09$
20	3	20 + 3	$\dfrac{3}{20+3} \doteq 13.04$
20	4.5	20 + 4.5	$\dfrac{4.5}{20+4.5} \doteq 18.37$
20	x	20 + x	$\dfrac{x}{20+x}$

There are at least two ways to formulate an equation from the last line of the chart that will answer the question, "How much salt must be added to 20 gallons of water to make a 22%-salt solution?"

a. Remember that 22% of the total solution must be salt. Thus, using columns 3 and 2, we need a value for x that makes

$$.22(20 + x) = x.$$

The equation is readily solved:

$$.22(20 + x) = x$$

$$.22(20) + .22x = x \qquad \text{(Distributive property)}$$

$$4.4 + .22x - .22x = x - .22x \qquad \text{(Subtract } .22x \text{ from each side)}$$

$$4.4 = .78x \qquad \text{(Combine like terms)}$$

$$\frac{4.4}{.78} = \frac{.78x}{.78} \qquad \text{(Divide both sides by .78)}$$

$$5.64 \doteq x.$$

We need to add approximately 5.64 gallons of salt to 20 gallons of water to make a 22%-salt solution.

b. A second way to use the information in the last line is to realize that the percent of salt (column 4) must equal 22%. Thus we need a value for x that makes

$$\frac{x}{20 + x} = .22.$$

This equation is not linear, but it can be changed to an equivalent linear equation by cross multiplication:

$$\frac{x}{20 + x} \diagdown \diagup \frac{.22}{1}$$

$$x = .22(20 + x).$$

This is the linear equation we wrote in part (a). Thus the two equations that look different have in fact the same solution. □

There are many questions that can be asked about mixing substances together, but the basic ideas are the same. The important step is to organize the information of the problem before writing an equation.

Example 3   The laboratory has received 40 gallons of a 25%-salt solution. How much 10% solution must be added to the original 40 gallons of 25% solution to dilute the concentration to 20%?

Again we start by analyzing some special cases, for example, when the amount of 10% solution added is 5 gallons, 10 gallons, 50 gallons, or x gallons. In each case we must record the amount of salt in the new solution and the amount of new solution so that we can compute the percent of salt. Remember that 25% of the original 40-gallon solution, or .25(40) gallons, is salt.

Amount of 10% solution added (gallons)	Amount of salt added (gallons)	Amount of salt in new solution (gallons)	Amount of new solution (gallons)	Percent salt (%)
5	.10(5)	.10(5) + .25(40)	40 + 5	$\dfrac{.10(5) + .25(40)}{40 + 5} \doteq 23.33$
10	.10(10)	.10(10) + .25(40)	40 + 10	$\dfrac{.10(10) + .25(40)}{40 + 10} = 22$
50	.10(50)	.10(50) + .25(40)	40 + 50	$\dfrac{.10(50) + .25(40)}{40 + 50} \doteq 16.67$
x	.10x	.10x + .25(40)	40 + x	$\dfrac{.10x + .25(40)}{40 + x}$

a. One way to formulate an equation is to realize that the salt in the new solution must be 20% of the new solution. Using columns 3 and 4, this means that we want to solve

$$.10x + .25(40) = .20(40 + x)$$

$$.10x + 10 = 8 + .20x \qquad \text{(Distributive property)}$$

$$.10x + 2 = .20x \qquad \text{(Subtract 8 from both sides)}$$

$$2 = .10x \qquad \text{(Subtract .10x from both sides)}$$

$$\frac{2}{.10} = x \qquad \text{(Divide both sides by .10)}$$

$$20 = x.$$

Thus we must add 20 gallons of 10% solution.

b. A second way to view the problem is to see that the percent of salt (column 5) must equal .20. Thus

$$\frac{.10x + .25(40)}{40 + x} = .20$$

or, using cross multiplication,

$$.10x + .25(40) = .20(40 + x).$$

This is the equation we solved in part (a). ☐

Not every mixture problem involves salt and water, of course. Here is an example involving nuts that is similar to the problem we have just solved.

Example 4 The manager of Nuts Galore has 50 pounds of roasted peanuts that ordinarily sell for $1.70 a pound. She proposes to mix cashews with the peanuts and create a mixture that will sell for $2.00 a pound. If her cashews are worth $3.20 a pound, how many pounds should she add to the 50 pounds of peanuts?

It may be possible for you to write an equation without first making a chart. The important idea is that the amount of money obtained from selling all the mixture should be the same as the owner would have received if she had sold the peanuts and cashews separately. Looking at what happens if she adds 5 pounds of cashews or 10 pounds of cashews can help us write an equation to describe the problem.

Pounds of cashews added	Money received from cashews	Money received from two kinds of nuts	Weight of mixture	Price per pound of mixture ($)
5	3.20(5)	3.20(5) + 1.70(50)	5 + 50	$\dfrac{3.20(5) + 1.70(50)}{5 + 50} = 1.84$
10	3.20(10)	3.20(10) + 1.70(50)	10 + 50	$\dfrac{3.20(10) + 1.70(50)}{10 + 50} = 1.95$
x	3.20(x)	$3.20x + 1.70(50)$	$x + 50$	$\dfrac{3.20x + 1.70(50)}{x + 50}$

One thing we see from the chart is that she can put in more than 10 pounds of cashews. Again, there are two valid ways to view the problem.

a. The money received from the two kinds of nuts sold separately must equal the money received from the mixture at $2.00 a pound. Thus we need x pounds of cashews, where

$$3.20x + 1.70(50) = 2.00(x + 50)$$
$$3.20x + 85 = 2.00x + 100$$
$$3.20x = 2.00x + 15$$
$$1.20x = 15$$
$$x = 12.5.$$

The owner should add 12.5 pounds of cashews.

b. Since the problem states that she wants the price per pound of the mixture to be $2.00, we can use column 5 to get the equation:

$$\frac{3.20x + 1.70(50)}{x + 50} = 2.00$$

or $3.20x + 1.70(50) = 2.00(x + 50)$. This is the equation of part (a). □

Exercises 5.3

Solve the equations in Exercises 1–8.

1. $x + .182x = 1477.5$

2. $x - .125x = 1225$

3. $.35(x - 26) = x$ 4. $.6x + .35(25) = .55(25 + x)$

5. $x + .04x = 260$ 6. $x - .095x = -16{,}742.5$

7. $.07x + .05(17{,}500 - x) = 1125$ 8. $2.5x + 3.75(2500 - x) = 7875$

9. A department store must add 5% sales tax to the price charged consumers for its goods.

   (a) Complete the following table:

Retail price ($)	Sales tax ($)	Consumer cost ($)
45	$.05(45) = 2.25$	$45 + .05(45) = 47.25$
57		
x		

   (b) Use the last line in the chart above to answer the question, "What is the retail price of an item whose cost to a consumer, including 5% sales tax, is $86.10?"

10. A department store is having a 24%-off sale.

   (a) Complete the following table:

Original price ($)	Discount ($)	Sale price ($)
40	$.24(40) = 9.60$	$40 - .24(40) = 30.40$
62.50		
x		

   (b) Use the last line of the chart above to answer the question, "What is the original price of an item whose sale price is $101.68?"

Write an equation and solve the problems in Exercises 11–25. (If you have trouble writing an equation, set up a chart and compute several special cases first.)

11. 72 is 25% of what number?

12. During a sale, a sofa that was originally marked $379 sold for $303. What was the percent reduction?

13. If a diner leaves a $.50 tip, which is 20% of the bill, how much is the bill?

14. The price of a suit, including 4% sales tax, is $159.75. What is the price of the suit excluding tax?

15. A salesman earns 30% commission on his sales. If he expects to earn $20,000 a year, what is the total amount in sales that he must make?

16. Find the number that, when increased by 22%, equals 45.

17. Find the number that, when decreased by 32%, equals 43.

18. A department store is having a 24%-off sale.

   (a) Find the sale price if the original price is $50.50.

   (b) Find the original price if the sale price is $27.74.

19. State employees receive a 9.5% pay raise.

   (a) Jane's salary was $14,500 before she received the raise. What is her new salary?

   (b) Bill's salary after the raise is $15,603.75. What was his salary before the raise?

20. Tickets to a football game cost $1.50 for students and $2.00 for adults. If the total amount of money collected from 1904 sales was $3395, how many of each type of ticket were sold?

21. $20,000 is invested, part at 9% and the remainder at 6.5%. If the annual income (simple interest) is $1637.75, how much is invested at each rate?

22. How much candy worth $.85 per pound should be mixed with candy worth $1.65 per pound to produce 40 pounds of a mixture worth $1.40 per pound?

23. How much pure alcohol should be added to 15 ounces of a 40%-alcohol solution to obtain a solution that is 66% alcohol?

24. How much water should be added to 6 liters of a 15%-salt solution to obtain a solution that is 9% salt?

25. How much 35%-alcohol solution should be mixed with a 65%-alcohol solution to form 30 gallons of a 47%-alcohol solution?

5.4 / USING LINEAR EQUATIONS TO SOLVE PROBLEMS ABOUT GEOMETRIC FIGURES, COINS, AND TRAVEL

In this section we will look at several other problems that can be described by linear equations. If you do not see immediately how to write an equation for a problem, remember to organize the information in the form of a chart and record what happens in several special cases.

Example 1   A rectangle is three times longer than it is wide. If the rectangle has a perimeter of 100 feet, what are the lengths of its sides?

Remember that the perimeter is the length of the boundary of the rectangle. If we denote the width by x, the length will be $3x$ (Fig. 5.1).

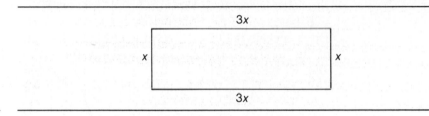

Figure 5.1

Thus the perimeter of the rectangle is $x + 3x + x + 3x$; and the perimeter must equal 100:

$$x + 3x + x + 3x = 100$$

$$8x = 100$$

$$x = 12.5.$$

Thus the width is 12.5 feet and the length is 3(12.5), or 37.5 feet.   □

If you did not see immediately how to write an equation for the above example, you could have made a chart first showing some specific cases.

Width	Length	Perimeter
5	3(5) = 15	5 + 15 + 5 + 15 = 40
10	3(10) = 30	10 + 30 + 10 + 30 = 80

Width	Length	Perimeter
15	$3(15) = 45$	$15 + 45 + 15 + 45 = 120$
x	$3x$	$x + 3x + x + 3x$

You should see from the last line that in order for the perimeter to be 100, we must have $x + 3x + x + 3x = 100$.

Example 2   Don has twice as many dimes as nickels and two fewer quarters than dimes. How many of each does he have if the value of his coins is $4.00?

The equation for this problem will be one which states that the value of the coins is $4.00. If the number of nickels is n, the number of dimes will be $2n$, and the number of quarters will be $2n - 2$. The value of Don's nickels is $(.05)n$; the value of his dimes is $(.10)(2n)$; and the value of his quarters is $(.25)(2n - 2)$. Thus the total value of his coins is $.05n + .10(2n) + .25(2n - 2)$. Since the total value equals $4.00, we have

$$.05n + .10(2n) + .25(2n - 2) = 4.00$$

$$.05n + .20n + .50n - .50 = 4.00$$

$$.75n - .50 = 4.00$$

$$.75n = 4.50$$

$$n = 6.$$

The number of nickels must be 6, the number of dimes must be $2(6) = 12$, and the number of quarters must be $12 - 2 = 10$.   □

The next example uses the relationship between distance, rate, and time. If we know how fast a vehicle is going and how long it travels, we can compute how far it goes. Sometimes we say

distance $=$ rate $\times$ time.

For example, a vehicle going an average of 60 mph for 3 hours goes a distance of $60(3)$, or 180 miles. The multiplication statement above can also be written as two division statements:

distance ÷ rate = time

distance ÷ time = rate.

For example, if a distance of 400 miles is traveled at an average speed of 40 mph, the time required must be $400 \div 40$, or 10 hours; if the distance is 400 miles and the time is 8 hours, the average rate must be $400 \div 8$, or 50 mph. Notice that there must be an agreement among the units of rate, time, and distance; if the time is given in hours and the distance in miles, the rate must be given in miles per hour.

Example 3   Louise lives 4 miles from school. Ordinarily, she rides her bicycle for 20 minutes to her friend's house and then walks with her friend another 20 minutes to school. If she rides three times as fast as she walks, how fast does she ride?

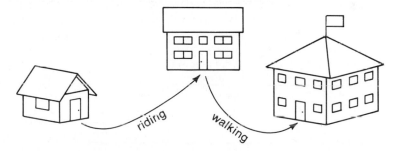

We can organize the information into a chart. The important headings are rate, time, and distance for both the walking and the riding parts of the trip. If x denotes the walking rate, $3x$ denotes the riding rate. Notice that we have expressed time in terms of hours. Distance is written in the chart as the product of rate and time.

	Rate (mph)	Time (hours)	Distance (miles)
Riding	$3x$	$\dfrac{20}{60} = \dfrac{1}{3}$	$3x\left(\dfrac{1}{3}\right) = x$
Walking	x	$\dfrac{20}{60} = \dfrac{1}{3}$	$x\left(\dfrac{1}{3}\right) = \dfrac{1}{3}x$

The information that provides an equation is that the total distance Louise travels is 4 miles. Thus from column 3

$$x + \frac{1}{3}x = 4$$

$$\frac{4}{3}x = 4 \qquad \text{(Combine terms)}$$

$$\frac{3}{4} \cdot \frac{4}{3}x = \frac{3}{4} \cdot 4 \qquad \left(\text{Multiply both sides by } \frac{3}{4}\right)$$

$$x = 3.$$

The walking rate x is 3 mph and the riding rate $3x$ is 9 mph.   □

Example 4   Each of the equal angles in an isosceles triangle is twice the third angle. Find the measure of the three angles of the triangle.

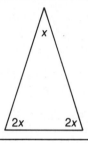

Figure 5.2

If x denotes the measure of the third angle, the two equal angles each have measure $2x$ (Fig. 5.2). Since the sum of the measures of the three angles in any triangle is 180°, we can write this equation:

$$x + 2x + 2x = 180$$

$$5x = 180$$

$$x = 36.$$

The smallest angle is 36°; the other two angles are each $2(36) = 72°$.

    □

Exercises   Solve the equations in Exercises 1–9.
5.4
   1. $4(x - 200) = 36x$ 2. $2x + 2(2x - 1) = 67$

3. $y + (3y + 10) + (y - 18) = 180$ 4. $.10y + .25(50 - y) = 9.95$

5. $.05x + .10(2x - 1) = 5.15$

6. $.05x + .10(x - 2) + .25(2x + 1) = 11.10$

7. $.05(3z + 1) + .10z + .25(59 - 4z) = 5.05$

8. $\dfrac{x}{15} + \dfrac{110 - x}{20} = 5$

9. $\dfrac{x}{12} + \dfrac{84 - x}{18} = \dfrac{11}{2}$

Set up an equation and solve the problems in Exercises 10–25.

10. A money box contains $10.55 in dimes and quarters. If the number of quarters is one less than twice the number of dimes, how many of each does the box contain?

11. Jane has 65 coins in nickels, dimes, and quarters. The number of nickels is two more than three times the number of dimes. How many of each does she have if the value of the coins is $6.85?

12. An 80-foot rope is cut into two pieces. If one piece is 4 feet shorter than the other, find the length of each piece.

13 A 75-foot tree trunk is cut into three pieces. The larger piece is three times the smaller piece, and the other piece is 5 feet longer than the smaller piece. Find the length of each piece.

14. Two cars start from the same point and travel in the same direction at average speeds of 45 mph and 55 mph. After how many hours are they 35 miles apart?

15. Sue leaves Dayton traveling at 55 mph. Two hours later a highway patrolman leaves Dayton along the same road traveling at 75 mph. How far from Dayton will the patrolman overtake Sue?

16. A man starts his journey in a boat and then completes his 297-mile trip by train. The entire trip takes 6 hours. If the boat averages 18 mph and the train averages 60 mph, how far does he go by boat?

17. Bill and Sue are 63 miles apart and are bicycling toward each other on the same road. Bill averages 10 mph and Sue averages 8 mph. In how many hours will they meet?

18. Tom's house is 22 miles from the place where he works. Each

morning he walks at the rate of 6 mph to a bus stop and then completes the trip on a bus that averages 30 mph. If the entire trip takes 1 hour, how much time is required for Tom to reach the bus stop?

19. A boat goes the same distance in 20 hours that a plane does in 3 hours. If the plane goes 140 mph faster than the boat, find the rate of each.

20. A rectangle is 4 feet longer than it is wide. Find the dimensions of the rectangle if the perimeter is 98 feet.

21. The width of a rectangle is 3 feet less than the length. Find the dimensions of the rectangle if the perimeter is 123 feet.

22. The length of a rectangle is 2 feet more than three times the width. Find the dimensions of the rectangle if the perimeter is 256 feet.

23. The largest angle of a triangle is 6° more than four times the smallest angle. The other angle is 24° larger than the smallest angle. Find the measure of each angle. (Remember that the sum of the measures of the angles of a triangle equals 180°.)

24. One of the acute angles of a right triangle is 3° less than twice the other acute angle. Find the measure of each angle.

25. Each of the equal angles of an isosceles triangle is 6° more than three times the third angle. Find the measure of each angle.

5.5 / RATIO AND PROPORTION

One way to compare the relative sizes of two sets of objects is to use the language of ratio. If there are 10 girls and 12 boys in a class, we say that the ratio of girls to boys is 10 to 12. In symbols we would write $10:12$. The ratio $5:6$ expresses the same information about the comparative numbers of girls and boys as does the ratio $10:12$, because $5:6$ means that 5 out of 11 students are girls; so of 22 students, 10 would be girls and 12 boys. On the other hand, $10:12$ indicates that 10 out of 22 students are girls, so of 11 students, 5 would be girls and 6 boys.

Even though ratios are not actually numbers (indeed they are pairs of whole numbers), we want to say that certain ratios are equal; for example,

$$5 : 6 = 10 : 12 = 15 : 18.$$

The criterion for equality of ratios is analogous to the definition of equal fractions. We can check that $10 : 12 = 15 : 18$ by comparing $10 \cdot 18$ and $12 \cdot 15$. (We might draw arrows this way to show the products, $10 : 12 = 15 : 18$.) Since $10 \cdot 18 = 180$ and $12 \cdot 15 = 180$, we conclude that $10 : 12 = 15 : 18$.

EQUAL RATIOS

For $b \neq 0$ and $d \neq 0$, we say

$$a : b = c : d \text{ if } ad = bc.$$

A statement that two ratios are equal is called a **proportion**. Some problems can be solved by setting up a proportion.

Example 1   The ratio of girls to boys in a school is $7 : 9$. If there are 2135 girls, how many boys are there?

If we let x denote the number of boys, then the statement that the ratio of girls to boys equals $7 : 9$ becomes the proportion

girls : boys   girls : boys

$$2135 : x = 7 : 9.$$

Using the definition of equality of ratios, we find that

$$2135 \cdot 9 = 7 \cdot x.$$

Thus we have a linear equation:

$$19{,}215 = 7x$$

$$2745 = x. \qquad \text{(Divide both sides by 7)}$$

There must be 2745 boys in the school.   □

You should notice that the proportion $2135 : x = 7 : 9$, used in the example above, and the equation $\frac{2135}{x} = \frac{7}{9}$ both give rise to the same linear equation, $2135 \cdot 9 = 7 \cdot x$.

Example 2   The fishery manager wishes to estimate the number of fish in a stocked pond. He tags 200 fish and returns them to the pond. Then he takes a sample of 50 fish from the pond and finds that 8 are tagged. How many fish can he predict are in the pond?

We assume that the ratio of tagged fish in the pond to the total number of fish in the pond is the same as the ratio of tagged fish in the sample to the total number of fish in the sample. If x denotes the number of fish in the pond, we get this proportion:

tagged : total   tagged : total,

$$200 : x = 8 : 50.$$

From the proportion, we have the linear equation

$$200 \cdot 50 = 8 \cdot x$$
$$10{,}000 = 8x$$
$$1250 = x.$$

The manager can predict that the pond contains 1250 fish.   □

In geometry, we may use ratios to compare the lengths of segments or areas of regions. Two triangles that have corresponding angles of equal measure need not be the same size, but they do have the same shape. Such triangles are said to be **similar**.

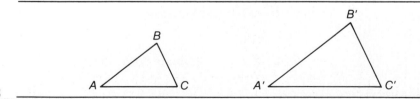

Figure 5.3

In Fig. 5.3 the angles at A and A' have the same measure, the angles at B and B' have the same measure, and the angles at C and C' have the same measure. When two figures are similar, one looks like a photographic enlargement of the other.

It is a theorem of geometry that for similar triangles the ratio of the lengths of two sides in one triangle equals the ratio of the lengths of the corresponding sides in the other triangle; for example, in Fig. 5.3, length $\overline{AB}$: length $\overline{BC}$ = length $\overline{A'B'}$: length $\overline{B'C'}$. If we know the lengths of the sides of one triangle and also one side of a triangle similar to it, then we can find the other two sides by setting up a proportion.

Example 3   In Fig. 5.4 two triangles are similar with the side lengths as shown. Find the length of the side labeled s.

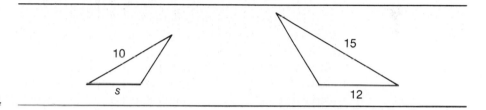

Figure 5.4

We need to observe that the side labeled s corresponds to the side labeled 12, and the side labeled 10 corresponds to the side labeled 15. Thus

$$s : 10 = 12 : 15$$

$$15s = 120$$

$$s = 8. \qquad\qquad \square$$

Example 4   A tree casts a 20-foot shadow at the same time that a 3-foot post casts a 5-foot shadow. How tall is the tree?

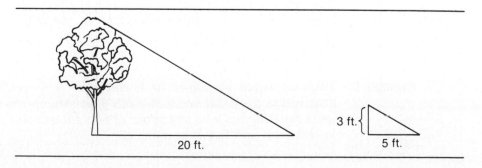

Figure 5.5

We assume that because the rays of the sun are parallel, two similar right triangles are formed by the tree with its shadow and the post with its shadow (Fig. 5.5). If the height of the tree is denoted by h, then

$$h : 20 = 3 : 5.$$

Thus

$$5h = 60$$
$$h = 12.$$

Without measuring the tree directly, we can compute its height as 12 feet.   □

You have undoubtedly heard the term *gear ratio* applied to the gears on a bicycle or a car. Two gears are used so that a shaft can turn at a different speed than a motor (or pedals). A driving gear is attached to the motor, a driven gear (usually smaller) is attached to the shaft of a wheel, and the two gears are either connected directly to each other or connected by a belt or a chain. The speeds of the two gears are related by a proportion that is given in terms of the number of teeth on the two gears. If one gear has T teeth and rotates at a speed S and the second gear has t teeth and rotates at a speed s, then

$$T : t = s : S.$$

Of course, the two speeds must be expressed in the same units. Notice that the ratio of the speeds is given in an order that is inverted compared to the ratio of the number of teeth. Sometimes we use the proportion $T : t = s : S$ in the form

$$TS = ts.$$

This equation states that the product of the number of teeth and the speed of one gear equals the product of the number of teeth and the speed for the other gear.

Example 5   Bill's ten-speed bicycle in its lowest gear setting has a 39-tooth gear attached to the pedal and a 28-tooth gear attached to the back wheel. If he is pedaling his bike at a speed of 60 rpm (revolutions per minute) in lowest gear, how fast is he moving?

The gear on the pedal has 39 teeth and a speed of 60 rpm. The back gear has 28 teeth; its speed is denoted by s. Then

$$T:t = s:S$$
$$39:28 = s:60$$

or

$$TS = ts$$
$$39(60) = 28s$$
$$2340 = 28s$$
$$83.57 \doteq s.$$

Thus the back gear, and hence the back tire, has a speed of approximately 83.6 rpm. It is interesting to change that speed to miles per hour. Assume that the wheel has a 26-inch diameter. Then the circumference of the tire ($c = \pi d$) is 26π, or approximately 81.7 inches. Each revolution of the gear results in the bicycle moving 81.7 inches. Thus the bicycle moves $(81.7)(83.6) = 6830.12$ inches each minute, or $(6830.12)(60) = 409{,}807.2$ inches each hour. The number of inches in a mile is 63,360 (since the number of feet in a mile is 5280). Thus the bicycle is moving $\dfrac{409{,}807.2}{63{,}360} \doteq 6.47$ miles each hour. The speed of the bicycle is approximately $6\frac{1}{2}$ mph. □

Example 6 If one gallon is 3.785 liters, how many gallons are 35 liters?

We can use a proportion to convert gallons to liters and liters to gallons. If x denotes the number of gallons in 35 liters, then the ratio of 1 to 3.785 must equal the ratio of x to 35.

gallons : liters   gallons : liters
$$1:3.785 = x:35$$
$$35 = 3.785x$$
$$\frac{35}{3.785} = x$$
$$9.247 \doteq x.$$

Thus 35 liters is the same as 9.247 gallons. □

Exercises 5.5

Solve for x in Exercises 1–8.

1. $2:3 = x:15$

2. $\dfrac{2}{3} = \dfrac{x}{15}$

3. $7:8 = 70:x$

4. $\dfrac{7}{8} = \dfrac{70}{x}$

5. $x:5 = 14:3$

6. $\dfrac{30}{x} = \dfrac{9}{-8}$

7. $6:7 = x:x-3$

8. $\dfrac{8}{9} = \dfrac{x}{2000-x}$

9. In order to estimate the number of deer in a park, 100 deer are tagged and then released. Later, in a sample of 75 deer, 30 are found to be tagged. Estimate the number of deer in the park.

10. To estimate the number of fish in a pond, 150 fish are tagged and then returned to the pond. Later, a sample of 100 fish shows 6 tagged. Estimate the number of fish in the pond.

11. The ratio of boys to girls in a school is $8:9$.

    (a) If there are 2480 boys, how many girls are there?

    (b) If there are 3240 girls, how many boys are there?

12. An architect makes a scale drawing of a soccer field that measures 70 yards by 120 yards. If the drawing is 3.5 inches wide, how long is it?

13. An enlargement of a 5-inch by 7-inch photograph is 10.5 inches long. How wide is it?

14. A triangle has sides 3, 4, and 5 feet. If the longest side of a similar triangle is 20 feet, find the lengths of the other two sides.

15. A triangle has sides 8, 15, and 17 feet. If the shortest side of a similar triangle is 12 feet, find the lengths of the other two sides.

16. If a 5-foot pole casts an 8-foot shadow, what length shadow will a 150-foot tree cast?

17. If a tree casts a 111-foot shadow at the same time that a 7-foot man casts a 12-foot shadow, how tall is the tree?

18. If a building casts a 200-foot shadow at the same time that a yardstick casts a 4-foot 6-inch shadow, how tall is the building?

19. If a motor that turns at 330 rpm is connected to a gear with 40 teeth, and this gear is connected to a gear with 55 teeth, how fast is the second gear turning?

20. If a motor that turns at 270 rpm is connected to a gear, and this gear is connected to a second gear with 21 teeth that turns at 900 rpm, how many teeth are on the first gear?

21. If a motor that turns at 360 rpm is connected to a gear with 75 teeth, and this gear is connected to a second gear that is turning at 500 rpm, how many teeth does the second gear have?

22. If a motor is connected to a gear with 45 teeth, and this gear is connected to a second gear with 80 teeth that is turning at 225 rpm, how fast is the motor turning?

23. Two people share $44,000 in the ratio $4:7$. How much does each receive?

24. Two people share $38,808 in the ratio $3:8$. How much does each receive?

25. One unit of length on a map represents 150,000 units in the real world.

    (a) How far apart are two cities if they are 2.64 inches apart on the map?

    (b) If two cities are 8 miles apart, how far apart are they on the map?

26. One inch on a map represents 10 miles in the real world.

    (a) How far apart are two cities if they are 3.67 inches apart on the map?

    (b) If two cities are 125 miles apart, how far apart are they on the map?

27. One kilometer is 0.6214 miles.

    (a) How many miles are there in 4.5 kilometers?

    (b) How many kilometers are there in 2 miles?

28. One liter is 0.2642 gallons.

    (a) How many gallons are there in 5 liters?

    (b) How many liters are there in 5 gallons?

29. If 60 mph is 88 feet per second, find how many feet per second equal

    (a) 50 mph   (b) x mph

30. If 60 mph is 88 feet per second, find how miles per hour equal

    (a) 66 feet per second   (b) x feet per second

5.6 / EQUATIONS WITH MORE THAN ONE VARIABLE

We have investigated problem situations that can be described by equations containing more than one variable. Here are three examples.

■ A car travels at 65 mph. If t denotes the number of hours for a trip and d the number of miles traveled, then

$$d = 65t.$$

■ Of two numbers, the second is three more than twice the first. If the first number is denoted by f and the second by s, then

$$s = 2f + 3.$$

■ A purse contains 30 coins in nickels and dimes. If n is the number of nickels and V the value of the money, then

$$V = .05n + .10(30 - n).$$

Each of these equations is a linear equation with two variables. A solution to any one of them is a pair of numbers. The values $t = 1$, $d = 65$, for example, give a solution to the equation $d = 65t$. There are many solutions to this equation: $t = 2, d = 130$ and $t = 3, d = 195$ are two more. You should find others. Often we write a pair of numbers that is a solution to an equation in two variables as an ordered pair: $(1, 65)$ or

(2, 130) or (3, 195). The order in which the numbers appear is important in an ordered pair; here we have used the order (t, d).

From the first equation above, if we know the number of hours, we can compute the number of miles. From the second, if we know the first number, we can compute the second. From the third, if we know the number of nickels, we can compute the value of the money. Sometimes we want to turn these around: we may want to write the number of hours in terms of the number of miles, the first number in terms of the second, and the number of nickels in terms of the value of the money.

Example 1 Solve the equation $d = 65t$ for t.

We must say what t equals in terms of d. We use the same procedures for solving this linear equation that we used in Section 5.2 for linear equations in one variable:

$$d = 65t$$

$$\frac{d}{65} = t. \qquad \text{(Divide both sides by 65)}$$

The equation $t = \frac{d}{65}$ gives us the number of hours in terms of the number of miles traveled. Now if we know the number of miles, we can find the number of hours. We divide the number of miles by 65. For example, if $d = 130$, then $t = \dfrac{130}{65} = 2$ hours. $\qquad\square$

Example 2 Solve the equation $s = 2f + 3$ for f.

Since we want to say what f equals, we collect the terms containing f on one side and all other terms on the other side:

$$s = 2f + 3$$

$$s - 3 = 2f \qquad \text{(Subtract 3 from both sides)}$$

$$\frac{s - 3}{2} = f. \qquad \text{(Divide both sides by 2)}$$

The equation states that the first number is found by subtracting 3 from the second number and dividing the result by 2. For instance, if the second number is 17, the first number is $\dfrac{17-3}{2} = 7$.   □

Example 3   Solve the equation $V = .05n + .10(30 - n)$ for n.

We want to show how n can be computed in terms of V:

$$V = .05n + .10(30 - n)$$

$$V = .05n + 3 - .10n \qquad \text{(Distributive property)}$$

$$V = -.05n + 3 \qquad \text{(Combine like terms)}$$

$$V - 3 = -.05 \qquad \text{(Subtract 3 from both sides)}$$

$$\dfrac{V - 3}{-.05} = n. \qquad \text{(Divide both sides by } -.05)$$

□

Now, in the example above, if we know the value of the money, we can compute the number of nickels among the 30 coins. For example, if $V = \$2.50$, then $n = \frac{V-3}{-.05} = \frac{2.50-3}{-.05} = 10$. The number of nickels is 10 and the number of dimes must be 20. What if we try $V = \$3.50$? Then $n = \frac{V-3}{-.05} = \frac{3.50-3}{-.05} = -10$. How can the number of nickels be a negative number? Clearly it cannot. Indeed, the value of 30 coins consisting of nickels and dimes can never be \$3.50!

Example 4   Solve $A = \dfrac{1}{2}bh$ for h.

This equation tells how to compute the area of a triangle in terms of the length of a base and the height. If we solve for h, we will be able to find the height in terms of the area and base. The equation has three variables; in the computation they behave just like numbers. A reasonable first step is to multiply both sides of the equation by 2 to clear the fractional coefficient:

$$A = \dfrac{1}{2}bh$$

$$2A = bh \qquad \text{(Multiply both sides by 2)}$$

$$\frac{2A}{b} = \frac{bh}{b} \qquad \text{(Divide both sides by } b\text{)}$$

$$\frac{2A}{b} = h.$$ □

Example 5   Solve $\frac{1}{3}x + \frac{1}{4}y = 1$ for y.

When fractions occur in equations, we can clear the fractions first by multiplying both sides by the LCM of the denominators:

$$\frac{1}{3}x + \frac{1}{4}y = 1$$

$$4x + 3y = 12 \qquad \text{(Multiply both sides by 12)}$$

$$3y = 12 - 4x \qquad \text{(Subtract } 4x \text{ from both sides)}$$

$$y = \frac{12 - 4x}{3}. \qquad \text{(Divide both sides by 3)}$$ □

Exercises 5.6   In Exercises 1–4, determine if the specified values of the variables give a solution of the equation.

1. $y = 2x + 3$

(a) $x = 0, y = 3$ (b) $x = -1, y = 1$

(c) $x = -2, y = -7$ (d) $x = -\frac{1}{2}, y = 2$

2. $V = .05x + .10(2x + 1) + .25(x + 3)$

(a) $x = 0, V = 85$ (b) $x = 12, V = 6.85$

3. $\frac{x}{2} - \frac{y}{3} = 1$

(a) $x = 6, y = 6$ (b) $x = 0, y = 3$

(c) $x = 2, y = 0$ (d) $x = 0, y = -3$

4. $z = 2x - 3y + 1$

    (a) $x = 3, y = 2, z = 1$ (b) $x = -1, y = -2, z = -7$

5. Let $y = \dfrac{1}{2}x$. Find

    (a) y if x is 4, 0, -3 (b) x if y is 2, 0, -1

6. Let $s = 3f - 2$. Find

    (a) s if f is $0, -\dfrac{1}{2}, 2$ (b) f if s is 1, 0, -5

7. Let $2x + 3y = 6$. Find

    (a) y if x is 3, -1, 0   (b) x if y is 2, 3, 0

8. Let $\dfrac{x}{2} - \dfrac{y}{3} = 1$. Find

    (a) y if x is 2, 0, 5   (b) x if y is 0, 3, 4

Find three different solutions to each of the equations in Exercises 9–11. (In each case a solution is a pair of numbers.)

9. $lw = 100$ 10. $3x + 4y = 12$

11. $\dfrac{1}{2}x + \dfrac{1}{3}y = 2$

Solve the equations in Exercises 12–32 for the variables specified.

12. $x = 15y$, for y 13. $1.5x = y$, for x

14. $y = \dfrac{2}{3}x$, for x 15. $x = \dfrac{8}{7}y$, for y

16. $y - 2x + 3 = 0$, for x 17. $2x + 3y = 1$, for y

18. $.05x + .25y = 2.75$, for y 19. $y = 3(x + 2)$, for x

20. $y = 2x + 3$, for x 21. $3x + 4y = 2$, for x

22. $10x + 15(2 - y) = 45$, for y 23. $2(y - 1) + 3(2x - 1) = 0$, for y

24. $\dfrac{x}{2} + \dfrac{y}{3} = 1$, for y 25. $\dfrac{1}{3}x - 2y = 1$, for y

26. $x + \dfrac{1}{2}y = 3$, for y 27. $y - .17x = 2$, for x

28. $\dfrac{x}{6} + \dfrac{10 - y}{15} = 2$, for y

29. $d = rt$, for t

30. $p = 2l + 2w$, for l

31. $s = \dfrac{2}{1 - r}$, for r

32. $y = mx + b$, for x

33. The formula for converting miles m to kilometers k is $k = (0.6214)m$. Find the formula for converting kilometers to miles.

34. The formula for converting gallons G to liters L is $L = (0.2642)G$. Find the formula for converting liters to gallons.

35. The formula for the circumference of a circle in terms of the radius is $C = 2\pi r$. Find the formula for the radius in terms of the circumference.

36. The formula for converting degrees Fahrenheit to degrees Celsius is $C = \dfrac{5}{9}(F - 32)$. Find the formula for converting degrees Celsius to degrees Fahrenheit.

5.7 / LINEAR INEQUALITIES IN ONE VARIABLE

In this chapter we have seen many problems that can be described by equations. An equation can be written when the information in the problem implies that two expressions are equal. Rather than being equal, sometimes the information in the problem implies that one expression is smaller than the other. Consider this problem.

> Bank Two charges a service charge on its checking accounts of $.75 a month plus $.10 for each check processed. If David wants to keep his monthly charges less than $2.00 a month, how many checks can he write?

If, in the above problem, n represents the number of checks that David writes in a month, then the expression $.75 + .10n$ represents the monthly bank charges. The fact that David wants to keep his monthly charges lower than $2.00 means that n must be a number such that

$.75 + .10n < 2.00$.

This statement is called an **inequality** (not-equal). Any number that replaces n to make $.75 + .10n < 2.00$ is called a **solution** to the inequality. There are several solutions to this inequality:

0 is a solution because $.75 + .10(0) \; < 2.00$;

1 is a solution because $.75 + .10(1) \; < 2.00$;

10 is a solution because $.75 + .10(10) < 2.00$.

However, 15 is not a solution because $.75 + .10(15) = 2.25$. In fact, the largest solution is 12; David will keep his bank charges lower than \$2.00 if the number of checks he writes is 0, 1, 2, 3, 4, 5, 6, 7, 8, 9, 10, 11, or 12.

Equivalent Inequalities

Although the example above could be solved by experimentation, there are techniques for solving linear inequalities that are similar to the techniques we already developed for solving linear equations. The general strategy is to replace the inequality with a simpler inequality that has the same solutions (that is, a simpler **equivalent** inequality). To simplify an inequality, we can add the same expression to both sides of the inequality or subtract the same expression from both sides; we can multiply both sides by the same *positive* number or divide by the same *positive* number. However, we must pay particular attention to what happens when we multiply or divide both sides of an inequality by a negative number. The following chart summarizes the results when these operations are performed on one inequality, $2 < 3$.

Original inequality:	$2 < 3$
Add 2 to both sides of the original inequality:	$4 < 5$
Subtract 3 from both sides of the original inequality:	$-1 < 0$
Multiply both sides of the original inequality by 5:	$10 < 15$
Divide both sides of the original inequality by 2:	$1 < \dfrac{3}{2}$
Multiply both sides of the original inequality by -1:	$-2 > -3$
Divide both sides of the original inequality by -2:	$-1 > -\dfrac{3}{2}$

Observe that adding or subtracting the same number leaves the left side of the inequality less than the right; so does multiplying or dividing by a positive number. However, when we multiply or divide by a negative number, the left side becomes *greater* than the right side. Multiplying or dividing an inequality by a negative number has the effect of reversing the direction of the inequality sign. Consider another example of this fact.

Original inequality: $\qquad\qquad\qquad\qquad\qquad\qquad\qquad\qquad$ $5 > -1$

Multiply both sides of the original inequality by -2: $\qquad$ $-10 < 2$

Divide both sides of the original inequality by -5: $\qquad$ $-1 < \dfrac{1}{5}$

Thus we get an equivalent inequality if we

- add the same algebraic expression to, or subtract the same algebraic expression from, both sides of an inequality;
- multiply or divide both sides of an inequality by the same positive number;
- multiply or divide both sides of an inequality by the same *negative* number *and reverse the direction of the inequality sign.*

Example 1   Solve the inequality $6x - 9 > 3x + 12$.

We want to replace the given inequality by a sequence of simpler equivalent inequalities until we get to one so simple that we can read its solutions:

$$6x - 9 > 3x + 12$$
$$6x > 3x + 21 \qquad \text{(Add 9 to both sides)}$$
$$3x > 21 \qquad \text{(Subtract } 3x \text{ from both sides)}$$
$$x > 7. \qquad \text{(Divide both sides by 3)}$$

Thus any number greater than 7 will be a solution to each of these inequalities, including the original. (For example, $\frac{15}{2}$ and 9 are both solutions; 7 and $\frac{5}{3}$ are not solutions.) $\qquad\qquad$ □

Example 2   Solve the inequality $7 - 2x < 9$.

Our strategy is to get the terms containing the variable on one side of the inequality and the terms without the variable on the other side:

$$7 - 2x < 9$$

$-2x < 2$ (Subtract 7 from both sides)

$x > -1.$ (Divide both sides by -2 *and reverse the inequality sign*)

Thus any number greater than -1 is a solution to this inequality.   □

Two New Symbols

Consider another problem that involves an inequality:

> Sue has \$175 to spend on four pairs of matching draperies. What prices of draperies can she afford?

If x represents the price of a pair of draperies, then Sue can afford the draperies if $4x < 175$ or if $4x = 175$. Rather than write these statements separately, we can combine them and write $4x \leq 175$. This is read, "$4x$ is less than or equal to 175." Sue can afford the draperies provided that $4x \leq 175$, or $x \leq \$43.75$.

    We also use the symbol $\geq$. This symbol means **greater than or equal to**.

Example 3   Solve the inequality $2x \leq 3x + 5$.

The solution to the inequality consists of the solutions to $2x < 3x + 5$ and the solutions to $2x = 3x + 5$.

$2x < 3x + 5$		$2x = 3x + 5$	
$-x < 5$	(Subtract $3x$ from both sides)	$-x = 5$	(Subtract $3x$ from both sides)
$x > -5$	(Multiply both sides by -1 *and reverse the inequality sign*)	$x = -5$	(Multiply both sides by -1)

Thus the solutions to $2x \leq 3x + 5$ consist of -5 and all numbers greater than -5. A more efficient way to display the solution is to combine the solving of the inequality and the equation this way:

$$2x \leq 3x + 5$$

$$-x \leq 5 \qquad \text{(Subtract } 3x \text{ from both sides)}$$

$$x \geq -5. \qquad \text{(Multiply both sides by } -1 \text{ \textit{and reverse the inequality sign}})$$ □

Example 4 Solve the inequality $x - 5 \leq 0$.

Here we are asking when the expression $x - 5$ is either 0 or negative:

$$x - 5 \leq 0$$
$$x \leq 5. \qquad \text{(Add 5 to both sides)}$$

The answer is that $x - 5 \leq 0$ for values of x less than 5 or equal to 5. □

Example 5 Solve the inequality $2x + 1 > 0$.

We want values of x for which $2x + 1$ is positive:

$$2x + 1 > 0$$

$$2x > -1 \qquad \text{(Subtract 1 from both sides)}$$

$$x > -\frac{1}{2}. \qquad \text{(Divide both sides by 2)}$$

Thus $2x + 1 > 0$ if $x > -\frac{1}{2}$. □

We can extend Example 5 by asking, "When is $2x + 1 = 0$? When is $2x + 1 < 0$?"

$$2x + 1 = 0 \qquad\qquad\qquad 2x + 1 < 0$$

$$2x = -1 \qquad\qquad\qquad\quad 2x < -1$$

$$x = -\frac{1}{2} \qquad\qquad\qquad\quad x < -\frac{1}{2}$$

For any value of x exactly one of the following is true: either $2x + 1 > 0$ or $2x + 1 = 0$ or $2x + 1 < 0$. From the computation we can tell the entire story:

$$2x + 1 > 0 \quad \text{if} \quad x > -\frac{1}{2},$$

$$2x + 1 = 0 \quad \text{if} \quad x = -\frac{1}{2},$$

$$2x + 1 < 0 \quad \text{if} \quad x < -\frac{1}{2}.$$

Using Graphs to Solve Inequalities

Earlier in this section we looked at two problems that involved inequalities, one about bank charges and one about drapery costs. When a problem situation can be represented graphically, the graph can be used to solve the inequality. For example, in Section 4.5 we drew the same graph as Fig. 5.6 to show the percent of alcohol in a mixture when varying amounts of alcohol are added to 20 gallons of water. That graph can be used to answer a question involving inequalities.

Example 6   How much alcohol must be added to 20 gallons of water to give a solution that is 60% alcohol or more.

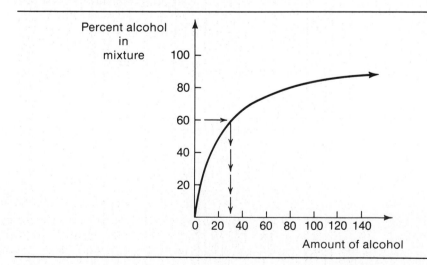

Figure 5.6

The percent of alcohol is 60% when 30 gallons of alcohol are added. The percent of alcohol is more than 60% when more than 30 gallons of alcohol are added. Thus to get a solution that is 60% alcohol or more, you must add 30 gallons of alcohol or more (that is, *at least* 30 gallons). □

Exercises 5.7

1. Find the value(s) of x for which

   (a) $x - 2 = 0$ (b) $x - 2 > 0$

   (c) $x - 2 < 0$

2. Find the value(s) of x for which

   (a) $2x + 5 = 0$ (b) $2x + 5 > 0$

   (c) $2x + 5 < 0$

Solve the inequalities in Exercises 3–19.

3. $3x + 1 < 5$

4. $4x + 3 \geq 2x - 1$

5. $5x - 6 < 0$

6. $5x - 3 \leq 2x + 12$

7. $3 - x \geq 2x + 9$

8. $3(2x - 1) - 2(x + 1) \leq x - 5$

9. $5(x + 3) > 45$

10. $\frac{1}{2}x + \frac{1}{3} < \frac{1}{3}x + \frac{1}{2}$

11. $\frac{x + 1}{3} - \frac{x}{2} \leq x + 1$

12. $x(x + 2) + 1 \geq x(x - 2) - 1$

13. $x - .135x < 1081.25$

14. $.09x + .07(18,500 - x) \geq 1459$

15. $\frac{x}{12} + \frac{84 - x}{18} < \frac{11}{2}$

16. $5x - (7x + 12) < x + 12$

17. $\frac{x}{5} - 1 \geq \frac{x}{2} + \frac{3}{5}$

18. $4(2x - 1) \leq 12 - 2(x + 4)$

19. $.05x + .10(2x + 1) + .25(59 - x) \leq 8.85$

20. The graph below shows how the amount of money in a savings account depends on the number of years the amount is left on deposit. Use the graph to find the number of years required for the amount in the account to reach $4500 or more.

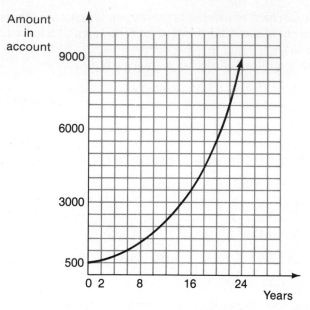

21. The graph below shows how 25% of a number depends on the number. Use the graph to find the numbers for which 25% of the number is

   (a) less than −20   (b) greater than or equal to 15.

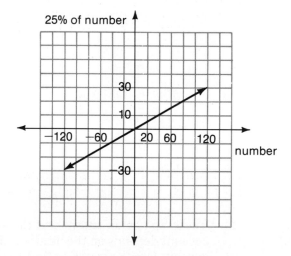

Write an inequality and solve each problem in Exercises 22–25.

22. Jim has $90.30 to spend on trophies for the 15 members of his little-league team. What prices of trophies can he afford?

23. Tickets to a football game cost $1.50 for students and $2.75 for adults. The number of student tickets sold was double that of the adult tickets. How many student tickets were sold if the total amount of money collected was more than $7350?

24. The cost of publication of a certain monthly magazine is $.45 per copy plus a monthly overhead of $1000. If the magazine sells for $1.25 a copy, how many copies must be sold for the company to realize a profit?

25. A money box contains 60 coins in dimes and quarters. Find the possible number of dimes in the money box if the value of the coins is $9.00 or more.

5.8 / CHAPTER 5 PROBLEM COLLECTION

In Exercises 1–4, find the value of the algebraic expression for the values of the variables specified.

1. $P\left(1 + \dfrac{r}{12}\right)^{12n}$; $P = 1200, r = .065, n = 3$

2. $\dfrac{b}{1 - r}$; $b = 1, r = \dfrac{1}{3}$

3. $2x - 3y$; $y = 2, x = -3$

4. $\dfrac{x}{2} + \dfrac{y}{3}$; $y = 4, x = -3$

State whether the equations in Exercises 5–8 are linear in one variable, linear in more than one variable, or not linear.

5. $\dfrac{2}{3}x + 6 = \dfrac{1}{2}x - 3$

6. $\sqrt{x + y} = 2$

7. $x^2 + y^2 = 4$

8. $\dfrac{x}{3} + \dfrac{y}{4} = 1$

Determine if the specified values of the variables give solutions of the equation or inequality in Exercises 9–11.

9. $x^3 = x; x = -1$ 10. $2x - 3y = 3; x = 3, y = -2$

11. $3x - 1 \le -4; x = -1$

12. Let $\frac{1}{2}x + 3y = -1$. Find

(a) y if x is 0, 2 (b) x if y is 0, 1

13. A 32%-alcohol solution is mixed with a 58%-alcohol solution to form 30 gallons of a new solution. Let x be the amount of 32% solution used.

(a) Write an algebraic expression that gives the percent of alcohol in the new solution.

(b) Write an equation that expresses the fact that the percent of alcohol in the new solution is 43%.

(c) Write an inequality that expresses the fact that the percent of alcohol in the new solution is less than 43%.

14. All workers must pay $3\frac{1}{2}$% state income tax. Let s represent a worker's salary.

(a) Write an algebraic expression that gives the amount of money a worker receives after state income tax is paid.

(b) Write an equation that expresses the fact that a worker's salary after state income tax is $21,616.

15. Write an equation that expresses the fact that the perimeter of a rectangle is 45 feet.

16. Write an equation that expresses the fact that the square root of one more than twice a number is equal to one more than the square root of the number.

17. Jack invests $19,000, part at 7.5% simple interest and the remainder at 5.6% simple interest.

(a) Complete the table on p. 301.

(b) Use the last line of the table in (a) to answer the question, "How much did Jack invest at 7.5% if he received $1219.80 interest?"

Amount at 7.5% ($)	Amount at 5.6% ($)	Total interest ($)
4000	$19{,}000 - 4000 = 15{,}000$	$.075(4000) + .056(19{,}000 - 4000) = 1140$
6000		
x		

18. Amy has 50 coins in dimes, quarters, and half-dollars. The number of quarters is one less than twice the number of dimes. Find the value of the coins if the number of dimes is

   (a) 6   (b) x

Solve the equations and inequalities in Exercises 19–43.

19. $.28x = 11.97$

20. $1.45x = 82.36$

21. $W + .35W = 17.01$

22. $R - .25R = 84.45$

23. $\dfrac{x}{3} + \dfrac{x}{4} = 14$

24. $\dfrac{1}{3}x - \dfrac{1}{2}x = \dfrac{2}{3}$

25. $7(3 - x) = 49$

26. $\dfrac{2z + 1}{3} = \dfrac{3z - 1}{7}$

27. $\dfrac{1}{2}y + \dfrac{1}{4}y + \dfrac{1}{8}y + \dfrac{1}{16}y = 15$

28. $2(3u + 1) - 3(u - 1) = u - 1$

29. $7(3x + 5) - 3(7x + 10) = 2$

30. $2t + 3(t - 3) = 5(t - 2) + 1$

31. $x(2x + 1) - 2 = 2x(x - 1) + 2$

32. $.075x + .052(18{,}000 - x) = 1217.75$

33. $\dfrac{x}{3} + \dfrac{2 - x}{2} = \dfrac{17}{3}$

34. $\dfrac{3}{8} = \dfrac{y}{14{,}000 - y}$

35. $3x - 2y = 4$, for y

36. $y - 3 = 2(x - 4)$, for x

37. $\dfrac{2x}{3} + \dfrac{5y}{7} = 1$, for y

38. $T = \dfrac{b}{1 - s}$, for s

39. $3x - 1 \le 0$

40. $2x + 3 \ge 5x - 1$

41. $2(3x + 1) - 2(2x - 1) < 2 - x$

42. $\dfrac{2}{3}x + 1 > \dfrac{3}{4}x - \dfrac{1}{2}$

43. $\dfrac{x}{6} - \dfrac{12 - 5x}{4} \le \dfrac{x}{4} + 1$

44. Part of $12,000 is invested at 5% and the remainder at 9%. The graph below shows how the interest received at the end of 1 year depends on the amount invested at 5%.

   (a) Use the graph to find the amount invested at 5% if the interest received is $720.

   (b) Use the graph to find the amount invested at 5% if the interest received is more than $720.

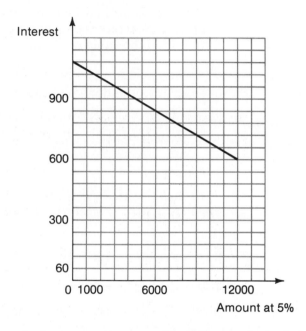

45. The formula for converting speed in miles per hour (m) to speed in feet per second (f) is $f = \frac{22}{15}m$. Find the formula for converting speed in feet per second to speed in miles per hour.

46. If a TV tower casts a 300-foot shadow at the same time that a 6-foot man casts a 12.5-foot shadow, how tall is the tower?

47. If a motor that turns at 380 rpm is connected to a gear with 45 teeth, and this gear is connected to a second gear with 50 teeth, how fast is the second gear turning?

48. If a motor is connected to a gear with 65 teeth, and this gear is connected to a second gear with 80 teeth that is turning at 234 rpm, how fast is the motor turning?

49. A motor that turns at 360 rpm is connected to a gear with 42 teeth, and this gear is connected to a gear with 48 teeth. If the second gear is connected to a third gear with 35 teeth, how fast is the third gear turning?

50. One unit of length on a map represents 250,000 units in the real world.

    (a) If two cities are 6.5 inches apart on the map, how far apart are they in the real world?

    (b) If two cities are 2.5 miles apart in the real world, how far apart are they on the map?

51. Two people share $76,500 in the ratio $13 : 21$. How much does each receive?

52. 75 fish are tagged and returned to a pond. Later, a sample of 100 fish has 12 tagged fish. Estimate the number of fish in the pond.

53. The ratio of girls to boys in a school is $12 : 13$. If there are 1157 boys in the school, how many girls are there in the school?

54. A triangle has sides of length 12, 13, and 18 feet. Assuming that the longest side of a similar triangle is 45 feet, find the lengths of the other two sides.

Write an equation or inequality and solve each problem in Exercises 55–70. (If you have trouble writing an equation, set up a chart and compute several special cases first.)

55. Mike has $1055 in 5-, 10- and 20-dollar bills. There are three more 10-dollar bills than 5-dollar bills. The number of 20-dollar bills is one less than twice the number of 5-dollar bills. Find the number of 5-, 10-, and 20-dollar bills that Mike has.

56. Gayle invests $22,000, part at 8.7% and the remainder at 6.5%. If her annual income (simple interest)is $1,726.56, how much has she invested at each rate?

57. A total of 4422 tickets to a concert were sold at $2 for children, $3.50 for adults, and $2.50 for senior citizens. There were twice as many $3.50 tickets sold as $2 tickets. If the total receipts were $12,930, how many of each ticket were sold?

58. A 117-foot chain is cut into three pieces. The largest piece is four

times the smallest, and the other piece is one-half the difference between the largest and the smallest. Find the length of each piece.

59. Two cars start from the same point and travel in the same direction at constant speeds of 55 mph and 60 mph. When will they be 48 miles apart?

60. Kyle traveled by car from his office to the airport and then took a plane to his destination. His trip was 895 miles in total and took $2\frac{3}{4}$ hours. If the car averaged 55 mph, and the plane averaged 650 mph, how far did he travel from his office to the airport?

61. The length of a rectangle is 3 feet less than twice the width. Find the dimensions of the rectangle if the perimeter is 99 feet.

62. The largest angle of a triangle is 3° less than three times the smallest angle. If the third angle is 12° larger than the smallest angle, find the measure of each angle.

63. How many gallons of a 12%-salt solution must be mixed with a 24%-salt solution to obtain 30 gallons of a 20%-salt solution?

64. How many pounds of candy worth $1.45 a pound must be mixed with candy worth $2.70 a pound to obtain 40 pounds of candy worth $2.10 a pound?

65. How many gallons must be drained from 30 gallons of a 10%-salt solution and replaced by a 22%-salt solution to obtain 30 gallons of a 15%-salt solution?

66. David's scores on the first three tests were 68, 76, and 91. What score does he need on the final exam, which is counted as two tests, to bring his average up to 85?

67. Find two numbers in the ratio 3 : 4 that have a sum of 77.

68. The ratio of the length of a rectangle to its width is 5 : 3. Find the dimensions of the rectangle if the perimeter is 88.

69. How much 10%-alcohol solution should be added to 40 gallons of a 60%-alcohol solution to produce a mixture that is at most 20% alcohol?

70. Jane's current annual salary is $15,000. What percent increase is required for her new salary to be at least $16,800 a year?

6

Graphing Equations
in Two Variables

6.1 / GRAPHS DETERMINED
BY CERTAIN CALCULATOR KEYS

We have used graphs to summarize information in problems, to solve problems, and to picture numerical relationships. Whenever one number is determined by another, a point on a graph can be used to represent the pair of numbers. There are keys on the calculator that determine for a first number a corresponding second number. To gain more practice with graphing, we will graph the points that correspond to these pairs of numbers for three different keys: $\boxed{1/x}$, $\boxed{\sqrt{x}}$, and $\boxed{\sin}$.

In the first example below, we want to graph points $(x, \frac{1}{x})$ for all numbers x except $x = 0$. On this graph the first coordinate of a point can be any number $x \neq 0$; the second coordinate must be its reciprocal $\frac{1}{x}$. The usual convention is to denote the first coordinate of a point by x and the second by y. In this example, $y = \frac{1}{x}$. Thus instead of saying, "Graph all points $(x, \frac{1}{x})$ for $x \neq 0$," we usually say, "Graph the equation $y = \frac{1}{x}$." This means that we should graph all points (x,y), where $y = \frac{1}{x}$. In general, the **graph of an equation** in two variables is the collection of points whose coordinates are solutions to the equation. In this chapter the y values are computed accurate to two decimal places.

Example 1   Graph the equation $y = \dfrac{1}{x}$.

The $\boxed{1/x}$ key assigns to any number its reciprocal, except for 0. If you key $\boxed{0}$ $\boxed{1/x}$, your calculator will give an error message, because 0 does not have a reciprocal. This means there will be no point on the graph corresponding to 0 on the horizontal axis. However, the calculator will give a value for every other number, positive and negative, that you key. To get representative points, we start with some whole numbers and their opposites.

x	-7	-6	-5	-4	-3	-2	-1	1	2	3	4	5	6	7
$y = \frac{1}{x}$	$-.14$	$-.17$	$-.2$	$-.25$	$-.33$	$-.5$	-1	1	.5	.33	.25	.2	.17	.14

The x value is measured on the horizontal axis and the corresponding $\dfrac{1}{x}$ value on the vertical axis to determine a point for each pair $\left(x, \dfrac{1}{x}\right)$ shown in Fig. 6.1.

It is possible to imagine many curves that contain the points we have graphed. Remember that there is no $\dfrac{1}{x}$ value for $x = 0$, so there can

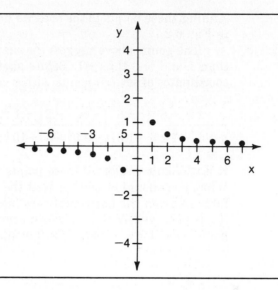

Figure 6.1

be no point of the graph on the vertical axis. To get a better indication of the shape of the graph, we choose several more values of x and compute the corresponding values for $y = \dfrac{1}{x}$:

x	-6.5	-5.5	-4.5	-3.5	-2.5	-1.5	$-.5$	$.5$	1.5	2.5	3.5	4.5	5.5	6.5
$y = \frac{1}{x}$	$-.15$	$-.18$	$-.22$	$-.29$	$-.4$	$-.67$	-2	2	$.67$	$.4$	$.29$	$.22$	$.18$	$.15$

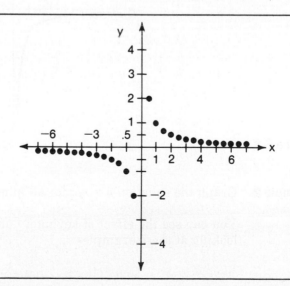

Figure 6.2

Adding these points to the ones we had before gives us the graph shown in Fig. 6.2.

The points above suggest the shape of the graph for x values greater than 1 and less than -1. Before sketching the full graph, we compute coordinates of several points with x values between -1 and 1.

x	$-.8$	$-.6$	$-.4$	$-.2$	$-.1$	$.1$	$.2$	$.4$	$.6$	$.8$
$y = \frac{1}{x}$	-1.25	-1.67	-2.5	-5	-10	10	5	2.5	1.67	1.25

It is difficult to plot all these points because they are so close together. What you should observe is that the x values closest to 0 have y values farthest from the horizontal axis (negative if x is negative and positive if x is positive). With this information, we assume that graphing all the points (x,y) for $y = \frac{1}{x}$ and $x \neq 0$ would give the curve shown in Fig. 6.3.

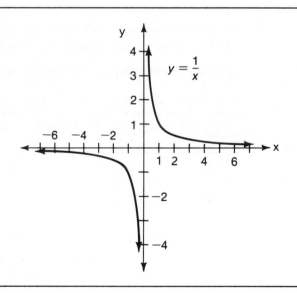

Figure 6.3

□

Example 2   Graph the equation $y = \sqrt{x}$ for all numbers $x \geq 0$.

You can see the effect of keying a number followed by the $\boxed{\sqrt{x}}$ key by looking at these examples:

Calculator	Display
25 $\boxed{\sqrt{x}}$	5
16 $\boxed{\sqrt{x}}$	4
$\boxed{4}$ $\boxed{\sqrt{x}}$	2

The $\boxed{\sqrt{x}}$ key, called the **square root key**, assigns to a given number a second number whose square equals the first number. Since no number squared equals a negative number, you will get an error message if you key a negative number followed by the square root key. (Try it!) To draw the graph we first locate several points for x as a whole number and graph them in Fig. 6.4. In a second step, we can fill in additional points.

x	0	1	2	3	4	5	6	7	8	9
$y = \sqrt{x}$	0	1	1.41	1.73	2	2.24	2.45	2.65	2.83	3

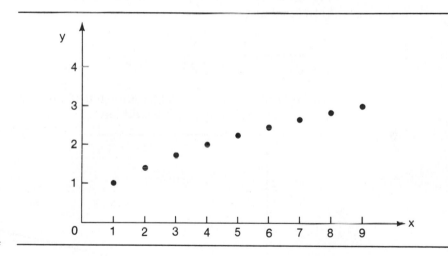

Figure 6.4

By taking x values halfway between those in the graph in Figure 6.4, we obtain the points shown in Fig. 6.5.

x	.5	1.5	2.5	3.5	4.5	5.5	6.5	7.5	8.5	9.5
$y = \sqrt{x}$	.71	1.22	1.58	1.87	2.12	2.35	2.55	2.74	2.92	3.08

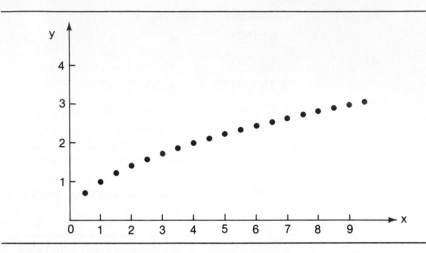

Figure 6.5

The part of the graph near the origin is still somewhat unclear. To see the shape of this part, we compute a few more points with x values between 0 and 1.

x	.2	.4	.6	.8
$y = \sqrt{x}$	.45	.63	.77	.89

Adding these points, we conjecture that if we could plot all solutions to $y = \sqrt{x}$ for $x \geq 0$, the graph would have the shape of Fig. 6.6.

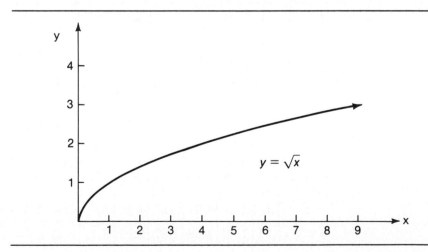

Figure 6.6

Since there is a point on the graph for every positive number x, the graph extends indefinitely to the right. $\qquad\qquad\qquad\qquad\qquad$ □

There are times when we wish to restrict attention to a portion of a graph. Inequalities can be used to describe the portion intended. For example, we might wish to look only at the part that has positive values of x. In this case we write $x > 0$. To indicate the part that has x values between -5 and 5, we could write $x > -5$ and $x < 5$. These two statements combine to state that x is a number greater than -5 and less than 5. A notation for this same set of numbers is $-5 < x < 5$. For example, the phrase $-180 \leq x \leq 180$ means that x represents the numbers between -180 and 180, including both -180 and 180.

Example 3 $\qquad$ Graph the equation $y = \sin x$ for $-180 \leq x \leq 180$.

We need numbers on the horizontal axis that extend from -180 to 180. Since the space is limited, we will label only the multiples of 20. To see the number associated with 100, for example, we key $\underline{100}$ ⎡sin⎤ . To see the number associated with -100, we key $\underline{100}$ ⎡+/−⎤ ⎡sin⎤ . Here are some points to get the graph started:

x	-180	-160	-140	-120	-100	-80	-60	-40	-20	0
$y = \sin x$	0	$-.34$	$-.64$	$-.87$	$-.98$	$-.98$	$-.87$	$-.64$	$-.34$	0

x	20	40	60	80	100	120	140	160	180
$y = \sin x$	.34	.64	.87	.98	.98	.87	.64	.34	0

Since the values for $\sin x$ are all between -1 and 1, we choose a different scale for the vertical axis than for the horizontal axis (Fig. 6.7).

There are many curves that go through the points graphed in Fig. 6.7. We need to add many more points before we guess the shape of the graph. Both the table of values above and the points we have already graphed suggest that the values associated with x when it is positive can be predicted from the values associated with x when it is negative. We compute additional values in the table on p. 312.

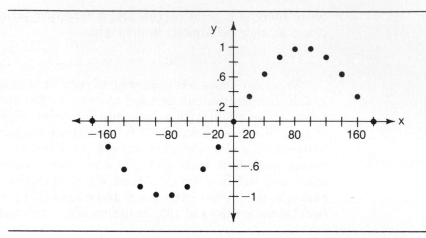

Figure 6.7

x	-170	-150	-130	-110	-90	-70	-50	-30	-10
$y = \sin x$	$-.17$	$-.5$	$-.77$	$-.94$	-1	$-.94$	$-.77$	$-.5$	$-.17$

x	10	30	50	70	90	110	130	150	170
$y = \sin x$	.17	.5	.77	.94	1	.94	.77	.5	.17

These points are graphed in Fig. 6.8.

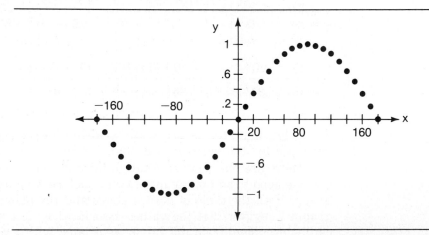

Figure 6.8

Even now there is some risk in predicting the shape of this graph. You may want to add more points to convince yourself that the graph looks like Fig. 6.9.   □

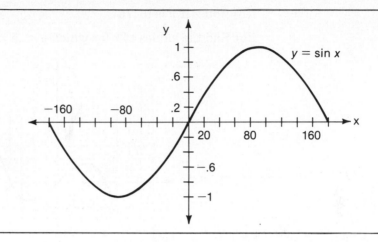

Figure 6.9

Exercises 6.1

1. Find all integers x that satisfy

    (a) $-2 < x \le 5$ (b) $-3 \le x < 2$

    (c) $4 < x < 5$ (d) $-1 \le \frac{x}{2} \le 3$

2. Write an inequality that expresses the fact that x is a number

    (a) greater than or equal to 4

    (b) greater than -2 and less than or equal to 3

    (c) between -1 and 3 but not equal to either -1 or 3

    (d) greater than or equal to 1 and less than or equal to 4

3. Use your calculator to find the value of

    (a) $\dfrac{1}{1.8}$ (b) $\dfrac{1}{-2.1}$ (c) $\sqrt{15}$ (d) $\sqrt{-10}$

    (e) $\sin 45$ (f) $\cos -85$ (g) $\tan 102$ (h) $\log 10$

    (i) $\ln 8$ (j) $\ln -2$

4. Below is the graph of y as a function of x. Use the graph to

(a) find y if x is .5, $-.5$

(b) find x if y is 0, .75

(c) find the values of x for which $y < .5$

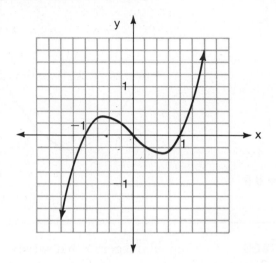

5. Let $y = x^2$.

(a) Compute the value of y for those values of x that divide the interval $-4 \le x \le 4$ into 16 pieces of equal length and for the endpoints $x = -4$ and $x = 4$.

(b) Graph the points (x,y) determined in (a).

(c) Describe the y values when the x values are greater than 4; describe the y values when x is less than -4.

(d) Use the information in (b) and (c) to give the complete graph of $y = x^2$.

6. Let $y = \dfrac{1}{x - 2}$.

(a) Compute the value of y for those values of x that divide the interval $-2 \le x \le 6$ into 16 pieces of equal length and for the endpoints $x = -2$ and $x = 6$.

(b) Graph the points (x,y) determined in (a).

(c) Describe the y values when the x values are very close to $x = 2$.

(d) Use the information in (b) and (c) to give the complete graph of
$$y = \frac{1}{x-2} \text{ for } -2 \le x \le 6.$$

7. Let $y = \cos x$.

(a) Compute the value of y for those values of x that divide the interval $-180 \le x \le 180$ into 36 pieces of equal length and for the endpoints $x = -180$ and $x = 180$.

(b) Graph the points (x,y) determined in (a).

(c) Use the information in (b) to give the complete graph of $y = \cos x$ for $-180 \le x \le 180$.

8. Let $y = \tan x$.

(a) Compute the value of y for those values of x that divide the interval $-90 \le x \le 90$ into 18 pieces of equal length and for the endpoints $x = -90$ and $x = 90$.

(b) Graph the points determined in (a).

(c) Describe the y values when the x values are very close to $x = -90$ and to $x = 90$.

(d) Use the information in (b) and (c) to give the complete graph of $y = \tan x$ for $-90 < x < 90$.

9. Let $y = \sqrt{x-2}$.

(a) Compute the value of y for those values of x that divide the interval $2 < x \le 12$ into 20 pieces of equal length and for the endpoints $x = 2$ and $x = 12$.

(b) Graph the points determined in (a).

(c) Describe the y values when the x values are close to $x = 2$; describe the y values when the x values are greater than 12.

(d) Use the information in (b) and (c) to give the complete graph of $y = \sqrt{x-2}$ for $x \ge 2$.

10. Let $y = \ln x$.

(a) Compute the value of y for those values of x that divide the interval $0 \le x \le 10$ into 20 pieces of equal length and for the endpoints $x = 0$ and $x = 10$.

(b) Graph the points determined in (a).

(c) Describe the y values when the x values are close to $x = 0$ or greater than 10.

(d) Use the information in (b) and (c) to give the complete graph of $y = \ln x$ for $x > 0$.

6.2 / GRAPHS OF POLYNOMIAL EQUATIONS

In the previous section we looked at graphs of equations in two variables, x and y, in which the y value was obtained from the x value by using a single calculator key. Now we study equations where computing the y value is somewhat more complicated. The particular type of equation that we investigate in this section is called a polynomial equation. A **polynomial** is either one term or a sum of terms. A typical example of a polynomial term in one variable is ax^n, where a is any number, x is the variable, and n is a whole number. For example, $7.5x^4$ could be a term. Another example is $\frac{1}{3}x^3$. A number without a variable can be a term; we can, for example, think of 5 as $5x^0$. Here are some polynomials in the variable x:

$$6 + 2x - 3x^2$$

$$\frac{1}{2} - x + \frac{3}{2}x^3 + 4x^5$$

$$x^4 - 2x^3 + .5x$$

$$7x - 11$$

$$3x^5.$$

The largest exponent that appears in a polynomial is called the **degree** of the polynomial. Thus the first polynomial above has degree 2, the second has degree 5, the third has degree 4, the fourth has degree 1, and the last has degree 5. Degree 1 polynomials are also called **linear polynomials.**

We can evaluate a polynomial for any value of x. In an equation in two variables, such as $y = 6 + 2x - 3x^2$, each value of x determines a value for y and we can graph the points with coordinates (x,y) to obtain a graph of the equation. You graphed the polynomial equation $y = x^2$ in the last set of exercises by plotting many points (x, x^2). Here are some pairs of numbers (x, y) for $y = x^2$.

x	-3	-2.5	-2	-1.5	-1	$-.5$	0	$.5$	1	1.5	2	2.5	3
y	9	6.25	4	2.25	1	.25	0	.25	1	2.25	4	6.25	9

Your graph for $y = x^2$ containing these points should look like Fig. 6.10.

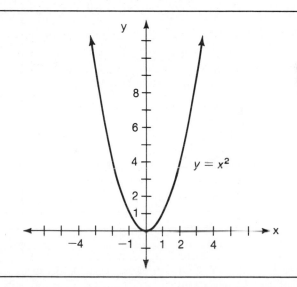

Figure 6.10

A graph shaped like Fig. 6.10 is called a **parabola**. As you graph other polynomials of degree 2, you will see that each one has a parabola for its graph. Can you explain why the left side of this curve has the same shape as the right side?

Example 1   Graph the equation $y = -x^2$.

To start this graph we use the same x values that we used for $y = x^2$, this time evaluating $y = -x^2$. To find y when $x = -2$, for example, we key ② ⊕⁄⊖ ⊗² ⊕⁄⊖ ⊜ (Display: -4).

x	-3	-2.5	-2	-1.5	-1	$-.5$	0	$.5$	1	1.5	2	2.5	3
y	-9	-6.25	-4	-2.25	-1	$-.25$	0	$-.25$	-1	-2.25	-4	-6.25	-9

These points are graphed in Fig. 6.11. The complete graph of $y = -x^2$ is shown in Fig. 6.12.

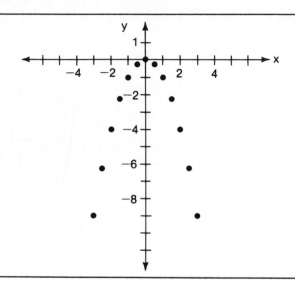

Figure 6.11

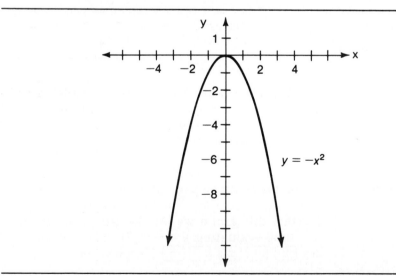

Figure 6.12

You should see that the y values in this example are the opposites of the y values in the $y = x^2$ table. (Why?) The graph in Fig. 6.12 has the same shape as the first parabola, but it opens downward instead of upward. (We sometimes say it is a "cap" rather than a "cup.") □

Example 2 Graph the equation $y = -x^2 + 3$.

If we again use the same x values that we used for $y = -x^2$, we can compute the y values for $y = -x^2 + 3$ readily. For each x value, the corresponding y value is 3 more than it was in Example 1.

x	-3	-2.5	-2	-1.5	-1	$-.5$	0	$.5$	1	1.5	2	2.5	3
$y = -x^2$	-9	-6.25	-4	-2.25	-1	$-.25$	0	$-.25$	-1	-2.25	-4	-6.25	-9
$y = -x^2 + 3$	-6	-3.25	-1	$.75$	2	2.75	3	2.75	2	$.75$	-1	-3.25	-6

In Fig. 6.13 we start the graph of $y = -x^2 + 3$ with the points whose coordinates are given in the table above.

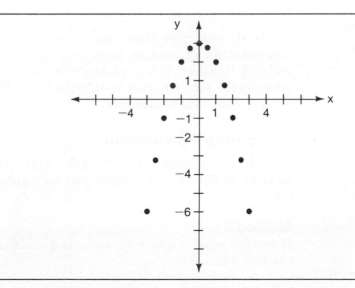

Figure 6.13

The graph in Fig. 6.13 has the same shape as the graph of $y = -x^2$, but it is positioned differently with respect to the coordinate axes. The difference is that each point on the graph $y = -x^2 + 3$ is 3 units above the corresponding point on the graph of $y = -x^2$. That is the effect of adding the number 3. Thus the graph looks like Fig. 6.14.

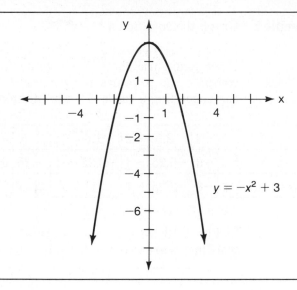

Figure 6.14

□

In the examples above, evaluating the polynomials for values of x was not difficult because these polynomials were short: x^2, $-x^2$, and $-x^2 + 3$. However, before we begin to graph a polynomial with several terms in the sum, we want to describe some efficient ways of evaluating a polynomial with a calculator.

Evaluating Polynomials

Consider the equation $y = 2x^3 - 5x^2 - 11x + 14$. We describe below two methods of finding the y value that corresponds to a particular value for x.

Method 1

If $y = 2x^3 - 5x^2 - 11x + 14$ and x is a positive number, say $x = 2$, we can find y by keying

⟨2⟩⟨×⟩⟨2⟩⟨yˣ⟩⟨3⟩⟨−⟩⟨5⟩⟨×⟩⟨2⟩⟨x²⟩⟨−⟩ 11 ⟨×⟩⟨2⟩⟨+⟩ 14 ⟨=⟩

The display shows that $y = -12$ when $x = 2$. Notice that at three different times in the keying sequence above, we keyed ⟨2⟩ to replace x in
the polynomial. If we are replacing x by a number with more than one
digit (say $x = 2.55$) we will save time in the keying sequence if we
STORE the value of x and RECALL it as needed:

2.55 ⟨STO⟩

⟨2⟩⟨×⟩⟨RCL⟩⟨yˣ⟩⟨3⟩⟨−⟩⟨5⟩⟨×⟩⟨RCL⟩⟨x²⟩⟨−⟩ 11 ⟨×⟩⟨RCL⟩⟨+⟩ 14 ⟨=⟩

Display: -13.39975

If we wish to evaluate $y = 2x^3 - 5x^2 - 11x + 14$ at a negative number, say $x = -2.55$, we recognize that the calculator cannot compute a
power of a negative number with the ⟨yˣ⟩ key. However, $2(-2.55)^3 -
5(-2.55)^2 - 11(-2.55) + 14 = -2(2.55)^3 - 5(2.55)^2 + 11(2.55) + 14$. Thus
to evaluate $y = 2x^3 - 5x^2 - 11x + 14$ for $x = -2.55$, we can evaluate
$y = -2x^3 - 5x^2 + 11x + 14$ for $x = 2.55$. Our method is to rewrite the
polynomial, changing the signs of the coefficients of odd powers of x,
and evaluate it at the opposite of the negative x value. Here is the
keying sequence:

2.55 ⟨STO⟩

⟨2⟩⟨⁺/₋⟩⟨×⟩⟨RCL⟩⟨yˣ⟩⟨3⟩⟨−⟩⟨5⟩⟨×⟩⟨RCL⟩⟨x²⟩⟨+⟩ 11 ⟨×⟩⟨RCL⟩⟨+⟩ 14 ⟨=⟩

Display: -23.62525

Method 2

The distributive property gives us a very efficient way of evaluating a
polynomial that works for both positive and negative values of x:

$$2x^3 - 5x^2 - 11x + 14 = (2x^2 - 5x - 11)x + 14$$

$$= [(2x - 5)x - 11]x + 14.$$

With the polynomial in this form, the value of x appears only with
exponent 1 and we do not need the ⟨yˣ⟩ key. If we want the value of the
polynomial for $x = -2.55$, we first STORE -2.55 to reduce the number
of key strokes and then use the polynomial in the form
$[(2x - 5)x - 11]x + 14$:

2.55 ⟨⁺/₋⟩⟨STO⟩

⟨2⟩⟨×⟩⟨RCL⟩⟨−⟩⟨5⟩⟨=⟩

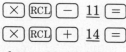

Display: -23.62525

This method works for any value of x and is so efficient that you may want to commit it to memory. Observe how the coefficients of the polynomial $(2, -5, -11, 14)$ occur in the keying sequence. The coefficients are taken in order, with the coefficient of the highest power coming first. If you have a polynomial that has 0 coefficient for some terms, the 0 must be regarded as a coefficient in this method of evaluating the polynomial. For example, if $y = x^4 - 2x^3 + 5x$, the sequence of coefficients is $1, -2, 0, 5, 0$. If we used the exact form above to evaluate this polynomial when $x = -1.1$, for example, the keying sequence would look like this:

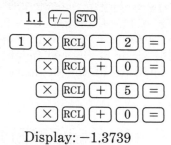

Display: -1.3739

The keying sequence can be shortened, since multiplying by 1 and adding 0 do not change the computation. You should see that the following sequence has the same result:

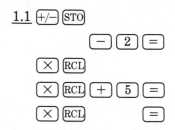

Example 3   Graph the equation $y = 2x^3 - 5x^2 - 11x + 14$.

We start by finding solutions of the equation that correspond to some small whole number values of x and their opposites. For positive values of x, we can use a keying sequence that directly parallels the terms of the polynomial. For example, if we want y when $x = 5$, we key

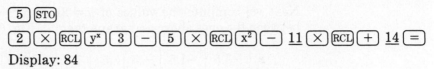

Display: 84

For negative values (and positive values also, if you wish) we might use method 2 discussed above. For $x = -5$, we key

5 +/- STO

2 × RCL − 5 =

× RCL − 11 =

× RCL + 14 =

Display: -306

For certain values of x, such as 0 and 1, you may choose not to use the calculator:

x	−3	−2	−1	0	1	2	3	4
y	−52	0	18	14	0	−12	−10	18

These points are shown in Fig. 6.15.

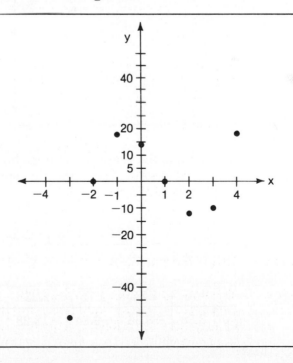

Figure 6.15

Next we compute the values of $y = 2x^3 - 5x^2 - 11x + 14$ for x values between the ones above and graph them in Fig. 6.16.

x	-2.5	-1.5	$-.5$	$.5$	1.5	2.5	3.5	4.5
y	-21	12.5	18	7.5	-7	-13.5	0	45.5

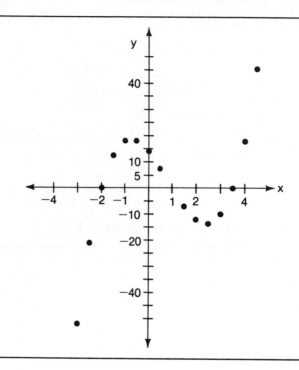

Figure 6.16

The holes in the graph are still too big for us to predict the shape of the graph accurately. Therefore, we compute more values and add more points (Fig. 6.17).

x	-2.75	-2.25	-1.75	-1.25	$-.75$	$-.25$	$.25$
y	-35.16	-9.34	7.22	16.03	18.59	16.41	10.97

x	$.75$	1.25	1.75	2.25	2.75	3.25	3.75	4.25
y	3.78	-3.66	-9.84	-13.28	-12.47	-5.91	7.91	30.47

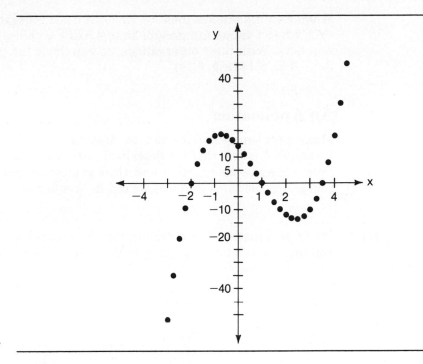

Figure 6.17

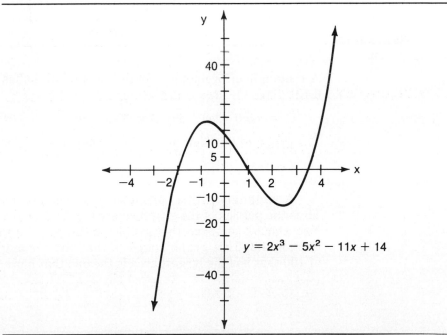

$y = 2x^3 - 5x^2 - 11x + 14$

Figure 6.18

When x values are less than -3, the y values are negative and are below -52; when x values are greater than 4.5, the y values are large positive numbers. With these observations we can draw the graph of $y = 2x^3 - 5x^2 - 11x + 14$ (Fig. 6.18). ☐

An Application

Many problem situations can be described by polynomial equations. Example 4 is one that we described with a graph in Chapter 4. This time we write an equation and then graph the equation. We will, of course, get the same graph as we did in Chapter 4.

Example 4　Write an equation that describes for all rectangles with a perimeter of 100 feet the relationship between the width of a rectangle and its area. Then graph the equation.

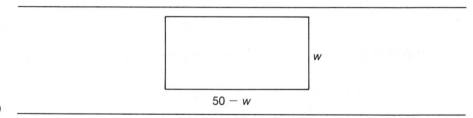

Figure 6.19

A rectangle of perimeter 100 that has width w has length $50 - w$ (Fig. 6.19). Thus the area A is given by

$$A = w(50 - w), \quad \text{or} \quad A = 50w - w^2.$$

The graph of this equation is the parabola shown in Fig. 6.20. ☐

A rectangle with a perimeter of 100 cannot have $w < 0$ or $w > 50$. Thus the portion of the graph where $0 \le w \le 50$ describes this problem. You should compare this graph to the one we drew in Example 1, Section 4.2. This graph suggests that the rectangle with a perimeter of 100 that has the largest area is the one that has a width of 25.

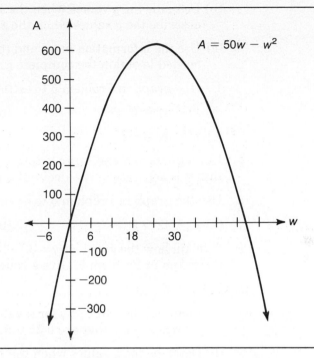

$$A = 50w - w^2$$

Figure 6.20

**Exercises
6.2**

Evaluate the polynomials in Exercises 1-4 for the values of the variables specified.

1. $x^3 - x$; $x = -1, 2.2, -2.2$

2. $3x^4 - 2x^3 + 5x^2 - x + 1$; $x = 1, 1.5, -1.5$

3. $3x^5 + 2x^3 - x$; $x = 1.8, -1.8, -2.35$

4. $2x^4 + x^2 - 4$; $x = -1, 2.5, -2.5$

5. Let $y = x^3 - 4x$.

    (a) Compute the value of y for x values in the interval $-3 \le x \le 3$ which are 0.5 apart.

(b) Describe the y values when the x values are greater than 3; describe the y values when the x values are less than -3.

(c) Use the information in (a) and (b) and plot additional points if needed to obtain the complete graph of $y = x^3 - 4x$.

6. Use the graph in Problem 5 to estimate the value(s) of x for which

   (a) $x^3 - 4x = 0$ (b) $x^3 - 4x = 5$

   (c) $x^3 - 4x = -2$

7. Use a guess-and-check procedure to refine your answer to 6(b) so that it is accurate to two decimal places.

8. Use the graph in Problem 5 to estimate the values of x for which

   (a) $x^3 - 4x > 5$ (b) $x^3 - 4x < -5$

9. Explain how the graph of $y = x^3 - 4x + 3$ can be obtained from the graph in Problem 5. Give a rough sketch of $y = x^3 - 4x + 3$.

10. Let $y = x^4 - 4x^2$.

    (a) Compute the value of y for x values in the interval $-2.5 \le x \le 2.5$ that are 0.25 units apart.

    (b) Describe the y values when the x values are greater than 2.5 and when the x values are less than -2.5.

    (c) Use the information in (a) and (b) and plot additional points if needed to give the complete graph of $y = x^4 - 4x^2$.

11. Use the graph in Problem 10 to estimate the value(s) of x for which

    (a) $x^4 - 4x^2 = 2$ (b) $x^4 - 4x^2 = -5$

12. Use a guess-and-check procedure to refine your answer to Problem 11(a) so that it is accurate to two decimal places.

13. Use the graph in Problem 10 to estimate the values of x for which

    (a) $x^4 - 4x^2 < 0$ (b) $x^4 - 4x^2 > 2$

14. Explain how the graph of $y = x^4 - 4x^2 - 2$ can be obtained from the graph in Problem 10. Give a rough sketch of $y = x^4 - 4x^2 - 2$.

15. Let $y = x^4 - 5x^2 + 4$.

(a) Compute the value of y for x values in the interval $-2.5 \le x \le 2.5$ that are 0.25 units apart.

(b) Describe the y values when the x values are greater than 2.5 or less than -2.5.

(c) Use the information in (a) and (b) and plot additional points if needed to give the complete graph of $y = x^4 - 5x^2 + 4$.

16. Let $y = x^4 - 5x^2 + 4$. Use the graph in Problem 15 to determine the values of x for which

(a) $y = 0$ (b) $y > 0$

(c) $y < 0$

17. Use the graphs of $y = x^2$ and $y = -x^2$ in the text to obtain the graphs of

(a) $y = x^2 + 2$ (b) $y = x^2 - 2$

(c) $y = -x^2 + 1$ (d) $y = -x^2 - 2$

18. A rectangular garden is enclosed by 200 feet of fence and one side of a barn. Let x denote the width of the garden.

(a) Write an equation for the area A of the garden.

(b) Give the complete graph of the equation in (a) and indicate which portion of the graph represents the problem.

6.3 / GRAPHS OF RATIONAL EQUATIONS

An algebraic expression that is a quotient of two polynomials is called a **rational expression**. Here are some rational expressions in one variable:

$$\frac{x^2 - 2x - 3}{x - 2}$$

$$\frac{x^5 + x}{x^2 - 2x + 1}$$

$$\frac{\frac{1}{2}x}{x^4 - 2x^2 + \frac{1}{4}}.$$

A polynomial can be thought of as a rational expression, since it can be written as itself divided by 1. A rational number (or fraction) like $\frac{3}{4}$ is also a rational expression, since the numerator is a polynomial (with one term, $3x^0$) and the denominator is a polynomial.

We can evaluate rational expressions for certain values of the variable. For example,

$$\text{if } x = 1, \text{ then } \frac{x^2 - 2x - 3}{x - 2} = \frac{1^2 - 2 \cdot 1 - 3}{1 - 2} = \frac{-4}{-1} = 4.$$

$$\text{if } x = 4, \text{ then } \frac{x^2 - 2x - 3}{x - 2} = \frac{4^2 - 2 \cdot 4 - 3}{4 - 2} = \frac{5}{2}.$$

There is an important difference between polynomials and rational expressions with respect to the numbers that the variable can represent. In a polynomial, any number replacing the variable will give a value for the polynomial. However, for a rational expression there may be numbers that, if they replaced the variable, would give no value for the rational expression. For example, consider the rational expression $\frac{x^2 - 2x - 3}{x - 2}$. If $x = 2$, then $x^2 - 2x - 3 = -3$ and $x - 2 = 0$. There is no number $(-3) \div 0$. Any number that replaces the variable to make the denominator equal 0 gives no value for a rational expression. Thus the variable cannot represent a number that makes the denominator equal to 0. If you attempt to evaluate a rational expression at a number that gives 0 in the denominator, the calculator will give an error message.

If we were being very careful, we would talk about the rational expression $\frac{x^2 - 2x - 3}{x - 2}$ for $x \neq 2$. However, we assume when we write a rational expression that the reader will exclude x values that make the denominator 0. The graph of $y = \frac{x^2 - 2x - 3}{x - 2}$ does not have a point corresponding to $x = 2$. Thus the behavior of the graph for x values near 2 is very important and must be investigated carefully.

Example 1   Graph the equation $y = \dfrac{x^2 - 2x - 3}{x - 2}$.

To use the calculator in finding y values for particular x values we can evaluate the numerator, key ⬚= and ⬚÷, then evaluate the denominator with parentheses. For example, we can use this sequence for $x = -5$.

⬚5 ⬚+/− ⬚STO
⬚RCL ⬚x² ⬚− ⬚2 ⬚× ⬚RCL ⬚− ⬚3 ⬚= ⬚÷ ⬚(⬚RCL ⬚− ⬚2 ⬚) ⬚=
Display: -4.5714286

Here are the y values for some whole number x values and their opposites. Notice that $x = 2$ is omitted.

x	−6	−5	−4	−3	−2	−1	0	1	3	4	5	6	7
y	−5.63	−4.57	−3.5	−2.4	−1.25	0	1.5	4	0	2.5	4	5.25	6.4

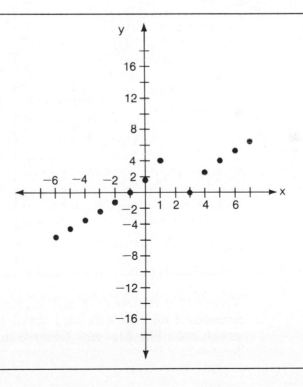

Figure 6.21

These points are graphed in Fig. 6.21 and give a good idea of the shape of the graph for x values less than 1 and greater than 3, but not for x values between 1 and 3. We compute $y = \dfrac{x^2 - 2x - 3}{x - 2}$ for additional x values in this interval but not for $x = 2$.

x	1.2	1.4	1.6	1.8	2.2	2.4	2.6	2.8
y	4.95	6.4	9.1	16.8	-12.8	-5.1	-2.4	$-.95$

Adding these points to the graph shows more clearly the behavior of the graph near the x value 2 (Fig. 6.22).

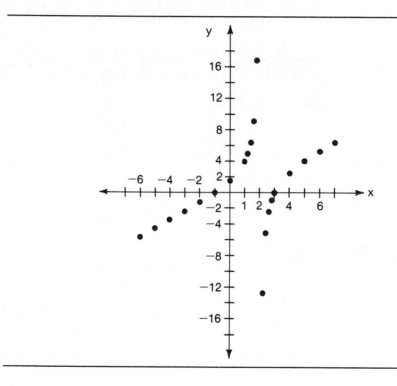

Figure 6.22

Values of x between 1.8 and 2 give large positive y values. Values between 2 and 2.2 give very small negative y values. The complete graph looks like Fig. 6.23. Observe that there is no point on the graph for $x = 2$.

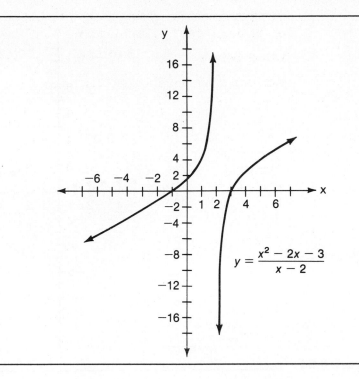

$$y = \frac{x^2 - 2x - 3}{x - 2}$$

Figure 6.23

□

Example 2 Graph the equation $y = \frac{x}{x-1}$.

When $x = 1$, the denominator of the rational expression $\frac{x}{x-1}$ equals 0.
Thus there is no y value for $x = 1$ and no point on the graph of the
equation for $x = 1$. To get the graph started we compute y for other
whole number values of x and their opposites, then graph the points in
Fig. 6.24.

x	-5	-4	-3	-2	-1	0	2	3	4	5
y	.83	.8	.75	.67	.5	0	2	1.5	1.33	1.25

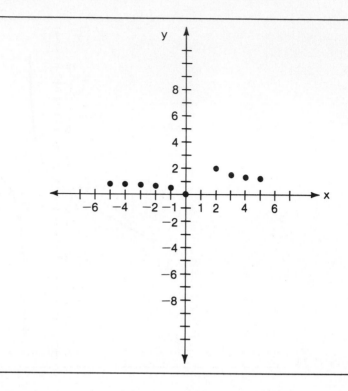

Figure 6.24

To get a better idea of the shape of the graph for x values between -5 and 5, we compute additional points. (Remember to key parentheses in the denominator.) These points are graphed in Fig. 6.25.

x	-4.5	-3.5	-2.5	-1.5	$-.5$	$.5$	1.5	2.5	3.5	4.5
y	.82	.78	.71	.6	.33	-1	3	1.67	1.4	1.29

It appears that more points are needed between $x = .5$ and $x = 1.5$. If we compute more y values in this interval, we can hope to infer the behavior of the graph.

x	.6	.7	.8	.9	1.1	1.2	1.3	1.4
y	-1.5	-2.33	-4	-9	11	6	4.33	3.5

What you should observe is that the x values close to 1 but slightly smaller than 1, that is to say, on the left side of 1, give negative y values;

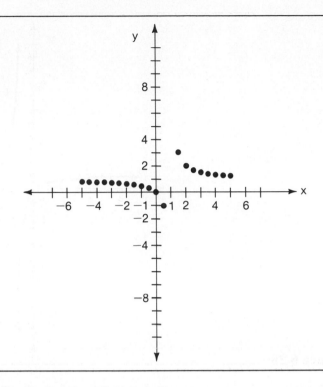

Figure 6.25

the closer the x value is to 1, the farther the y value is from 0. Thus the graph goes down on the left side of $x = 1$. On the right side of 1, the y values are positive; x values close to 1 give large y values, so the graph goes up to the right of $x = 1$.

We can predict y values for x values greater than 5. If x takes the values 10 or 100 or 1000, for example, then y correspondingly equals 1.11 or 1.01 or 1.001. These numbers are close to the whole number 1, but are always a little larger than 1. Consider x values less than -5, say x equals -10 or -100 or -1000, for example; then, correspondingly, y equals .91 or .99 or .999. These numbers are again close to the whole number 1, but are always a little less than 1. These observations help in sketching the extremes of the graph (Fig. 6.26).   □

Look again at the graphs of the two rational equations in Figs. 6.23 and 6.26. We are able to infer the behavior of the graphs near the values of x that make the denominator equal to 0. First of all, there are

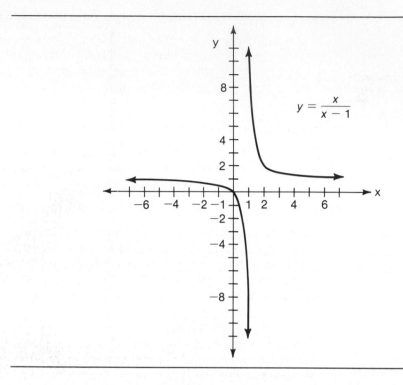

$$y = \frac{x}{x - 1}$$

Figure 6.26

no points on the graph corresponding to these values of x. Furthermore, values of x close to these make the denominator of the rational expression almost equal to 0; thus the y values will either be positive and far from 0 or negative and far from 0.

In Summary

From the examples we learn that in graphing rational expressions, the values of x that make the *denominator* 0 are especially important. First of all, there are no points on the graph corresponding to these values of x. Furthermore, values of x close to these make the denominator of the rational expression almost equal to 0; thus, if the numerator is not 0, the y values will either be positive and very large or negative and very small. The graph, therefore, appears either to shoot up or to shoot down near the x values that make the denominator of a rational expression 0.

The values of x that make the numerator of a rational expression equal to 0 are also important. If there is an x value that makes the numerator 0 but the denominator not 0, then the corresponding y value is 0. A point with y coordinate 0 lies on the horizontal axis. An x value that gives a corresponding y value of 0 determines a point where the graph crosses the horizontal axis. Look back now at Examples 1 and 2 and find such points.

An Application

Among the problem situations that can be described with a rational equation is one in which speed is computed in terms of time for a given distance, say 200 miles: $s = \frac{200}{t}$. Another is one involving gear ratios.

Example 3 Dan's two-speed bike has 39 teeth on the gear attached to the pedal. He generally pedals at 60 rpm (revolutions per minute). Write an equation that shows the relationship between the speed of the back wheel and the number of teeth on the gear attached to the back wheel; then graph the equation.

You need to remember the way the speed and number of teeth on one gear are related to the speed and number of teeth on a second gear connected to it. If S and T denote the speed and number of teeth on the first gear and s and t denote the speed and number of teeth on the second gear, then $s : S = T : t$. In this problem $T = 39$ and $S = 60$ rpm. Thus the proportion $s : 60 = 39 : t$ gives the relationship between the speed of the second gear (which is the same as the number of revolutions per minute of the back wheel) and the number of teeth on the second gear. Now $s : 60 = 39 : t$ means that

$$s \cdot t = 60 \cdot 39$$

$$s = \frac{60 \cdot 39}{t}$$

$$s = \frac{2340}{t}.$$

Now we can see that if the number of teeth is large, the speed is small; and if the number of teeth is small, the speed is large.

The graph that shows the relationship between the number of teeth and the speed is part of the graph of $s = \dfrac{2340}{t}$. It is not the entire graph since in this problem the variable t must be a positive whole number. Before we can draw the graph, we must decide what numbers t can represent in this situation. The smallest gear on a bicycle usually has about 14 teeth; the largest gear has about 30 teeth. Thus we might decide to let t represent the whole numbers from 10 to 42. We compute the corresponding s values for some values of t and draw the graph of these points in Fig. 6.27.

t	10	14	18	22	26	30	34	38	42
s (rpm)	234	167.14	130	106.36	90	78	68.82	61.58	55.71

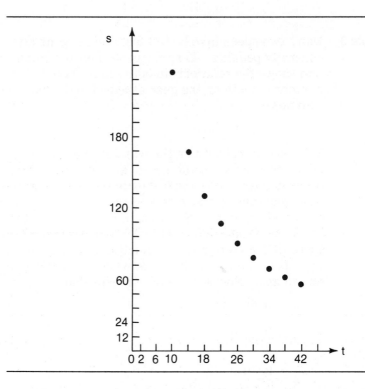

Figure 6.27

If we look at this problem for gears with the number of teeth from 10 to 42, then the graph describing the relationship between the number of teeth and the speed of the wheel will have 33 points on it (one for each number from 10 to 42). Each point lies on the positive part of the graph (Fig. 6.28) of the rational equation $s = \dfrac{2340}{t}$.

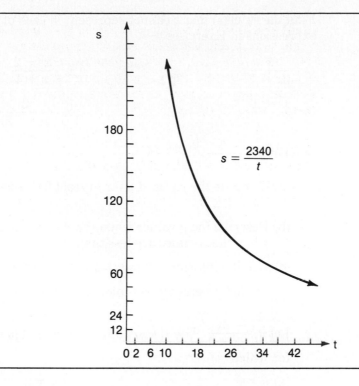

Figure 6.28

Actually the equation $s = \dfrac{2340}{t}$ can be graphed for all t except $t = 0$. You may want to draw the complete graph. □

Exercises 6.3

Evaluate the rational expressions in Exercises 1-3 for the specified values of the variable.

1. $\dfrac{2x + 1}{x^2 + 1}$; $x = -.5, 1.6, 100$

2. $\dfrac{x^2 - x - 6}{x + 1}$; $x = -2, 3.5, 100$

3. $\dfrac{2x^2 - x + 3}{x^2 - 2x + 1}$; $x = 1, -3.2, -100$

Find the value(s) that x cannot represent in each of the rational expressions in Exercises 4–7.

4. $\dfrac{x}{2x - 1}$

5. $\dfrac{x^2 - 2}{x + 4}$

6. $\dfrac{x^2 + 1}{x}$

7. $\dfrac{2}{x(x - 2)}$

8. Let $y = \dfrac{2x}{x + 1}$.

(a) Compute the value of y for at least 24 values of x in the interval $-3 \le x \le 3$.

(b) Describe the y values when the x values are greater than 3; when the x values are less than -3.

(c) Use the information in (a) and (b), plotting additional points if needed, to give the complete graph of $y = \dfrac{2x}{x + 1}$.

9. Let $y = \dfrac{2x}{x + 1}$. Use the graph in Problem 8 to find the value(s) of x for which

(a) $y = 0$

(b) $y = 2$

(c) $y > 0$

(d) $y < 0$

10. Let $y = \dfrac{1}{(x - 1)^2}$.

(a) Compute the value of y for at least 16 values of x in the interval $-4 \le x \le 4$.

(b) Describe the y values when the x values are greater than 4; when the x values are less than -4.

(c) Use the information in (a) and (b), plotting additional points if needed, to give the complete graph of $y = \dfrac{1}{(x - 1)^2}$.

11. Let $y = \dfrac{1}{(x - 1)^2}$. Use the graph in Problem 10 to find the value(s) of x for which

(a) $y = 0$ (b) $y = 4$

(c) $y > 0$ (d) $y < 0$

12. Let $y = \dfrac{x^2 - x}{x + 2}$.

(a) Compute the value of y for at least 20 values of x in the interval $-6 \le x \le 4$.

(b) Describe the y values when the x values are greater than 4; when the x values are less than -6.

(c) Use the information in (a) and (b), plotting additional points if needed, to give the complete graph of $y = \dfrac{x^2 - x}{x + 2}$.

13. Let $y = \dfrac{x^2 - x}{x + 2}$. Use the graph in Problem 12 to find the value(s) of x for which

(a) $y = 1$ (b) $y = -2$ (c) $y > 1$

14. Let $y = \dfrac{1}{x^2 - 4}$.

(a) Compute the value of y for at least 16 values of x in the interval $-4 \le x \le 4$.

(b) Describe the y values when the x values are greater than 4; when the x values are less than -4.

(c) Use the information in (a) and (b), plotting additional points if needed, to give the complete graph of $y = \dfrac{1}{x^2 - 4}$.

15. Let $y = \dfrac{1}{x^2 - 4}$. Use the graph in Problem 14 to find the value(s) of x for which

(a) $y = -1$ (b) $y < -1$

16. A motor that turns at 360 rpm is connected to a gear with 42 teeth. This gear is connected to a second gear.

(a) Write an equation that shows how the speed of the second gear depends on the number of teeth on the second gear.

(b) Give the complete graph of the equation in (a) and indicate which portion of the graph represents this problem.

6.4 / GRAPHS OF EXPONENTIAL EQUATIONS

We have seen powers of algebraic expressions in equations that arise from compound interest, inflation, population growth, and carbon dating. In this section we will study graphs of this type of equation and use graphs to answer questions about problems the equations describe.

Example 1 Draw the graph of the equation $y = 2^x$.

We have discussed the meaning of the expression 2^x in cases where x is a fraction or an integer. For example,

$$2^3 = 2 \cdot 2 \cdot 2, \ 2^{-3} = \frac{1}{2^3}, \ 2^0 = 1, \ \text{and } 2^{1/2} \doteq 1.414.$$

To look at the graph of $y = 2^x$ for x between -5 and 5, we compute the following values. (A keying sequence for the value of 2^{-4}, for example, is $\boxed{2}\ \boxed{y^x}\ \boxed{4}\ \boxed{+/-}\ \boxed{=}$.)

x	-5	-4	-3	-2	-1	0	1	2	3	4	5
$y = 2^x$	.03	.06	.13	.25	.5	1	2	4	8	16	32

These points are shown in Fig. 6.29.

All powers of 2 are positive numbers. The negative powers get near 0; the positive powers get very large. To obtain additional points for the graph, we choose values of x halfway between the ones we already have.

x	-4.5	-3.5	-2.5	-1.5	$-.5$	.5	1.5	2.5	3.5	4.5
y	.04	.09	.18	.35	.71	1.41	2.83	5.66	11.31	22.63

These values give the points in Fig. 6.30.

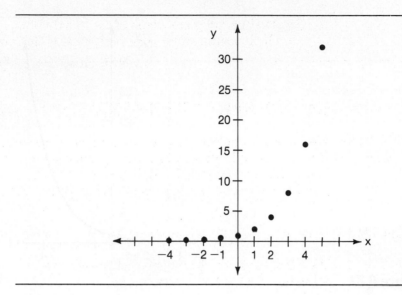

Figure 6.29

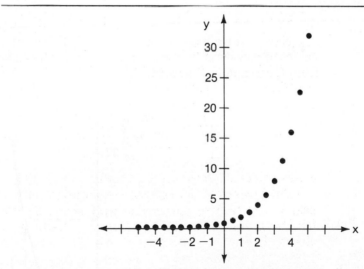

Figure 6.30

If we were able to fill in all the points on the graph of $y = 2^x$, it would look like Fig. 6.31.

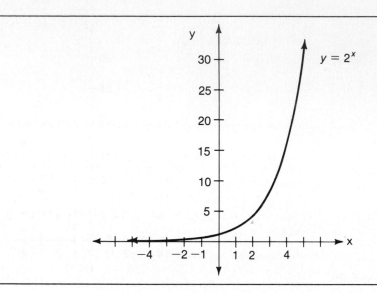

Figure 6.31

Now that we have the graph of $y = 2^x$, we can use it to estimate the values of powers of 2. For example, see if you can read the values of $2^{4.25}$ from the graph in Fig. 6.32.

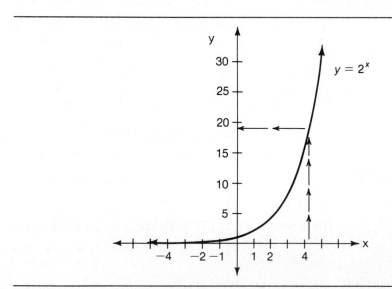

Figure 6.32

According to the calculator, $2^{4.25} \doteq 19.027314$. Is your estimate close to the actual value?

Example 2   Draw the graph of $y = \left(\dfrac{1}{2}\right)^{x}$.

We can start by computing y for certain values of x between -5 and 5.

x	-5	-4	-3	-2	-1	0	1	2	3	4	5
y	32	16	8	4	2	1	$.5$	$.25$	$.13$	$.06$	$.03$

Compare these values to the values we obtained for 2^{x} in Example 1. You should observe that

$$2^{-n} = \left(\frac{1}{2}\right)^{n}.$$

This is not surprising because

$$2^{-n} = \frac{1}{2^{n}} = \left(\frac{1}{2}\right)^{n}.$$

When we plot these points, we will see graphically the relationship between 2^{x} and $\left(\dfrac{1}{2}\right)^{x}$ for each value of x (Fig. 6.33).

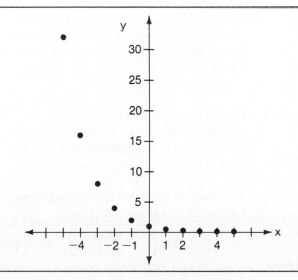

Figure 6.33

The complete graph of $y = \left(\frac{1}{2}\right)^x$ is shown in Fig. 6.34.

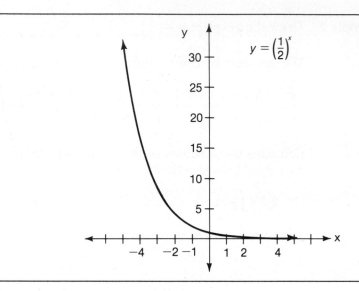

Figure 6.34

One of the problems that involves exponents is a compound interest problem. We described this type of problem with a graph in Chapter 4. We can also write an equation that describes the problem and then graph the equation. We will, of course, get the same graph as we did in Chapter 4.

Example 3   Nancy deposits $1000 in an account that pays 18% interest compounded quarterly. Write an equation that shows the amount of money in the account after interest has been paid t times. Then graph the equation for $t \geq 0$.

To compute the total amount of money M in the account after t interest payments, we must compute

$$M = 1000 \left(1 + \frac{.18}{4}\right)^t, \quad \text{or} \quad M = 1000(1.045)^t \text{ for } t \geq 0.$$

Look again at the computation in Example 1 of Section 4.3. You should see that the graph there is, in fact, the graph of $M = 1000(1.045)^t$ for $t \geq 0$. The graph of $M = 1000(1.045)^t$ for all values of t is shown in Fig. 6.35.

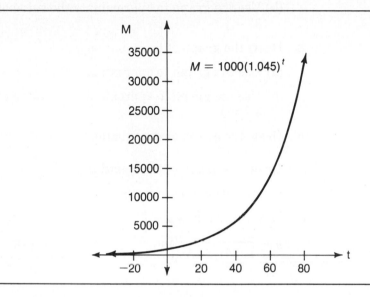

Figure 6.35

However, the portion of this graph for $t < 0$ does not apply to the problem situation.   □

Exercises 6.4

Use your calculator to evaluate the expressions in Exercises 1–4.

1. $3^{2.75}$

2. $3^{-1.25}$

3. $(.45)^{2.7}$

4. $\dfrac{5^{1.5} + 5^{-1.5}}{2}$

5. Find the smallest whole number n for which

   (a) $2^n \geq 1000$

   (b) $(1.5)^n \geq 1000$

   (c) $(1.2)^n \geq 1000$

   (d) $(1.005)^n \geq 1000$

6. Find the smallest whole number n for which

(a) $(.1)^n \le .001$ (b) $(.5)^n \le .001$ (c) $(.9)^n \le .001$ (d) $(.99)^n \le .001$

7. Draw the graph of the equation $y = 3^x$.

(a) Use the graph to approximate the value of $3^{2.75}$ and $3^{-1.25}$.

(b) Use the graph to approximate the value of x for which $3^x = 7$, $3^x = \frac{1}{4}$.

8. Draw the graph of the equation $y = 3^{-x}$.

(a) Use the graph to approximate the value of $3^{3.75}$ and $3^{-2.25}$.

(b) Use the graph to approximate the value of x for which $3^{-x} = 7$, $3^{-x} = \frac{1}{4}$.

9. Draw the graph of the equation $y = 5^x$.

10. Draw the graph of the equation $y = \left(\frac{1}{5}\right)^x$. Describe how you can obtain this graph from the graph of $y = 5^x$.

11. Use the graph of $y = 2^x$ and $y = 2^{-x}$ below to draw the graph of $y = \dfrac{2^x + 2^{-x}}{2}$. $\left(\text{Hint: } \dfrac{2^x + 2^{-x}}{2} \text{ is the average of } 2^x \text{ and } 2^{-x}.\right)$

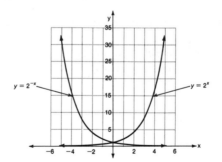

12. Use the graphs in Problems 7 and 8 to graph $y = \dfrac{3^x + 3^{-x}}{2}$.

13. The number of bacteria in a culture doubles every 4 hours. There are 500 present initially.

(a) Write an equation that gives the number of bacteria present for any time t.

(b) Draw the complete graph of the equation in (a) and indicate which portion is the graph of the problem.

6.5 / GRAPHS OF SPECIAL LINEAR EQUATIONS

In this section and the next we want to study equations whose graphs are lines. It should not surprise you that *linear* equations turn out to be the ones that have lines for their graphs. You have already graphed some linear equations by plotting many points. We will identify certain characteristics of linear equations that make it possible to graph them without plotting a large number of points.

Equations of the Form $y = mx$

One of the simplest examples of a linear equation in two variables is $y = 2x$. Consider these points on the graph of $y = 2x$:

x	-5	-4	-3	-2	-1	0	1	2	3	4	5
$y = 2x$	-10	-8	-6	-4	-2	0	2	4	6	8	10

These points are shown in Fig. 6.36.

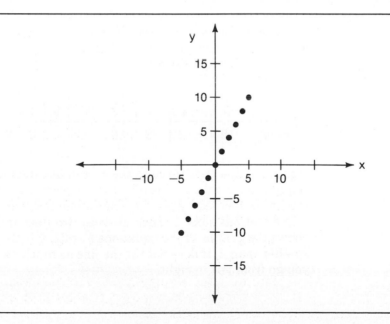

Figure 6.36

In each case the y value of a point is obtained by doubling the x value. The result is the line through the origin in Fig. 6.37.

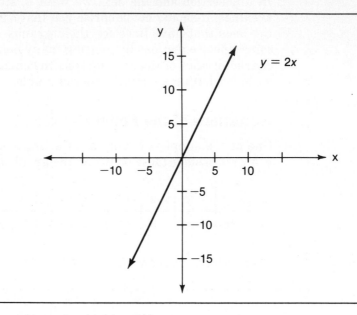

Figure 6.37

The multiplier of x, that is to say, the coefficient 2 in the example $y = 2x$, determines how the line slants with respect to the horizontal axis. We call it the **slope** of the line. Compare the graph of $y = 2x$ to the graph of $y = 3x$.

x	-5	-4	-3	-2	-1	0	1	2	3	4	5
$y = 3x$	-15	-12	-9	-6	-3	0	3	6	9	12	15

Again the line goes through the origin, but the y values are obtained by tripling the x values. The effect is to describe a line that is steeper than the line described by $y = 2x$. Thus a line that has slope 3 is steeper than a line that has slope 2. Look at these two lines in Fig. 6.38. Figure 6.39 shows the graphs of the equations $y = 3x$, $y = 2x$, and $y = \frac{1}{2}x$. Since $\frac{1}{2}$ is smaller than 2, it does not tilt the line as much as the coefficient 2 when you go from left to right.

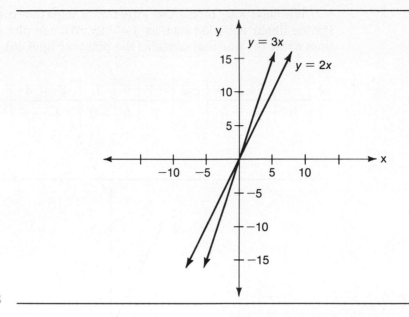

Figure 6.38

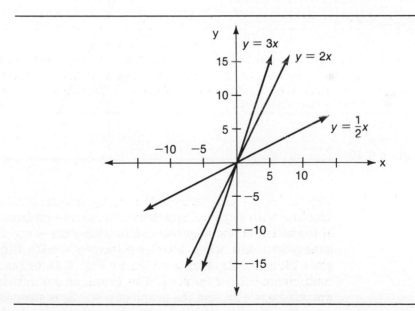

Figure 6.39

We need also to see the effect of a negative coefficient of x in a simple linear equation such as $y = -2x$. We can plot several points and then draw the line that contains the points (Fig. 6.40).

x	-5	-4	-3	-2	-1	0	1	2	3	4	5
y	10	8	6	4	2	0	-2	-4	-6	-8	-10

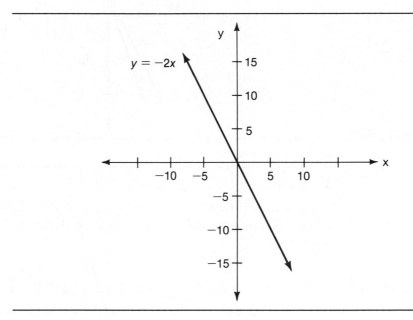

Figure 6.40

The line with slope -2 appears to fall as you go from left to right. This is the effect of the negative coefficient: since $y = -2x$, when $x = 0$ we have $y = 0$, but when $x = 5$ we have $y = -10$. Study the graphs of $y = -3x$, $y = -2x$, and $y = -\frac{1}{2}x$ in Fig. 6.41 to see the effects of the coefficients -3, -2, and $-\frac{1}{2}$. The graph of $y = -3x$ is steeper than the graph of $y = -2x$, and the graph of $y = -2x$ is steeper than the graph of $y = -\frac{1}{2}x$. Observe that the origin is a point on each of these lines and is, in fact, a point on the graph of $y = mx$ for any number m.

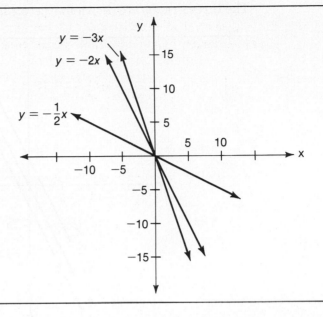

Figure 6.41

Equations of the Form $y = mx + b$

Linear equations such as $y = 2x + 4$ or $y = 2x - 3$ have graphs that are related to the graph of $y = 2x$. You need to understand what this relationship is.

Example 1   On the same coordinate system, graph $y = 2x$, $y = 2x + 4$, and $y = 2x - 3$.

The important thing to observe is that for each x, the y value, when $y = 2x + 4$, is 4 more than it is for $y = 2x$; for each x value, the y value for $y = 2x - 3$ is 3 less than $y = 2x$:

x	-4	-2	0	2	4
$y = 2x$	-8	-4	0	4	8
$y = 2x + 4$	-4	0	4	8	12
$y = 2x - 3$	-11	-7	-3	1	5

The three graphs are shown in Fig. 6.42.

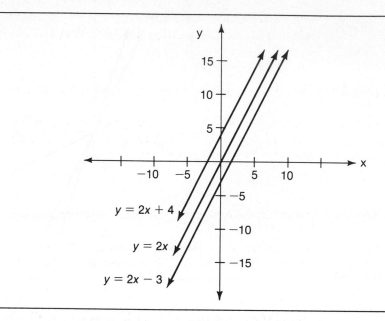

Figure 6.42

☐

The graph of $y = 2x + 4$ is obtained from the graph of $y = 2x$ by moving each point up 4 units; the graph of $y = 2x - 3$ is obtained from the graph of $y = 2x$ by moving each point down 3 units. Thus the three lines are parallel. In fact, the graph of any equation of the form $y = 2x + b$ is obtained from the graph of $y = 2x$ by moving each point up b units if b is positive or down b units if b is negative. The result is the same for any coefficient of x. Therefore, the graph of equations of the form $y = mx + b$ and $y = mx + c$ are parallel.

Observe that the graph of $y = 2x + 4$ crosses the vertical axis at 4. This happens because $y = 4$ when $x = 0$. The graph of $y = 2x - 3$ crosses the vertical axis at -3 because $y = -3$ when $x = 0$.

The graph of a linear equation of the form $y = mx + b$ is determined by the two numbers m and b. The coefficient m (for multiplier) determines how much the line slants and whether it goes up or down; m

is called the **slope** of the line. In the example we have seen that *lines with the same slope are parallel.* Also, parallel lines have the same slope. The number b in the equation $y = mx + b$ tells where the line crosses the vertical axis; b is called the **y-intercept** of the line.

Example 2   Give the slope and y-intercept of the graph of $y = -x + 5$.

We must remember that an equation $y = mx + b$ has for its graph a line with slope m and y-intercept b. Thus the graph of $y = -x + 5$ is a line with slope -1 and y-intercept 5.   □

Example 3   Write the equation of a line with slope -4 and y-intercept 1.5.

A line with slope m and y-intercept b has equation $y = mx + b$. Thus a line with slope -4 and y-intercept 1.5 has equation $y = -4x + 1.5$ □

Graphing a Line with Given Slope and y-Intercept

If the slope and y-intercept of a line are known, the line can be graphed efficiently.

Example 4   Graph the line that has slope $\frac{2}{3}$ and y-intercept 0.

a. One way to graph this line is first to write the equation $y = \frac{2}{3}x + 0$ and then to graph the equation by plotting several points.

b. A second approach is to plot the y-intercept 0 first and then analyze what it means for the line to have slope $\frac{2}{3}$. Since the equation of the line is $y = \frac{2}{3}x$, if $x = 3$, then $y = 2$. To move from (0,0) to (3,2) you move 3 units right and 2 units up (or 2 units up and 3 units right) as shown in

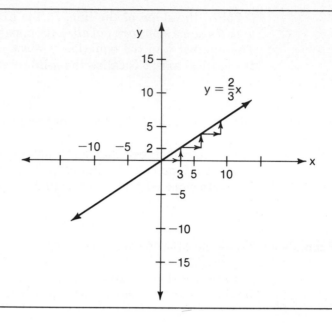

Figure 6.43

□

Fig. 6.43. [Notice that if you start at the point (3,2) on the line and again go 3 units right and 2 units up you locate the point (6,4), another point on the line. If you start at (6,4) and go 3 units right and 2 units up, you locate (9,6).] Thus we can graph the line with slope $\frac{2}{3}$ and y-intercept 0 by starting at (0,0), then moving 3 units right and 2 units up to locate a second point. Once we have two points, we can draw the line through them.

Example 5   Graph $y = -\frac{4}{3}x$.

If we want to graph this line without plotting several points, we can recognize first that the origin (0,0) is a point on the line. Notice that if $x = 3$, then $y = -4$. This means that by starting at the origin and moving 3 units *right* and then moving 4 units *down*, we locate a second point (Fig. 6.44).   □

    The interpretation of slope that we used in the examples above can be used in graphing any line. The graph of $y = -\frac{4}{3}x + 3$ is a line with slope $-\frac{4}{3}$ and y-intercept 3. Viewing the slope as $\frac{-4}{3}$, we can graph the line as in Fig. 6.45 by starting at (0,3) and moving 3 units right and then 4 units down to locate the point (3,−1). If we had viewed the slope

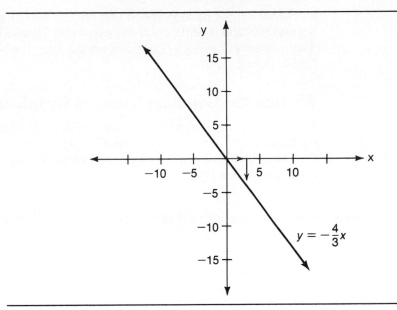

Figure 6.44

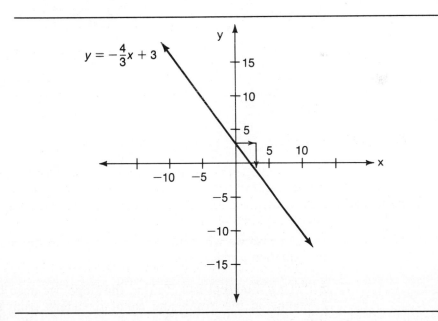

Figure 6.45

as $\frac{4}{-3}$, we might have graphed the line by starting at (0,3) and moving 3 units left and 4 units up to the point (−3,7), another point on the line. The y-intercept gives a first point on the line; the slope indicates how to find a second point.

Writing the Equation from the Graph of a Line

We have been concerned with the problem of finding the graph when we know an equation. We can turn the problem around and attempt to find an equation when we know the graph. This problem is not difficult if the graph is a line.

Example 6   Find the slope, y-intercept, and equation for lines L_1 and L_2 in Fig. 6.46.

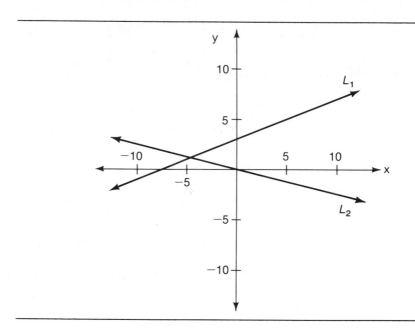

Figure 6.46

In each case we will find the slope m, the y-intercept b, and then write the equation $y = mx + b$.

L_1: The y-intercept is 3. To find the slope, we need to identify a second point on the graph and count the steps from (0,3) to the second point. Say we identify (5,5). To move from (0,3) to (5,5) we go 5 units right and 2 units up. Thus the slope is $\frac{2}{5}$. The equation is $y = \frac{2}{5}x + 3$.

L_2: The y-intercept is 0. To find the slope we need a second point, say (4,−1). Moving from (0,0) to (4,−1) requires us to move 4 units right and 1 unit down. Thus the slope is $\frac{-1}{4}$. The equation is $y = -\frac{1}{4}x + 0$

or $y = -\frac{1}{4}x$.   □

Exercises
6.5

1. Let $y = 2x + 3$.

(a) Complete the following table:

x	−2	−1	0	1	2	3
y						

(b) Graph the points in (a) and use a straightedge to verify that these points lie on a line.

(c) Draw the line $y = 2x + 3$.

(d) Use the graph in (c) to find the value of y if x is $-\frac{3}{2}, -\frac{1}{2}, \frac{1}{2}, \frac{3}{2}$.

(e) Use the graph in (c) to find the value of x if y is $-2, \frac{5}{2}, 8$.

Draw the graph for each of the equations in Exercises 2–9.

2. $y = \frac{5}{2}x$

3. $y = -\frac{3}{2}x$

4. $y = x$

5. $y = -x$

6. $y = \frac{4}{3}x - 2$

7. $y = \frac{5}{4}x - 1$

8. $y = -\frac{1}{4}x + 2$

9. $y = -\frac{4}{5}x + 2$

10. On the same coordinate system, graph $y = -\frac{1}{2}x$, $y = -\frac{1}{2}x + 2$, and $y = -\frac{1}{2}x - 3$.

Give the slope and y-intercept of the graphs of each of the equations in Exercises 11–13.

11. $y = \frac{5}{3}x$

12. $y = -3x + 4$

13. $y = \frac{x}{4} - 2$

Write the equations for the lines with slopes and y-intercepts as given in Exercises 14–16.

14. Slope $\frac{3}{4}$, y-intercept -2

15. Slope -1, y-intercept 3

16. Slope $-\frac{1}{2}$, y-intercept 2.5

17. Write equations for three distinct lines parallel to $y = -2x + 1$.

In Exercises 18–23, fill in the blanks to describe how to move from point A to point B.

18. $A = (0,0)$, $B = (3,4)$. Move _____ units right and then _____ units up.

19. $A = (0,0)$, $B = (2,-3)$. Move _____ units right and then _____ units down.

20. $A = (0,2)$, $B = (4,5)$. Move _____ units right and then _____ units up.

21. $A = (0,2)$, $B = (5,-1)$. Move _____ units right and then _____ units down.

22. $A = (0,-3)$, $B = (1,0)$. Move _____ units right and then _____ units up.

23. $A = (0,-3)$, $B = (3,-5)$. Move _____ units right and then _____ units down.

24. Find the slope and write an equation for each line below:

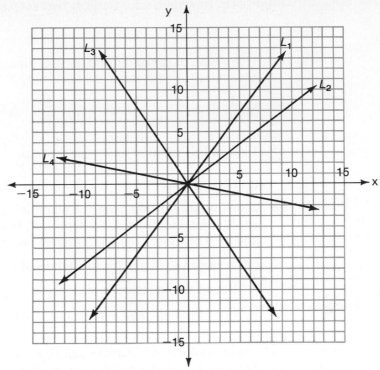

25. Find the slope and y-intercept, and write an equation for each line below:

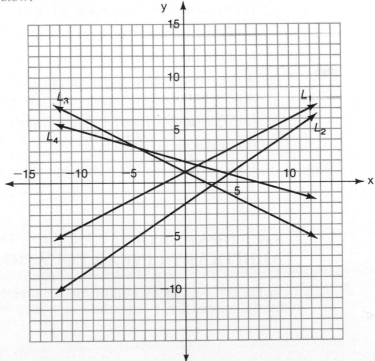

26. Use graph paper and the same coordinate system to draw the lines through (0,0) with slope

   (a) $\dfrac{1}{3}$ (b) $\dfrac{3}{2}$ (c) -1 (d) $-\dfrac{2}{3}$ (e) 1

27. Draw a line parallel to $y = \dfrac{2}{3}x$ through

   (a) $(1,2)$ (b) $(-2,-1)$

28. Jane invests \$18,000, part at 8% annual interest and the remainder at 5% annual interest. Let x represent the amount invested at 8%, and let y represent the total interest received at the end of one year.

   (a) Write an equation for y in terms of x.

   (b) Graph the equation in (a).

   (c) Find the slope and y-intercept of the graph in (b).

29. A 22%-alcohol solution is mixed with a 54%-alcohol solution to produce 40 gallons of a new solution. Let x represent the amount of 22% solution used, and let y represent the amount of alcohol in the new solution.

   (a) Write an equation for y in terms of x.

   (b) Graph the equation in (a).

   (c) Find the slope and y-intercept of the graph in (b).

30. A department store marks up wholesale prices 60%. Let x represent the wholesale price, and let y represent the retail price.

   (a) Write an equation for y in terms of x.

   (b) Graph the equation in (a).

   (c) Find the slope and y-intercept of the graph in (b).

6.6 / GRAPHS OF ARBITRARY LINEAR EQUATIONS

In the last section we studied linear equations in two variables that have a special form—equations such as $y = mx + b$ for two numbers

m and b. Graphs of these equations are lines that have slope m and y-intercept b. For example, the graph of the equation $y = -3x + \frac{1}{2}$ is a line with slope -3 and y-intercept $\frac{1}{2}$; the graph of the equation $y = \frac{3}{4}x - 7$ is a line with slope $\frac{3}{4}$ and y-intercept -7.

However, not every linear equation in two variables is written in the form $y = mx + b$. Here are some examples of equations that have different forms:

$$3x + 4y = 8$$

$$x = -2y + \frac{1}{2}$$

$$-5x + 2y + 6 = 0.$$

Even though these equations do not look like the special equation $y = mx + b$, they can be put in that form by solving for the variable y. Then it becomes clear that the graph of each is a line, and the slope and y-intercept can be read directly from the equation.

Example 1 Give the slope and y-intercept of the graph of $3x + 4y = 8$; then graph the equation.

First we solve the equation for the variable y:

$$3x + 4y = 8$$

$$-3x + 3x + 4y = -3x + 8 \qquad \text{(Add } -3x \text{ to both sides)}$$

$$4y = -3x + 8$$

$$y = -\frac{3}{4}x + 2. \qquad \text{(Divide both sides by 4)}$$

Now we have an equation equivalent to the original equation from which we can read the slope and the y-intercept. Remember that the graph of an equation in the form $y = mx + b$ is a line with slope m and y-intercept b. Thus the graph of the equation $y = -\frac{3}{4}x + 2$ has slope $-\frac{3}{4}$ and y-intercept 2. We can graph the equation (Fig. 6.47) by first locating the y-intercept $(0,2)$. Then using the fact that the slope is $-\frac{3}{4}$, move 4 units right and 3 units down to the point $(4, -1)$ to find a second point on the line. □

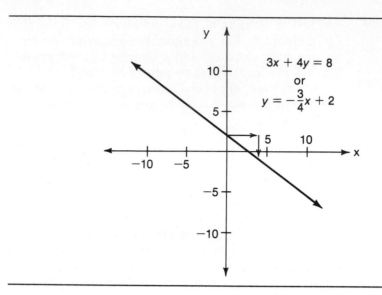

Figure 6.47

Example 2   Give the slope and y-intercept of the graph of $x = -2y + \dfrac{1}{2}$; then graph the equation.

We change the form of the equation by solving for the variable y:

$$x = -2y + \frac{1}{2}$$

$$x - \frac{1}{2} = -2y \qquad \text{(Subtract } \frac{1}{2} \text{ from both sides)}$$

$$-\frac{1}{2}x + \frac{1}{4} = y. \qquad \left(\begin{array}{l} \text{Divide both sides by } -2, \text{ or} \\ \text{multiply by } -\frac{1}{2} \end{array} \right)$$

In this form the slope is $-\dfrac{1}{2}$ and the y-intercept is $\dfrac{1}{4}$. We can graph the equation by locating $\left(0, \dfrac{1}{4}\right)$ on the vertical axis and then moving 2 units right and 1 unit down to the point $\left(2, -\dfrac{3}{4}\right)$, as is shown in Fig. 6.48.   □

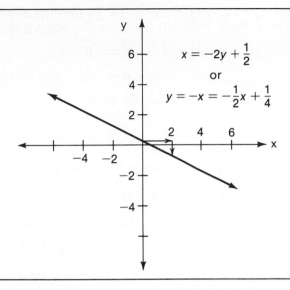

Figure 6.48

Intercepts

If our goal in Examples 1 and 2 were only to graph the equations, we could have done that more efficiently than to find the slope and the y-intercept. We can graph a line by locating any two points on the line. Two easy points to locate are the points on the horizontal and vertical axes if the line crosses both of these axes. The y-coordinate of the point where a line crosses the vertical axis is called its **y-intercept**. (Observe that this point has x-coordinate 0.) The x-coordinate of the point where a line crosses the horizontal axis is called its **x-intercept**. (Observe that this point has y-coordinate 0.)

Example 3 Find the x-intercept and the y-intercept for the graph of $-5x + 2y + 6 = 0$. Then graph the equation.

To find the y-intercept we must answer the question, "What is the value of y when $x = 0$?"

$$-5(0) + 2y + 6 = 0$$
$$2y + 6 = 0$$
$$2y = -6$$
$$y = -3. \quad \text{(Divide both sides by 2)}$$

The y-intercept is -3. To find the x-intercept we must answer the question, "What is the value of x when $y = 0$?"

$$-5x + 2(0) + 6 = \;\; 0$$
$$-5x + 6 = \;\; 0$$
$$-5x = -6$$
$$x = \frac{-6}{-5} = \frac{6}{5}. \quad \text{(Divide both sides by } -5\text{)}$$

The x-intercept is $\dfrac{6}{5}$. We can graph the equation by plotting the two intercepts (Fig. 6.49).

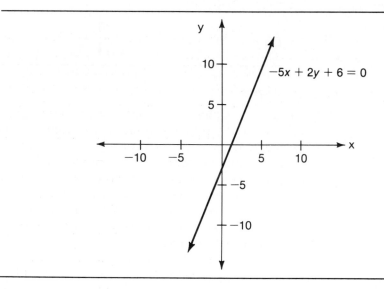

Figure 6.49

□

Horizontal and Vertical Lines

Notice in the above example that finding the y-intercept required that we divide by the y-coefficient 2 in the computation; finding the x-intercept required that we divide by the x-coefficient -5 in the computation. If either of these coefficients had been 0, we could not have found the x- or y-intercept. We need to investigate separately the graphs of equations $ax + by = c$, where either a or b is 0. Here are some equations of that form:

$$x = 5$$
$$2x = -7$$
$$y = 15$$
$$\frac{1}{2}y = -3.$$

We can think of these as equations in two variables, where one variable has coefficient 0. With this interpretation, each equation has many solutions, and we can graph the equation by plotting points that have the solutions as coordinates.

Example 4   Graph the equation $x = 5$.

We want to locate the points whose coordinates are solutions to $x = 5$. Here are some: $(5, -12)$, $(5, -3)$, $(5,0)$, $(5,2)$, $(5,5)$, and $(5,7)$. Any point with x-coordinate 5 will give a solution; the y-coordinate can be any number. The graph consists of all points with x-coordinates 5. The x-intercept is 5; there is no y-intercept. The graph is shown in Fig. 6.50.

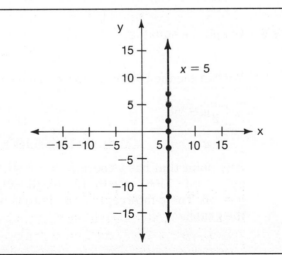

Figure 6.50

□

The graph of $x = -5$ is parallel to the graph of $x = 5$ and has x-

intercept -5. In fact, the graphs of $x = 5$, $x = -5$, $x = \frac{1}{2}$, and $x = -2\frac{1}{3}$ are all parallel and are all vertical. Their graphs are in Fig. 6.51.

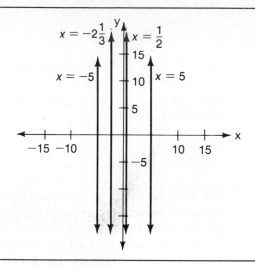

Figure 6.51

Example 5   Graph the equation $\frac{1}{2}y = -3$.

We can see the points on this graph easily if we solve for y:

$$\frac{1}{2}y = -3$$

$$y = -6. \qquad \text{(Multiply both sides by 2)}$$

Any point that has y-coordinate -6 will lie on this graph. For example, $(-5,-6)$, $(-1,-6)$, $(0,-6)$, $(2,-6)$, $(3,-6)$, and $(5,-6)$ are all solutions to $y = -6$. The y-intercept is -6; there is no x-intercept. Figure 6.52 shows the graph.   □

You should see that the graphs of $y = 6$, $y = \frac{1}{4}$, and $y = -2$ are all horizontal and are parallel to the graph of $y = -6$. Their graphs are shown in Fig. 6.53.

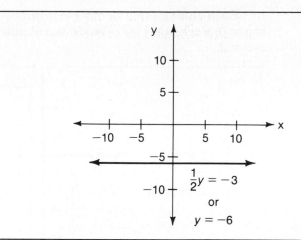

Figure 6.52

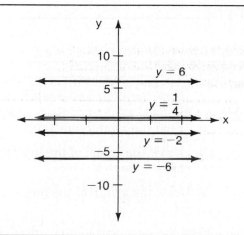

Figure 6.53

Example 6   Write an equation for the line that contains (2,3) and is

(a) parallel to the x-axis,

(b) parallel to the y-axis.

The two lines are shown in Fig. 6.54.

Each point on line L_1 has y-coordinate 3. An equation describing L_1 is $y = 3$. Each point on L_2 has x-coordinate 2. An equation describing L_2 is $x = 2$.

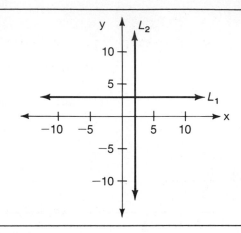

Figure 6.54

□

Exercises 6.6

1. Find the slope and y-intercept of the graph of $\dfrac{x}{3} - y = 4$.

2. Find the x-intercept and y-intercept of the graph of $3x + 2y = 6$.

3. Draw the graph of the line with y-intercept -2 that is parallel to $2x + 3y = 4$.

4. Draw the graph of the line with x-intercept -1 and y-intercept -2.

Find the slope and y-intercept of the graphs of the equations in Exercises 5–10. Then graph the equation.

5. $x + 2y = 1$ 6. $3x - y = 2$

7. $4x + 3y + 2 = 0$ 8. $2x - 3y = 0$

9. $x + y = 5$ 10. $3x - 4y + 6 = 0$

Find the x-intercept and y-intercept of the graphs of the equations in Exercises 11–18. Then graph the equation.

11. $x - 2y = 4$ 12. $2x + 3y = 0$

13. $\dfrac{x}{2} + \dfrac{y}{3} = 1$

14. $2x - 3y = 1$

15. $x + y = 4$

16. $3x - 4y + 6 = 0$

17. $2x + 3 = 0$

18. $3y - 1 = 0$

Draw the graphs of the equations in Exercises 19–25.

19. $x + \dfrac{1}{2}y + 1 = 0$

20. $2x - 4 = 0$

21. $5x - 3y + 10 = 0$

22. $\dfrac{2}{3}y - 3 = 0$

23. $\dfrac{1}{2}x + 2 = 0$

24. $\dfrac{1}{2}x - \dfrac{1}{3}y = 1$

25. $4y + 3 = 0$

26. Write an equation for the line parallel to the y-axis that passes through

  (a) $(-1,0)$

  (b) $(-3,1)$

  (c) $(2,4)$

  (d) $(3,0)$

27. Write an equation for the line parallel to the x-axis that passes through

  (a) $(0,-2)$

  (b) $(-1,-1)$

  (c) $(1,2)$

  (d) $(2,4)$

28. Write an equation for the line with y-intercept 3 that is parallel to $5x - 3y = 1$.

6.7 / CHAPTER 6 PROBLEM COLLECTION

1. Find the value of

  (a) $\cos 75$

  (b) $\log 8$

  (c) $7^{1.6}$

  (d) $10^{-2.1}$

2. Find all integers x that satisfy

  (a) $-2 < x \leq 1$

  (b) $-\dfrac{1}{2} \leq x \leq 3$

3. Write the points that divide the interval $-1 \le x \le 3$ into 16 pieces of equal length.

Evaluate the expressions in Exercises 4 and 5 for the values of the variable specified.

4. $2x^5 - 3x^3 + x^2 - 2x + 1$; $x = -1.5, 2.5$

5. $\dfrac{x^2 - 3x}{2x + 1}$; $x = -.5, 1.8$

6. The graph below represents y as a function of x.

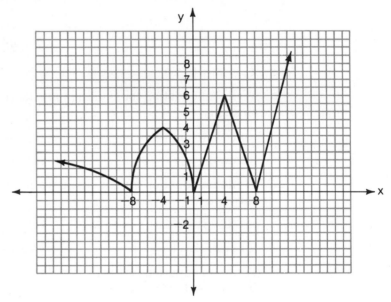

(a) Find y if $x = -2, -.5, 8, 12$.

(b) Find x if $y = -1, 0, 4, 6$.

(c) Find the value(s) of x for which $y > 4$.

(d) Find the value(s) of x for which $y < 0$.

7. Use graph paper to draw the line through $(0,-2)$ with slope

    (a) 2   (b) -2

    (c) $\dfrac{2}{3}$ (d) $-\dfrac{2}{3}$

8. Graph $y = \frac{3}{4}x$, $y = \frac{3}{4}x - 4$, and $y = \frac{3}{4}x + 3$ on the same coordinate system.

Find the slope and y-intercept of the graph of each equation in Exercises 9–13. Then graph the equation.

9. $5x + 3y = 4$

10. $\frac{1}{3}x - \frac{1}{4}y = 1$

11. $2x - 5 = 0$

12. $2y + 5 = 0$

13. $2y - 3x = 0$

Find the x-intercept and y-intercept of the graphs of the equations in Exercises 14–18. Then graph the equations.

14. $4x - 5y = 8$

15. $\frac{1}{3}x + \frac{1}{5}y = 1$

16. $y - x = 0$

17. $2y - 3 = 0$

18. $3x + 4 = 0$

19. Let $y = \dfrac{1}{\cos x}$.

   (a) Compute the value of y for x values in the interval $-90 \le x \le 270$ that are 10 units apart.

   (b) Graph the points determined in (a).

   (c) Give the complete graph of $y = \dfrac{1}{\cos x}$ in the interval $-90 \le x \le 270$.

Draw the graph of the equations in Exercises 20–33.

20. $y = \frac{2}{3}x + 1$

21. $y = -\frac{3}{4}x - 2$

22. $\frac{x}{2} - \frac{y}{5} = 1$

23. $\frac{2}{3}x + \frac{3}{4}y = 6$

24. $5x + 3y = 12$

25. $3x - 5 = 0$

26. $4y + 7 = 0$

27. $y = x^2 - x$

28. $y = x^3 + x$

29. $y = \sqrt{2x - 1}$

30. $y = \dfrac{1}{x - 4}$

31. $y = \dfrac{x}{x^2 - 1}$

32. $y = (1.8)^x$

33. $y = (.3)^x$

34. Below is the graph of $y = \dfrac{1}{8}x^3$. On the same coordinate system, draw the graph of $y = \dfrac{1}{8}x^3 - 2$.

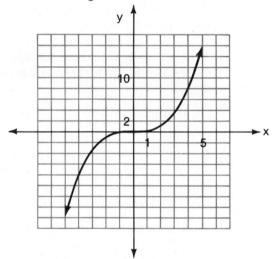

35. Find the slope and y-intercept, and write an equation for the lines L_1 and L_2 below:

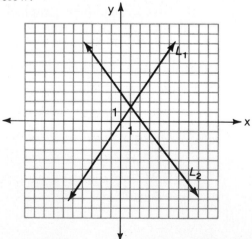

Write an equation for each line in Exercises 36–39.

36. Slope $-\dfrac{1}{2}$ and y-intercept 3

37. y-intercept -2 and parallel to $y = 3x - 1$

38. Through $(3,1)$ and parallel to the y-axis

39. Through $(-2,-1)$ and parallel to the x-axis

40. A department store marks up wholesale prices 25%. Let x denote the wholesale price and let y represent the corresponding retail price.

   (a) Write an equation for y in terms of x.

   (b) Give the complete graph of the equation in (a) and indicate which portion of the graph represents the problem.

   (c) Use the graph to find the wholesale price if the retail price is $90.

41. Bill drives to a train station at an average speed of 50 mph and then completes his 300-mile trip on a train that averages 80 mph. Let x represent the distance Bill travels by car, and let y represent the total time for the trip.

   (a) Write an equation for y in terms of x.

   (b) Give the complete graph of the equation in (a) and indicate which portion of the graph represents the problem.

   (c) Use the graph to find how far Bill travels by car if the trip takes 4 hours.

42. An open rectangular box is to be made from a piece of cardboard 8 inches wide and 15 inches long by cutting a square from each corner and bending up the sides. Let x denote the side of the square, and let y represent the volume of the box.

   (a) Write an equation for y in terms of x.

   (b) Give the complete graph of the equation in (a) and indicate which portion of the graph represents the problem.

   (c) Use the graph to find the dimensions of the box of largest volume.

Linear Equations in Two Variables and Their Graphs

7.1 / SLOPE OF A LINE

If the equation of a line is known, we can find its slope by solving the equation for y and then reading the coefficient of x. It is also possible to find the slope of a line without knowing the equation if two points on the line are known.

To see how this is done, remember how we can sketch the line that is the graph of $y = \frac{2}{3}x + 4$. First, we locate the y-intercept $(0,4)$ (Fig. 7.1). Since the slope is $\frac{2}{3}$, we move 3 units right and 2 units up to the point $(3,6)$.

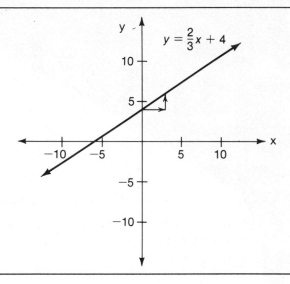

Figure 7.1

The coordinates $(3,6)$ are obtained from the coordinates $(0,4)$ by adding the numerator of the slope to the y-coordinate 4 and the denominator of the slope to the x-coordinate 0:

$$6 = 4 + 2$$

$$3 = 0 + 3.$$

Thus if we start with the coordinates of the two points $(3,6)$ and $(0,4)$, we can compute the numerator of the slope by taking the difference of the y-coordinates, and we can find the denominator of the slope by taking the difference of the x-coordinates:

$$6 - 4 = 2$$
$$3 - 0 = 3.$$

Consider two other points on the graph of $y = \frac{2}{3}x + 4$: $(1, 4\frac{2}{3})$ and $(-1, 3\frac{1}{3})$. (These points are labeled C and A in Fig. 7.2. D and F are the two points we considered above.)

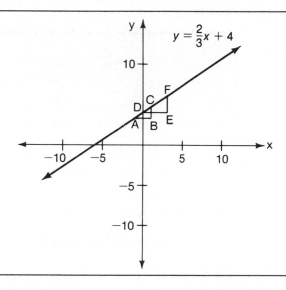

Figure 7.2

If we compute the difference of the y-coordinates and the difference of the x-coordinates, we can find the slope:

$$4\frac{2}{3} - 3\frac{1}{3} = 1\frac{1}{3}$$

$$1 - (-1) = 2.$$

The quotient $1\frac{1}{3} \div 2$ simplifies to $\frac{2}{3}$, the slope of the line:

$$\frac{1\frac{1}{3}}{2} = \frac{\frac{4}{3}}{2} = \frac{4}{3} \cdot \frac{1}{2} = \frac{2}{3}.$$

In Fig. 7.2 where the line has positive slope, the difference between the y-coordinates is the length of the vertical leg of the triangle, and the difference of the x-coordinates is the length of the horizontal leg. The two triangles ABC and DEF have corresponding angles that are congruent. Thus the triangles are similar. The ratio of the lengths of the legs in one triangle equals the ratio of the lengths of the legs in the other. In each, the length of the vertical leg divided by the length of the horizontal leg is the slope $\frac{2}{3}$. If we take any two points on the line and form a right triangle as we did above, the triangle will be similar to the other two, and thus the result will be the same. The slope of the line, therefore, can be computed from the coordinates of any two points on the line.

> If (a,b) and (c,d) are two points on a line, the slope of the line is $\dfrac{b-d}{a-c}$.

Sometimes this statement is summarized by saying that the slope can be computed from the coordinates of two points using the ratio

$$\frac{\text{change in } y}{\text{change in } x}.$$

Example 1   Find the slope of the line containing the points (6,4) and (2,1).

The graph of the line is shown in Fig. 7.3.

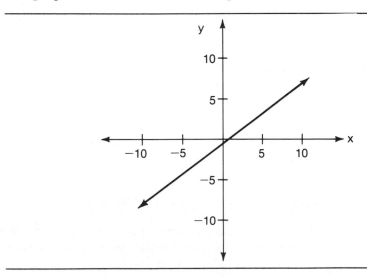

Figure 7.3

The slope equals $\dfrac{4-1}{6-2} = \dfrac{3}{4}$. What would happen if we had taken the two points in the opposite order?

$$\frac{1-4}{2-6} = \frac{-3}{-4} = \frac{3}{4}.$$

The order in which we use the points in the computation does not matter because there is a sign change in both the numerator and the denominator.   □

Example 2 Find the slope of the line that contains the points $(1,-3)$ and $(-2,4)$.

The graph of the line in Fig. 7.4 suggests that the slope is a negative number.

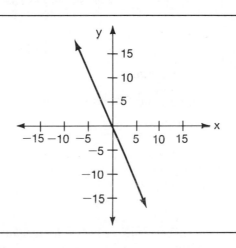

Figure 7.4

The slope is $\dfrac{-3-4}{1-(-2)} = \dfrac{-7}{3}$. If we take the points in the opposite order, the quotient is $\dfrac{4-(-3)}{-2-1} = \dfrac{7}{-3}$. Both computations show that the slope is $-\dfrac{7}{3}$.   □

Slopes of Lines Parallel to Axes

We want to analyze what happens when we attempt to compute the slope for a line that is parallel to the horizontal axis and for a line that is parallel to the vertical axis.

Example 3   Find the slope of the line through the points $(-2,3)$ and $(5,3)$.

These are two points on the line parallel to the horizontal axis with y-intercept 3 (Fig. 7.5).

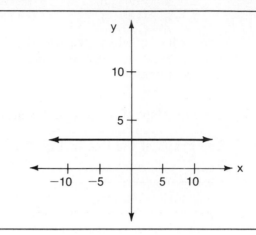

Figure 7.5

Computing the slope gives

$$\text{slope} = \frac{3-3}{-2-5} = \frac{0}{-7} = 0.$$

Thus the line has slope 0. If we recall that the equation of this line is $y = 3$ (or $y = 0x + 3$), we are not surprised it has slope 0.   □

Any line parallel to the horizontal axis has slope 0 because two points on the line have the same y-coordinates. When the quotient is computed to give the slope of the line, the numerator is always 0.

Example 4   Find the slope of the line through $(2,5)$ and $(2,-3)$.

This line is parallel to the vertical axis with x-intercept 2 (Fig. 7.6). When we attempt to compute the slope of this line, the numerator is $5 - (-3) = 8$ and the denominator is $2 - 2 = 0$. But there is no number $8 \div 0$. Thus this line has no slope.   □

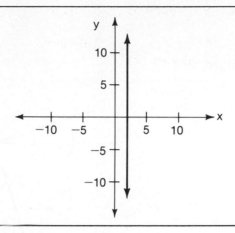

Figure 7.6

If a line is parallel to the vertical axis, any two points on the line have the same x-coordinate. When the quotient is computed to give the slope of the line, the denominator is 0. Thus *lines parallel to the vertical axis have no slope*; we often say their slope is *undefined*.

Slopes of Perpendicular Lines

We have observed that parallel lines have the same slope. We can also make a statement about the slopes of perpendicular lines. To see what this statement is, consider these examples.

Example 5   Compare the graphs of $y = 3x$ and $y = -\dfrac{1}{3}x$.

The graphs are shown in Fig. 7.7 on the same coordinate system. These two lines go through the origin and intersect in a right angle.   □

In the above example the graphs of $y = 3x$ and $y = -\frac{1}{3}x$ are perpendicular. A pair of lines, one parallel to each of these, will also be perpendicular to each other. For example, the graphs of $y = 3x - 2$ and $y = -\frac{1}{3}x + 4$ are perpendicular lines.

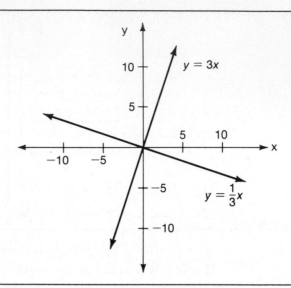

Figure 7.7

Example 6   Compare the graphs of $y = -\dfrac{4}{3}x + 1$ and $y = \dfrac{3}{4}x + 1$.

The two graphs are shown in Fig. 7.8 on the same coordinate system.

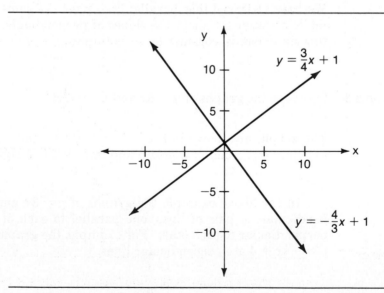

Figure 7.8

The two lines have the same y-intercept and intersect at right angles.

□

In fact, any line perpendicular to a line with slope $-\frac{4}{3}$ has slope $\frac{3}{4}$; any line perpendicular to a line with slope $\frac{3}{4}$ has slope $-\frac{4}{3}$. We can summarize these observations as follows:

Two lines are perpendicular whenever the slope of one is the opposite of the reciprocal of the slope of the other.

Using Slopes of Lines

The facts that two lines are parallel if they have the same slope and that two lines are perpendicular if the slope of one is the opposite of the reciprocal of the slope of the other help us to analyze some geometric situations.

Example 7   Determine whether the points with coordinates $(-1,-3)$, $(1,-1)$, $(2,3)$, and $(0,1)$ are vertices of a parallelogram.

The graph of these points in Fig. 7.9 suggests that they are vertices of a parallelogram, but we must check that the opposite sides are actually on parallel lines.

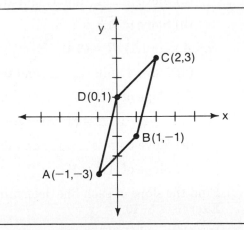

Figure 7.9

The symbol $\overleftrightarrow{AB}$ denotes the line through points A and B. Then

$$\text{Slope of } \overleftrightarrow{AB} = \frac{-3 - (-1)}{-1 - 1} = \frac{-2}{-2} = 1$$

$$\text{Slope of } \overleftrightarrow{DC} = \frac{1 - 3}{0 - 2} = \frac{-2}{-2} = 1$$

$$\text{Slope of } \overleftrightarrow{AD} = \frac{-3 - 1}{-1 - 0} = \frac{-4}{-1} = 4$$

$$\text{Slope of } \overleftrightarrow{BC} = \frac{-1 - 3}{1 - 2} = \frac{-4}{-1} = 4$$

Since the slopes of the opposite sides are equal, we can conclude that this four-sided figure is a parallelogram.   □

Exercises 7.1

In part (a) of Exercises 1–4 fill in the blanks to describe how to move from point A to point B.

1. $A = (2,-3)$, $B = (7,2)$

   (a) Move _____ units right and then _____ units up.

   (b) Slope of $\overleftrightarrow{AB}$ = ?

2. $A = (-1,3)$, $B = (4,-2)$

   (a) Move _____ units right and then _____ units down.

   (b) Slope of $\overleftrightarrow{AB}$ = ?

3. $A = (-1,1)$, $B = (3,4)$

   (a) Move 4 units _____ and then 3 units _____.

   (b) Slope of $\overleftrightarrow{AB}$ = ?

4. $A = (-2,-1)$, $B = (3,-3)$

   (a) Move 5 units _____ and then 2 units _____.

   (b) Slope of $\overleftrightarrow{AB}$ = ?

Find the slope of each line determined by the two points in Exercises 5–10.

5. $(1,1)$ and $(5,6)$

6. $(2,1)$ and $(6,-2)$

7. $(0,1)$ and $(5,-2)$

8. $(1,2)$ and $(-2,-1)$

9. $(2,-1)$ and $(2,6)$ 10. $(-1,2)$ and $(5,2)$

11. Find c so that the slope of the line determined by $(1,2)$ and $(3,c)$ is $\frac{2}{3}$.

12. Find c so that the slope of the line determined by $(-1,2)$ and $(3,c)$ is $-\frac{1}{2}$.

13. Find c so that the slope of the line determined by $(-1,3)$ and $(c, c + 2)$ is 2.

In Exercises 14–19, draw the line through each given point with the given slope.

14. $(1,2)$, slope $= \frac{2}{3}$ 15. $(0,-2)$, slope $= \frac{3}{4}$

16. $(-1,2)$, slope $= -2$ 17. $(-2,3)$, slope $= -\frac{3}{5}$

18. $(-4,-3)$, slope $= 0$ 19. $(-4,-3)$, slope undefined

20. Graph $y = 2x$ and $y = -\frac{1}{2}x$ on the same coordinate system.

21. Graph $y = \frac{2}{3}x + 1$ and $y = -\frac{3}{2}x - 2$ on the same coordinate system.

22. Write equations for three distinct lines perpendicular to $y = \frac{4}{5}x - 1$.

23. Write an equation for the line with y-intercept 3 that is perpendicular to the line that is the graph of $3x - 2y = 1$.

Show that the four-sided figures $ABCD$ in Exercises 24 and 25 are parallelograms.

24. $A = (0,0)$, $B = (4,0)$, $C = (7,6)$, and $D = (3,6)$

25. $A = (-4,-3)$, $B = (3,-1)$, $C = (6,6)$, and $D = (-1,4)$

Decide if the triangles determined by the points A, B, and C in Exercises 26 and 27 are right triangles.

26. $A = (0,1)$, $B = (2,-2)$, and $C = (3,3)$

27. $A = (0,-1)$, $B = (2,-5)$, and $C = (3,5)$

Determine whether the diagonals of the four-sided figures $ABCD$ in Exercises 28 and 29 are perpendicular.

28. $A = (0,0)$, $B = (2,0)$, $C = (2,3)$, and $D = (0,3)$

29. $A = (3,4)$, $B = (-2,4)$, $C = (-2,-1)$, and $D = (3,-1)$

30. Let $A = (-1,-1)$, $B = (1,3)$, and $C = (4,9)$. Find the slope of each line determined by the points

(a) A and B (b) B and C

(c) A and C

What conclusion can you draw about the three points A, B, and C?

7.2 / WRITING EQUATIONS OF LINES

Every linear equation in two variables has a line for its graph. You have had considerable practice sketching the graphs of linear equations. Going the other way, if we have a line, we should be able to write the linear equation that describes the line. The examples below show typical cases in which equations are written to describe particular lines. Remember that a line with slope m and y-intercept b has equation $y = mx + b$.

Example 1   Write an equation for the line that has x-intercept 6 and y-intercept 2.

The equation can be written as $y = mx + b$ if we find the slope m and the y-intercept b. The slope can be computed from the coordinates of the x-intercept $(6,0)$ and the y-intercept $(0,2)$: $m = \dfrac{0 - 2}{6 - 0} = \dfrac{-2}{6} = -\dfrac{1}{3}$. The y-intercept of this line is given as 2. Thus the equation is $y = -\dfrac{1}{3}x + 2$.

$\square$

Example 2 Write an equation for the line containing the points $(-2,3)$ and $(1,5)$.

Again, we want to find both the slope and the y-intercept of this line so that we can write the equation as $y = mx + b$. The slope can be computed from the coordinates of the two points: $m = \dfrac{3-5}{-2-1} = \dfrac{-2}{-3} = \dfrac{2}{3}$.

Thus the equation is $y = \dfrac{2}{3}x + b$ for a number b that still must be determined. The information that permits us to find b is that the point $(-2,3)$ lies on the line. [The point $(1,5)$ lies on the line also, but one point is enough to find b.] Since $(-2,3)$ is on the line with equation $y = \dfrac{2}{3}x + b$, we can replace x by -2 and y by 3 and get an equation that yields the value of b:

$$3 = \frac{2}{3}(-2) + b$$

$$3 = -\frac{4}{3} + b$$

$$3 + \frac{4}{3} = b$$

$$\frac{13}{3} = b.$$

Now we know that $m = \dfrac{2}{3}$ and $b = \dfrac{13}{3}$. Thus the equation of the line is $y = \dfrac{2}{3}x + \dfrac{13}{3}$. Notice that if we had used the point $(1,5)$ to solve for b in the equation $y = \dfrac{2}{3}x + b$, we would have obtained the same y-intercept:

$$5 = \frac{2}{3}(1) + b$$

$$5 - \frac{2}{3} = b$$

$$\frac{13}{3} = b. \qquad \qquad \square$$

Example 3 Write an equation for the line through the two points $(5,4)$ and $(5,-3)$.

Again we attempt to compute the slope and y-intercept. Using the coordinates of the two points, we find that the difference $4 - (-3) = 7$ and the difference $5 - 5 = 0$. Thus the line has no slope. If you look at the graph of the line in Fig. 7.10, you should see that it is parallel to the vertical axis.

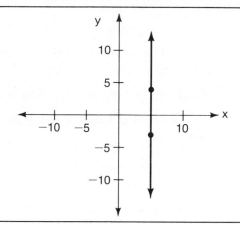

Figure 7.10

The equation that describes this line is $x = 5$. Notice that there is no y-intercept in this case.   □

Example 4   Write an equation for the line that contains the point (1,6) and is parallel to the graph of $2x + y = 4$.

In order to write an equation for a line, we need to find the slope and y-intercept of the line. Remember that parallel lines have the same slope. We can find the slope from the equation $2x + y = 4$:

$$2x + y = 4$$

$$y = -2x + 4.$$

Thus both lines have slope -2. An equation describing the line through (1,6) has the form $y = -2x + b$ where b is the y-intercept. We can find the value of b because we know that the point (1,6) is on the line. Replacing x by 1 and y by 6 in the equation gives

$$6 = -2(1) + b$$

$$2 + 6 = b$$

$$8 = b.$$

Thus the equation is $y = -2x + 8$. □

Example 5 Write an equation for the line perpendicular to $\frac{x}{6} + \frac{y}{4} = \frac{1}{3}$ that contains the point $(-1,2)$.

The slope m of the line we want is the negative reciprocal of the slope of $\frac{x}{6} + \frac{y}{4} = \frac{1}{3}$. We can find that slope by solving for y. (Notice that we first multiply by 12 to clear the fractions in the equation.)

$$\frac{x}{6} + \frac{y}{4} = \frac{1}{3}$$

$$2x + 3y = 4 \qquad \text{(Multiply both sides by 12)}$$

$$3y = -2x + 4 \qquad \text{(Subtract } 2x \text{ from both sides)}$$

$$y = \frac{-2}{3}x + \frac{4}{3}. \qquad \text{(Divide both sides by 3)}$$

Thus the line $\frac{x}{6} + \frac{y}{4} = \frac{1}{3}$ has slope $\frac{-2}{3}$. The slope of a line perpendicular to it is $\frac{3}{2}$. The line with slope $\frac{3}{2}$ through $(-1,2)$ has the form $y = \frac{3}{2}x + b$; replacing y by 2 and x by -1 gives the value for b:

$$2 = \frac{3}{2}(-1) + b$$

$$2 = -\frac{3}{2} + b$$

$$\frac{7}{2} = b. \qquad \text{(Add } \frac{3}{2} \text{ to both sides)}$$

Thus the equation we seek is $y = \frac{3}{2}x + \frac{7}{2}$. □

In the examples above when two points were given, we wrote the equation of a line by finding its slope and y-intercept. The next example illustrates a slightly different way of determining an equation for a line if two points on the line are given.

Example 6   Write an equation for the line through the two points (1,2) and (5,8).

We start by computing the slope of the line using the two points (1,2) and (5,8): $m = \dfrac{2 - 8}{1 - 5} = \dfrac{-6}{-4} = \dfrac{3}{2}$. We want to describe a relationship between the coordinates x and y of any point on the line.

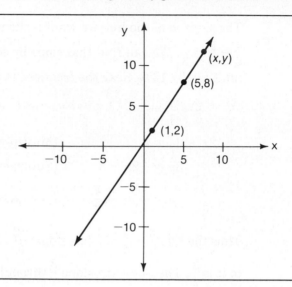

Figure 7.11

If (x,y) is a point on the line, then computing the slope with (x,y) and (5,8) [or with (x,y) and (1,2)] must give $\dfrac{3}{2}$ (Fig. 7.11). Thus

$$\frac{y - 8}{x - 5} = \frac{3}{2}$$

$$y - 8 = \frac{3}{2}(x - 5) \qquad \text{(Multiply both sides by } x - 5)$$

$$y = \frac{3}{2}x - \frac{15}{2} + 8$$

$$y = \frac{3}{2}x + \frac{1}{2}.$$

If the slope $\dfrac{3}{2}$ is written in terms of (x,y) and (1,2), the equation has the form $\dfrac{y-2}{x-1}=\dfrac{3}{2}$. Check to see that this equation simplifies to $y=\dfrac{3}{2}x+\dfrac{1}{2}$.

□

Exercises 7.2

Find the x-intercept and the y-intercept of the graphs of the equations in Exercises 1 and 2.

1. $2x + 3y = 4$

2. $2y - 3x = 6$

Find the slope and the y-intercept of each of the graphs of the equations in Exercises 3 and 4.

3. $3x + 4y = 12$

4. $\dfrac{y}{5} - \dfrac{x}{4} = 1$

Determine k in the equation of each line in Exercises 5–11 so that the graph of the equation satisfies the given condition.

5. $y = \dfrac{2}{3}x + k$; has y-intercept -2.

6. $y = kx + 2$; has slope $\dfrac{1}{2}$.

7. $y = 3x + k$; goes through $(-1,1)$.

8. $2x + 3y = k$; has x-intercept 2.

9. $kx + 2y - 2 = 0$; goes through $(-4,2)$.

10. $x + ky - 3 = 0$; the line is vertical.

11. $kx + 2y + 5 = 0$; the line is horizontal.

Write an equation for each line passing through the given points in Exercises 12–17.

12. $(-1,2)$ and $(3,4)$

13. $(-2,1)$ and $(3,-2)$

14. $(0,-2)$ and $(4,0)$

15. $(-3,0)$ and $(0,4)$

16. $(-2,-2)$ and $(3,-2)$

17. $(2,-3)$ and $(2,4)$

Write an equation for each line described in Exercises 18–28.

18. Slope $-\dfrac{3}{2}$, y-intercept 3

19. Slope 2, y-intercept $\dfrac{3}{2}$

20. Slope -2, through $(-2,1)$

21. Slope $\dfrac{1}{2}$, x-intercept 5

22. Through $(-1,-2)$, x-intercept 2

23. Through $(3,1)$, parallel to $y = \dfrac{-3}{4}x + 1$

24. Through $(2,-3)$, perpendicular to $2x + 4y = 1$

25. Through $(0,2)$, parallel to the line determined by $(1,-2)$ and $(4,3)$

26. x-intercept -1, y-intercept -2

27. Through $(1,2)$, perpendicular to the y-axis

28. Through $(1,2)$, perpendicular to the x-axis

29. Show that the three points $(-2,-3)$, $(2,5)$, and $(4,9)$ lie on a line.

30. B lies on the line determined by the points A $(2,1)$ and $C(6,9)$. Find the coordinates of B so that

   (a) the length of $\overline{AB}$ is $\dfrac{1}{2}$ of the length of $\overline{AC}$

   (b) the length of $\overline{AB}$ is $\dfrac{1}{3}$ of the length of $\overline{AC}$

   (c) the length of $\overline{AB}$ is $\dfrac{3}{2}$ of the length of $\overline{AC}$

7.3 / USING GRAPHS TO SOLVE TWO EQUATIONS IN TWO VARIABLES

Some problem situations can be described naturally using two variables. Consider this problem, for example.

Tickets to the Thespians production of *Our Town* are priced at $4.00 for adults and $1.50 for children. As of last Monday, 1232 tickets had been sold and a total of $3898 was received. How many of the tickets were for adults and how many were for children?

In Chapter 6 we solved problems like this by letting x represent the number of adult tickets sold and $1232 - x$ the number of children's tickets sold. However, since there are two quantities in the problem that are not known, the number of adult tickets and the number of children's tickets, it is natural to represent each with a different variable. Let

x denote the number of adult tickets sold, and

y denote the number of children's tickets sold.

Observe that there are two pieces of information in the problem that describe the relationship between x and y:

- The number of adult tickets plus the number of children's tickets sold is 1232;
- The money received from the ticket sales is $3898.

These two statements can be translated directly into equations in the two variables x and y:

$$x + y = 1232$$

$$4x + 1.5y = 3898.$$

You should see that the first equation states that the number of adult tickets plus the number of children's tickets is 1232, and the second equation states that the money received from the adult tickets plus the money received from the children's tickets is $3898.

In order to solve this problem, we must find a pair of numbers that give a solution to $x + y = 1232$ and at the same time give a solution to $4x + 1.5y = 3898$. A pair of numbers, one value for x and one value for y, that solves both of the equations is called a **simultaneous solution**.

Finding a Solution Graphically

Each of the equations, $x + y = 1232$ and $4x + 1.5y = 3898$, has a line for its graph. The point where the two lines intersect lies on both lines, and hence its coordinates give a solution to both equations. Thus one way to find the simultaneous solution is to graph the two lines and find the coordinates of the point of intersection.

Equations in this form are graphed easily by finding the x-intercept and the y-intercept.

$$x + y = 1232$$

x-intercept: $x = 1232$

y-intercept: $y = 1232$

$$4x + 1.5y = 3898$$

x-intercept: $x = 974.5$

y-intercept: $y = 2598.\overline{6}$

The two lines are graphed in Fig. 7.12.

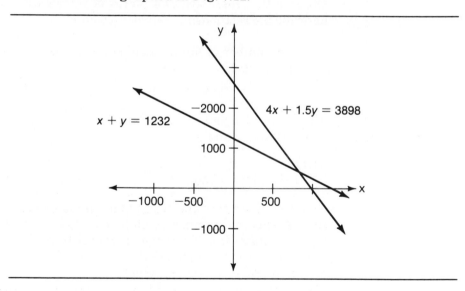

Figure 7.12

The graphs of the two equations show that there is a simultaneous solution to the two equations. Although it is not possible to read exactly the coordinates of the point of intersection, we can approximate the solution. It appears that x (the number of adult tickets) is slightly more than 800 and that y (the number of children's tickets) is slightly more than 400. If we need the exact solution, we can use a guess-and-check approach to find it remembering that the values for x and y in this example must be whole numbers.

Example 1   Find the simultaneous solutions to the two equations

$$x + y = 9$$

$$x - 2y = -6.$$

The graph of the first equation has x-intercept 9 and y-intercept 9. The graph of the second equation has x-intercept -6 and y-intercept 3 (Fig. 7.13).

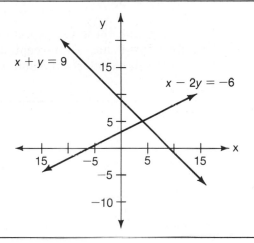

Figure 7.13

It is clear from the graphs that there is exactly one point common to both lines. Its coordinates are (4,5). We can check to see that $x = 4$ and $y = 5$ is a solution to both of the equations. Thus the two equations have one simultaneous solution.   □

Systems of Equations that Have No Solution or More Than One Solution

In the two examples above, the system of equations has exactly one solution because the graphs are lines that intersect. However, it is possible to have two linear equations whose graphs are different lines that are parallel. Then the lines have no point in common and the equations have no simultaneous solution.

Another thing that can happen is that two linear equations can appear to be different but actually be equivalent. In this case the graphs of the two equations will be the same line, and every pair of coordinates that is a solution to one equation will also be a solution to the other. It is easy to recognize these two special cases if the equations are graphed.

Example 2   Find the simultaneous solutions to the two equations

$$2x - y = 6$$

$$x - .5y = 5.$$

First we graph these two equations using their intercepts. The equation $2x - y = 6$ has x-intercept 3 and y-intercept -6. The equation $x - .5y = 5$ has x-intercept 5 and y-intercept -10 (Fig. 7.14).

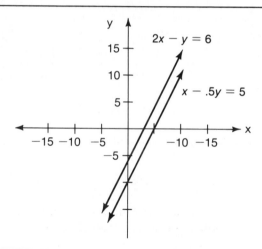

Figure 7.14

The graphs suggest that the two lines are parallel and have no points in common. Hence the equations have *no* simultaneous solution. To check that the lines are indeed parallel, we can compute the slope of each using the x- and y-intercepts.

$$2x - y = 6: \quad \text{slope} = \frac{0 - (-6)}{3 - 0} = \frac{6}{3} = 2$$

$$x - .5y = 5: \quad \text{slope} = \frac{0 - (-10)}{5 - 0} = \frac{10}{5} = 2. \qquad \square$$

Example 3   Find the simultaneous solutions to the two equations

$$3x + 2y = 8$$

$$9x + 6y = 24.$$

Again, we compute the intercepts so that we can sketch the graphs of the equations. The equation $3x + 2y = 8$ has x-intercept $\frac{8}{3}$ and y-intercept 4; the equation $9x + 6y = 24$ has x-intercept $\frac{8}{3}$ and y-intercept 4. We see that the two equations have the same x- and y-intercepts. In fact, the two equations describe the same line as shown in Fig. 7.15.

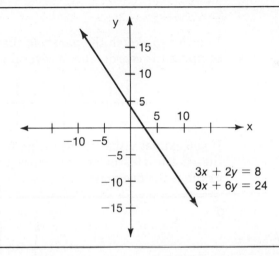

Figure 7.15

Any solution to one equation is a solution to the other. We have a whole line of simultaneous solutions.   □

What the discussion and examples above have shown is that three different things can happen with two simultaneous linear equations:

1. The graphs of the equations can intersect in one point, and there will be one simultaneous solution to the two equations.
2. The graphs of the equations can be different parallel lines, and there will be no simultaneous solutions to the two equations.
3. The graphs of the equations can be the same line, and there will be infinitely many simultaneous solutions to the two equations.

Simultaneous Solutions to Equations that Are Not Linear

If we have any two equations in two variables, not necessarily linear, we can graph the equations and determine whether they have any points in common. The coordinates of common points are the simultaneous solutions to the two equations.

Example 4   Find the simultaneous solutions to the two equations

$$y = x^2 - x - 6$$

$$y = -2x - 4.$$

In order to sketch the parabola that is the graph of $y = x^2 - x - 6$, we compute the coordinates of several points:

x	-3	-2	-1	0	1	2	3	4
y	6	0	-4	-6	-6	-4	0	6

If you cannot visualize the graph from these points, you should compute the coordinates of several more points to convince yourself that the graph is the one given in Fig. 7.16.

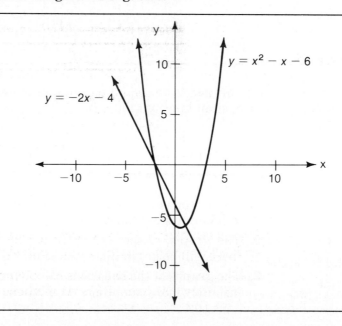

Figure 7.16

The graph of $y = -2x - 4$ is a line with slope -2 and y-intercept -4. We can see that these graphs have two points in common. The graphs suggest that the two points are $(-2,0)$ and $(1,-6)$. You should check that $x = -2$ and $y = 0$ is a solution to both equations and that $x = 1$ and $y = -6$ is also a solution to both. We conclude that these equations have two simultaneous solutions. ☐

Exercises 7.3

In Exercises 1–12, find the simultaneous solutions to each of the equations graphically.

1. $2x + 3y = -1$
   $3x - 2y = -8$

2. $\frac{1}{2}x - y = 1$
   $\frac{1}{3}x - \frac{2}{3}y = 2$

3. $2x + \frac{1}{3}y = 2$
   $3x + \frac{1}{2}y = 3$

4. $x + 2y = 5$
   $2x - 3y = 3$

5. $x + 3y = 5$
   $3x - y = -5$

6. $3x - 4y = 18$
   $x + 3y = -7$

7. $2x - 3y - -1$
   $-4x + 6y = 2$

8. $2x + 3y = -17$
   $2x - 5y = 7$

9. $y = x^2$
   $y = 2x + 3$

10. $y = -x^2 + 6$
    $y = x^2 - 2$

11. $y = x^3$
    $y = 4x$

12. $y = \frac{1}{x - 2}$
    $x - y = 2$

Set up two equations in two variables for the problems in Exercises 13–16, and then find their solutions graphically.

13. Jim invests \$20,000, part at 11% simple interest and the remainder at 6% simple interest. If his annual income is \$1800, how much is invested at each rate?

14. How many gallons of a 20%-alcohol solution must be mixed with a 60%-alcohol solution to obtain 50 gallons of a 32%-alcohol solution?

15. How much candy worth $1.20 per pound must be mixed with candy worth $1.80 per pound to obtain 100 pounds of candy worth $1.35 per pound?

16. Find two numbers whose sum is 110 and whose difference is 36.

7.4 / ALGEBRAIC METHODS OF SOLVING TWO EQUATIONS IN TWO VARIABLES

In the last section we used graphing to find simultaneous solutions to two equations. In some cases we were able to read the exact value of a simultaneous solution from the graphs; in other cases we could only approximate the value of the solution and then use guess-and-check methods to find the actual solution. In this section we will develop two algebraic methods that always give exact solutions for simultaneous linear equations.

Solving Two Equations by Substitution

A pair of equations in the last section for which we found approximate solutions by graphing was the pair describing the number of adult tickets sold and the number of children's tickets sold to the Thespian's production of *Our Town*:

$$x + y = 1232$$

$$4x + 1.5y = 3898.$$

In the first equation of the pair, it is easy to solve for y and write $y = 1232 - x$. Now, in the *second* equation, we can replace y by the expression $1232 - x$ and write $4x + 1.5(1232 - x) = 3898$. This procedure gives a linear equation in only one variable x, which can be solved:

$$4x + 1.5(1232 - x) = 3898$$

$$4x + 1848 - 1.5x = 3898 \qquad \text{(Distributive property)}$$

$$2.5x + 1848 = 3898 \qquad \text{(Combine like terms)}$$

$$2.5x = 2050 \qquad \text{(Subtract 1848 from both sides)}$$

$$x = 820.$$

Thus we know that the number of adult tickets sold is 820. We also need the number of children's tickets sold. We can find the value of y by replacing x in either of the original equations by 820; using the first equation gives

$$820 + y = 1232$$

$$y = 412.$$

Thus the number of children's tickets sold is 412. $\qquad\qquad$ □

This procedure for finding a simultaneous solution to two linear equations is called **substitution**. We solve for one variable in one of the equations and then make a substitution for that variable in the other equation. Remember that a simultaneous solution to the two original equations has two parts; a value for x and a value for y.

Example 1 A boat travels 70 miles downstream on a river in 5 hours and makes the return trip in 7 hours. Find the rate of the boat in still water and the rate of the water.

It is an important observation from physics that the actual rate that a boat travels downstream is its rate in still water plus the rate of the water; and the actual rate at which a boat travels upstream is its rate in still water minus the rate of the water. Let x denote the rate of the boat in still water and y the rate of the water. Then $x + y$ is the rate at which the boat travels downstream, and $x - y$ is the rate at which the boat travels upstream. Since the distance traveled is computed by multiplying rate by time ($d = rt$), we have two equations:

downstream: $70 = (x + y)5$

upstream: $70 = (x - y)7.$

If we divide both sides of the first equation by 5 and both sides of the second equation by 7, we get the following pair of equivalent equations:

$$x + y = 14$$

$$x - y = 10.$$

If we solve the *second* equation for x, we get $x = y + 10$. Now, in the first equation we can replace x by the expression $y + 10$ and write $(y + 10) + y = 14$. Solving this equation, we find that

$$(y + 10) + y = 14$$
$$2y + 10 = 14 \qquad \text{(Combine like terms)}$$
$$2y = 4 \qquad \text{(Subtract 10 from both sides)}$$
$$y = 2. \qquad \text{(Divide both sides by 2)}$$

Thus the rate of the water is 2 mph. We can find the rate at which the boat travels in the still water by replacing y by 2 in either of the original equations. Using the first equation, we find that

$$5(x + 2) = 70$$
$$x + 2 = 14 \qquad \text{(Divide both sides by 5)}$$
$$x = 12. \qquad \text{(Subtract 2 from both sides)}$$

Thus the rate of the boat in still water is 12 mph. ◻

Solving Two Equations by Eliminating a Variable

We may have two linear equations in which solving one equation for one of the variables would be difficult. Here is an example:

$$5x + 2y = 7$$
$$3x - 2y = 9.$$

Solving for x or y in either equation gives an expression with fractions for coefficients. Substitution, though possible, would be a messy computation. A more efficient procedure is possible. Notice that the y terms have opposite coefficients. We add together the two left sides and the two right sides of the original equations to get a new equation:

$$5x + 2y + (3x - 2y) = 7 + 9$$
$$5x + 2y + 3x - 2y = 16$$
$$8x = 16$$
$$x = 2.$$

The y term disappears in the sum because the first equation contains the term $2y$ and the second equation, $-2y$. Since the y term disappears we are able to find the value of x. Now, replacing x by 2 in either of the original equations, say the first, gives the corresponding value for y:

$$5(2) + 2y = 7$$
$$10 + 2y = 7$$
$$2y = -3$$
$$y = -\frac{3}{2}.$$

This procedure, called **eliminating a variable**, worked in the example above because adding the coefficients for one variable gave 0. Although two equations may not have coefficents with this property, it is possible to replace each of the original equations by an equivalent equation so that the new pair will have this property.

Example 2 Find the simultaneous solutions to the two equations

$$2x + 6y = 14$$
$$3x - 4y = -5$$

using the procedure of eliminating a variable.

If we want to eliminate the variable x, we replace the two original equations by two equivalent equations where the coefficient of x in one is the opposite of the coefficient of x in the other. We can do this by multiplying both sides of the first equation by 3 and multiplying both sides of the second equation by -2:

$$6x + 18y = 42$$
$$-6x + 8y = 10.$$

Adding the equations together gives an equation only in y:

$$26y = 52$$
$$y = 2.$$

To find the corresponding value for x, we replace y by 2 in either of the original equations, say the first one:

$$2x + 6(2) = 14$$

$$2x + 12 = 14$$
$$2x = 2$$
$$x = 1.$$

□

Equations that Have
No Simultaneous Solutions or More Than One Solution

The examples above are equations that have one simultaneous solution because their graphs are lines that intersect in one point. We have seen that some simultaneous equations have no solutions and others have infinitely many. It is possible to detect these two situations when solving algebraically as well as when solving graphically. For illustration we will investigate algebraically two examples that we solved by graphing in the last section. You should compare these different approaches.

Example 3   Find the simultaneous solutions to the two equations

$$2x - y = 6$$
$$x - .5y = 5.$$

We can either use substitution or eliminate a variable. Since it is easy to solve for x in the second equation, we will do that and then replace x in the first equation:

$$x = 5 + .5y \qquad \text{(Solving for } x \text{ in the second equation)}$$
$$2(5 + .5y) - y = 6 \qquad \text{(Substituting for } x \text{ in the first equation)}$$
$$10 + y - y = 6$$
$$10 = 6.$$

Because $10 = 6$ can never happen, we are able to infer that there is no simultaneous solution to this pair of equations. If we had eliminated a variable instead of using substitution, we would have come to the same conclusion.   □

Example 4   Find the simultaneous solutions to the two equations

$$3x + 2y = 8$$
$$9x + 6y = 24.$$

This time we might eliminate the variable y by multiplying both sides of the first equation by -3 and adding it to the second equation:

$$-9x - 6y = -24$$
$$9x + 6y = 24.$$

Adding gives $0 = 0$, a statement that is always true. By examining the equations, we find that they are in fact equivalent (their graphs are the same line). Any solution to one equation is also a solution to the other. Thus we have a whole line of points whose coordinates are solutions to both of the equations.   □

Exercises 7.4

Find the simultaneous solutions to each pair of equations in Exercises 1–11. Use the algebraic method to solve the problems.

1. $3x - 4y = 5$
   $5x + 4y = 3$

2. $-2x + 7y = 6$
   $2x - 3y = 2$

3. $4x + 3y = 5$
   $5x + 3y = 2$

4. $3x + 2y = 10$
   $3x - 5y = -4$

5. $3x - 5y = 4$
   $4x + 3y = 15$

6. $2x + 5y = 1$
   $3x + 2y = 0$

7. $3x + 4y = -11$
   $7x - 5y = 3$

8. $2x + 3y = 4$
   $.5x + .75y = 2$

9. $5x + 15y = 23$
   $35x + 105y = 161$

10. $x + y = 12,000$
    $.08x + .05y = 843$

11. $x - \dfrac{1}{2}y = 2$
    $1.5x - .75y = 3$

Set up a system of two equations in two variables and solve the problems in Exercises 12–24.

12. Chris invested $18,800, part at 9.3% simple interest and the remainder at 6.7% simple interest. If her annual income is $1465, how much does she have invested at each rate?

13. How many gallons of a 24%-alcohol solution must be mixed with a 52%-alcohol solution to obtain 40 gallons of a $36\frac{1}{4}$%-alcohol solution?

14. How much candy worth $1.20 per pound must be mixed with candy worth $1.80 per pound to obtain 50 pounds of candy worth $1.35 per pound?

15. A money box contains $26.95 in dimes and quarters. If the number of quarters is one less than three times the number of dimes, how many of each coin are in the box?

16. A total number of 2025 tickets to a football game were sold at $2.75 for children and $4.25 for adults. If the total receipts were $6921.75, how many of each kind of ticket were sold?

17. The length of a rectangle is 5 feet less than twice its width. Find the dimensions of the rectangle if its perimeter is 146 feet.

18. In an isosceles triangle, each of the equal angles is 9° more than the third angle. Find the measure of each angle.

19. A 20-foot pipe is cut into two pieces so that one piece is 3 feet longer than the other. Find the length of each piece.

20. A plane flies 2800 miles with the wind in 4 hours and makes the return trip against the wind in 5 hours. What is the rate of the plane in still air and what is the rate of the wind?

21. A boat travels 45 miles downstream on a river in 3 hours and makes the return trip in 5 hours. Find the rate of the boat in still water and the rate of the river.

22. One number is five more than one-half the other number and their sum is 68. Find the numbers.

23. Determine m and b so that the graph of $y = mx + b$ passes through the points $(2,-1)$ and $(-1,5)$.

24. Determine a and b so that the graph of $ax + by = 5$ passes through the points $(1,2)$ and $(-2,1)$.

7.5 / SOLVING SYSTEMS OF EQUATIONS IN MORE THAN TWO VARIABLES

Some problem situations can be described by writing two equations in two variables. Other problems can be more naturally described with three variables and three equations. Here is an example:

A coin bank contains 50 coins in nickels, dimes, and quarters. The value of the nickels is the same as the value of the quarters. The total value of the money in the bank is $4.40. How many coins of each type are in the bank?

Since there are three types of coins, it is natural to assign a variable to the number of each. Let

n represent the number of nickels,

d represent the number of dimes, and

q represent the number of quarters.

These are the equations in the variables n, d, and q that describe the problem:

$$n + d + q = 50 \qquad \text{(Number of coins is 50)}$$

$$.05n = .25q \qquad \text{(Value of nickels equals the value of quarters)}$$

$$.05n + .10d + .25q = 4.40 \qquad \text{(Total value of the coins is \$4.40)}$$

What remains is to find a simultaneous solution for the three equations, that is, a value for n, a value for d, and a value for q, that gives a solution to each of the three equations. We can use the same techniques to find solutions to systems of three linear equations that we used in the last section to find solutions to two linear equations.

Example 1 Find the simultaneous solutions to the equations

$$n + d + q = 50$$

$$.05n = .25q$$

$$.05n + .10d + .25q = 4.40.$$

In this example the second equation $.05n = .25q$ is equivalent to $n = 5q$. Thus we can replace n by $5q$ in the first and third equations to obtain two equations in two variables:

$$(5q) + d + q = 50$$

$$.05(5q) + .10d + .25q = 4.40.$$

We have now reduced three equations in three variables to two equations in two variables. These can be solved by using the techniques of the last section. First, we simplify the two equations:

$$d + 6q = 50$$

$$.10d + .50q = 4.40.$$

Since the first equation of this pair can readily be solved for d, we write $d = 50 - 6q$, substitute $50 - 6q$ for d in the second equation, and simplify:

$$.10(50 - 6q) + .50q = 4.40$$

$$5 - .60q + .50q = 4.40$$

$$5 - .10q = 4.40$$

$$-.10q = -.60$$

$$q = 6.$$

Now we know that q (the number of quarters) must be 6. Since $d = 50 - 6q$, we can find the value of d:

$$d = 50 - 6(6)$$

$$d = 14.$$

Using the fact that $q = 6$ and $d = 14$, we can find n from any of the original equations. The second is an easy one:

$$.05n = .25q$$

$$.05n = .25(6)$$

$$n = 30.$$

The three values $n = 30$, $d = 14$, and $q = 6$ give a solution for each of the three equations.   □

In the example above we used the technique of substitution to reduce three equations in three variables to two equations in two variables. This worked because one equation in the example could be solved easily for one of the variables. If substitution had not been efficient, we might have used the technique of adding equations to eliminate one variable. In general, the strategy is to replace three equations in three variables with two equations in two variables. Solving the two equations will give values for two of the variables; the third can then be found from one of the original equations.

Example 2 Find the simultaneous solutions to the equations

$$x + y + 2z = 12$$

$$x + y - 2z = 0$$

$$2x - 3y + z = 5.$$

We could use the technique of substitution, solving the first equation for either x or y and then substituting in the second and third equations. You should try that technique. Another approach is to add the first two equations together to get an equation in x and y: $2x + 2y = 12$. We want another equation in the variables x and y, so we take the third equation and either one of the other original equations and combine them to remove the variable z:

$$\left.\begin{array}{l} x + y + 2z = 12 \\[2em] x + y - 2z = 0 \\[2em] 2x - 3y + z = 5 \end{array}\right\} \quad \begin{array}{l} \longrightarrow \left.\begin{array}{l} x + y + 2z = 12 \\ x + y - 2z = 0 \end{array}\right\} \longrightarrow 2x + 2y = 12 \\[2em] \longrightarrow \left.\begin{array}{l} x + y - 2z = 0 \\ 4x - 6y + 2z = 10 \end{array}\right\} \longrightarrow 5x - 5y = 10. \end{array}$$

Now we have two equations in two variables. Combining them removes the y variable and gives a value for x:

$$\left.\begin{array}{l} 2x + 2y = 12 \\ 5x - 5y = 10 \end{array}\right\} \longrightarrow \left.\begin{array}{l} x + y = 6 \\ x - y = 2 \end{array}\right\} \longrightarrow 2x = 8.$$

Thus $x = 4$. We can find y by substituting 4 for x in the equation $2x + 2y = 12$:

$$2(4) + 2y = 12$$
$$2y = 12 - 8$$
$$2y = 4$$
$$y = 2.$$

To get a value for z, we can return to one of the original equations, say the first:

$$x + y + 2z = 12$$
$$4 + 2 + 2z = 12$$
$$2z = 12 - 6$$
$$2z = 6$$
$$z = 3.$$

Thus $(4,2,3)$ is a simultaneous solution to the three equations.   □

It is possible to have a system of three equations that uses three variables, but each equation contains only two of the three variables. The same methods of substitution and elimination of a variable can be used.

Example 3   Find the simultaneous solutions to the equations

$$x + 2z = 5$$
$$2x - 3y = 5$$
$$y - z = -3.$$

If we elect to use the method of eliminating a variable, we can note that the third equation only contains y and z, and then combine the first and second equations to eliminate x. This procedure will give two equations in y and z:

$$\left.\begin{array}{r} x + 2z = 5 \\ 2x - 3y = 5 \end{array}\right\} \longrightarrow \left.\begin{array}{r} -2x - 4z = -10 \\ 2x - 3y = 5 \end{array}\right\} \longrightarrow -3y - 4z = -5$$

$$y - z = -3 \longrightarrow \quad y - z = -3 \longrightarrow y - z = -3.$$

Now we combine these two equations to get values for y and z. We can use substitution. From the second equation, we have $y = z - 3$ so

$$-3(z - 3) - 4z = -5 \qquad \text{(From the first equation)}$$

$$-3z + 9 - 4z = -5$$

$$-7z = -14$$

$$z = 2.$$

Thus $y = z - 3 = 2 - 3 = -1$ and $x + 2z = 5$, so $x = 1$. If we had elected to use substitution, we might have left the third equation as is, written $x = -2z + 5$ from the first equation and substituted for x in the second. That procedure would give two equations in y and z. Try this approach.

□

Not all sets of three equations in three variables will have a simultaneous solution. The situation is analogous to what we found for two linear equations in two variables. Sometimes there will be no common solution; other times there could be infinitely many common solutions. In attempting to find a solution algebraically, you will be able to tell if either of these situations occurs.

Example 4 Find the simultaneous solutions to the equations

$$2x + y - z = 12$$

$$2x + z = 4$$

$$2x - y + 3z = -4.$$

We might use substitution, solving the second equation for $z = -2x + 4$ and then replacing z in the first and third equations by $-2x + 4$:

$$2x + y - (-2x + 4) = 12$$

$$2x - y + 3(-2x + 4) = -4.$$

Simplifying these two equations gives

$$\left. \begin{array}{l} 2x + y + 2x - 4 = 12 \\ 2x - y - 6x + 12 = -4 \end{array} \right\} \longrightarrow \begin{array}{l} 4x + y = 16 \\ -4x - y = -16 \end{array}$$

These two equations are equivalent; any pair of numbers (x,y) that is a solution to one equation is also a solution to the other. What we see is that if x is any number, we get a solution by letting $y = -4x + 16$; we also know that $z = -2x + 4$. Thus we can generate any number of solutions to this system of equations. For example, we might let $x = 5$ and $y = -4x + 16 = -4(5) + 16 = -4$ and $z = -2x + 4 = -2(5) + 4 = -6$. Check to see that $(5,-4,-6)$ is a solution. Or, we might let $x = 0$; then $y = -4x + 16 = -4(0) + 16 = 16$ and $z = -2x + 4 = -2(0) + 4 = 4$. Check to see that $(0,16,4)$ is a solution and then write another solution yourself. There are infinitely many solutions to this system of equations.

□

Example 5   Find the simultaneous solutions to the equations

$$x + 2y - z = 1$$

$$2x - y + 3z = 2$$

$$2x + y + z = -1.$$

We can combine the first and second equations and then the second and third equations to eliminate the y-variable.

$$\left. \begin{array}{l} x + 2y - z = 1 \\ \\ 2x - y + 3z = 2 \\ \\ 2x + y + z = -1 \end{array} \right\} \longrightarrow \left. \begin{array}{l} x + 2y - z = 1 \\ 4x - 2y + 6z = 4 \end{array} \right\} \longrightarrow 5x + 5z = 5 \longrightarrow x + z = 1$$

$$\longrightarrow \left. \begin{array}{l} 2x - y + 3z = 2 \\ 2x + y + z = -1 \end{array} \right\} \longrightarrow 4x + 4z = 1 \longrightarrow x + z = \frac{1}{4}.$$

We are looking for values of x and z such that both $x + z = 1$ and

$x + z = \frac{1}{4}$. But the sum $x + z$ cannot be both 1 and $\frac{1}{4}$. We can only conclude that there are no simultaneous solutions to the equations. □

Graphing an equation in three variables requires three-dimensional space, just as graphing an equation in two variables requires two dimensions. The graph of a linear equation in three variables is a plane (think of a sheet of paper that extends infinitely far in every direction). Thus when we look for simultaneous solutions to three linear equations in three variables, we are looking for points common to three planes in three-dimensional space. Three planes may have no point that lies on all three of them. This is the situation in Example 5 above. Or, three planes can intersect in exactly one point. Think of a corner where two walls and the ceiling come together. This is the situation in Examples 1, 2, and 3. Three planes can also intersect in a line. Think of a revolving door. This is the situation in Example 4. There are several other possibilities for the intersection of three planes. See how many of them you can describe.

Exercises 7.5

Solve the systems of equations in Exercises 1–12. If there are infinitely many solutions, give at least three different ones.

1. $x + y + z = 3$
   $2x - y + 3z = 4$
   $x + 2y + 2z = 5$

2. $2x - y + 2z = -2$
   $x + 2y - z = 1$
   $3x + y + z = -1$

3. $x + y + z = 50$
   $.05x + .10y + .25z = 7.35$
   $.25x + .10y + .05z = 5.95$

4. $x + y + z = 5$
   $2x - y = 1$
   $2y = -1$

5. $2x + y = 0$
   $3x - z = -4$
   $y + 2z = 4$

6. $x - y + 2z = 3$
   $x + 2y - 3z = 1$
   $2x + y - z = 2$

7. $3x + 2y - z = 2$
$2x - y + 2z = 1$
$7x + 3z = 4$

8. $2x - y + z = -1$
$3x + \frac{1}{2}y + 2z = 0$
$4x + 2y - z = 5$

9. $2x + y - 3z = -1$
$2y + z = 2$
$3z = 6$

10. $x + y + z = 2500$
$-x + 4z = 70$
$1.5x + 2.5y + 2z = 4880$

11. $x + y + z = 1$
$-x + 2y - 3z = -1$
$2x + 3y + z = 3$

12. $x + 2z = 5$
$2x + 3y = -1$
$y - z = -3$

Write a system of three equations in three variables and solve for each of the problems in Exercises 13–20.

13. A money box contains $8.85 in nickels, dimes, and quarters. If the number of dimes is one more than twice the number of nickels and there are 60 coins in all, how many of each coin are in the box?

14. A money box contains $1490 in 5-, 10-, and 20-dollar bills. There are 150 bills in all. If the number of 5-dollar bills and the number of 20-dollar bills are interchanged, the value of the bills is $1895. How many of each are in the box?

15. A total of 2000 tickets to a concert were sold at $2.50 for students, $3.75 for adults, and $3.00 for senior citizens. The number of adult tickets sold was 250 more than the number of student tickets. If the total receipts were $6328.75, now many of each ticket were sold?

16. Sue invests $20,000, one part at 9.5% simple interest, another part at 7% simple interest, and the remainder at 6.5% simple interest. If she invests twice as much at 9.5% as 6.5% and her annual income is $1634, how much does she invest at each rate?

17. A 40-foot pipe is cut into 3 pieces. The largest piece is three times longer than the smallest piece and the other piece is 2 feet shorter than the largest piece. Find the length of each piece.

18. One angle of a triangle is 10° more than twice a second angle. The

third angle is the sum of the other two angles. Find the measure of each angle.

19. The sum of three numbers is 60. The ratio of the sum of the first two numbers to the third number is $2:1$. The first number minus the second number is 4. Find the three numbers.

20. Determine a, b, and c so that the graph of $y = ax^2 + bx + c$ passes through the points $(-1,4)$, $(1,2)$, and $(2,7)$.

7.6 / LINEAR INEQUALITIES IN TWO VARIABLES

Consider again the coordinate system with the graph of the equation $y = x + 4$, as shown in Fig. 7.17.

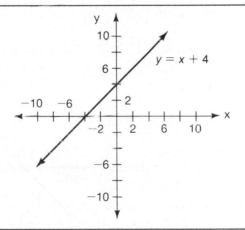

Figure 7.17

Think about all the points in the coordinate system that have a given x-coordinate; for example, focus on all the points that have 2 for their x-coordinate. Of all such points, only one lies on the graph of $y = x + 4$; it is the point where the y-coordinate equals the x-coordinate plus 4.

Those points that have y-coordinate greater than the x-coordinate plus 4 lie above the line $y = x + 4$; all points that have y-coordinate less

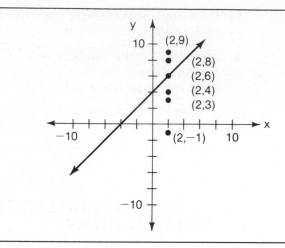

Figure 7.18

than the x-coordinate plus 4 lie below the line $y = x + 4$. In Fig. 7.18, we have illustrated this fact by looking at the points with x-coordinate 2. However, the same is true no matter what the x-coordinate is. Look at the points in Fig. 7.19 with x-coordinate -1.

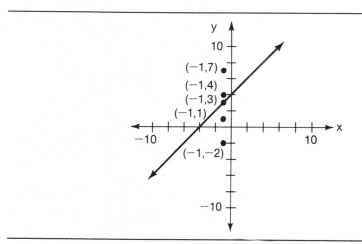

Figure 7.19

Only one of these points lies on the line $y = x + 4$; its y-coordinate is equal to the x-coordinate plus 4. The points that have y-coordinate greater than the x-coordinate plus 4 are above the line; those with y-coordinate less than the x-coordinate plus 4 are below the line. Thus the points on the line are described by the equation $y = x + 4$; the points above the line are described by the inequality $y > x + 4$ (Fig. 7.20); the points below the line are described by the inequality $y < x + 4$ (Fig. 7.21).

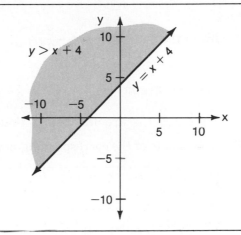

Figure 7.20

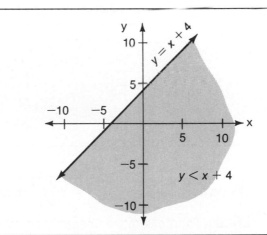

Figure 7.21

Every nonvertical line separates the plane into three parts: the line itself, the region "above" the line, and the region "below" the line. The line is described by a linear equation. The region above the line and the region below the line (both of which are called **half-planes**) are described by inequalities. A vertical line also separates the plane into three parts: the line itself, the region to the "left" of the line, and the region to the "right" of the line. Again, the two regions are called half-planes.

Example 1   Graph the inequality $y < -\frac{2}{3}x + 1$.

First we graph the line $y = -\frac{2}{3}x + 1$. The points that are solutions to $y < -\frac{2}{3} + 1$ are the points that have y-coordinate less than the y-coordinate of the corresponding point on the line (Fig. 7.22).

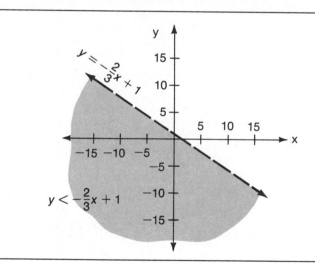

Figure 7.22

We have dashed the line $y = -\frac{2}{3}x + 1$ to indicate that points on the line are not solutions to the inequality $y < -\frac{2}{3}x + 1$. The solutions are the points below the line.   □

Example 2   Graph the inequality $x - 2y < 4$.

a. To use the same reasoning that we applied in the examples above, we need to see what the inequality says about the y-coordinate of a point, so we solve for y:

$$x - 2y < 4$$

$$-2y < -x + 4 \qquad \text{(Subtract } x \text{ from both sides)}$$

$$y > \frac{1}{2}x - 2 \qquad \text{(Divide both sides by } -2 \text{ and } \textit{reverse} \text{ the inequality sign)}$$

We can graph the inequality $y > \frac{1}{2}x - 2$ by graphing the equation $y = \frac{1}{2}x - 2$ and marking the points above the line (Fig. 7.23). Again, we dash the line to indicate that points on the line are *not* solutions to $y > \frac{1}{2}x - 2$.

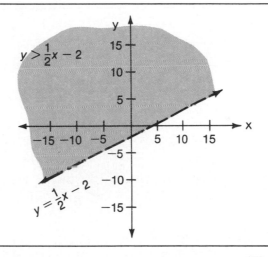

Figure 7.23

b. We know that the equation $x - 2y = 4$ describes a line; the line determines two half-planes, one described by $x - 2y < 4$ and the other by $x - 2y > 4$. To see which region is the graph of $x - 2y < 4$, we can graph $x - 2y = 4$ (Fig. 7.24) and check one point that is not on the line.

Say we check the easy point $(0,0)$ to see if it is a solution to $x - 2y < 4$ or to $x - 2y > 4$. If $x = 0$ and $y = 0$, then $x - 2y = 0 - 2(0) = 0 < 4$. Thus $(0,0)$ is a solution to $x - 2y < 4$. We can conclude that the half-plane containing $(0,0)$ is described by $x - 2y < 4$ (Fig. 7.25).

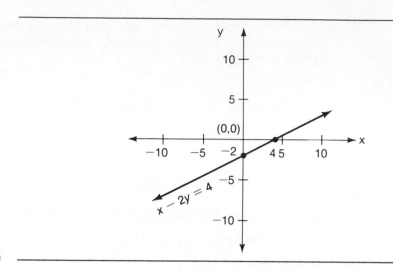

Figure 7.24

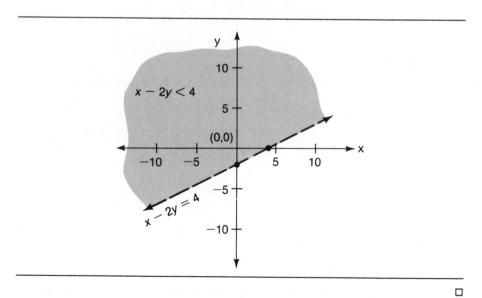

Figure 7.25

☐

Example 3   Write an inequality that describes the half-plane below the line through (−3,2) and (4,−5).

The half-plane below the line through (−3,2) and (4,−5) is shown in

Fig. 7.26. The line is dashed to indicate that points on the line are not "below the line."

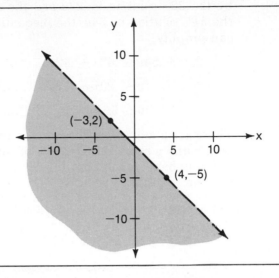

Figure 7.26

To write the inequality describing the half-plane, we first write the equation of the line through $(-3,2)$ and $(4,-5)$. This line has slope $\frac{2-(-5)}{-3-4} = \frac{7}{-7} - -1$. Thus the equation is $y = -x + b$; to find b, we replace x by -3 and y by 2:

$$2 = 3 + b$$

$$-1 = b.$$

The line has equation $y = -x - 1$. The half-plane below the line is described by the inequality $y < -x - 1$.   □

Example 4 When a 22%-alcohol solution and a 54%-alcohol solution are combined, the mixture will have an alcohol concentration between 22% and 54%, depending on the amounts of the two solutions used. Write and graph an inequality that describes all possible amounts of 22% solution and 54% solution that give a mixture of not more than 42% alcohol.

If x represents the amount of 22% solution and y the amount of 54%

solution, then $x + y$ represents the amount of the mixture. The problem states that x and y should be chosen so that the mixture contains no more alcohol than $.42(x + y)$. Since the amount of alcohol contributed by the 22% solution is $.22x$ and the amount of alcohol contributed by the 54% solution is $.54y$, the inequality is $.22x + .54y \leq .42(x + y)$. We can simplify:

$$.22x + .54y \leq .42x + .42y$$

$$0 \leq .20x - .12y$$

$$0 \leq 5x - 3y. \qquad \text{(Multiply both sides by 25)}$$

There are many values for x and y that make $5x - 3y \geq 0$. In our problem, x and y represent amounts of solutions; thus they cannot be negative numbers. A solution to $5x - 3y \geq 0$ in which neither the x value nor the y value is negative is a solution to the problem. To see the solutions graphically, we first graph $5x - 3y = 0$ (or $y = \frac{5}{3}x$), as shown in Fig. 7.27.

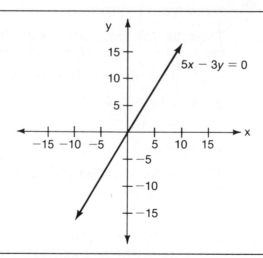

Figure 7.27

A point not on the line is $(0,1)$. Computing the value of $5x - 3y$ when $x = 0$, $y = 1$ gives $5(0) - 3(1) = -3 < 0$. Thus the region above the line contains solutions to $5x - 3y < 0$. The region below the line contains solutions to $5x - 3y > 0$. This region is shown in Fig. 7.28.

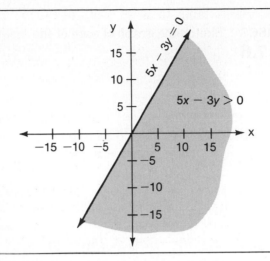

Figure 7.28

The solutions to $5x - 3y \geq 0$ are the points on the line and the points in the shaded region. However, the points that have a negative coordinate are not solutions to our problem. To graph the solutions to the problem, we take the points whose coordinates are positive or 0 (Fig. 7.29).

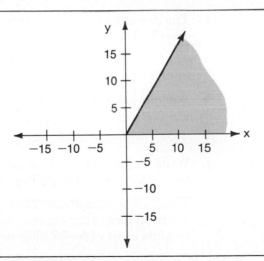

Figure 7.29

Each of these points has coordinates that give a quantity of 22%-alcohol solution and a quantity of 54%-alcohol solution that can be combined to yield a mixture of not more than 42% alcohol.   □

Exercises 7.6

Sketch the graph of each of the inequalities in Exercises 1–15.

1. $y < 2x + 3$
2. $y \geq 2x + 3$
3. $y \leq 2x$
4. $y > 2x$
5. $3x - 2y \leq 4$
6. $3x - 2y > 4$
7. $4y + 5x < 10$
8. $x \leq 2$
9. $4(x + y) \geq 8 + x - y$
10. $x > 0$
11. $y \geq 4$
12. $y < 0$
13. $\dfrac{x}{2} + \dfrac{y}{3} \leq 1$
14. $\dfrac{x}{3} - \dfrac{y}{4} \geq 1$
15. $.12x + .26y \leq .15(x + y)$

Write an inequality whose solution is the region described in each of Exercises 16–21.

16. Above the line determined by (1,2) and (4,6)

17. Above the line determined by (0,4) and (4,0)

18. The line and the region below the line determined by (−1,2) and (3,−1)

19. The line and the region below the line with slope 2 and y-intercept −3

20. The line and the region to the right of the line determined by (2,1) and (2,4)

21. Above the line determined by (−2,−3) and (3,−3)

22. Write an inequality that expresses the requirement that the perimeter of a rectangle is at least 100 feet, and draw its graph.

23. Tickets to a concert cost $2.50 for students and $3.75 for adults. Write an inequality expressing the requirement that the total receipts must exceed $1500, and draw its graph.

24. Candy worth $.95 per pound is mixed with candy worth $1.35 per pound. Write an inequality expressing the requirement that the value of the new mixture cannot exceed $1.15 per pound, and draw its graph.

25. Jim has some money invested in stocks that yield 10.5% annually and the rest in an account that pays 16.7% annually. Write an inequality expressing the requirement that his investments must yield at least 13% annually, and draw its graph.

26. An alcohol solution is formed by mixing a 34%-alcohol solution with a 62%-alcohol solution. Write an inequality expressing the requirement that the percent of alcohol in the new mixture cannot exceed 44%, and draw its graph.

27. A fast-food restaurant makes $.25 profit on hamburgers, $.15 profit on French fries, and breaks even on the other items. If the weekly overhead is $600, write an inequality that expresses the requirement that the weekly profits must exceed $400, and draw its graph.

7.7 / SYSTEMS OF LINEAR INEQUALITIES IN TWO VARIABLES

In the last section, we graphed the inequality $5x - 3y \geq 0$ (Fig. 7.30).

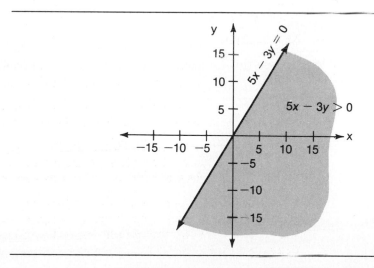

Figure 7.30

Because the values in that problem were quantities of alcohol solution, we restricted the graph to those points whose coordinates are positive or 0 (Fig. 7.31).

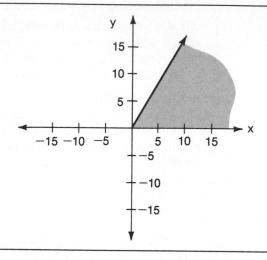

Figure 7.31

One way to say that the coordinates of the points are positive or 0 is to use the inequalities $x \geq 0$, $y \geq 0$. In fact, the region in Fig. 7.31 contains the points that are solutions to all three inequalities:

$$x \geq 0$$

$$y \geq 0$$

$$5x - 3y \geq 0.$$

We can see this by looking first at the graph of each of the inequalities (Fig. 7.32).

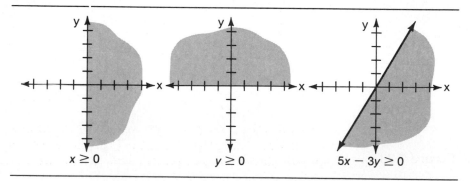

Figure 7.32

The points that are solutions to all three inequalities simultaneously are the points that are shaded in all of the graphs, that is, the region shown in Fig. 7.33.

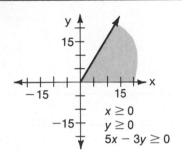

$x \geq 0$
$y \geq 0$
$5x - 3y \geq 0$

Figure 7.33

Example 1 Graph the solutions of the simultaneous inequalities

$$y \geq 0$$

$$y > x$$

$$y \leq -\frac{1}{2}x + 5.$$

Each inequality has its own graph, as shown in Fig. 7.34.

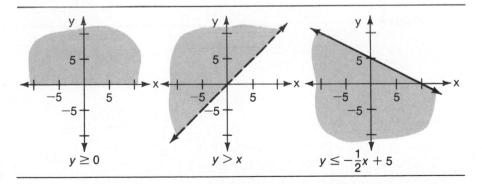

$y \geq 0$ $y > x$ $y \leq -\frac{1}{2}x + 5$

Figure 7.34

If we put the three graphs on the same coordinate system, it is possible to see which points are solutions to all three inequalities (Fig. 7.35). The points that are solution to all three inequalities form the region shown in Fig. 7.36.

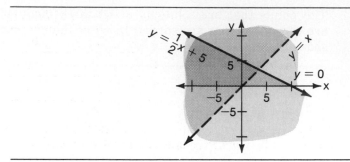

Figure 7.35

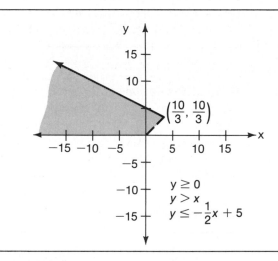

Figure 7.36

The dashed line indicates that points on that line are not solutions. The solid line indicates that point on that line are solutions. The point of intersection of the two lines is found by solving simultaneously the equations $y = x$ and $y = -\dfrac{1}{2}x + 5$. □

Example 2   A bakery uses the same mixing, baking, and packaging machines to make its bread and its cakes. The table below shows the amount of time each machine must be used to make a batch of bread and to make a batch of cakes. It also shows the maximum amount of time each machine can be operated in a day.

	Bread	Cakes	Maximum
Mixing (hours)	1	2	10
Baking (hours)	2	1	11
Packaging (hours)	1	1	6

Write and graph inequalities that describe the possible combinations of batches of bread and batches of cakes that can be baked in one day.

Let x denote the number of batches of bread and let y denote the number of batches of cakes. The fact that the time for mixing bread plus the time for mixing cakes must be less than or equal to 10 is expressed by the inequality $x + 2y \le 10$. The fact that the baking time for bread plus the baking time for cakes must be less than or equal to 11 is expressed by the inequality $2x + y \le 11$. The fact that the sum of the times required for the packaging machine must be less than or equal to 6 is expressed by the inequality $x + y \le 6$. Thus the three simultaneous inequalities are the conditions that determine how many batches of bread and batches of cakes can be baked in a day. The graph in Fig. 7.37 shows the possibilities.

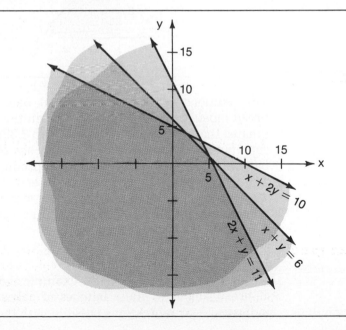

Figure 7.37

The solutions to the problem about batches of cakes and bread cannot be negative. The points in Fig. 7.37 that satisfy the three inequalities and have coordinates that are not negative are shown in Fig. 7.38. These points are the simultaneous solutions to the five inequalities

$$x \geq 0$$

$$y \geq 0$$

$$x + 2y \leq 10$$

$$2x + y \leq 11$$

$$x + y \leq 6.$$

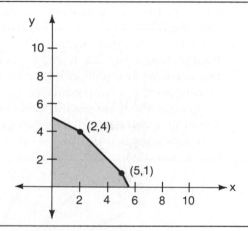

Figure 7.38

The solutions to the problem about batches of cakes and batches of bread must be whole numbers. Within this region, 25 points have coordinates that are whole numbers (Fig. 7.39). Thus there are 25 possible combinations of the number of batches of bread and the number of batches of cakes that the bakery can produce in a day.

    The point (2,4) is the point of intersection of the line with equation $x + 2y = 10$ and the line with equation $x + y = 6$; the point (5,1) is the point of intersection of the lines that have equations $x + y = 6$ and $2x + y = 11$.   □

    How will the bakery in the example above decide how many batches of bread and how many batches of cakes to bake a day? Our graph indicates the combinations possible with the machines the bakery owns.

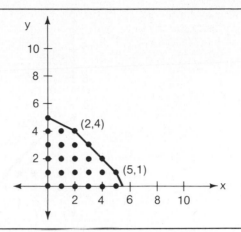

Figure 7.39

For example, they could make 1 batch of bread and 4 batches of cakes [that is the solution (1,4) on the graph], or they could make 3 batches of bread and 3 batches of cakes, or 4 batches of bread and 2 batches of cakes, and so on. How will they decide? The factors that influence their decision will be things such as the buying patterns of their customers, whether they make more money on cakes or bread, and which of their personnel can handle different jobs. The information on the graph is helpful in making the decision because it shows what the possibilities are.

Example 3 Using Example 2, assume that the bakery makes a clear profit of $30 on each batch of bread and $50 on each batch of cakes. Of all the possible combinations of batches of bread and cakes that can be baked in one day, which combination gives the maximum profit for the bakery?

We have seen that there are 25 possibilities (Fig. 7.40). We could compute the profit in each case and see which is the greatest. However, we can reduce that computation by observing the following. Of the possibilities represented by points in the first column—(0,5), (0,4), (0,3), (0,2), (0,1), and (0,0)—the bakery does best with 0 bread and 5 cake batches. That is to say, if the bakery is baking no bread, it should bake the maximum number of cakes, namely 5. In the second column of possible points—(1,4), (1,3), (1,2), (1,1), and (1,0)—the most profit comes from 1 bread and 4 cake batches. In fact, if attention is restricted to only the points in one column, the one at the top gives the most profit.

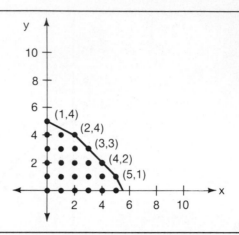

Figure 7.40

Thus we only need to compute the profit possible for the combinations represented by (0,5), (1,4), (2,4), (3,3), (4,2), and (5,1). Remember that the first coordinate designates the number of batches of bread and the second coordinate, the number of batches of cakes.

Point	Profit ($)
(0,5)	$0 \cdot 30 + 5 \cdot 50 = 250$
(1,4)	$1 \cdot 30 + 4 \cdot 50 = 230$
(2,4)	$2 \cdot 30 + 4 \cdot 50 = 260$
(3,3)	$3 \cdot 30 + 3 \cdot 50 = 240$
(4,2)	$4 \cdot 30 + 2 \cdot 50 = 220$
(5,1)	$5 \cdot 30 + 1 \cdot 50 = 200$

The computation shows that the greatest profit results when the bakery prepares 2 batches of bread and 4 batches of cakes a day.   □

Exercises 7.7   Graph the system of inequalities in Exercises 1–8.

1. $y \leq x + 2$

   $y \geq 3x - 2$

2. $y \geq \frac{1}{3}x + 1$

   $y \leq \frac{-3}{4}x + 3$

3. $3y - 2x < 6$

   $4x + 3y \geq 12$

4. $3y - x > 3$

   $5x + 3y > 15$

5. $x + 2y \leq 3$

   $x - y \leq 3$

   $x \geq 1$

   $y \geq 0$

6. $5x - 3y \geq 0$

   $5x + 3y \geq 0$

   $y \geq 0$

7. $y - 2x > 0$

   $y - 2x \leq 4$

   $x \geq 0$

   $x \leq 5$

8. $x - 2y \leq 4$

   $2y - x \leq 10$

   $2x + y \leq 13$

   $2x + y \geq -7$

Write a system of inequalities whose simultaneous solutions form the region described in each of Exercises 9–14.

9. The region that is above the line $y = 2x + 1$ and below the line $3x + y = 2$

10. The region that is above both the line $\frac{x}{2} + \frac{y}{3} = 1$ and the line $\frac{y}{2} - \frac{x}{3} = 1$

11. The region that is below the line determined by (0,1) and (2,5) and below the line determined by (1,4) and (3,0)

12. The region that is below the line determined by (−3,−1) and (4,5) and below the line determined by (−2,6) and (7,−2)

13. The interior of the triangle determined by (0,0), (0,4), and (4,2)

14. The interior of the four-sided figure determined by (−1,−2), (1,6), (5,−1), and (8,2)

15. Timber University's basketball arena seats 10,000. Student and faculty tickets sell for $5 each while tickets to the general public sell for $8. The overhead for each game is $2000. Write and graph inequalities that describe the possible combinations of ticket sales that yield a profit for the university.

16. Mrs. Jones has 4 cups of mix and 6 cups of milk to make pancakes and waffles. To make one pancake requires $\frac{1}{8}$ cup of mix and $\frac{1}{4}$ cup of milk. To make one waffle requires $\frac{2}{3}$ cup of mix and $\frac{1}{2}$ cup of milk. Write and graph inequalities that describe the possible combinations of pancakes and waffles that can be made.

17. The Keystone Company has 800 pounds of peanuts, 500 pounds of cashews, and 300 pounds of pecans to use in producing two different kinds of nut mixtures. The proportions used for each mixture are shown in the chart below:

Mixture	Peanuts (pounds)	Cashews (pounds)	Pecans (pounds)
Party Pack	$1\frac{1}{4}$	$\frac{1}{2}$	$\frac{1}{4}$
Fancy Mix	$\frac{3}{4}$	$\frac{3}{4}$	$\frac{1}{2}$

Write and graph inequalities that describe the possible combinations of Party Pack and Fancy Mix packages that can be made.

18. A tailor has 14 square yards of cotton and 10 square yards of silk on hand. To make a suit he needs 3 square yards of cotton and 1 square yard of silk. An evening gown requires 1 square yard of cotton and 3 square years of silk. Write and graph inequalities that describe the possible combinations of suits and evening gowns that the tailor can make with the material on hand.

19. If the tailor in Problem 18 earns $40 per evening gown but only $25 per suit, which of the combinations in Problem 18 earns him the most money?

7.8 / CHAPTER 7 PROBLEM COLLECTION

In Exercises 1–4, draw the line through the given point with the given slope.

1. (0,3), slope $\dfrac{2}{3}$

2. (−2,4), slope $-\dfrac{3}{4}$

3. (2,−3), slope 0

4. (2,−3) slope undefined

5. Find the slope and y-intercept, and graph the equation $2x - 3y + 4 = 0$.

6. Find the x-intercept and y-intercept, and graph the equation $3x + 4y + 12 = 0$.

7. On the same coordinate system, graph $y = \dfrac{3}{5}x$, $y = -\dfrac{5}{3}x + 1$,

   $y = -\dfrac{5}{3}x - 2$.

Determine k in the equation of the line in Exercises 8 and 9 so that the graph of the equation satisfies the given condition.

8. $y = \dfrac{3}{4}x + k$, goes through (−2,1)

9. $kx + 3y + 1 = 0$, goes through (−1,1)

Find the slope of the line determined by the two points in Exercises 10–13.

10. (1,−2) and (5,1)

11. (−2,1) and (3,−2)

12. (−2,3) and (4,3)

13. (1,−2) and (1,5)

14. Find c so that the slope of the line determined by (−3,1) and (2,c) is

    (a) 2

    (b) $-\dfrac{3}{4}$

    (c) 0

Write an equation for each of the lines in Exercises 15–22.

15. With slope $\dfrac{2}{3}$ and y-intercept −1

16. Through $(1,1)$ and $(5,7)$

17. Through $(-2,1)$ and $(2,-5)$

18. Through $(-1,2)$, parallel to $3x - 2y = 1$

19. Through $(1,-1)$, perpendicular to $3x - 2y = 1$

20. With x-intercept 2 and y-intercept -1

21. Through $(-1,1)$, no x-intercept

22. Through $(1,-2)$, no y-intercept

23. Show that the four-sided figure $ABCD$ determined by the points $A = (-2,-2)$, $B = (2,8)$, $C = (8,17)$, and $D = (4,7)$ is a parallelogram.

24. Show that the triangle determined by the points $(-3,-3)$, $(3,-12)$, and $(6,3)$ is a right triangle.

Find the simultaneous solutions of the equations in Exercises 25 and 26 graphically.

25. $2x - 4y = 5$

      $4x + 6y = 3$

26. $y + x = 3$

      $y = x^2 + 1$

Find the simultaneous solutions of the equations in Exercises 27–36.

27. $3x - 4y = 10$

      $2x + 5y = -1$

28. $x - 2y = -8$

      $2x - 3y = -13$

29. $6x - 3y = 2$

      $12x + 3y = 7$

30. $x - 2y = 2$

      $2x - 4y = -4$

31. $2x + y = 3$

      $5x + 2.5y = 7.5$

32. $2x - 3y + z = -5$

      $x - 2y + 2z = 1$

      $3x + 2y - 2z = -5$

33. $2x + y - z = 2$

      $2y - 3z = -5$

      $x - 3y = 5$

34. $3x + y - 2z = 1$

      $3x + 8y - 7z = -4$

      $x - 2y + z = 2$

35. $3x - y + z = 2$
     $-5x + 4y - 2z = -2$
     $4x + y + z = 1$

36. $x + y + z + w = 0$
     $2x - y + 2z - w = 0$
     $3x + 2y - z + 2w = -4$
     $x - 2y + 3z - w = 0$

37. Find two numbers whose sum is 62 and whose difference is 12.

38. How many gallons of a 32%-acid solution must be mixed with 30 gallons of a 58%-acid solution to obtains a 44%-acid solution?

39. Tim invests $22,000, part at 6.7% simple interest and the remainder at 5.2% simple interest. If his annual income is $1349.50, how much does he have invested at each rate?

40. A money box contains $2040 in 20-dollar bills and 50-dollar bills. If there are 60 bills, how many of each kind of bill are in the box?

41. The length of a rectangle is one foot more than three times its width. Find the dimensions if the perimeter of the rectangle is 178 feet.

42. Determine m and b so that the graph of $y = mx + b$ passes through the points $(3,-1)$ and $(6,-3)$.

43. A money box contains $16.25 in nickels, dimes, and quarters. There are twice as many dimes as nickels and 2 more quarters than dimes. Find the number of each in the box.

44. Determine a, b, and c so that the graph of $y = ax^2 + bx + c$ passes through the points $(-1,-4)$, $(0, 2)$, and $(3,-8)$.

Sketch the graph of the inequalities in Exercises 45–50.

45. $2x + 3y + 1 < 4$

46. $x - 2y \geq 1 + y$

47. $\dfrac{1}{3}x - \dfrac{1}{4}y \leq 1$

48. $2(x - 1) + 3(y - 2) > -2$

49. $x + 1 \leq 0$

50. $2y - 3 \geq 1$

Write an inequality whose solutions form the region described in Exercises 51 and 52.

51. Above the line $2y = x + 4$

52. Below the line determined by the points $(-1,2)$ and $(3,5)$

53. Hank walks at the rate of 5 mph to a bus stop. He then travels by bus at 30 mph. Write and graph an inequality that expresses the requirement that the time for the entire trip is less than or equal to 3 hours.

54. An acid solution is formed by mixing a 32%-acid solution with a 74%-acid solution. Write and graph an inequality that expresses the requirement that the percent of acid in the new mixture does not exceed 52%.

Graph each system of inequalities in Exercises 55–59.

55. $y \geq \dfrac{2}{3}x + 2$

   $3x + 5y \leq 3$

56. $3y - 5x < 5$

   $y \leq 4$

57. $y \geq x$

   $y \leq 4$

   $y \geq 0$

58. $4x + y \leq 12$

   $2x + y \leq 8$

   $3x + 5y \leq 30$

   $y \geq 0$

59. $4y - 5x \leq 32$

   $y \leq -\dfrac{5}{4}x + 8$

   $x \leq 4$

   $x \geq -4$

   $y \geq 0$

Write a system of inequalities whose simultaneous solutions form the regions described in Exercises 60 and 61.

60. The region above the lines $y = x$ and $x + y = 0$

61. The region that is below the line $y = \dfrac{8}{5}x + 1$, below the line

   $y = -\dfrac{1}{2}x + 4$, and above the line $5x - 11y = 19$

62. A tailor has 14 square yards of cotton and 10 square yards of silk on hand. To make a suit he needs 2 square yards of cotton and 1

square yard of silk. A dress requires 1 square yard of cotton and 1 square yard of silk. Write and graph inequalities that describe the possible combinations of suits and dresses that the tailor can make with the material on hand.

63. If the tailor in Problem 62 earns $50 per suit and $40 per dress, which one of the combinations in Problem 62 earns him the most money?

Polynomial Arithmetic and Factoring

8.1 / ADDING, SUBTRACTING, AND MULTIPLYING POLYNOMIALS

In Section 6.2 we graphed equations containing special kinds of algebraic expressions called **polynomials**. In that section we considered polynomials in one variable. Here are some examples of polynomials in one variable:

$$6x^2 - 8x + 4$$

$$\frac{1}{2}x - \frac{1}{3}$$

$$7x^7$$

$$x^5 - .5x^3 + 4.$$

Polynomials can also be written in more than one variable, for example,

$$x^3 + y^2$$

$$3r^3 - r^2s + 4rs^2 - s^3$$

$$12xyz^2.$$

Problem situations are frequently described by polynomials. We have already studied techniques for solving equations that contain *linear* polynomials (highest exponent 1) in one variable. Before we study techniques for solving equations that contain polynomials in one variable with highest exponent 2 (that is, degree 2 or **quadratic polynomials**), we need more practice computing with polynomials. In particular, we need to be able to add, subtract, multiply, and divide polynomials.

Adding and Subtracting Polynomials

Polynomials are added and subtracted like other algebraic expressions. These examples should remind you of the procedures:

$$(x^2 + 4x - 3) + (-2x^2 + 5) = x^2 + 4x - 3 - 2x^2 + 5 = -x^2 + 4x + 2$$

$$(x^2 + 4x - 3) - (-2x^2 + 5) = x^2 + 4x - 3 + 2x^2 - 5 = 3x^2 + 4x - 8$$

$$(3xy - 5) + (y^2 - xy + 2) = 3xy - 5 + y^2 - xy + 2 = y^2 + 2xy - 3$$

$$(3xy - 5) - (y^2 - xy + 2) = 3xy - 5 - y^2 + xy - 2 = -y^2 + 4xy - 7.$$

Terms in polynomials that contain the same powers of the variables are called **like terms**; the distributive property permits like terms to be combined by addition and subtraction:

$$6x^2y - 2x^2y = (6 - 2)x^2y = 4x^2y$$

$$7z^3 + 3z^3 = (7 + 3)z^3 = 10z^3.$$

Terms that do not contain the same powers of the variables cannot be simplified in this way. For example, $3y - y^2$ cannot be reduced to a one-term polynomial.

In the examples above, the two polynomials that are added or subtracted have been written on the same line. In some computations involving addition or subtraction of polynomials, one polynomial is written below the other. When we do this, we position the terms of the lower polynomial directly under the corresponding like terms of the upper polynomial.

Example 1   Add the polynomials

$$\begin{array}{r} x^2 + 4x - 3 \\ -2x^2 \quad\;\; + 5 \\ \hline \end{array}.$$

This computation has been worked out in a horizontal manner in the discussion above. There it has the form $(x^2 + 4x - 3) + (-2x^2 + 5)$. To do it vertically we need only remember that adding two polynomials requires adding like terms in the polynomials.

$$\begin{array}{r} x^2 + 4x - 3 \\ -2x^2 \quad\;\; + 5 \\ \hline -x^2 + 4x + 2 \end{array}$$

There are actually three steps in the computations above:

$$x^2 + (-2x^2) = -x^2,$$

$$4x + 0x = 4x,$$

$$-3 + 5 = 2. \qquad\qquad\qquad \square$$

Example 2   Subtract the polynomials

$$\begin{array}{r} x^2 + 4x - 3 \\ -2x^2 \qquad + 5 \\ \hline \end{array}.$$

This computation is the same as the one above that was written $(x^2 + 4x - 3) - (-2x^2 + 5)$. To perform the computation, we must subtract each term of the second polynomial from the corresponding like term of the first polynomial.

$$\begin{array}{r} x^2 + 4x - 3 \\ -2x^2 \qquad + 5 \\ \hline 3x^2 + 4x - 8 \end{array}.$$

Again there are three parts to this computation:

$$x^2 - (-2x^2) = 3x^2$$

$$4x - 0x = 4x$$

$$-3 - 5 = -8. \qquad\qquad\qquad \square$$

Multiplying Polynomials

The procedure for multiplying polynomials is a consequence of the distributive property. We already know how to multiply by a one-term polynomial using the distributive property:

$$3x^2(x^3 - 2x + 7) = 3x^2(x^3) + 3x^2(-2x) + 3x^2(7) = 3x^5 - 6x^3 + 21x^2.$$

If the first of two polynomials has more than one term, the distributive property (used more than once) again determines the product:

$$\begin{aligned} (3x^2 + x)(x^3 - 2x + 7) &= (3x^2 + x)(x^3) + (3x^2 + x)(-2x) \\ &\quad + (3x^2 + x)(7) \\ &= (3x^2)(x^3) + x(x^3) + (3x^2)(-2x) + (x)(-2x) \\ &\quad + (3x^2)(7) + (x)(7). \end{aligned}$$

The product above still needs to be simplified, but before doing that, we should observe the result so far: each term of the first polynomial is multiplied by each term of the second polynomial to give six partial products that are then added. This is the effect of the distributive property; it is the procedure for multiplying any two polynomials.

Example 3 Find the product of $(x - 2y)(x^2 + xy - 4y^2)$.

a. Using the observation above, namely, that each term of the first polynomial multiplies each term of the second polynomial and then the partial products are added, we can write the product this way:

$$(x - 2y)(x^2 + xy - 4y^2) = x(x^2) + x(xy) + x(-4y^2) + (-2y)(x^2)$$
$$+ (-2y)(xy) + (-2y)(-4y^2)$$
$$= x^3 + x^2y - 4xy^2 - 2x^2y - 2xy^2 + 8y^3$$
$$= x^3 - x^2y - 6xy^2 + 8y^3.$$

We might call this a horizontal method for writing the product of two polynomials.

b. Sometimes products of polynomials are set up in a vertical manner analogous to the long multiplication process for large numbers. In the computation below, we first multiply $x^2 + xy - 4y^2$ by x and then multiply $x^2 + xy - 4y^2$ by $(-2y)$. Finally the partial products are added. Notice that the partial products are arranged to help with the addition.

$$
\begin{array}{r}
x^2 + xy - 4y^2 \\
x - 2y \\
\hline
x^3 + x^2y - 4xy^2 \\
- 2x^2y - 2xy^2 + 8y^3 \\
\hline
x^3 - x^2y - 6xy^2 + 8y^3 \, .
\end{array}
$$

□

Example 4 Perform the indicated operations and then simplify $2y^2 - (x + y)(x - 2y)$.

We first compute the product $(x + y)(x - 2y)$. This product has four partial products arising as each term of the first polynomial is multiplied by each term of the second:

$$2y^2 - (x + y)(x - 2y) = 2y^2 - (x^2 - 2xy + xy - 2y^2)$$
$$= 2y^2 - (x^2 - xy - 2y^2) \quad \text{(Combine like terms)}$$
$$= 2y^2 - x^2 + xy + 2y^2$$
$$= -x^2 + xy + 4y^2.$$

□

Example 5 A rectangular garden with dimensions of 16 feet by 12 feet is bordered by a concrete walk of uniform width. If the width of the walk is x feet, give an algebraic expression for its area (Fig. 8.1).

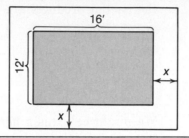

Figure 8.1

One way to express the area of the walkway is to subtract the area of the inner rectangle from the area of the outer rectangle. The inner area has dimensions 16 by 12. The outer rectangle has dimensions $(16 + 2x)$ by $(12 + 2x)$. Thus the walkway has area

$$(16 + 2x)(12 + 2x) - (16)(12) = [(16)(12) + 16(2x) + (2x)(12) \\ + (2x)(2x)] - (16)(12)$$

$$= (16)(12) + 32x + 24x + 4x^2 \\ - (16)(12)$$

$$= 56x + 4x^2$$

Thus the walk of width x has area $56x + 4x^2$.   □

Evaluating Sums, Differences, and Products of Polynomials

If we have the sum of two polynomials that needs to be evaluated for some value of the variable, it is natural to ask whether we should first add the two polynomials and then evaluate the sum, or first evaluate each of the two polynomials and then add their values. Actually, it does not matter. The two approaches give the same value because, in computing with polynomials, we treat the variable as though it were a number.

Example 6 Find the value of $(2x + 5)(x - 4) + 10$ for $x = -3$.

a. We can evaluate each polynomial in the expression for $x = -3$ and then perform the operations:

$$[2(-3) + 5](-3 - 4) + 10 = (-1)(-7) + 10$$
$$= 7 + 10$$
$$= 17.$$

b. We can perform the operations on the polynomials in the expression and then evaluate for $x = -3$:

$$(2x + 5)(x - 4) + 10 = 2x^2 - 8x + 5x - 20 + 10$$
$$= 2x^2 - 3x - 10$$
$$= 2(-3)^2 - 3(-3) - 10 \quad (\text{Letting } x = -3)$$
$$= 17.$$

You can see that method (a) is a somewhat shorter way to evaluate the expression.   □

Exercises 8.1

Subtract the second polynomial from the first polynomial in Exercises 1–4.

1. $3x^3 - 2x^2 + 5x - 4$
   $3x^2 - 2x - 3$

2. $2x^2 - 3xy + 5y^2$
   $2x^2 + xy + 2y^2$

3. $-2x^3y + 4x^2y^2 + 3xy - 2$
   $-2x^3y + 4x^2y^2 \quad\quad - 1$

4. $x^4 \quad\quad\quad - 2x^2 \quad\quad + 1$
   $\quad\quad 2x^3 - x^2 + 2x$

Perform the indicated operations and simplify the resulting algebraic expressions in Exercises 5–39.

5. $(3x - 4y) + (2x + 3y)$

6. $(7x^2y + xy^2 + x - y) - (2x^2y - 2xy^2 + xy - x + 1)$

7. $x(2y - x) - y(2x - y)$

8. $(2x^3y^2 - xy^3 + 4x^2y^2) + (x^3y^2 + xy^3 - 2x^2y^2) - (3x^3y^2 + 2x^2y^2)$

9. $(2x + 1)(3x - 2)$

10. $(x + 4)(x + 4)$

11. $(2x - 3)(2x - 3)$

12. $(x - 2)(x + 2)$

13. $(y - 3)(y + 3)$

14. $(2x + 3)(x - 4)$

15. $(2x^2 - 1)(3x^2 + 2)$

16. $(4x^2 + 3)(x^2 - 4)$

17. $(x + 1)(x + 1) - (x - 1)(x - 1)$

18. $(y - x)(y + x) + (x - y)(x + y)$

19. $(a + b + c)(a + b + c)$

20. $(a - b + c)(a - b + c)$

21. $(a + b + c)(a - b - c)$

22. $(a - b + c)(a - b - c)$

23. $x(x - 1)(x + 1)$

24. $(x - 1)(x + 1)(x^2 + 1)$

25. $(x^2 + x + 1)(x^2 - x + 1)$

26. $(x^2 + 2x + 4)(x^2 - 2x + 4)$

27. $(x - 1)(x^2 + x + 1)$

28. $(x + 1)(x^2 - x + 1)$

29. $(x + 2y)(x^2 - 2xy + 4y^2)$

30. $(x - 2y)(x^2 + 2xy + 4y^2)$

31. $(x + 1)(x + 1)(x + 1)$

32. $(x - 1)(x - 1)(x - 1)$

33. $(x - 1)(x^3 + x^2 + x + 1)$

34. $(x + 1)(x^3 - x^2 + x - 1)$

35. $(x + 1)(x^4 - x^3 + x^2 - x + 1)$

36. $(x - 1)(x^4 + x^3 + x^2 + x + 1)$

37. $(x^2 + x + 1)(2x^2 - x + 3) - (3x + 1)$

38. $y(3x - 2y) - (x + 2y)(2x - y)$

39. $(x^3 - x^2 y)(x^2 + xy + y^2) - (y^3 - xy^2)(xy + y^2)$

In Exercises 40 and 41, find the value of each algebraic expression for the specified value of the variable.

40. $(x - 1)(2x^2 - 3x + 1) - 2$; $x = -2$

41. $(x^2 - 2)(x^2 - x + 1) - (2x - 1)$; $x = 3$

42. A rectangular garden of dimensions 15 feet by 20 feet is bordered by a concrete walk of uniform width. If the width of the walk is x feet, write an algebraic expression for the area of the walk.

43. A rectangle is 2 feet longer than it is wide. If the width is x feet, write an algebraic expression for the perimeter of the rectangle and an algebraic expression for the area of the rectangle.

44. An open-topped box is constructed by cutting a square of side x from each corner of a 10-foot by 15-foot sheet of cardboard and folding up the sides. Write the volume of the box as a function of x.

45.   A piece of wire x feet long is bent to form a square. Write the area of the square as a function of x.

8.2 / SPECIAL POLYNOMIAL PRODUCTS

Some polynomial products occur frequently. Rather than multiply them out each time, it is helpful to observe patterns in the product so that the product can be written directly. These patterns will be useful when we want to write a polynomial as a product of factors in the next two sections.

Product of Like Binomials

A polynomial that has two terms is called a **binomial**. In general, the product of two binomials contains four terms. However, in the special case that the two binomials are made up of like terms, two of the terms in the product can be combined. Study this example:

$$(5x + 2)(3x - 7) = 5x(3x) + 5x(-7) + 2(3x) + 2(-7)$$
$$= 5x(3x) + [5x(-7) + 2(3x)] + 2(-7)$$
$$= 5x(3x) + [-35x + 6x] + 2(-7).$$

The product can, of course, be simplified to $15x^2 - 29x - 14$; however, the important thing to observe from the last line above is that the first term in the product is the product of $5x$ and $3x$, the first terms in the two binomials. The last term of the product is 2 times -7, the last terms of the two binomials. The middle term is a sum; we can indicate it this way:

$$(5x + 2)(3x - 7) \quad \text{to mean the sum} \quad (5x)(-7) + (2)(3x).$$

Here is a second example of multiplying a binomial by a like binomial:

$$(3x + y^2)(x - 2y^2) = 3x(x) + (3x)(-2y^2) + y^2(x) + y^2(-2y^2)$$
$$= 3x(x) + [3x(-2y^2) + y^2x] + y^2(-2y^2)$$

$$= 3x(x) + [-6xy^2 + xy^2] + y^2(-2y^2)$$
$$= 3x^2 - 5xy^2 - 2y^4.$$

Again we see that the first term in the product comes from multiplying the two first terms of the binomials; the last term comes from multiplying the last terms of the binomials; the middle term is a sum:

$$(3x + y^2)(x - 2y^2).$$

Squaring a Binomial

When we square a binomial, we multiply it by itself, thus computing the product of a binomial times a binomial:

$$(3x + 5)^2 = (3x + 5)(3x + 5) = (3x)^2 + 3x(5) + 5(3x) + 5^2$$
$$= (3x)^2 + [15x + 15x] + 5^2$$
$$= (3x)^2 + 2(15x) + 5^2$$
$$= 9x^2 + 30x + 25.$$

The middle term in this product is twice the product of the two terms of the binomial:

$$(3x + 5)(3x + 5).$$

If we represent the first term of the binomial by a and the second by b, we get this expression:

$$(a + b)^2 = (a + b)(a + b) = a^2 + (ab + ba) + b^2$$
$$= a^2 + 2ab + b^2.$$

To help remember that $(a + b)^2 = a^2 + 2ab + b^2$, we can draw a square of side length $a + b$ (Fig. 8.2). Then $(a + b)^2$ is the area of the square. Another way to express the area is to add the areas of the four pieces shown. Thus

$$(a + b)^2 = a^2 + ab + ab + b^2$$
$$= a^2 + 2ab + b^2.$$

We get an analogous result when we square a binomial that is a difference:

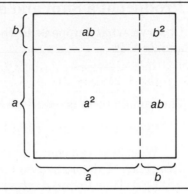

Figure 8.2

$$(a - b)^2 = (a - b)(a - b) = a^2 - ab - ba + b^2$$
$$= a^2 - 2ab - b^2.$$

We can also draw a picture of this result starting with a square that has side length a (Fig. 8.3). The shaded square has area $(a - b)^2$. This area can also be represented as the area of the large square minus the three unshaded areas:

$$a^2 - b(a - b) - b(a - b) - b^2 = a^2 - ba + b^2 - ba + b^2 - b^2$$
$$= a^2 - 2ab + b^2.$$

Thus $(a - b)^2 = a^2 - 2ab + b^2$.

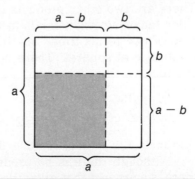

Figure 8.3

Product of a Sum and a Difference

Observe what happens to the middle term in the product of $3x + 2$ and $3x - 2$:

$$(3x + 2)(3x - 2).$$

The middle term is $-6x + 6x = 0$. Thus the product has only two terms:

$$(3x + 2)(3x - 2) = (3x)(3x) + (2)(-2) = 9x^2 - 4.$$

Whenever you compute the product of a sum of two terms times the difference of the same two terms, you can expect that the middle term will be 0. Thus the product will be the square of the first term minus the square of the second. Study these examples:

$$(2x + y)(2x - y) = 4x^2 - 2xy + 2xy - y^2$$
$$= 4x^2 + 0xy - y^2$$
$$= 4x^2 - y^2$$

$$(x^2 - 3)(x^2 + 3) = x^4 + 3x^2 - 3x^2 - 9$$
$$= x^4 + 0x^2 - 9$$
$$= x^4 - 9.$$

In these cases we note that the product is the difference of two squares.

Products that Equal the Difference and Sum of Two Cubes

There are two other products that we need to compute before we consider factoring polynomials in the next section. We have seen that the sum of two terms times the difference of the same two terms gives a difference of squares. There is also a product that gives a difference of cubes:

$$(x - y)(x^2 + xy + y^2) = x^3 + x^2y + xy^2 - x^2y - xy^2 - y^3$$
$$= x^3 - y^3.$$

Similarly, $(2x - 3)(4x^2 + 6x + 9) = (2x)^3 - 3^3 = 8x^3 - 27$.

Although there is no product that gives the sum of two squares, there is a product that gives the sum of two cubes:

$$(x + y)(x^2 - xy + y^2) = x^3 - x^2y + xy^2 + x^2y - xy^2 + y^3$$
$$= x^3 + y^3.$$

Similarly, $(2x + 3)(4x^2 - 6x + 9) = (2x)^3 + 3^3 = 8x^3 + 27.$

Summary of Special Products

The following examples summarize the special products we will use in the next two sections. Be sure you can compute each of them.

Example 1   Product of like binomials:

$$(4x - 3)(x + 5) = 4x^2 + 4x(5) - 3(x) - 15$$
$$= 4x^2 + 17x - 15$$
$$(2x - y)(7x + 3y) = 14x^2 + 2x(3y) - y(7x) - 3y^2$$
$$= 14x^2 - xy - 3y^2. \qquad \square$$

Example 2   Square of a sum:

$$(x + y)^2 = x^2 + 2xy + y^2$$
$$(3x^2 + y)^2 = 9x^4 + 6x^2y + y^2. \qquad \square$$

Example 3   Square of a difference:

$$(x - y)^2 = x^2 - 2xy + y^2$$
$$(3x^2 - y)^2 = 9x^4 - 6x^2y + y^2. \qquad \square$$

Example 4   Product equal to difference of squares:

$$(x - y)(x + y) = x^2 - y^2$$
$$(x^2 - 5y)(x^2 + 5y) = x^4 - 25y^2. \qquad \square$$

Example 5　Product equal to difference of cubes:

$$(x - y)(x^2 + xy + y^2) = x^3 - y^3$$
$$(7x - y^3)(49x^2 + 7xy^3 + y^6) = (7x)^3 - (y^3)^3 = 343x^3 - y^9.$$　□

Example 6　Product equal to sum of cubes:

$$(x + y)(x^2 - xy + y^2) = x^3 + y^3$$
$$(7x + y^3)(49x^2 - 7xy^3 + y^6) = (7x)^3 + (y^3)^3 = 343x^3 + y^9.$$　□

Exercises 8.2

Find the products in Exercises 1–28.

1. $(x - 2)(x^2 + 2x + 4)$
2. $(4x + 3y^2)(2x - 4y^2)$
3. $(x - 4)(x + 4)$
4. $(2x + 3)^2$
5. $(x + 2)(x^2 - 2x + 4)$
6. $(2y - x^3)(y + 2x^3)$
7. $(2x - 3)(2x + 3)$
8. $(3x - 2)^2$
9. $(x - 2y)(x^2 + 2xy + 4y^2)$
10. $(3z + 4y^2)(y^2 - 2z)$
11. $(3y - 2)(3y + 2)$
12. $(4x + 1)^2$
13. $(x + 2y)(x^2 - 2xy + 4y^2)$
14. $(4y - 3)(y + 4)$
15. $(1 - x)(1 + x)$
16. $(3x + 2y)^2$
17. $(x^2 - 3y)(x^4 + 3x^2y + 9y^2)$
18. $(2r + 1)(3r + 4)$
19. $(1 - x^2)(1 + x^2)$
20. $(1 - x)^2$
21. $(x^2 + y^2)(x^4 - x^2y^2 + y^4)$
22. $(3x - 2)(4x - 3)$
23. $(2 - xy)(2 + xy)$
24. $(2x^2 - 3)^2$
25. $(x^2 + 1)(x^4 - x^2 + 1)$
26. $(x^2 - 1)(x^4 + x^2 + 1)$
27. $(2y - 5)(y - 3)$
28. $(a + b)^2(a - b)^2$

Draw a diagram to illustrate the products in Exercises 29 and 30.

29. $(a + b)(c + d) = ac + ad + bc + bd$
30. $(a + b)(a - b) = a^2 - b^2$

8.3 / REMOVING COMMON FACTORS AND FACTORING TRINOMIALS

In Chapter 1 we observed that any whole number greater than 1 can be factored into a product of prime numbers. In fact, for each whole number there is essentially one way to write the number as a product of primes. There are at least two situations when factoring numbers into primes simplifies computation. One is when we want to find a common denominator to add or subtract two fractions. Another is when we want to remove common factors from the numerator and denominator to reduce a fraction.

In this section we investigate factoring polynomials into products of other polynomials. The discussion is limited to polynomials that have integers for coefficients (no fractions or decimals). If a polynomial is written as a product of other polynomials, we say it is **factored**. A polynomial that cannot be factored is said to be **irreducible**. An irreducible polynomial is similar to a prime number.

Although every polynomial can be written as a product of irreducible polynomials, it may be difficult to find the irreducible factors. Only in a few cases are we able to describe methods for writing the factors of a given polynomial; for the most part, these methods are consequences of the special products we computed in the last section. Factoring polynomials, like factoring whole numbers, simplifies adding and subtracting fractions and reducing fractions when the fractions have polynomials in their numerators and denominators. We will see that factoring helps us to solve certain equations. Even though we can only develop methods for factoring certain types of polynomials, these techniques are used frequently.

Removing Common Factors

The distributive property gives these products:

$$4x^2 - x = x(4x - 1)$$

$$3x + 9y - 24z = 3(x + 3y - 8z)$$

$$12x^2z - 10xz^2 = 2xz(6x - 5z)$$

$$(x + y) + 2(x + y)z - 3(x + y)z^2 = (x + y)(1 + 2z - 3z^2).$$

In each of these examples, there is a factor common to every term of the polynomial. Each polynomial can be written as this factor times another polynomial. The first step in factoring a polynomial is to look for common factors in the terms of the polynomial and write the polynomial as the common factor times a new polynomial. Then we attempt to factor the new polynomial.

Factoring Trinomials of Degree 2

We have seen that the product of two like binomials is a polynomial of three, rather than four, terms. (A polynomial with three terms is called a **trinomial**). This is the pattern:

$$(ax + b)(cx + d) = acx^2 + (ad + bc)x + bd.$$

This equation shows the relationship between the coefficients in the trinomial and the coefficients in the two binomials. If we want to write $x^2 + 7x + 12$, for example, as the product of two binomials, we need to find coefficients a, b, c, d so that

$$x^2 + 7x + 12 = (ax + b)(cx + d).$$

Since $ac = 1$ and a and c are integers, we can choose a and c both to be 1. We must guess values for b and d so that both $bd = 12$ and the middle term in the product has coefficient 7. If we guess $b = 2$, $d = 6$, for example, we have

$$(x + 2)(x + 6) = x^2 + 8x + 12.$$

The first and third terms are correct in the product, but the middle term is not. Say we next try $b = 3$, $c = 4$. Then

$$(x + 3)(x + 4) = x^2 + 7x + 12.$$

Good choice! The polynomial $x^2 + 7x + 12$ can be factored as $(x + 3)(x + 4)$.

In the example above, the number of choices for coefficients in the factors was not large. Sometimes there are many choices and the process becomes one of guessing and checking. Fortunately, the number of possibilities is finite; perseverance will pay off! There is always the possibility that the trinomial is irreducible. You have to exhaust all possibilities before you can conclude that none work for your polynomial.

Example 1   Factor $6x^2y + 22xy - 8y$ into irreducible factors.

First we remove the common factors:

$$6x^2y + 22xy - 8y = 2y(3x^2 + 11x - 4).$$

Next we attempt to factor $3x^2 + 11x - 4$. If $3x^2 + 11x - 4 = (ax + b)$ $(cx + d)$, we know that $ac = 3$, so we can let $a = 3$ and $c = 1$ (or $a = 1$ and $c = 3$). Then $3x^2 + 11x - 4 = (3x + b)(x + d)$. Since $bd = -4$, there are six possibilities for b and d. We can have $b = 2$, $d = -2$ or $b = -2$, $d = 2$ or $b = -4$, $d = 1$ or $b = 1$, $d = -4$ or $b = 4$, $d = -1$ or $b = -1$, $d = 4$. Only one of these gives us the product $3x^2 + 11x - 4$:

$$(3x + 2)(x - 2) = 3x^2 - 4x - 4,$$
$$(3x - 2)(x + 2) = 3x^2 + 4x - 4,$$
$$(3x - 4)(x + 1) = 3x^2 - x - 4,$$
$$(3x + 1)(x - 4) = 3x^2 - 11x - 4,$$
$$(3x + 4)(x - 1) = 3x^2 + x - 4,$$
$$(3x - 1)(x + 4) = 3x^2 + 11x - 4.$$

Thus $3x^2 + 11x - 4 = (3x - 1)(x + 4)$, and the complete factoring of the original polynomial is

$$6x^2y + 22xy - 8y = 2y(3x^2 + 11x - 4) \qquad \text{(Remove common factor first!)}$$

$$= 2y(3x - 1)(x + 4). \qquad \qquad \square$$

Example 2   Factor $-6x^2 + 7x - 2$ into irreducible factors.

We can start by first writing $-6x^2 + 7x - 2 = -(6x^2 - 7x + 2)$ and then factoring $6x^2 - 7x + 2$. In factoring this polynomial, if we write 6 as the product of two positive numbers, then we write 2 as the product of two negative numbers so that the coefficient of the middle term will be negative:

$$6x^2 - 7x + 2 = (ax - 2)(bx - 1).$$

The values for a and b that give a coefficient of -7 in the middle term are $a = 3$, $b = 2$:

$$6x^2 - 7x + 2 = (3x - 2)(2x - 1)$$

Thus $-6x^2 + 7x - 2 = -(3x - 2)(2x - 1)$.   □

Example 3   Factor $9x^2 + 12x + 4$ into irreducible factors.

If we are guided by our observations in Example 2, we will write

$$9x^2 + 12x + 4 = (ax + b)(cx + d)$$

and choose a, b, c, and d so they are all positive. The combination that works is

$$9x^2 + 12x + 4 = (3x + 2)(3x + 2) = (3x + 2)^2.$$

Remember that a binomial squared has a special form:

$$(nx + m)^2 = n^2x^2 + 2nmx + m^2.$$

This example is the case where $n = 3$, $m = 2$.   □

Example 4   Factor $2x^2 + 3x - 1$ into irreducible factors.

If $2x^2 + 3x - 1 = (ax + b)(cx + d)$, there are not many choices for a, b, c, and d. We check all possibilities that make $ac = 2$ and $bd = -1$, choosing a and c positive:

$$(2x - 1)(x + 1) = 2x^2 + x - 1,$$
$$(2x + 1)(x - 1) = 2x^2 - x - 1.$$

Since none of these give a middle term of $3x$, we conclude that $2x^2 + 3x - 1$ is an irreducible polynomial.   □

Exercises 8.3

Complete each factorization in Exercises 1–4.

1. $8x^2 - 18x + 9 = (4x - \quad)(2x - \quad)$

2. $6x^2 + x - 12 = (3x - \quad)(2x + \quad)$

3. $4x^2 + 11x - 3 = (4x - \quad)(x + \quad)$

4. $12x^2 + 17x + 6 = (3x + \quad)(4x + \quad)$

Factor the expressions in Exercises 5–25 into irreducible factors.

5. $-8x + 24y$

6. $x^2 - 7x + 12$

7. $x^2 + 2x - 15$

8. $x^2 + 9x + 20$

9. $6x^2 + x - 1$

10. $4x^2 + 10x - 6$

11. $3x^3 + 11x^2 - 4x$

12. $6x^2 + 23x + 20$

13. $5x^2 - 17x + 6$

14. $x^2 + 8x + 16$

15. $x^2 - 6x + 9$

16. $x^2y + y^3$

17. $x^3 - 4x^2y + 4xy^2$

18. $12x + 15x^2 - 18x^3$

19. $6x^4y + 13x^3y^2 - 5x^2y^3$

20. $x^3y^3 - x^2y^2 - xy^3$

21. $9x^2 + 6x + 1$

22. $16x^2 - 24xy + 9y^2$

23. $10y^4 - 7y^2 - 12$

24. $10x^4 + 11x^2y - 6y^2$

25. $3x^3y + 12x^2y + 12xy$

Exhaust all possibilities to show that the expressions in Exercises 26–30
are irreducible.

26. $x^2 + y^2$

27. $2x + 3$

28. $y^2 + yz + z^2$

29. $y^2 - yz + z^2$

30   $2x^2 + 2x + 1$

8.4 / FACTORING DIFFERENCES OF SQUARES, AND SUMS AND DIFFERENCES OF CUBES

In the last section we observed that the first step in factoring any
polynomial is to remove common factors. We also practiced factoring
trinomials. In this section we use the other special products we have
computed to factor the difference of squares and both sums and differences of cubes.

Factoring the Difference of Two Squares

Remember that $(a - b)(a + b) = a^2 - b^2$. Thus whenever you see a difference of two perfect squares, $a^2 - b^2$, you can factor it as $(a - b)(a + b)$. Study the following examples. Each is a perfect square minus a perfect square.

$$x^2 - 4 = x^2 - 2^2 = (x - 2)(x + 2),$$

$$36 - 25t^2 = 6^2 - (5t)^2 = (6 - 5t)(6 + 5t),$$

$$(x - 7)^2 - y^2 = [(x - 7) - y][(x - 7) + y],$$

$$x^4 - 1 = (x^2)^2 - 1^2 = (x^2 - 1)(x^2 + 1) = (x - 1)(x + 1)(x^2 + 1).$$

The sum of two squares cannot be factored. Experiment with $a^2 + b^2$ to convince yourself that it is an irreducible polynomial.

Factoring Differences of Cubes and Sums of Cubes

The key to factoring a difference of two cubes is given by the equation $(a - b)(a^2 + ab + b^2) = a^3 - b^3$. The key to factoring a sum of cubes is given by $(a + b)(a^2 - ab + b^2) = a^3 + b^3$. Actually, you will want to memorize these two equations in this form:

$$a^3 - b^3 = (a - b)(a^2 + ab + b^2),$$

$$a^3 + b^3 = (a + b)(a^2 - ab + b^2).$$

Look carefully at the signs in these factors and at the middle term in the second factor.

In factoring differences and sums of cubes, an important step is to recognize what is being cubed in each term. Be sure that is clear to you in each of the following examples:

$$x^3 - 8 = x^3 - 2^3 = (x - 2)(x^2 + 2x + 4),$$

$$125x^3 - y^6 = (5x)^3 - (y^2)^3 = (5x - y^2)(25x^2 + 5xy^2 + y^4),$$

$$125x^3 + y^6 = (5x)^3 + (y^2)^3 = (5x + y^2)(25x^2 - 5xy^2 + y^4)$$

$$(x + y)^3 + z^3 = [(x + y) + z][(x + y)^2 - (x + y)z + z^2].$$

Often, factoring polynomials requires a combination of the procedures we have developed. To get started, remember to remove common factors first.

Example 1   Factor $8x^3 - 2xy^2$ into a product of irreducible factors.

$$8x^3 - 2xy^2 = 2x(4x^2 - y^2) \qquad \text{(Common factors)}$$
$$= 2x(2x - y)(2x + y). \qquad \text{(Difference of squares)} \qquad \square$$

Example 2   Factor $27x - x^4$ into a product of irreducible factors.

$$27x - x^4 = x(27 - x^3) \qquad \text{(Common factor)}$$
$$= x(3^3 - x^3)$$
$$= x(3 - x)(9 + 3x + x^2). \qquad \text{(Difference of cubes)} \qquad \square$$

Example 3   Factor $y^6 - z^6$ into a product of irreducible factors.

$$y^6 - z^6 = (y^3 - z^3)(y^3 + z^3) \qquad \text{(Difference of squares)}$$

$$= (y - z)(y^2 + yz + z^2)(y^3 + z^3) \qquad \text{(Difference of cubes)}$$

$$= (y - z)(y^2 + yz + z^2)(y + z)(y^2 - yz + z^2). \qquad \text{(Sum of cubes)}$$

Each of the factors in the last product is irreducible. (In Problems 28 and 29 of Section 8.3, you showed that $y^2 + yz + z^2$ and $y^2 - yz + z^2$ are irreducible polynomials.) $\qquad \square$

In the example above, we factored $y^6 - z^6$ as the difference of squares and wrote

$$y^6 - z^6 = (y^3 - z^3)(y^3 + z^3)$$
$$= (y - z)(y^2 + yz + z^2)(y + z)(y^2 - yz + z^2).$$

It is also true that $y^6 - z^6$ is the difference of cubes. Thus we might have factored this way:

$$y^6 - z^6 = (y^2)^3 - (z^2)^3$$
$$= (y^2 - z^2)(y^4 + y^2z^2 + z^4) \qquad \text{(Difference of cubes)}$$
$$= (y - z)(y + z)(y^4 + y^2z^2 + z^4). \qquad \text{(Difference of squares)}$$

We must not stop here because $y^4 + y^2z^2 + z^4$ is not irreducible. If we had not already factored $y^6 - z^6$ as the difference of squares, we might not have recognized that, in fact,

$$y^4 + y^2z^2 + z^4 = (y^2 + yz + z^2)(y^2 - yz + z^2).$$

(You check!) Thus factoring as a difference of cubes gives the same complete factorization as factoring as a difference of squares:

$$y^6 - z^6 = (y - z)(y + z)(y^2 + yz + z^2)(y^2 - yz + z^2).$$

Grouping Terms Before Factoring

There are times when a polynomial can be fit into one of the forms we have discussed if its terms are grouped properly. Recognizing how to group terms so that a polynomial can be factored is a skill that requires considerable practice. Generally, in the exercises, polynomials that have more than three terms will require grouping of terms before factoring. Here is one example:

Example 4   Factor $6ax - 2by - 3bx + 4ay$ into a product of irreducible factors.

a. We can put together the two terms that contain x and the two terms that contain y, in each case removing common factors:

$$6ax - 2by - 3bx + 4ay = (6ax - 3bx) + (-2by + 4ay)$$
$$= 3x(2a - b) + 2y(-b + 2a).$$

Each of these terms contains the factor $2a - b$; we can remove it to complete the factoring:

$$6ax - 2by - 3bx + 4ay = (2a - b)(3x + 2y).$$

b. We can group together the two terms that contain a and the two terms that contain b, in each case removing common factors:

$$6ax - 2by - 3bx + 4ay = (6ax + 4ay) + (-2by - 3bx)$$
$$= 2a(3x + 2y) - b(2y + 3x).$$

Now both these terms contain the factor $3x + 2y$ and we can remove it to complete the factoring:

$$6ax - 2by - 3bx + 4ay = (3x + 2y)(2a - b). \qquad \square$$

Exercises
8.4

Complete the factorization in Exercises 1–4.

1. $4x^2 - 9 = (2x - \quad)(2x + \quad)$

2. $y^3 + 27 = (y + \quad)(y^2 - \quad + 9)$

3. $8x^3 - 1 = (\quad - 1)(4x^2 + \quad + 1)$

4. $3x(a - 2b) - 2y(a - 2b) = (a - 2b)(\quad - \quad)$

Factor the expressions in Exercises 5–30 into irreducible factors.

5. $x^2 - 25$

6. $x^2y^2 - 9$

7. $1 - x^3$

8. $8x^3 + y^3$

9. $8y^4 - y$

10. $16a^3 + 54$

11. $2xy + 4xz + 3yz + 6z^2$

12. $ac - bc - ad + bd$

13. $x^4 - x^2$

14. $2a^3b - 2ab^3$

15. $125r^3 - 8s^3$

16. $27y^3 + 8$

17. $x^4 - y^2$

18. $16 - y^2$

19. $3r^3 + 24$

20. $27y - y^4$

21. $5ax^2 + 10ay - 3bx^2 - 6by$

22. $x^6 - 64$

23. $x^4y^4 - 16$

24. $x^8 - y^8$

25. $x^5 - xy^4$

26. $x^6 - x^2y^4$

27. $x^2y - 2x^2z - y + 2z$

28. $2a^4 + a^2b^2 - 3b^4$

29. $x^4 - 13x^2 + 36$

30. $x^6 - 9x^3 + 8$

8.5 / DIVIDING POLYNOMIALS

The polynomial $x^3 + 3x^2 - 6x - 8$ can be factored this way:

$$x^3 + 3x^2 - 6x - 8 = (x^3 - 8) + (3x^2 - 6x)$$
$$= (x - 2)(x^2 + 2x + 4) + 3x(x - 2)$$

$$= (x - 2)(x^2 + 2x + 4 + 3x)$$
$$= (x - 2)(x^2 + 5x + 4).$$

Since $x^3 + 3x^2 - 6x - 8 = (x - 2)(x^2 + 5x + 4)$, it is possible to make these two division statements:

$$(x^3 + 3x^2 - 6x - 8) \div (x - 2) = x^2 + 5x + 4,$$
$$(x^3 + 3x^2 - 6x - 8) \div (x^2 + 5x + 4) = x - 2.$$

There is a procedure that gives the quotient of one polynomial divided by another and does not require factoring first. The procedure is similar to long division for whole numbers. We find partial quotients, multiply each times the divisor, and subtract, repeating until we get either a remainder of 0 or a polynomial remainder whose degree is less than the degree of the divisor. The steps are shown in this example:

$$
\begin{array}{r}
x \\
x^2 + 5x + 4 \overline{\smash{\big)}\, x^3 + 3x^2 - 6x - 8} \\
\underline{x^3 + 5x^2 + 4x } \\
-2x^2 - 10x - 8
\end{array}
$$

In this first step, the partial quotient x is chosen because it multiplies $x^2 + 5x + 4$ to give a polynomial with first term x^3. Subtraction then gives a polynomial of degree 2, which is less than the degree of $x^3 + 3x^2 - 6x - 8$.

$$
\begin{array}{r}
x - 2 \\
x^2 + 5x + 4 \overline{\smash{\big)}\, x^3 + 3x^2 - 6x - 8} \\
\underline{x^3 + 5x^2 + 4x } \\
-2x^2 - 10x - 8 \\
\underline{-2x^2 - 10x - 8} \\
0
\end{array}
$$

In this second step, the term -2 is chosen because it multiplies $x^2 + 5x + 4$ to give a polynomial with first term $-2x^2$. Subtraction gives a remainder of 0.

Example 1   If one factor of $4x^3 + 3x - 18$ is $2x - 3$, find another factor.

To find a polynomial that multiplies $2x - 3$ to give $4x^3 + 3x - 18$, we divide $4x^3 + 3x - 18$ by $2x - 3$. Notice that $4x^3 + 3x - 18$ is written as

$4x^3 + 0x^2 + 3x - 18$ so that there will be a place for the x^2 terms in the computation.

$$
\begin{array}{r}
2x^2 + 3x \; + \; 6 \\
2x - 3 \;\overline{)\,4x^3 + 0x^2 + \; 3x - 18} \\
4x^3 - 6x^2 \qquad\qquad \\
6x^2 + \; 3x - 18 \\
6x^2 - \; 9x \qquad \\
12x - 18 \\
12x - 18 \\
0
\end{array}
$$

$\longleftarrow$ From $2x^2(2x - 3)$

$\longleftarrow$ From $3x(2x - 3)$

$\longleftarrow$ From $6(2x - 3)$

Since the remainder is 0, we know that $(4x^3 + 3x - 18) \div (2x - 3) = 2x^2 + 3x + 6$. Thus $4x^3 + 3x - 18 = (2x - 3)(2x^2 + 3x + 6)$. It is important to recognize that the 0 remainder guarantees that $2x - 3$ is a factor of $4x^3 + 3x - 18$. $\square$

Example 2   Determine if $x^3 - 2x + 4$ is a factor of $3x^4 - x^3 - 6x^2 + 14x - 4$.

In order to decide if $x^3 - 2x + 4$ is a factor of $3x^4 - x^3 - 6x^2 + 14x - 4$, we perform the division and see if the remainder is 0.

$$
\begin{array}{r}
3x \; - 1 \\
x^3 - 2x + 4 \;\overline{)\,3x^4 - x^3 - 6x^2 + 14x - 4} \\
3x^4 \qquad - 6x^2 + 12x \\
-x^3 \qquad + 2x - 4 \\
-x^3 \qquad + 2x - 4 \\
0
\end{array}
$$

$\longleftarrow$ From $3x(x^3 - 2x + 4)$

$\longleftarrow$ From $-1(x^3 - 2x + 4)$

Thus $x^3 - 2x + 4$ *is* a factor of $3x^4 - x^3 - 6x^2 + 14x - 4$. In fact, $3x^4 - x^3 - 6x^2 + 14x - 4 = (3x - 1)(x^3 - 2x + 4)$. $\square$

Dividing one polynomial by another gives a remainder different from 0 if the first polynomial is not a factor of the second. In this case the division procedure continues until we get a polynomial of degree less than the degree of the divisor. This polynomial is the remainder.

Example 3   Find the quotient and remainder when $3x^4 + 3x^2 + 5$ is divided by $x^2 + 2x - 1$.

$$
\begin{array}{r}
3x^2 - 6x + 18 \\
x^2 + 2x - 1 \overline{\smash{)}3x^4 + 0x^3 + 3x^2 + 0x + 5} \\
\underline{3x^4 + 6x^3 - 3x^2} \qquad \text{From } 3x^2(x^2 + 2x - 1) \\
-6x^3 + 6x^2 + 0x + 5 \\
\underline{-6x^3 - 12x^2 + 6x} \qquad \text{From } -6x(x^2 + 2x - 1) \\
18x^2 - 6x + 5 \\
\underline{18x^2 + 36x - 18} \qquad \text{From } 18(x^2 + 2x - 1) \\
-42x + 23
\end{array}
$$

Notice that the procedure stops when we get a polynomial of degree 1. The quotient is $3x^2 - 6x + 18$ and the remainder is $-42x + 23$. $\qquad \square$

Writing an Equation to Summarize the Division Process

Remember how long-division problems are checked by elementary school children.

$$
\begin{array}{r}
18 \\
38 \overline{\smash{)}705} \\
\underline{38} \\
325 \\
\underline{304} \\
21
\end{array}
$$

To be sure the computation is correct, they would check that $18(38) + 21 = 705$. The quotient times the devisor plus the remainder must equal 705 (sometimes called the dividend). The same relationship is true for polynomial division. In Example 3 above, we performed this computation:

$$
\begin{array}{r}
3x^2 - 6x + 18 \\
x^2 + 2x - 1 \overline{\smash{)}3x^4 + 0x^3 + 3x^2 + 0x + 5} \\
\underline{3x^4 + 6x^3 - 3x^2} \\
-6x^3 + 6x^2 + 0x + 5 \\
\underline{-6x^3 - 12x^2 + 6x} \\
18x^2 - 6x + 5 \\
\underline{18x^2 + 36x - 18} \\
-42x + 23
\end{array}
$$

The computation can be summarized in the equation

$$(3x^2 - 6x + 18)(x^2 + 2x - 1) + (-42x + 23) = 3x^4 + 3x^2 + 5.$$

The statement that one polynomial can always be divided into another to give a quotient and a remainder that is either 0 or a polynomial of smaller degree than the divisor is called the **division algorithm for polynomials.** In Section 2.7 we started a division algorithm for whole numbers this way: If a and b are positive whole numbers, there is a quotient q and a remainder r (where q and r are whole numbers) such that $a = qb + r$ and $r < b$. From the division algorithm it follows that one positive whole number can always be divided into another to give a quotient that is a whole number and a remainder that is a whole number less than the divisor. In order to state the division algorithm for polynomials in a formal way, we need some notation for abbreviating polynomials.

Some Polynomial Shorthand

We can refer to a polynomial written in the variable x as $p(x)$. For example, we can denote $4x^2 - x + 3$ by $p(x)$. This notation does not mean p times x; it stands for a polynomial in the variable x. When we evaluate $p(x)$ for some value of the variable, say $x = 2$, we indicate the value of the polynomial at $x = 2$ by $p(2)$:

If $p(x) = 4x^2 - x + 3$, then $p(2) = 4(2)^2 - (2) + 3 = 17$,

$$p(-2) = 4(-2)^2 - (-2) + 3 = 21,$$

$$p(.5) = 4(.5)^2 - (.5) + 3 = 3.5.$$

If we want to talk about more than one polynomial in the variable x, we can denote them in this manner using different letters for each polynomial: $p(x), q(x), r(x), s(x)$, and so forth. With this notation we are able to state the division algorithm for polynomials.

DIVISION ALGORITHM FOR POLYNOMIALS

If $a(x)$ and $b(x)$ are polynomials, we can always find polynomials $q(x)$ and $r(x)$ such that
$$a(x) = q(x) \cdot b(x) + r(x)$$
and either $r(x)$ is the zero polynomial or has degree less than the degree of $b(x)$.

Although our primary interest in this chapter is with polynomials that have integers for coefficients, the polynomials $q(x)$ and $r(x)$ in the statement above may turn out to have fractions for coefficients. You will see examples of this in the exercises.

In Example 3, we divided $3x^4 + 3x^2 + 5$ by $x^2 + 2x - 1$ to get the quotient $3x^2 - 6x + 18$ and remainder $-42x + 23$. This example illustrates the division algorithm with

$$a(x) = 3x^4 + 3x^2 + 5$$

$$b(x) = x^2 + 2x - 1$$

$$q(x) = 3x^2 - 6x + 18$$

$$r(x) = -42x + 23.$$

Example 4   If a polynomial is divided by the polynomial $x^2 - 3x + 2$, what are the possible degrees for the remainder?

If $x^2 - 3x + 2$ is the divisor, it follows from the division algorithm that the remainder is either 0 or is a polynomial of degree less than 2. Thus if the remainder is not 0, the remainder will either be of degree 1 (that is, a linear polynomial such as $2x + 7$) or of degree 0 (that is, a constant polynomial such as the number 54).   □

Example 5   If $a(x) = 2x^4 - x^3 + 3x + 20$ and $b(x) = x^2 - 3x + 2$, find $q(x)$ and $r(x)$ so that $a(x) = q(x)b(x) + r(x)$.

The problem asks us to find the quotient and remainder when $2x^4 - x^3 + 3x + 20$ is divided by $x^2 - 3x + 2$.

$$
\require{enclose}
\begin{array}{r}
2x^2 + 5x + 11 \\
x^2 - 3x + 2 \enclose{longdiv}{2x^4 - x^3 + 0x^2 + 3x + 20} \\
\underline{2x^4 - 6x^3 + 4x^2} \\
5x^3 - 4x^2 + 3x + 20 \\
\underline{5x^3 - 15x^2 + 10x} \\
11x^2 - 7x + 20 \\
\underline{11x^2 - 33x + 22} \\
26x - 2
\end{array}
$$

From this computation, $q(x) = 2x^2 + 5x + 11$ and $r(x) = 26x - 2$. □

**Exercises
8.5**

Write two division statements given by the equations in Exercises 1–3.

1. $6x^3 + 5x^2 - 2x - 1 = (3x + 1)(2x^2 + x - 1)$

2. $x^5 - 2x^3 + 2x^2 - x - 1 = (x^2 - x + 1)(x^3 + x^2 - 2x - 1)$

3. $x^4 - 1 = (x^2 - 1)(x^2 + 1)$

The second polynomial in Exercises 4–7 is a factor of the first polynomial. Find another factor.

4. $x^2 - 9; x + 3$ 5. $x^3 + 1; x + 1$

6. $2x^3 - 3x^2 + 3x - 1; 2x - 1$ 7. $2x^5 - x^3 - x; 2x^2 + 1$

List the possible degrees for the remainder if any polynomial is divided by the particular polynomial in Exercises 8–10.

8. $2x^3 - x + 1$ 9. $x^2 - 3x - 1$

10. $2x - 5$

Find each quotient and remainder in Exercises 11–14 when the polynomial on the left side of the equation is divided by the specified divisor.

11. $2x^5 + x^3 + x^2 + x = (x^2 + 1)(2x^3 - x + 1) + 2x - 1$,

divisor $= x^2 + 1$.

12. $3x^3 - 10x^2 + 2x + 3 = (x - 3)(3x^2 - x - 1) + 5$, divisor $= x - 3$.

13. $2x^3 + x^2 - 2x - 1 = (x^2 - 1)(2x + 1)$, divisor $= x^2 - 1$.

14. $x^3 - x^2 - 7x - 2 = (x + 2)(x^2 - 3x - 1)$, divisor $= x + 2$.

Find each quotient and remainder in Exercises 15–26 when the first polynomial is divided by the second polynomial.

15. $2x^4 - 2x^3 + 2x^2; 2x^2$

16. $4x^5 - 4x^4 + 3x^3 - 2x^2 + 2x; x^2 - x + 1$

17. $x^4 + 5x^3 + 3x^2 + 3x + 9; x + 1$

18. $2x^4 - 4x^3 + x^2 + 2x - 1; 2x^2 - 1$

19. $4x^2 - 2x + 5; 2x^2 - x + 1$

20. $3x^3 + x^2 - 5x + 1; x^3 - 2x$

21. $x^2 - x; 2x^3 - 2x^2 + 1$

22. $2x - 1; x^2 - x$

23. $x^4 - 16; x - 2$

24. $2x^3 - 7x^2 + 5x - 3; 2x - 1$

25. $5x^3 - 3x^2 + 2x - 1; 2x + 1$

26. $x^3 - 3x^2 + x + 5; 2x^2 - x - 1$

Find $q(x)$ and $r(x)$ so that $a(x) = b(x)q(x) + r(x)$ for $a(x)$ and $b(x)$ as given in Exercises 27–31.

27. $a(x) = 2x^5 - 2x^4 + x^3 + 2x - 1; b(x) = x^2 - x$

28. $a(x) = x^3 - 6x^2 + 7x + 11; b(x) = x - 3$

29. $a(x) = x^5 - 1; b(x) = x - 1$

30. $a(x) = x^5 + 2x^2 - 2x + 2; b(x) = x^3 - x + 1$

31. $a(x) = 3x^4 - 2x^2 + x - 1; b(x) = 2x^2 - 3x + 1.$

32. Let $p(x) = 2x^3 - x^2 + 2x - 1$. Find

   (a) $p(0)$ (b) $p(-2)$ (c) $p(.7)$ (d) $p(c)$

33. Let $g(x) = (x + 1)^2(4x - 5)$. Find

   (a) $g(-1)$ (b) $g\left(\dfrac{5}{4}\right)$ (c) $g\left(\dfrac{1}{4}\right)$ (d) $g(0)$

34. If a polynomial is divided by $x + 2$, explain why the remainder is a number.

8.6 / CONSEQUENCES OF THE DIVISION ALGORITHM

The division algorithm for polynomials states that we can divide any polynomial $a(x)$ by any other polynomial $b(x)$ to get a quotient $q(x)$ and a remainder $r(x)$:

$$a(x) = q(x) \cdot b(x) + r(x).$$

The remainder is either 0 or is a polynomial of degree less than the degree of $b(x)$. If the variable x is replaced by a number, say 3, then we have the equation $a(3) = q(3) \cdot b(3) + r(3)$.

This observation leads to an important result when the divisor is a polynomial of degree 1, such as $x - 3$. In this case the remainder must be a polynomial of degree 0. Therefore, the remainder must be a number. If $b(x) = x - 3$, then the division algorithm gives

$$a(x) = (x - 3)q(x) + r(x)$$

$$a(3) = (3 - 3)q(3) + r(3)$$

$$a(3) = 0 \cdot q(3) + r(3)$$

$$a(3) = r(3).$$

Since $r(x)$ is a number and $r(3) = a(3)$, it must follow that $r(x)$ is the number $a(3)$. This example illustrates the following general statement.

REMAINDER THEOREM

If a polynomial $a(x)$ is divided by $x - c$, then the remainder is the number $a(c)$; that is to say, in the equation $a(x) = (x - c)q(x) + r(x)$, the remainder $r(x)$ equals the value of $a(x)$ when $x = c$.

Example 1 Find the remainder when the polynomial $2x^3 + 4x^2 - x + 5$ is divided by $x - 2$ and when it is divided by $x + 2$.

From the remainder theorem we see that dividing $2x^3 + 4x^2 - x + 5$ by $x - 2$ produces a remainder equal to the value of $2x^3 + 4x^2 - x + 5$ when $x = 2$. Thus the remainder is $2(2)^3 + 4(2)^2 - (2) + 5 = 35$. We are not surprised to see this same result in the long division:

$$
\begin{array}{r}
2x^2 + 8x + 15 \\
x - 2 \overline{)2x^3 + 4x^2 - x + 5} \\
\underline{2x^3 - 4x^2} \\
8x^2 - x + 5 \\
\underline{8x^2 - 16x} \\
15x + 5 \\
\underline{15x - 30} \\
35
\end{array}
$$

To find the remainder when $2x^3 + 4x^2 - x + 5$ is divided by $x + 2$, we can regard $x + 2$ as $x - (-2)$ and again use the remainder theorem. The remainder in this case is the value of $2x^3 + 4x^2 - x + 5$ when $x = -2$:

$$2(-2)^3 + 4(-2)^2 - (-2) + 5 = 7. \qquad \square$$

Roots of Polynomials

If a polynomial $p(x)$ has $x - 2$ for a factor, then $p(x)$ can be written as $p(x) = (x - 2)q(x)$ for some polynomial $q(x)$. What is the value of $p(x)$ when $x = 2$?

$$p(2) = (2 - 2)q(2) = 0 \cdot q(2) = 0.$$

When $x - 2$ is a factor of $p(x)$, then $p(2) = 0$.

Example 2   For what values of x does the polynomial $(x - 2)(x - 3)(x + 5)$ equal 0?

There are three values of x that cause the polynomial to equal 0:

if $x = 2$, then $(x - 2)(x - 3)(x + 5) = (2 - 2)(2 - 3)(2 + 5) = 0$;

if $x = 3$, then $(x - 2)(x - 3)(x + 5) = (3 - 2)(3 - 3)(3 + 5) = 0$;

if $x = -5$, then $(x - 2)(x - 3)(x + 5) = (-5 - 2)(-5 - 3)(-5 + 5) = 0$.

$\square$

Values of a variable that make a polynomial equal to 0 are called **roots** of the polynomial. For example, 2, 3, and -5 are roots of $(x - 2)(x - 3)(x + 5)$. The number 2 is a root of $x^3 + 5x^2 - 2x - 24$ because $2^3 + 5(2)^2 - 2(2) - 24 = 0$. You should see that *roots* of the polynomial $x^3 + 5x^2 - 2x - 24$ are *solutions* to the equation $x^3 + 5x^2 - 2x - 24 = 0$. In general, roots of a polynomial $p(x)$ are solutions to the equation $p(x) = 0$. The example above shows that each linear factor of a polynomial gives a root of the polynomial.

The remainder theorem gives an additional result: if a polynomial has root 2, then $x - 2$ must be a factor of the polynomial. To see that this is true, we denote the polynomial by $p(x)$. Then $p(x) = (x - 2)q(x) + r(x)$. According to the remainder theorem, the remainder $r(x) = p(2)$.

Since $p(2) = 0$, it must follow that $r(x) = 0$ and $p(x) = (x-2)q(x)$, demonstrating that $x-2$ is a factor of $p(x)$.

Of course, the above reasoning applies for any root of a polynomial: each whole number root c gives a linear factor $x-c$. These observations can be summarized this way.

FACTOR THEOREM (for whole number roots)

If $x-c$ is a factor of a polynomial, then c is a root of the polynomial.

If c is a root of a polynomial, then $x-c$ is a factor of the polynomial.

It is important to realize that we now have at least four ways of saying the same thing. Each of the following statements contains the same information:

- c is a root of $p(x)$;
- c is a solution to $p(x) = 0$;
- $x-c$ is a factor of $p(x)$;
- $p(x)$ divided by $x-c$ has remainder 0.

Example 3 Solve the equation $x^2 - x - 6 = 0$.

A solution to $x^2 - x - 6 = 0$ corresponds to a factor of $x^2 - x - 6$. Since $x^2 - x - 6 = (x-3)(x+2)$, there are two solutions: $x = 3$ and $x = -2$.
□

Example 4 Use the fact that 2 is a root of the polynomial $x^3 + 5x^2 - 2x - 24$ to factor $x^3 + 5x^2 - 2x - 24$. Then solve $x^3 + 5x^2 - 2x - 24 = 0$.

Since 2 is a root of $x^3 + 5x^2 - 2x - 24$, it follows from the factor theorem that $x - 2$ is a factor. We can use long division to find another factor:

$$\begin{array}{r}
x^2 + 7x + 12 \\
x - 2 \overline{\smash{)}x^3 + 5x^2 - 2x - 24} \\
\underline{x^3 - 2x^2} \\
7x^2 - 2x - 24 \\
\underline{7x^2 - 14x} \\
12x - 24 \\
\underline{12x - 24} \\
0
\end{array}$$

Thus $x^3 + 5x^2 - 2x - 24 = (x^2 + 7x + 12)(x - 2)$. We can factor $x^2 + 7x + 12$ as $(x + 3)(x + 4)$. Now we have

$$x^3 + 5x^2 - 2x - 24 = (x + 3)(x + 4)(x - 2).$$

Each linear factor gives a root of $x^3 + 5x^2 - 2x - 24$, that is to say, each linear factor gives a solution to $x^3 + 5x^2 - 2x - 24 = 0$. The solutions are -3, -4, and 2.   □

Example 5   Write a polynomial of degree 2 with roots 5 and -3.

Since 5 is a root of the desired polynomial, $x - 5$ must be a factor. Also, $x - (-3) = x + 3$ is a second factor. Thus the polynomial $(x - 5)(x + 3) = x^2 - 2x - 15$ is a degree 2 polynomial with roots 5 and -3.   □

**Exercises
8.6**

1. Let $p(x) = x^4 - 2x^3 - 2x^2 - 2x - 3$. Find

   (a) $p(0)$ (b) $p(-1)$ (c) $p(3)$ (d) $p(-1.5)$

2. Let $g(x) = (x - 2)^2(x + 3)(x - 5)$. Find

   (a) $g(-3)$ (b) $g(2)$ (c) $g(2.5)$ (d) $g(5)$

Use the remainder theorem to find each remainder when the first polynomial is divided by the second polynomial in Exercises 3–6.

3. $x^4 - x^3 + 2x^2 + x - 1, x + 2$ 4. $2x^3 + 3x^2 - 6x - 2, x - 2$

5. $x^{21} - 2x^{12} + 3x^4 - 2,\ x - 1$ 6. $x^3 - 4x^2 + 4x - 3,\ x - 3$

Determine if the polynomials given in Exercises 7 and 8 have the specified value of x for a root. If so, write the polynomials in factored form.

7. $x^4 - 8x^3 + 6x^2 + 40x + 25,\ x = 5$

8. $x^5 - x^4 + 2x^2 - 2x,\ x = -1$

Find all the roots of the polynomials in Exercises 9–12.

9. $(x + 3)(x + 8)$ 10. $(x + 6)(x + 4)(x - 5)$

11. $(x + 2)^2(x - 1)^3$ 12. $2x^3(x^2 - x)(x + 2)$

For each of the Exercises 13–20, use the factor theorem to determine if the second polynomial is a factor of the first polynomial.

13. $2x^4 - 5x^3 + 4x^2 - 4x + 3,$ 14. $x^3 - 3x + 2,\ x + 2$
     $x - 1$

15. $x^3 - 2x^2 - 2x + 1,\ x - 3$ 16. $2x^3 + 5x^2 - 3x - 2,\ x + 3$

17. $x^{43} + 2x^{20} - x^9 - 2,\ x + 1$ 18. $x^3 + 2x^2 - 2x + 3,\ x + 3$

19. $x^6 - 1,\ x - 1$ 20. $x^6 + 1,\ x + 1$

Solve for x in Exercises 21–24.

21. $(x - 7)(x + 3) = 0$ 22. $x(x + 1)(x - 3)(x + 4) = 0$

23. $x^2 + 3x - 10 = 0$ 24. $x^2 + 9x + 20 = 0$

Write each polynomial with degree and roots as specified in Exercises 25–27.

25. Roots 2, 3; degree 2   26. Roots $-1, -2, 3$; degree 3

27. Roots -2, 1; degree 3

28. Determine a value of k that makes $x - 2$ a factor of $2x^3 - 7x^2 + kx - 2$.

29. Determine a value of k that makes $x + 2$ a factor of $x^4 + kx^3 + 3kx - 1$.

30. Determine a value of k so that 4 is the remainder when $x^3 + 2x^2 + kx + 1$ is divided by $x - 1$.

8.7 / FRACTIONS AS ROOTS OF POLYNOMIALS

Sometimes polynomials with integers for coefficients have fractions for roots. For example, $\frac{3}{2}$ is a root of $2x - 3$ because $2(\frac{3}{2}) - 3 = 0$. Furthermore, any polynomial that has $2x - 3$ for a factor has $\frac{3}{2}$ for a root. In general, there is the same correspondence between factors and roots of polynomials that we saw in the last section. Here are some examples:

$$\text{Root} \xleftrightarrow{\quad\text{corresponds to}\quad} \text{Factor}$$

Root	Factor
$\dfrac{3}{2}$	$2x - 3$
$\dfrac{4}{5}$	$5x - 4$
$-\dfrac{2}{7}$	$7x + 2$
$-\dfrac{9}{16}$	$16x + 9$
$\dfrac{n}{m}$	$mx - n$

The factor theorem that we stated in the last section can be extended to include fractions as roots. Assume that m and n are integers.

FACTOR THEOREM

If $mx - n$ is a factor of a polynomial, then $\dfrac{n}{m}$ is a root of the polynomial.

If the reduced fraction $\dfrac{n}{m}$ is a root of a polynomial, then $mx - n$ is a factor of the polynomial.

Example 1   Write a polynomial with integer coefficients that has degree 3 and roots 2, $-\dfrac{1}{3}$, and $\dfrac{5}{2}$.

Each root provides one factor of the polynomial we want to write:

corresponds to

Root	⟷	Factor
2		$x - 2$
$-\dfrac{1}{3}$		$3x + 1$
$\dfrac{5}{2}$		$2x - 5$

The product of these three factors is a degree-3 polynomial with roots $2, -\dfrac{1}{3}$, and $\dfrac{5}{2}$:

$$(x - 2)(3x + 1)(2x - 5).$$

If you wish, you can perform the multiplication to put this polynomial in the form $6x^3 - 25x^2 + 21x + 10$. □

One way to find roots of a polynomial is to first find linear factors. But this may be difficult. Fortunately, we can tell by inspecting the coefficients of a polynomial what fractions can be roots of the polynomial. For example, $\frac{3}{2}$ can be a root of a polynomial only if $2x - 3$ is a factor. So a polynomial that has $\frac{3}{2}$ for a root can be written $(2x - 3) \cdot q(x)$ for another polynomial $q(x)$. When these two polynomials are multiplied together, the term of highest degree in the product contains the factor 2 and the term of lowest degree contains the factor 3. (Think of the long multiplication.) Thus a polynomial that looks like $ax^4 + bx^3 + cx^2 + dx + e$ (where $e \neq 0$) can have $\frac{3}{2}$ for a root only if 2 is a factor of a and 3 is a factor of e.

The same thing can be said for any fraction.

RATIONAL ROOTS THEOREM

The reduced fraction $\dfrac{n}{m}$ can be a root of a polynomial with integer coefficients only if m is a factor of the term of highest degree and n is a factor of the term of lowest degree.

Example 2   What fractions are roots of the polynomial $2x^4 + 5x^3 - 4x^2 - 6x - 9$?

We list the possible numerators and denominators and then form the fractions that could be roots:

Numerators (factors of -9): $1, -1, 3, -3, 9, -9$;

Denominators (factors of 2): $1, -1, 2, -2$;

Candidates for rational roots: $1, -1, \dfrac{1}{2}, -\dfrac{1}{2}, 3, -3,$

$$\frac{3}{2}, -\frac{3}{2}, 9, -9, \frac{9}{2}, -\frac{9}{2}.$$

Now we must check each of these to see if any are actually roots. An easy way to do this is to remember the calculator method for evaluating a polynomial. For example, to evaluate $2x^4 + 5x^3 - 4x^2 - 6x - 9$ for $x = \frac{9}{2}$, we can use this keying sequence:

<u>4.5</u> STO

$\boxed{2}$ $\boxed{\times}$ $\boxed{\text{RCL}}$ $\boxed{+}$ $\boxed{5}$ $\boxed{=}$

$\boxed{\times}$ $\boxed{\text{RCL}}$ $\boxed{-}$ $\boxed{4}$ $\boxed{=}$

$\boxed{\times}$ $\boxed{\text{RCL}}$ $\boxed{-}$ $\boxed{6}$ $\boxed{=}$

$\boxed{\times}$ $\boxed{\text{RCL}}$ $\boxed{-}$ $\boxed{9}$ $\boxed{=}$ Display: 1158.75

To evaluate the polynomial for $x = -3$, we can key

$\boxed{3}$ $\boxed{+/-}$ STO

$\boxed{2}$ $\boxed{\times}$ $\boxed{\text{RCL}}$ $\boxed{+}$ $\boxed{5}$ $\boxed{=}$

$\boxed{\times}$ $\boxed{\text{RCL}}$ $\boxed{-}$ $\boxed{4}$ $\boxed{=}$

$\boxed{\times}$ $\boxed{\text{RCL}}$ $\boxed{-}$ $\boxed{6}$ $\boxed{=}$

$\boxed{\times}$ $\boxed{\text{RCL}}$ $\boxed{-}$ $\boxed{9}$ $\boxed{=}$ Display: 0

We conclude that -3 is a root and that 4.5 is not a root. Testing all the candidates will yield two roots: $\frac{3}{2}$ and -3. No other fraction is a root of $2x^4 + 5x^3 - 4x^2 - 6x - 9$.   □

Example 3   Factor the polynomial $2x^4 + 5x^3 - 4x^2 - 6x - 9$ into a product of irreducible factors.

In Example 2, we found two roots: $\frac{3}{2}$ and -3. Thus $2x - 3$ is a factor, $x + 3$ is a second factor, and

$$2x^4 + 5x^3 - 4x^2 - 6x - 9 = (2x - 3)(x + 3)q(x)$$

for some polynomial $q(x)$. To find $q(x)$, we can divide $2x^4 + 5x^3 - 4x^2 - 6x - 9$ by $(2x - 3)(x + 3) = 2x^2 + 3x - 9$:

$$
\begin{array}{r}
x^2 + x + 1 \\
2x^2 + 3x - 9 \overline{\smash{\big)}\ 2x^4 + 5x^3 - 4x^2 - 6x - 9} \\
\underline{2x^4 + 3x^3 - 9x^2} \\
2x^3 + 5x^2 - 6x - 9 \\
\underline{2x^3 + 3x^2 - 9x} \\
2x^2 + 3x - 9 \\
\underline{2x^2 + 3x - 9} \\
0
\end{array}
$$

Thus $2x^4 + 5x^3 - 4x^2 - 6x - 9 = (x^2 + x + 1)(2x^2 + 3x - 9)$
$= (x^2 + x + 1)(2x - 3)(x + 3)$. Since $x^2 + x + 1$ is irreducible, this is the complete factorization.   □

In factoring the polynomial above we might have been more efficient if, when we found the first root -3, we had divided the polynomial by the factor $x + 3$. Then we could have looked for a second root among the candidates for roots to the quotient, a somewhat shorter list of candidates in general. In fact, since a linear factor can be repeated in the factorization, it is important to check the quotient for roots that have already been identified.

Example 4   Factor the polynomial $x^4 - 6x^3 + 10x^2 - 6x + 9$ into a product of irreducible factors.

We begin by listing possible numerators and denominators and then all candidates for rational roots.

Numerators (factor of 9): $1, -1, 3, -3, 9, -9$;
Denominators (factors of 1): $1, -1$;
Candidates for rational roots: $1, -1, 3, -3, 9, -9$.

Evaluating the polynomial for $x = 1$ and $x = -1$ shows that neither is a root. Evaluating for $x = 3$ gives

$$3^4 - 6(3)^3 + 10(3)^2 - 6(3) + 9 = 0.$$

Thus 3 is a root, so $x - 3$ is a factor. Division gives a second factor:

$$
\begin{array}{r}
x^3 - 3x^2 + x - 3 \\
x - 3 \overline{)x^4 - 6x^3 + 10x^2 - 6x + 9} \\
\underline{x^4 - 3x^3} \\
-3x^3 + 10x^2 - 6x + 9 \\
\underline{-3x^3 + 9x^2} \\
x^2 - 6x + 9 \\
\underline{x^2 - 3x} \\
-3x + 9 \\
\underline{-3x + 9} \\
0
\end{array}
$$

Thus $x^4 - 6x^3 + 10x^2 - 6x + 9 = (x - 3)(x^3 - 3x^2 + x - 3)$. The polynomial $x^3 - 3x^2 + x - 3$ has a new list of candidates for rational roots: 1, -1, 3, -3. Since 1 and -1 were not roots of the original polynomial, they cannot be roots of a factor. However, evaluating $x^3 - 3x^2 + x - 3$ for $x = 3$ gives $3^3 - 3(3)^2 + 3 - 3 = 0$. Thus $x - 3$ is a factor of $x^3 - 3x^2 + x - 3$. To get another factor, we divide:

$$
\begin{array}{r}
x^2 + 1 \\
x - 3 \overline{)x^3 - 3x^2 + x - 3} \\
\underline{x^3 - 3x^2} \\
x - 3 \\
\underline{x - 3} \\
0
\end{array}
$$

Since $x^3 - 3x^2 + x - 3 = (x - 3)(x^2 + 1)$, we have

$$x^4 - 6x^3 + 10x^2 - 6x + 9 = (x - 3)(x^3 - 3x^2 + x - 3)$$
$$= (x - 3)(x - 3)(x^2 + 1).$$

The factorization is complete, since $x^2 + 1$ is irreducible. Notice that $x - 3$ occurs twice in the factorization.   □

Exercises 8.7

Write a complete list of candidates for rational roots of the polynomials in Exercises 1 and 2.

1. $x^4 - 3x^2 + x + 4$

2. $2x^3 - x^2 + x - 3$

Find all roots of the polynomials in Exercises 3 and 4.

3. $x(x + 2)^2(3x - 5)$

4. $x^2(4x + 7)(2x + 1)(x - 5)$

Determine if each polynomial in Exercises 5 and 6 has the specified value of x for a root. If so, write the polynomial in factored form.

5. $5x^3 - 7x^2 + 7x - 2$, $x = \dfrac{2}{5}$

6. $2x^4 + 5x^3 + 2x + 5$, $x = -\dfrac{5}{2}$

Find all rational roots of the polynomials in Exercises 7–10.

7. $6x^3 - 4x^2 - 11x + 12$

8. $x^3 - 3x^2 + 4$

9. $x^3 + 6x^2 + 12x + 8$

10. $3x^4 + 5x^3 - 5x^2 - 5x + 2$

Factor the polynomials in Exercises 11–14 into a product of irreducible factors.

11. $2x^3 + 3x^2 - 11x - 6$

12. $9x^3 + 30x^2 - 23x + 4$

13. $x^4 - 6x^3 + 13x^2 - 12x + 4$

14. $3x^4 - 4x^3 + 6x^2 - 8x$

Write a polynomial with integer coefficients that has degree and roots as specified in each of Exercises 15–18.

15. Roots $-\dfrac{1}{2}, \dfrac{2}{3}$; degree 2

16. Roots $0, \dfrac{1}{2}, -1$; degree 3

17. Roots $\dfrac{3}{5}, -\dfrac{2}{3}, -1$; degree 4

18. Root $\dfrac{5}{6}$; degree 4

19. Determine the value of k that makes $3x + 4$ a factor of $3x^3 + 7x^2 + kx - 4$.

20. Determine the value of k that makes $2x - 5$ a factor of $2x^3 + kx^2 + 2x - 5$.

8.8 / CHAPTER 8 PROBLEM COLLECTION

Subtract the second polynomial from the first polynomial in
Exercises 1 and 2.

1. $3x^4 - 2x^3 + x^2 + 2x - 1$
   $\underline{4x^4 - 2x^3 + 2x^2 - 3x + 5}$

2. $2x^4 - x^3 - 2x + 2$
   $\underline{2x^4 + x^2 + 3x - 3}$

Perform the indicated operations and simplify the resulting algebraic
expressions in Exercises 3–18.

3. $(3x^2y - 2xy + 1) + (3 - x^2y + 4xy)$

4. $(3a^2b + ab^2 + 2ab) - (ab^2 - 2a^2b - a)$

5. $(5x - 2y)(3x + 4y)$
6. $(x + 5)(x - 7)$

7. $(3x - 2)(4x + 3)$
8. $(a + b)^2 - (a - b)^2$

9. $(a - b - c)^2$
10. $(2x - 1)(4x^2 + 2x + 1)$

11. $(3y^2 + 1)(2y^2 + 1)$
12. $(3x + 5)(3x - 5)$

13. $(2 - ab)(2 + ab)$
14. $(a^2 - 2b)(2a^2 + 3b)$

15. $xy(x + 1) - xy(1 + y)$
16. $(x + 2r)(x^2 - 2rx + 4r^2)$

17. $(3x - 1)^2$
18. $(1 + 4x)^2$

Factor the expressions in Exercises 19–40 into irreducible factors.

19. $3x^3 + 3x$
20. $a^2bc + ab^2c - abc^2$

21. $x^2 + x - 12$
22. $8x^2 + 6x - 9$

23. $15x^2 - 22x + 8$
24. $2x^2 - x + 2$

25. $r^2 + s^2$
26. $4y^2 - 1$

27. $6ac + 9bc - 4ab - 6b^2$
28. $x^3 + 27$

29. $x^4 + 8x$
30. $4x^3 - 4x^2y + xy^2$

31. $4a^4 - 5a^2b^2 - 6b^4$
32. $12a^2 + 36ab + 27b^2$

33. $2a^4b + a^3b^2 - 6a^2b^3$
34. $4x^3y - 9xy^3$

35. $2r^3 - 54$
36. $2r^2 - 18$

37. $3x^4 + x^2y^2 - 4y^4$
38. $2y^4 - 20y^2 + 18$

39. $x^6 + x^5 + x^4 + x^3$
40. $x^2y^2 - w^2y^2 - x^2z^2 + w^2z^2$

Complete the factorization in Exercises 41 and 42.

41. $6x^4 - 5x^3 + 11x^2 - 3x - 1 = (2x - 1)(\qquad)$

42. $2x^5 + x^4 - x^3 - 2x^2 - x + 1 = (x^3 - 1)(\qquad)$

Find the quotient and remainder when the first polynomial is divided by the second polynomial in Exercises 43–47.

43. $x^5 - 1, x - 1$
44. $x^5 + 1, x + 1$

45. $2x^4 + x^3 - 3x^2 + 3x + 3, 2x - 1$
46. $x^3 - 5x, x^2 + x$

47. $2x^5 + 8x^4 - 9x^3 - 10x^2 - 7x - 4, 2x^2 + 1$

48. Find $q(x)$ and $r(x)$ so that $a(x) = b(x)q(x) + r(x)$, where $a(x) = 2x^4 - x^3 - 2x^2 + 3x - 1$ and $b(x) = 2x^2 - 3x + 1$.

49. Let $p(x) = (3x - 1)(3x + 1)(x - 2)(x^2 + 1)$. Find

   (a) $p(0)$ (b) $p\left(-\dfrac{1}{3}\right)$ (c) $p\left(\dfrac{1}{3}\right)$ (d) $p(-2)$ (e) $p(2)$

50. Find the remainder when $(x - 1)(x + 2)(x^2 - x + 1)$ is divided by

   (a) $x - 1$ (b) $x - 2$ (c) $x + 1$

Determine if the second polynomial is a factor of the first polynomial in Exercises 51–57.

51. $3x^3 - 5x^2 + 4x + 2, 3x + 1$

52. $x^4 - 3x^3 + x^2 - 2x - 1, x - 3$

53. $x^4 - 4x^3 + 5x^2 - 3x + 2, x - 2$

54. $3x^3 - 5x^2 - 5x - 1, 3x + 1$
55. $x^{12} - 2x^{11} + 2x - 4, x - 2$

56. $2x^3 - 2x^2 - 8x + 5, 2x - 3$
57. $2x^3 + 4x^2 - 5x + 1, x + 3$

Find all roots of the polynomials in Exercises 58 and 59.

58. $(2x - 5)(x - 3)(x + 1)$
59. $x^2(3x + 5)(x - 7)$

Find all rational roots of the polynomials in Exercises 60 and 61.

60. $6x^3 + 11x^2 - 4x - 4$
61. $2x^4 + 3x^3 + 6x^2 + 12x - 8$

Factor the polynomials in Exercises 62 and 63 into a product of irreducible factors.

62. $2x^4 - 3x^3 - 8x^2 + 12x$

63. $2x^4 + 9x^3 + 11x^2 - 4$

Solve for x in Exercises 64–67.

64. $(2x + 3)(3x - 4) = 0$

65. $x(x - 2)(x + 5) = 0$

66. $4x^2 + 7x - 2 = 0$

67. $3x^3 - 8x^2 - 5x + 6 = 0$

68. Write a polynomial with integer coefficients that has degree 2 and roots $\dfrac{3}{4}, -\dfrac{1}{2}$.

69. Write a polynomial with integer coefficients that has degree 3 and roots $-3, 0, 2$.

70. Determine a value of k that makes $2x - 1$ a factor of $2x^4 - kx^3 + (k + 1)x^2 - 7x + 2$.

71. The length of a rectangle is one foot less than twice its width. If the width is x feet, write the area and the perimeter of the rectangle as functions of x.

72. A piece of wire x feet long is bent to form a rectangle that is twice as long as it is wide. Write the area of the rectangle as a function of x.

Quadratic Equations
and Inequalities

9.1 / GRAPHS OF QUADRATIC EQUATIONS IN TWO VARIABLES

In Chapter 6 we saw that the graph of a quadratic equation of the form $y = ax^2 + bx + c$ is a parabola. If the coefficient a is a positive number, the parabola opens up (a *cup*); if the coefficient a is a negative number, the parabola opens down (a *cap*). In this section we will study other characteristics of graphs of quadratic equations. Our goal is to graph quadratic equations as efficiently as possible without plotting a large number of points.

Consider the parabola that is the graph of $y = x^2$ (Fig. 9.1).

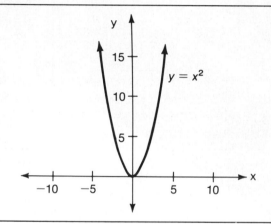

Figure 9.1

On the graph of $y = x^2$, the point (0,0) is called the **vertex** of the parabola. It is the lowest point on this graph. Every parabola has a vertex. If the parabola is a cup, the vertex is the lowest point on the graph. If the parabola is a cap, the vertex is the highest point on the graph. Observe that if we make a fold along the vertical axis, the two branches of the graph of $y = x^2$ match exactly. When this happens, we say that the graph is symmetric with respect to the line along which we make the fold. In this example, the **line of symmetry** is the vertical axis. Every parabola has a line of symmetry; the vertex of the parabola always lies on this line. What we shall see is that the equation of a parabola can be written in a form that permits us to read the coordinates of the vertex. Once we know the vertex, we can find the line of symmetry for the parabola. This is a good way to begin the graph.

Example 1   Graph the equation $y = (x - 1)^2 + 3$.

Instead of simplifying the right side of this equation, we want to analyze it in the form given to determine its vertex and line of symmetry. Since the coefficient of x^2 is positive, we know that the graph is a parabola that opens up. Its vertex is the lowest point on the graph, that is, the point with the smallest y value. In the equation $y = (x - 1)^2 + 3$, if $x - 1 = 0$, then $y = 3$; if $x - 1 \neq 0$, then $y > 3$. Thus the smallest y value is 3. It occurs when $x - 1 = 0$, that is, when $x = 1$. Thus (1,3) is the vertex. The line of symmetry is the vertical line through (1,3); its equation is $x = 1$. Here are some other points on the graph of $y = (x - 1)^2 + 3$.

x	-3	-2	-1	0	1	2	3	4	5
y	19	12	7	4	3	4	7	12	19

Observe in the chart above that two values of x the same distance from the number 1 give the same y values. This fact explains the symmetry of the graph across the line $x = 1$.

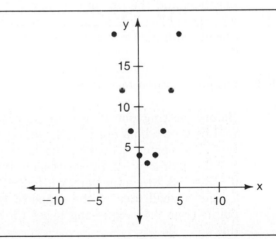

Figure 9.2

The points in Fig. 9.2 are sufficient to draw the complete graph. In Fig. 9.3 we have dashed the line of symmetry $x = 1$ to indicate that it is not part of the parabola.   □

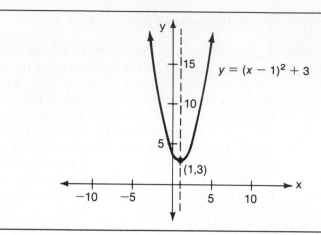

Figure 9.3

Look again at the equation $y = (x - 1)^2 + 3$ in the example above. The coordinates of the vertex are $x = 1$ and $y = 3$. The symmetry of the parabola is explained by observing that two values of x the same distance from 1 give the same value for $(x - 1)^2$ and, hence, the same value for y. Thus the two numbers 1 and 3 identify the vertex of the parabola and the line of symmetry. In general, for $a > 0$, the equation $y = a(x - h)^2 + k$ is a parabola that opens up, has vertex (h,k), and has line of symmetry $x = h$.

Example 2   Graph the equation $y = -(x + 2)^2 - 3$.

Before plotting points, we can observe that the graph of this equation will be a parabola opening down, because the coefficient of x^2 is negative. The vertex occurs at the largest possible y value, that is, when $x + 2 = 0$ or $x = -2$; the corresponding y value is $y = -(-2 + 2)^2 - 3 = 0 - 3 = -3$. Since the vertex is $(-2,-3)$, the line of symmetry is $x = -2$. To get a good idea of the shape of the parabola, we can plot several points (Fig. 9.4), expecting to see the symmetry of the graph about the line $x = -2$.

x	-5	-4	-3	-2	-1	0	1
y	-12	-7	-4	-3	-4	-7	-12

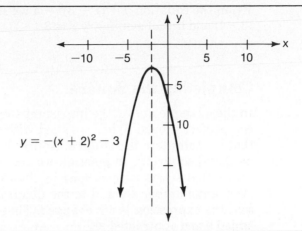

$$y = -(x + 2)^2 - 3$$

Figure 9.4

□

In Example 2, the equation $y = -(x + 2)^2 - 3$ has its vertex at $(-2, -3)$. In general, for $a > 0$, the graph of the equation $y = -a(x - h)^2 + k$ is a parabola opening down with vertex (h,k) and line of symmetry $x = h$.

These observations suggest that if we can write a quadratic equation in the form $y = a(x - h)^2 + k$ we are able to read the line of symmetry and the coordinates of the vertex directly from the equation. An important thing to remember in writing a quadratic equation in the form $y = a(x - h)^2 + k$ is that $(x - h)^2 = x^2 - 2hx + h^2$.

Example 3 Write the equation $y = x^2 - 6x + 5$ in a form that gives its vertex and its line of symmetry.

If we write the equation in the form $y = (x - h)^2 + k$, we will be able to read the vertex and the line of symmetry directly from the equation. The quadratic $x^2 - 6x + 5$ is not a binomial squared. However, $x^2 - 6x + 9$ is; in fact, $x^2 - 6x + 9 = (x - 3)^2$. Thus we can rewrite the equation this way:

$$y = x^2 - 6x + 5$$
$$= x^2 - 6x + 9 - 9 + 5$$
$$= (x - 3)^2 - 4.$$

From the equation $y = (x - 3)^2 - 4$ we see that the parabola has vertex $(3,-4)$ and line of symmetry $x = 3$.   □

Completing the Square

In the example above, the important step is to observe that $x^2 - 6x + 9$ is a perfect square. When we start with $x^2 - 6x$ and then add 9, we say that we **complete the square**. Notice that 9 is the result of dividing -6 by 2 and squaring. In general, we can complete the square for a quadratic that starts with $x^2 + nx$ by dividing n by 2 and squaring. However, any number we add to the quadratic also must be subtracted so that the expression is not changed. Thus in the example above we both added 9 and subtracted 9.

Not every quadratic polynomial has x^2 terms with coefficient 1, of course. We can complete the square for any polynomial by first factoring the coefficient of x^2 from the x^2 and x terms and then working on the resulting quadratic that has x^2 term with coefficient 1. Remember that any number added to complete the square must also be subtracted.

Example 4   Write the equation $y = 2x^2 + 8x + 9$ in a form that gives its vertex and its line of symmetry.

To complete the square efficiently, we first factor 2 from the first two terms so that the coefficient of x^2 is 1; then we proceed to complete the square:

$$y = 2x^2 + 8 + 9$$
$$= 2(x^2 + 4x) + 9$$
$$= 2(x^2 + 4x + 4 - 4) + 9 \quad \text{(Divide coefficient of } x \text{ by 2 and}$$
$$\text{square; then add and subtract 4)}$$
$$= 2(x^2 + 4x + 4) - 8 + 9 \quad \text{(Distributive property)}$$
$$= 2(x + 2)^2 + 1.$$

From this equation, we see that the parabola has vertex $(-2,1)$ and line of symmetry $x = -2$.   □

**Exercises
9.1**

Find the value of the quadratic expressions in Exercises 1–4 for the specified values of the variable.

1. $(x - 2)^2 - 2, x = 1, 3$

2. $-(x + 3)^2 + 1, x = -4, -2$

3. $2(x + 1)^2 - 3, x = -3, 1$

4. $-3(x - 4)^2 + 2, x = 1, 7$

Part of the graph of a parabola is given in Exercises 5–8. Use the specified line of symmetry to complete the graph.

5. Line of symmetry: $x = 2$

6. Line of symmetry: $x = 0$

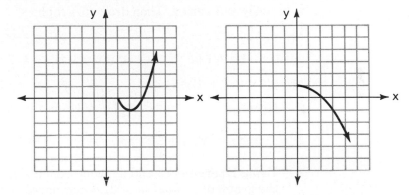

7. Line of symmetry: $x = 0$

8. Line of symmetry: $x = -1$

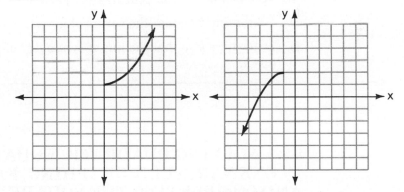

Give the coordinates of the vertex, the equations of the line of symmetry, and draw the graphs of the equations in Exercises 9–14.

9. $y = (x - 3)^2 + 2$

10. $y = (x + 3)^2 - 1$

11. $y = -(x - 1)^2 + 3$

12. $y = -(x + 1)^2 + 3$

13. $y = 2(x + 2)^2 - 1$

14. $y = -\dfrac{1}{2}(x - 2)^2 - 1$

Supply the missing terms in Exercises 15–18 to make each expression the square of a binomial.

15. $x^2 - 4x + (?)$

16. $x^2 + 2x + (?)$

17. $x^2 - 6x + (?)$

18. $x^2 - (?)x + 16$

Write each equation in Exercises 19–24 in a form that gives its line of symmetry and vertex. Then draw its graph.

19. $y = x^2 - 2x - 1$

20. $y = x^2 + 6x + 5$

21. $y = -x^2 + 6x$

22. $y = 2x^2 - 12x + 10$

23. $y = -3x^2 + 12x - 10$

24. $y = x^2 - 8$

25. Graph $y = x^2$, $y = 2x^2$, and $y = 3x^2$ on the same coordinate system.

26. Graph $y = x^2$, $y = \dfrac{1}{2}x^2$, and $y = \dfrac{1}{3}x^2$ on the same coordinate system.

27. Using Exercises 25 and 26, what conclusions can you draw about the graph of $y = ax^2$ ($a > 0$) as compared to $y = x^2$?

28. Write equations for four distinct parabolas with vertex (3,2).

29. Determine k so that the vertex of $y = x^2 - 4x + k$ is (2,3).

30. Determine k so that the line of symmetry of $y = x^2 - kx + 2$ is $x = 1$.

9.2 / SOLVING QUADRATIC EQUATIONS IN ONE VARIABLE: GRAPHING, FACTORING, AND COMPLETING THE SQUARE

We have seen that the graph of an equation of the form $y = ax^2 + bx + c$ is a parabola. By completing the square we can identify the vertex and the line of symmetry of the parabola. Another natural question to ask is where the graph crosses the horizontal axis; that is to say, what are the x values that have corresponding y values equal to 0. We need to find values for x that make $0 = ax^2 + bx + c$. Consider these three different examples.

Example 1   Use the graph of $y = x^2 - 2x - 3 = (x - 1)^2 - 4$ to determine the solutions of $0 = x^2 - 2x - 3$.

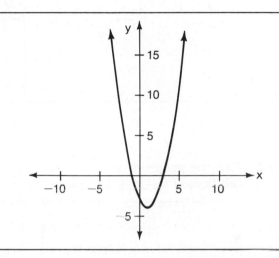

Figure 9.5

The graph in Fig. 9.5 contains two points of the horizontal axis, namely, where $x = -1$ and $x = 3$. These values are the solutions to the equation $0 = x^2 - 2x - 3$.   □

Example 2   Use the graph of $y = x^2 - 2x + 1 = (x - 1)^2$ to determine the solutions of $0 = x^2 - 2x + 1$.

There is one point of the horizontal axis on the graph of this equation (Fig. 9.6). Thus the equation $0 = x^2 - 2x + 1$ has only one solution: $x = 1$.

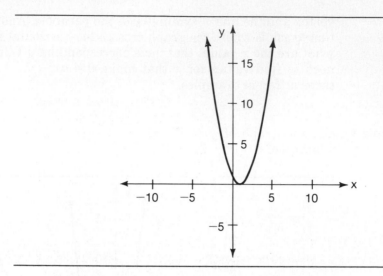

Figure 9.6

□

Example 3   Use the graph of $y = x^2 - 2x + 5 = (x - 1)^2 + 4$ to determine the solutions of $0 = x^2 - 2x + 5$.

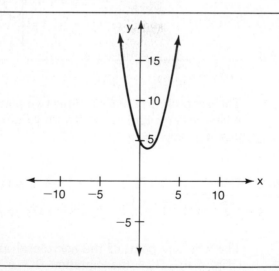

Figure 9.7

No point of the horizontal axis lies on the graph shown in Fig. 9.7. Thus there is no value of x on the number line that gives a solution to $0 = x^2 - 2x + 5$. (The numbers on the number line are called **real numbers**. We can say there is no real number solution to $0 = x^2 - 2x + 5$.)   □

The examples above illustrate that there are three possibilities for solutions to a quadratic equation in one variable, $0 = ax^2 + bx + c$:

2 solutions   (Example 1)

1 solution   (Example 2)

0 solutions   (Example 3)

If we have the graph of $y = ax^2 + bx + c$, we can approximate solutions to $0 = ax^2 + bx + c$. However, problem situations frequently require exact solutions to quadratic equations of the form $0 = ax^2 + bx + c$. We need techniques that permit us to solve these equations algebraically without first drawing a graph. If the quadratic polynomial $ax^2 + bx + c$ can be factored, then finding solutions to $0 = ax^2 + bx + c$ is done easily. This is the first case we consider.

Solving Quadratic Equations by Factoring

In the first example above, we looked for solutions to $0 = x^2 - 2x - 3$. Since $x^2 - 2x - 3 = (x - 3)(x + 1)$, we want the values of x that make $(x - 3)(x + 1) = 0$. Each factor determines a solution:

if $x = 3$, then $(x - 3)(x + 1) = (3 - 3)(3 + 1) = 0 \cdot 4 = 0$,

if $x = -1$, then $(x - 3)(x + 1) = (-1 - 3)(-1 + 1) = -4 \cdot 0 = 0$.

Example 4   Find solutions to the equation $2x^2 - 8x + 6 = 0$.

The polynomial $2x^2 - 8x + 6$ can be factored and the equation rewritten this way:

$$2x^2 - 8x + 6 = 0$$

$$2(x^2 - 4x + 3) = 0$$

$$2(x - 3)(x - 1) = 0.$$

The factor $x - 3$ gives the solution $x = 3$. The factor $x - 1$ gives the solution $x = 1$.   □

What we see in the example above is that if the quadratic expression can be factored into two different linear expressions, then we have two solutions to the equation.

Example 5   Find solutions to the equation $2x^2 - x - 10 = 0$.

Again we have a quadratic polynomial that can be factored:

$$2x^2 - x - 10 = 0$$

$$(2x - 5)(x + 2) = 0.$$

We have solutions whenever

$$2x - 5 = 0 \quad \text{or} \quad x + 2 = 0$$

$$2x = 5 \qquad\qquad x = -2.$$

$$x = \frac{5}{2}$$

Thus there are two solutions: $x = -2$, $x = \dfrac{5}{2}$.   □

It is worthwhile to observe that if we want to graph the equation $y = 2x^2 - x - 10$, Example 5 tells us that the graph crosses the horizontal axis at $x = -2$ and $x = \frac{5}{2}$. Furthermore, the symmetry of the graph assures us that the vertex of the parabola will be halfway between these x values, that is to say, when

$$x = \frac{1}{2}\left[-2 + \frac{5}{2}\right] = \frac{1}{2}\left(\frac{1}{2}\right) = \frac{1}{4}$$

and

$$y = 2\left(\frac{1}{4}\right)^2 - \frac{1}{4} - 10 = -10\frac{1}{8}.$$

Thus knowing where a parabola crosses the x-axis permits us to find the x-coordinate of the vertex without completing the square, and then we can find the y-coordinate. The graph of $y = 2x^2 - x - 10$ is shown in Fig. 9.8.

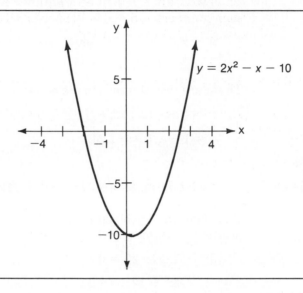

$$y = 2x^2 - x - 10$$

Figure 9.8

Solving Quadratic Equations by Completing the Square

If a quadratic polynomial can be factored, we can easily solve the quadratic equation consisting of the polynomial equal to 0. Unfortunately, many quadratic polynomials with integer coeffcents cannot be factored into polynomials with integer coefficients. A technique that gives a solution for any quadratic equation in one variable uses the process of completing the square.

Example 6   Find solutions to the equation $x^2 - 2x - 4 = 0$.

The quadratic polynomial $x^2 - 2x - 4$ cannot be factored. However, we can use the process of completing the square:

$$x^2 - 2x - 4 = 0$$

$$x^2 - 2x = 4 \qquad \text{(Add 4 to both sides)}$$

$$x^2 - 2x + 1 = 4 + 1 \qquad \text{(Divide coefficient of } x \text{ by 2,}$$
$$\text{square, and add to both sides)}$$

$$(x - 1)^2 = 5.$$

Thus we want $x - 1$ to be a number whose square is 5. There are two such numbers: $\sqrt{5}, -\sqrt{5}$. We need either

$$x - 1 = \sqrt{5} \qquad \text{or} \qquad x - 1 = -\sqrt{5}$$
$$x = 1 + \sqrt{5} \qquad\qquad x = 1 - \sqrt{5}.$$

There are two solutions to the equation:

$$x = 1 + \sqrt{5} \doteq 3.24$$
$$x = 1 - \sqrt{5} \doteq -1.24. \qquad\qquad\qquad \square$$

Example 7   Find solutions to the equation $3x^2 + 12x + 20 = 0$.

The quadratic polynomial does not factor. However, we can use the process of completing the square:

$$3x^2 + 12x + 20 = 0$$

$$3x^2 + 12x = -20 \qquad \text{(Subtract 20 from both sides)}$$

$$x^2 + 4x = \frac{-20}{3} \qquad \text{(Divide both sides by 3 to make the}$$
$$\text{coefficient of } x^2 \text{ equal to 1)}$$

$$x^2 + 4x + 4 = \frac{-20}{3} + 4 \qquad \text{(Divide the coefficient of } x \text{ by 2,}$$
$$\text{square, and add to both sides)}$$

$$(x + 2)^2 = -\frac{8}{3}.$$

We need values of x that make $(x + 2)^2 = -\frac{8}{3}$. But no real number squared is a negative number. Thus we conclude that this equation has no real solution. $\qquad\qquad \square$

Completing the square is a reliable procedure, but a clumsy one. In the next section we will refine it to get a highly efficient method of finding solutions for any quadratic equation.

Exercises 9.2 Find the solutions of the equations in Exercises 1–19 by the method of factoring.

1. $x^2 - 2x = 0$

2. $4x^2 - 4x + 1 = 0$

3. $6x^2 - 7x - 3 = 0$

4. $x^2 - 9 = 0$

5. $4x^2 - 9 = 0$

6. $2x^2 + 5x = 3$

7. $x^2 + x = 12$

8. $\left(\dfrac{2x - 1}{2}\right)^2 - 4 = 0$

9. $\left(\dfrac{3x + 1}{3}\right)^2 = 1$

10. $9x^2 + 12x + 4 = 0$

11. $4x^2 + 5x = 6$

12. $9x^2 - 4 = 0$

13. $x^2 - 6x = 16$

14. $x^2 = -3x$

15. $x^2 - 4x - 21 = 0$

16. $2x(x + 1) - 1 = 3 - 5x$

17. $x(4x - 3) - 7 = (x + 1)^2 + 4$

18. $x(2x + 8) = (x + 4)(x - 1)$

19. $(2x + 1)^2 = (x + 3)^2$

Find the values of x for which $y = 0$. Then find each vertex and line of symmetry, and graph the equations in Exercises 20–23.

20. $y = (x - 2)(x - 5)$

21. $y = (x + 3)(2x - 1)$

22. $y = x(3x + 4)$

23. $y = (2x + 1)(x + 4)$

Use the process of completing the square to find the solutions of the equations in Exercises 24–29.

24. $x^2 + 4x + 1 = 0$

25. $x^2 + 6x + 2 = 0$

26. $x^2 - 2x - 2 = 0$

27. $x^2 + 4x + 5 = 0$

28. $4x^2 - 12x - 11 = 0$

29. $9x^2 + 6x - 26 = 0$

30. Find two numbers whose sum is 20 and whose product is 96.

31. Find two numbers whose difference is 5 and whose product is 84.

32. One side of a rectangle is 4 feet shorter than the other side. If the area is 21 square feet, find the dimensions of the rectangle.

33. The length of a rectangle is one foot more than twice its width. If the area of the rectangle is 45 square feet, find the dimensions of the rectangle.

34. Determine k so that 3 is a solution of $x^2 + x - 3k = 0$, and then find the other solution.

35. Write a quadratic equation with integer coefficients whose solutions are

(a) $-3, 4$ (b) $-\dfrac{1}{2}, 3$

(c) $-\dfrac{1}{3}, \dfrac{5}{2}$

9.3 / SQUARE ROOT AND SOLUTIONS TO QUADRATIC EQUATIONS

We have seen that quadratic equations may have solutions that contain the square roots of numbers. We will review the properties of square root in this section before pursuing efficient methods of solving quadratic equations.

Remember that $\sqrt{9}$ and $9^{1/2}$ denote the positive number whose square is 9. Thus

$$\sqrt{9} = 9^{1/2} = (3^2)^{1/2} = 3.$$

The symbol $\sqrt{9}$ is read as the **square root** of 9, or **radical** 9. If a number equals another number squared, we call it a **perfect square**; for example, 9, 16, and .25 are perfect squares because $9 = 3^2$, $16 = 4^2$, and $.25 = (.5)^2$. If n is a positive number, it is always the case that $\sqrt{n^2} = n$.

Positive numbers that are not perfect squares also have square roots. For example, there is a number whose square is 5. We denote the number by $\sqrt{5}$. This number is not a finite decimal; it is not even a repeating infinite decimal. However, we can approximate $\sqrt{5}$ by a finite decimal. We can give decimal approximations correct to tenths   2.2

correct to hundredths   2.24

correct to thousandths   2.236

correct to ten-thousandths   2.2361.

Each of these decimals squared equals a number close to 5. However, none of the decimals exactly equals $\sqrt{5}$.

A negative number cannot have a real number square root because no real number squared is negative. Thus symbols such as $\sqrt{-9}$ and $\sqrt{-5}$ do not represent real numbers.

Properties of Square Root

Two properties of square root can be used to simplify computation. The first property is illustrated by the fact that $\sqrt{12} \cdot \sqrt{3} = \sqrt{36}$. This is true because

$$\sqrt{12} \cdot \sqrt{3} = 12^{1/2} \cdot 3^{1/2} = (12 \cdot 3)^{1/2} = 36^{1/2} = \sqrt{36}.$$

The general property can be stated this way:

For any two positive numbers n and m, $\sqrt{n} \cdot \sqrt{m} = \sqrt{nm}$.

This property of square root, read from right to left, permits us to simplify the square root of a number by first writing the number as a product of factors, some of which are perfect squares:

$$\sqrt{450} = \sqrt{2 \cdot 9 \cdot 25} = \sqrt{2} \cdot \sqrt{9} \cdot \sqrt{25} = \sqrt{2} \cdot 3 \cdot 5 = 15\sqrt{2}$$

$$\sqrt{180} = \sqrt{5 \cdot 36} = \sqrt{5} \cdot \sqrt{36} = \sqrt{5} \cdot 6 = 6\sqrt{5}$$

$$\sqrt{a^2 b^3} = \sqrt{a^2 \cdot b^2 \cdot b} = \sqrt{a^2} \cdot \sqrt{b^2} \cdot \sqrt{b} = ab\sqrt{b}.$$

When we have rewritten the square root of a number so that no perfect squares remain under the radical sign, we say we have **simplified the radical**.

Example 1   Simplify each term in the following sum and then combine terms:

$$\sqrt{252} + \sqrt{28} + \sqrt{175}.$$

To simplify each term, we first factor and then remove perfect squares from the radical:

$$\begin{aligned}
\sqrt{252} + \sqrt{28} + \sqrt{175} &= \sqrt{2^2 \cdot 3^2 \cdot 7} + \sqrt{2^2 \cdot 7} + \sqrt{5^2 \cdot 7} \\
&= 2 \cdot 3 \cdot \sqrt{7} + 2\sqrt{7} + 5\sqrt{7} \\
&= 6\sqrt{7} + 2\sqrt{7} + 5\sqrt{7} \\
&= 13\sqrt{7}.
\end{aligned}$$ □

A second property of square root concerns quotients. It is analogous to the property for products that we stated above.

> For any two positive numbers n and m,
> $$\frac{\sqrt{n}}{\sqrt{m}} = \sqrt{\frac{n}{m}}.$$

Using this property we can simplify some expressions that contain quotients:

$$\sqrt{\frac{25}{9}} = \frac{\sqrt{25}}{\sqrt{9}} = \frac{5}{3}$$

$$\sqrt{\frac{b^2 - 4ac}{4a^2}} = \frac{\sqrt{b^2 - 4ac}}{\sqrt{4a^2}} = \frac{\sqrt{b^2 - 4ac}}{2a}.$$

Numbers that contain square roots often occur in numerical computation. The usual properties of arithmetic apply. For example, the distributive property guarantees that *like* radicals can be combined under addition and subtraction:

$$\sqrt{5} + 3\sqrt{5} = (1 + 3)\sqrt{5} = 4\sqrt{5}$$
$$3\sqrt{2} - 5\sqrt{2} = (3 - 5)\sqrt{2} = -2\sqrt{2}.$$

Not every expression containing radicals can be simplified. Consider $\sqrt{5} - 3\sqrt{2} + 15$. We can approximate its value with a finite decimal, but the exact value is denoted $\sqrt{5} - 3\sqrt{2} + 15$.

Study these computations involving multiplication and division with radicals:

$$\frac{4 - 2\sqrt{3}}{2} = \frac{2(2 - \sqrt{3})}{2} = 2 - \sqrt{3},$$

$$(1 + \sqrt{3})^2 = (1 + \sqrt{3})(1 + \sqrt{3})$$
$$= 1 + 2\sqrt{3} + (\sqrt{3})^2$$
$$= 1 + 2\sqrt{3} + 3$$
$$= 4 + 2\sqrt{3},$$

$$(1 + \sqrt{3})(1 - \sqrt{3}) = 1 - (\sqrt{3})^2$$
$$= 1 - 3$$
$$= -2.$$

Writing Quadratic Equations with Given Solutions

Usually we have an equation, and our job is to find its solutions. However, sometimes the problem is the opposite: we have two numbers and we want an equation that has the numbers for its solutions.

Example 2   Find a quadratic equation with integer coefficients whose solutions are

$$\frac{-1 + \sqrt{3}}{2} \text{ and } \frac{-1 - \sqrt{3}}{2}.$$

One way to write an equation that has the numbers

$$\frac{-1 + \sqrt{3}}{2} \text{ and } \frac{-1 - \sqrt{3}}{2}.$$

for solutions is to write

$$\left(x - \frac{-1 + \sqrt{3}}{2}\right)\left(x - \frac{-1 - \sqrt{3}}{2}\right) = 0.$$

Multiplying the two binomials gives the equation in standard form:

$$0 = \left(x - \frac{-1 + \sqrt{3}}{2}\right)\left(x - \frac{-1 - \sqrt{3}}{2}\right)$$

$$= x^2 - \left(\frac{-1 + \sqrt{3}}{2} + \frac{-1 - \sqrt{3}}{2}\right)x + \left(\frac{-1 + \sqrt{3}}{2}\right)\left(\frac{-1 - \sqrt{3}}{2}\right)$$

$$= x^2 - \left(\frac{-1 + \sqrt{3} - 1 - \sqrt{3}}{2}\right)x + \left(\frac{(-1)^2 - (\sqrt{3})^2}{4}\right)$$

$$= x^2 - \left(\frac{-2}{2}\right)x + \left(\frac{1 - 3}{4}\right)$$

$$= x^2 + x - \frac{1}{2}.$$

Thus a quadratic equation with solutions $\dfrac{-1 + \sqrt{3}}{2}$ and $\dfrac{-1 - \sqrt{3}}{2}$ is $0 = x^2 + x - \dfrac{1}{2}$ or, multiplying both sides by 2, $0 = 2x^2 + 2x - 1$. ☐

Example 3   Verify that $\dfrac{-1 + \sqrt{3}}{2}$ is a solution to the equation $2x^2 + 2x - 1 = 0$.

a. If we replace x by $\dfrac{-1 + \sqrt{3}}{2}$ in the quadratic polynomial $2x^2 + 2x - 1$, we get

$$2\left(\frac{-1 + \sqrt{3}}{2}\right)^2 + 2\left(\frac{-1 + \sqrt{3}}{2}\right) - 1 = 2\left(\frac{(-1)^2 - 2\sqrt{3} + (\sqrt{3})^2}{4}\right) + (-1 + \sqrt{3}) - 1$$

$$= \left(\frac{1 - 2\sqrt{3} + 3}{2}\right) - 1 + \sqrt{3} - 1$$

$$= \left(\frac{4 - 2\sqrt{3}}{2}\right) - 2 + \sqrt{3}$$

$$= 2 - \sqrt{3} - 2 + \sqrt{3}$$

$$= 0.$$

b. We might want to use a calculator to check that $\dfrac{-1 + \sqrt{3}}{2}$ is a solution to the equation $2x^2 + 2x - 1 = 0$. Of course, the calculator will approximate $\dfrac{-1 + \sqrt{3}}{2}$. A keying sequence could look like this:

Keying sequence:

$\boxed{1}\ \boxed{+/-}\ \boxed{+}\ \boxed{3}\ \boxed{\sqrt{x}}\ \boxed{=}\ \boxed{\div}\ \boxed{2}\ \boxed{=}\ \boxed{\text{STO}}$

$\boxed{2}\ \boxed{\times}\ \boxed{\text{RCL}}\ \boxed{x^2}\ \boxed{+}\ \boxed{2}\ \boxed{\times}\ \boxed{\text{RCL}}\ \boxed{-}\ \boxed{1}\ \boxed{=}$

Display: $-1.3\ -09$

Thus instead of the value 0, the calculator gives -1.3×10^{-9}, or $-.0000000013$, a number that is very close to 0. The error occurs because the calculator uses an approximation to the exact value of $\dfrac{-1 + \sqrt{3}}{2}$. However, the error is so small that the computation suggests that $\dfrac{-1 + \sqrt{3}}{2}$ is in fact a solution.   □

Exercises 9.3

Simplify the radicals in Exercises 1–18 by computing the square roots of perfect square factors.

1. $\sqrt{18}$ 2. $\sqrt{50}$

3. $\sqrt{450}$ 4. $\sqrt{.16}$

5. $\sqrt{.81}$ 6. $\sqrt{\dfrac{25}{16}}$

7. $\sqrt{\dfrac{50}{9}}$ 8. $\dfrac{-2 - \sqrt{8}}{2}$

9. $\dfrac{-4 + \sqrt{28}}{6}$ 10. $\dfrac{6 - \sqrt{45}}{6}$

11. $\sqrt{a^3 b^2}$ 12. $\sqrt{a^5 b^4}$

13. $\sqrt{12 a^7 b^7}$ 14. $\sqrt{8 x^3 y^7}$

15. $\sqrt{\dfrac{a^4 b^3}{c^4}}$

16. $\sqrt{\dfrac{a^2 - b}{c^2}}$

17. $\sqrt{(a + b)^4}$

18. $\sqrt{x^2 + 4x + 4}$

Simplify each term in the expressions in Exercises 19–27 and then combine like terms. Give answers in radical form.

19. $4\sqrt{15} - \sqrt{15}$

20. $2\sqrt{21} + 3\sqrt{21}$

21. $\sqrt{24} + 2\sqrt{54}$

22. $\sqrt{18} - 2\sqrt{50} + 3\sqrt{98}$

23. $4\sqrt{12} - 2\sqrt{75} + 5\sqrt{27}$

24. $\sqrt{242} - \sqrt{32} + 3\sqrt{48}$

25. $\sqrt{44} + 2\sqrt{99} - \sqrt{175}$

26. $2\sqrt{20} - \sqrt{28} + 2\sqrt{45}$

27. $2\sqrt{x^3 y} - 3\sqrt{xy^3} + \sqrt{x^3 y^3}$

Perform the indicated operations and simplify the resulting expressions in Exercises 28–37. Give answers in radical form.

28. $(\sqrt{2})^3$

29. $\sqrt{2}(\sqrt{2} - \sqrt{3})$

30. $(\sqrt{2} + \sqrt{3})(\sqrt{2} - \sqrt{3})$

31. $\left(\dfrac{2 - 3\sqrt{3}}{2}\right)\left(\dfrac{2 + 3\sqrt{3}}{2}\right)$

32. $\dfrac{2 - 3\sqrt{3}}{2} + \dfrac{2 + 3\sqrt{3}}{2}$

33. $(\sqrt{2} + \sqrt{3})^2$

34. $(\sqrt{5} - \sqrt{7})^2$

35. $(2\sqrt{3} + \sqrt{5})(\sqrt{3} - 2\sqrt{5})$

36. $\left(\dfrac{2 - 3\sqrt{3}}{2}\right)^2$

37. $(\sqrt{x} + \sqrt{y})(\sqrt{x} - \sqrt{y})$

Compute with radicals to show that the numbers in Exercises 38 and 39 are solutions of the given equations.

38. $2 - \sqrt{3}$; $x^2 - 4x + 1 = 0$

39. $\dfrac{1 + \sqrt{5}}{2}$; $x^2 - x - 1 = 0$

Use your calculator to show that the numbers in Exercises 40 and 41 are solutions of the given equations.

40. $3 + \sqrt{2}$; $x^2 - 6x + 7 = 0$

41. $\dfrac{3 - \sqrt{5}}{7}$; $49x^2 - 42x + 4 = 0$

Write quadratic equations with integer coefficients that have solutions as specified in Exercises 42–44.

42. $5 + \sqrt{3}, 5 - \sqrt{3}$

43. $\dfrac{-1 + \sqrt{7}}{2}, \dfrac{-1 - \sqrt{7}}{2}$

44. $\dfrac{2 + \sqrt{5}}{3}, \dfrac{2 - \sqrt{5}}{3}$

9.4 / SOLVING QUADRATIC EQUATIONS IN ONE VARIABLE: THE QUADRATIC FORMULA

The solutions to a quadratic equation in one variable do not depend on the symbol that is used for the variable. For example, the equations $x^2 + 2x - 5 = 0$ and $y^2 + 2y - 5 = 0$ have the same solutions. The solutions depend on what the coefficients are in the equation. In this section we will find expressions for solutions to quadratic equations given in terms of the coefficients of the equation.

A Formula for Solutions of a Quadratic Equation

We can use the process of completing the square to find solutions to the equation $ax^2 + bx + c = 0$ for any coefficients a, b, and c. Before completing the square, we divide both sides of the equation by a to give an equivalent equation that has 1 for the coefficient of x^2:

$$ax^2 + bx + c = 0$$

$$ax^2 + bx = -c \qquad \text{(Subtract } c \text{ from both sides)}$$

$$x^2 + \frac{b}{a}x = -\frac{c}{a} \qquad \text{(Divide both sides by } a)$$

$$x^2 + \frac{b}{a}x + \frac{b^2}{4a^2} = -\frac{c}{a} + \frac{b^2}{4a^2} \qquad \text{(Divide coefficient of } x \text{ by 2, square, and add to both sides)}$$

$$\left(x + \frac{b}{2a}\right)^2 = -\frac{c}{a} + \frac{b^2}{4a^2} \qquad \text{(Factor left side)}$$

$$\left(x + \frac{b}{2a}\right)^2 = \frac{b^2 - 4ac}{4a^2}. \qquad \text{(Combine fractions on right side over common denominator)}$$

In order for $x + \dfrac{b}{2a}$ to be a number whose square is $\dfrac{b^2 - 4ac}{4a^2}$, it must be that either

$$x + \frac{b}{2a} = \sqrt{\frac{b^2 - 4ac}{4a^2}} \quad \text{or} \quad x + \frac{b}{2a} = -\sqrt{\frac{b^2 - 4ac}{4a^2}}$$

$$x + \frac{b}{2a} = \frac{\sqrt{b^2 - 4ac}}{2a} \qquad x + \frac{b}{2a} = -\frac{\sqrt{b^2 - 4ac}}{2a}$$

$$x = \frac{-b + \sqrt{b^2 - 4ac}}{2a} \qquad x = \frac{-b - \sqrt{b^2 - 4ac}}{2a}.$$

These expressions for x look horrendous, but they are very helpful. They permit us to write the solutions to a quadratic equation in terms of its coefficients a, b, and c. These expressions are referred to as the **quadratic formula**.

Example 1   Find the solutions to the equation $2x^2 + 3x - 1 = 0$.

If the polynomial $2x^2 + 3x - 1$ could be factored, we could use factoring to solve the equation. Since it cannot, we will write the solutions using the quadratic formula:

$$x = \frac{-b + \sqrt{b^2 - 4ac}}{2a} \quad \text{and} \quad x = \frac{-b - \sqrt{b^2 - 4ac}}{2a}.$$

In this example, $a = 2$, $b = 3$, $c = -1$. Thus

$$x = \frac{-3 + \sqrt{3^2 - 4(2)(-1)}}{2(2)} \quad \text{and} \quad x = \frac{-3 - \sqrt{3^2 - 4(2)(-1)}}{2(2)}$$

$$x = \frac{-3 + \sqrt{17}}{4} \qquad x = \frac{-3 - \sqrt{17}}{4}.$$

There are two solutions to the equation. If you replace x in the polynomial $2x^2 + 3x - 1$ by either of these values, you will get the number 0. Remember if you use a decimal approximation to these values, you may not get the number 0 exactly. ☐

Example 2   Use the quadratic formula to find solutions to

$$2x(x + 10) + 17 = 2x(2 - x).$$

To use the quadratic formula, we must first write the equation in the form $ax^2 + bx + c = 0$:

$$2x(x + 10) + 17 = 2x(2 - x)$$

$$2x^2 + 20x + 17 = 4x - 2x^2 \qquad \text{(Distributive property)}$$

$$4x^2 + 20x + 17 = 4x \qquad \text{(Add } 2x^2 \text{ to both sides)}$$

$$4x^2 + 16x + 17 = 0. \qquad \text{(Subtract } 4x \text{ from both sides)}$$

Now we can use the quadratic formula, letting $a = 4$, $b = 16$, and $c = 17$. Thus

$$x = \frac{-16 + \sqrt{16^2 - 4(4)(17)}}{2(4)} \qquad \text{or} \qquad x = \frac{-16 - \sqrt{16^2 - 4(4)(17)}}{2(4)}$$

$$= \frac{-16 + \sqrt{-16}}{8} \qquad\qquad\qquad = \frac{-16 - \sqrt{-16}}{8}.$$

But there is no real number $\sqrt{-16}$. We conclude that the equation has no real number solutions. □

In Example 2, the expression $\sqrt{b^2 - 4ac}$ became $\sqrt{-16}$ and we concluded that the equation had no solutions. Indeed, whenever $b^2 - 4ac$ is a negative number, the equation $ax^2 + bx + c = 0$ will not have real number solutions. On the other hand, if $b^2 - 4ac$ is a positive number, the equation $ax^2 + bx + c = 0$ has two different solutions:

$$x = \frac{-b + \sqrt{b^2 - 4ac}}{2a} \qquad \text{and} \qquad x = \frac{-b - \sqrt{b^2 - 4ac}}{2a}.$$

If $b^2 - 4ac$ equals 0, the two solutions provided by the quadratic formula are equal; both are $x = \frac{-b}{2a}$.

The number $b^2 - 4ac$ is called the discriminant of the equation $ax^2 + bx + c = 0$. If the discriminant is negative, the equation $0 = ax^2 + bx + c$ has no solutions. If the discriminant is 0, the equation has one solution. If the discriminant is positive, the equation has two solutions. Since a solution to $0 = ax^2 + bx + c$ gives a value of x where the graph of $y = ax^2 + bx + c$ crosses the x-axis, the discriminant also indicates the number of x-intercepts for $y = ax^2 + bx + c$. The chart below provides a summary.

Value of discriminant	Number of solutions	Number of x-intercepts
Negative	0	0
Zero	1	1
Positive	2	2

Example 3   How many times does the graph of $y = 4x^2 - 4x + 1$ cross the x-axis?

The question is the same as asking for the number of x-intercepts for $y = 4x^2 - 4x + 1$. Since $a = 4$, $b = -4$, and $c = 1$, the discriminant is $b^2 - 4ac = (-4)^2 - 4(4)(1) = 0$. Thus the graph has one intercept; that is to say, it is tangent to the x-axis.   □

Example 4   For what values of c does the equation $x^2 - 2x + c = 0$ have two different solutions?

In order for the equation to have two solutions, the coefficients must be such that the discriminant $b^2 - 4ac$ is positive. In this equation $a = 1$ and $b = -2$. Thus c must be a number such that

$$(-2)^2 - 4(1)(c) > 0$$

$$4 - 4c > 0$$

$$-4c > -4 \qquad \text{(Subtract 4 from both sides)}$$

$$c < 1. \qquad \text{(Divide both sides by } -4 \text{ and } \textit{reverse} \text{ the inequality sign)}$$

If c is any number less than 1, the equation $x^2 - 2x + c = 0$ will have two different solutions.   □

A Formula for the Vertex and the Line of Symmetry of a Parabola

Just as we are able to write the solutions to a quadratic equation $0 = ax^2 + bx + c$ in terms of its coefficients, we can also write the line of symmetry and the coordinates of the vertex of the graph of $y = ax^2 + bx + c$ in terms of a, b, and c. We need to use again the technique of completing the square:

$y = ax^2 + bx + c$

$y = a\left(x^2 + \dfrac{b}{a}x\right) + c$ (Factor a out of x^2 and x terms)

$y = a\left(x^2 + \dfrac{b}{a}x + \dfrac{b^2}{4a^2}\right) - \dfrac{b^2}{4a} + c$ (Divide the coefficient of x by 2, square; then add and subtract the same quantity)

$y = a\left(x + \dfrac{b}{2a}\right)^2 + c - \dfrac{b^2}{4a}.$

Now the equation has the form $y = a(x - h)^2 + k$, where h is $\dfrac{-b}{2a}$ and k is $c - \dfrac{b^2}{4a}$. The vertex is (h,k) or $\left(\dfrac{-b}{2a}, c - \dfrac{b^2}{4a}\right)$. It is enough to remember the x-coordinate of the vertex because the y-coordinate can then be computed from the equation. Here is a simple summary of the computation.

The graph of the equation $y = ax^2 + bx + c$ is a parabola whose vertex has x-coordinate $-\dfrac{b}{2a}$. The line of symmetry has equation $x = -\dfrac{b}{2a}$.

Example 5   Find the vertex and the line of symmetry of the graph of $y = 2x^2 - 12x + 3$.

In this equation $a = 2$, $b = -12$, and $c = 3$. The x-coordinate of the vertex is

$$x = -\dfrac{b}{2a} = -\dfrac{-12}{2(2)} = 3.$$

Substituting 3 for x in the equation gives the y-coordinate of the vertex:

$$y = 2x^2 - 12x + 3 = 2(3)^2 - 12(3) + 3 = -15.$$

Thus the vertex is $(3,-15)$. The equation of the line of symmetry is $x = 3$. □

**Exercises
9.4**

Use the quadratic formula to find all real solutions of the equations in Exercises 1–11.

1. $2x^2 + 9x - 5 = 0$
.2 $2x^2 + 2x - 3 = 0$

3. $2x^2 - 2x - 1 = x^2 - 2x + 2$
4. $25x^2 - 30x + 4 = 0$

5. $x(3x + 1) - 3 = 3(x - 1)$
6. $5x(x + 3) - 2 = x(3 - 4x)$

7. $2x^2 + 2x + 5 = 0$
8. $9x^2 - 12x + 5 = 0$

9. $4x(x - 1) = 13$
10. $9x^2 + 30x + 13 = 0$

11. $4x^2 - 16x - 11 = 0$

Determine if the equations in Exercises 12–15 have 0, 1, or 2 solutions without actually solving the equations.

12. $4x^2 - 4x + 1 = 0$
13. $9x^2 + 6x + 5 = 0$

14. $x^2 + 4x - 1 = 0$
15. $9x^2 + 6x - 2 = 0$

Find each vertex and line of symmetry, and graph the equations in Exercises 16–19.

16. $y = 2x^2 - 4x - 1$
17. $y = 3x^2 - 12x - 11$

18. $y = 4x^2 + 16x + 15$
19. $y = -2x^2 + 12x - 17$

20. A rectangle is 1 foot longer than it is wide. If the area of the rectangle is $\frac{1}{2}$ square foot, find the dimensions of the rectangle.

21. A number plus its square is equal to 1. Find the number(s) accurate to two decimal places.

22. A number subtracted from its square is equal to 1. Find the number(s) accurate to two decimal places.

23. The square of a number is equal to one more than twice the number. Find the number(s) accurate to two decimal places.

24. Determine the value(s) of b so that the equation $2x^2 + bx + 2 = 0$ has exactly one real solution.

25. Determine the value(s) of c so that the equation $x^2 - 2x + c = 0$ has no real solutions.

26. Determine the number of x-intercepts of the graph of $y = 2x^2 + 7x - 4$.

9.5 / SOLVING PROBLEMS WITH QUADRATIC EQUATIONS

There are many problem situations that can be described by quadratic equations. Some typical examples follow.

Example 1 The sum of the squares of two consecutive even integers is 340. Find the two integers.

If the smaller of the two even integers is denoted by x, the next larger *even* integer is $x + 2$. The problem states that $x^2 + (x + 2)^2 = 340$. The equation can be rewritten as follows:

$$x^2 + (x + 2)^2 = 340$$
$$x^2 + x^2 + 4x + 4 = 340$$
$$2x^2 + 4x + 4 = 340$$
$$2x^3 + 4x - 336 = 0$$
$$x^2 + 2x - 168 = 0$$
$$(x + 14)(x - 12) = 0.$$

Thus x can be either -14 or 12. If the smaller even integer is -14, the next larger is $-14 + 2$, or -12. If the smaller even integer is 12, the next larger is $12 + 2$, or 14. Thus the problem has two solutions: the integers -14 and -12, and the integers 12 and 14.   □

Remember, if you do not see how to factor a quadratic expression readily, you can use the quadratic formula to solve a quadratic equation. We use that method in the next example.

Example 2   A rectangular poster is three times as high as it is wide. It contains a picture of 204 square inches surrounded by a border 4 inches wide on all four sides. What are the dimensions of the poster?

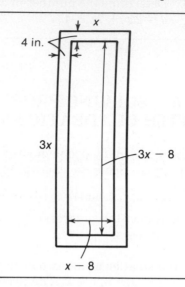

Figure 9.9

If the width of the rectangle is denoted by x, the height is $3x$. Since the border is 4 inches wide, the picture part has width $x - 8$ and height $3x - 8$ (Fig. 9.9). The area of the picture is 204 square inches. Thus

$$(x - 8)(3x - 8) = 204$$

$$3x^2 - 32x + 64 = 204$$

$$3x^2 - 32x - 140 = 0.$$

We use the quadratic formula to solve this equation:

$$x = \frac{32 + \sqrt{(-32)^2 - 4(3)(-140)}}{2(3)} \quad \text{or} \quad x = \frac{32 - \sqrt{(-32)^2 - 4(3)(-140)}}{2(3)}$$

$$= \frac{32 + \sqrt{2704}}{6} \qquad\qquad\qquad = \frac{32 - \sqrt{2704}}{6}$$

$$= \frac{32 + 52}{6} \qquad\qquad\qquad\qquad = \frac{32 - 52}{6}$$

$$= 14 \qquad\qquad\qquad\qquad\qquad = -3\frac{1}{3}.$$

Although the equation describing this problem has two solutions, x represents the width of a rectangle and cannot be a negative number. Thus $x = -3\frac{1}{3}$ is not a solution to the problem. The width x is 14 inches and the height is 3(14), or 42 inches.   □

The first two examples give problems that can be solved by writing a quadratic equation and solving it. Occasionally, writing an equation and graphing it provides a solution to a problem.

Example 3   A farmer wants to fence a rectangular garden. One side of the garden will be against the side of his house; the other three sides must be fenced with 100 feet of fence. Find the dimensions of the rectangular garden of largest area that the farmer can make with the 100 feet of fence.

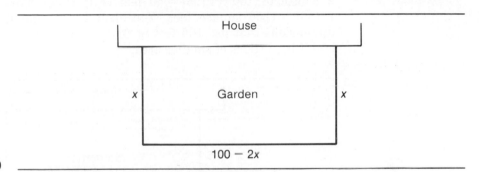

Figure 9.10

If the width of the garden is denoted by x, the length is $100 - 2x$, since the farmer has 100 feet of fencing to use (Fig. 9.10). A garden with these dimensions has area $A = x(100 - 2x)$ or $A = 100x - 2x^2$.

We want to find the value of x that makes A as large as possible. One way to do this is to consider the graph of the equation and see what value of x makes A maximal. Rewriting the equation as $A = -2x^2 + 100x$, we can observe that the graph is a parabola opening down because the coefficient of x^2 is negative. Thus the maximum value for A occurs at the vertex. Remember that the vertex of a parabola has x-coordinate $-\frac{b}{2a}$ or, for this equation, $\frac{-100}{2(-2)} = 25$. This reasoning shows that the greatest area A occurs when $x = 25$. You can also see this by examining the graph shown in Fig. 9.11.

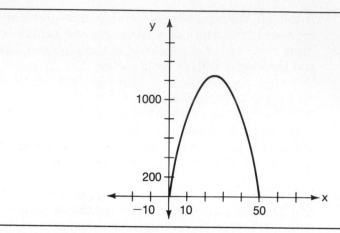

Figure 9.11

The farmer gets the largest garden by taking a width of 25 feet and a length of $100 - 2(25) = 50$ feet (Fig. 9.12). This garden has area $25(50) = 1250$ square feet. You should experiment with some other dimensions that use 100 feet of fence and convince yourself that they result in gardens of smaller area.

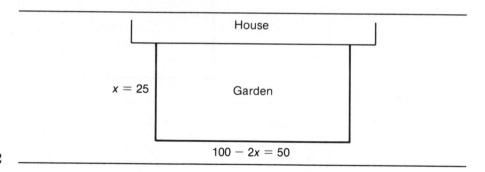

Figure 9.12

An Application from Physics

We learn from physics that an object shot straight up from the ground at an initial speed of 1000 feet per second will reach a height of $-16t^2 + 1000t$, t seconds after being shot. Thus we can compute the height y of the object after it has been in the air t seconds by using the equation $y = -16t^2 + 1000t$. For example, after 2 seconds, the height of the object is $y = -16(2)^2 + 1000(2) = 1936$ feet.

In general, the equation that gives the height y of an object, t seconds after it has been shot into the air, has the form $y = -16t^2 + \square t + \triangle$. In this equation, $\square$ is filled by the initial speed of the object; $\triangle$ is filled by the height of the object above the ground when it is launched; and the number -16 gets into the equation because it is half the acceleration due to gravity.

Example 4 An object is shot into the air from ground level at a speed of 1000 feet per second. When will the object be 13,600 feet above the ground?

After t seconds the object is at height $-16t^2 + 1000t$; the problem asks for the time t at which $-16t^2 + 1000t = 13,600$. Using the quadratic formula to solve the quadratic equation gives

$$-16t^2 + 1000t - 13,600 = 0$$

$$-2t^2 + 125t - 1700 = 0$$

$$t = \frac{-125 + \sqrt{(125)^2 - 4(-2)(-1700)}}{2(-2)} \quad \text{or} \quad t = \frac{-125 - \sqrt{(125)^2 - 4(-2)(-1700)}}{2(-2)}$$

$$= \frac{-125 + \sqrt{2025}}{-4} \qquad\qquad\qquad = \frac{-125 - \sqrt{2025}}{-4}$$

$$= \frac{-125 + 45}{-4} \qquad\qquad\qquad\quad = \frac{-125 - 45}{-4}$$

$$= 20 \qquad\qquad\qquad\qquad\qquad = 42.5.$$

There are two times when the object is 13,600 feet above ground: at time 20 seconds when the object is going up, and at time 42.5 seconds when it is coming down. □

Exercises 9.5

1. The sum of the squares of two consecutive integers is 421. Find the integers.

2. The sum of the squares of two consecutive odd integers is 290. Find the integers.

3. The sum of the squares of three consecutive integers is 245. Find the integers.

4. The sum of the squares of three consecutive even integers is 440. Find the integers.

5. If 2 times an integer is added to the square of the integer, the result is 195. Find the integer(s).

6. The square of a number is 2 more than the number. Find the number(s).

7. A rectangle is 3 feet longer than it is wide. If each side is increased by 1 foot, the area of the new rectangle is 208 square feet. Find the dimensions of the original rectangle.

8. A page of a book is 4 inches longer than it is wide. The margin at top and bottom is 2 inches, and the margin on each side is 1 inch. If the area of the printed matter is 35 square inches, find the dimensions of the page.

9. A rectangular poster is twice as high as it is wide. It contains a picture of 312 square inches bordered by a margin of 3 inches on the top and the bottom and 2 inches on each side. Find the dimensions of the poster.

10. A rectangular pool of dimensions 25 feet by 40 feet is surrounded by a walk of uniform width. If the area of the walk is 504 square feet, find the width of the walk.

11. Find the dimensions of a rectangle with a perimeter of 37 feet and an area of 85 square feet.

12. Among the rectangles with perimeters of 100 feet, find the one with maximum area.

13. An open-topped box is to be made from a rectangular piece of cardboard whose length is 1 inch more than twice its width. This box is made by cutting a 4-inch square from each corner and folding up the sides. If the box is to hold 272 cubic inches, find the dimensions of the piece of cardboard.

14. A piece of wire 20 feet long is cut into two pieces so that the sum of the squares of the length of each piece is 202. Find the length of each piece.

15. $1000 is deposited in an account that compounds interest annually. Two years later, the amount in the account is $1115.14. Find the annual interest rate.

16. Write the quadratic equations with integer coefficients whose solutions are

    (a) 1, −1

    (b) −2, 3

    (c) $\dfrac{1}{2}, -\dfrac{1}{3}$

    (d) $\dfrac{-2 + \sqrt{3}}{2}, \dfrac{-2 - \sqrt{3}}{2}$

17. If one solution of $2x^2 + 11x + k = 0$ is $x = -5$, find the other solution.

18. The sum of the solutions of a quadratic equation is −2 and their product is −15. Find the solutions and write the quadratic equation.

19. An object is shot into the air from ground level at a speed of 800 feet per second.

    (a) When will the object be 7500 feet above ground?

    (b) When will the object reach its maximum height?

    (c) What is the maximum height attained by the object?

    (d) When will the object strike the ground?

20. An object is shot into the air from the top of a 200-foot-tall building at a speed of 960 feet per second.

    (a) When will the object be 13,000 feet above ground?

    (b) When will the object be 200 feet above ground?

    (c) When will the object reach its maximum height?

    (d) What is the maximum height attained by the object?

    (e) When will the object strike the ground?

9.6 / EQUATIONS QUADRATIC IN FORM

We have developed methods for solving equations that are quadratic in one variable. These methods also can be used to solve equations that are not necessarily quadratic in the technical sense that the highest exponent of the variable is 2, but which are quadratic in some algebraic expression containing the variable. Such equations have the form $a\,\boxed{}^2 + b\,\boxed{} + c = 0$, where $\boxed{}$ is filled with an algebraic expression in one variable. We say the expression is quadratic in the expression $\boxed{}$. Consider these equations:

$2x^4 - 10x^2 + 8 = 0$ This equation can also be written as $2(x^2)^2 - 10x^2 + 8 = 0$. It is quadratic in the expression x^2.

$x + \sqrt{x} = 1$ Rewriting as $(\sqrt{x})^2 + \sqrt{x} - 1 = 0$ shows that this equation is quadratic in the expression $\sqrt{x}$.

$6y^{-2} + y^{-1} - 2 = 0$ This equation can be written as $6(y^{-1})^2 + (y^{-1}) - 2 = 0$. It is quadratic in the expression y^{-1}.

Once we recognize an equation as quadratic in form, we can solve it by factoring or by using the quadratic formula.

Example 1   Solve $2x^4 - 10x^2 + 8 = 0$ for x.

The polynomial on the left side is quadratic in x^2 and can be factored:

$$2x^4 - 10x^2 + 8 = 0$$
$$2(x^4 - 5x^2 + 4) = 0$$
$$2(x^2 - 4)(x^2 - 1) = 0.$$

We conclude that either

$$x^2 - 4 = 0 \qquad \text{or} \qquad x^2 - 1 = 0$$
$$(x - 2)(x + 2) = 0 \qquad\qquad (x - 1)(x + 1) = 0$$
$$x = 2,\ x = -2 \qquad\qquad x = 1,\ x = -1.$$

Thus the equation $2x^4 - 10x^2 + 8 = 0$ has four solutions:

$$x = 1,\ x = -1,\ x = 2,\ x = -2. \qquad\qquad \square$$

Example 2   Solve $x + \sqrt{x} = 1$ for x.

a. We have seen that this equation is quadratic in the expression $\sqrt{x}$ since the equation can be written as $(\sqrt{x})^2 + \sqrt{x} - 1 = 0$. Thus the quadratic formula with $a = 1$, $b = 1$, $c = -1$ gives values for $\sqrt{x}$. Once we know $\sqrt{x}$, we will find x.

$$\sqrt{x} = \frac{-1 + \sqrt{1^2 - 4(1)(-1)}}{2(1)} \qquad \text{or} \qquad \sqrt{x} = \frac{-1 - \sqrt{1^2 - 4(1)(-1)}}{2(1)}$$

$$= \frac{-1 + \sqrt{1 + 4}}{2} \qquad\qquad\qquad = \frac{-1 - \sqrt{1 + 4}}{2}$$

$$= \frac{-1 + \sqrt{5}}{2} \qquad\qquad\qquad = \frac{-1 - \sqrt{5}}{2}$$

$$\doteq .618 \qquad\qquad\qquad\qquad \doteq -1.618.$$

We must examine these results carefully. We are looking for values of x that make $\sqrt{x} \doteq .618$ and $\sqrt{x} \doteq -1.618$. But $\sqrt{x}$ denotes the positive square root of x, so $\sqrt{x}$ cannot equal a negative number. Thus there is no solution to $\sqrt{x} \doteq -1.618$. We can get a solution to $\sqrt{x} \doteq .618$, namely, $x \doteq .618^2$ or $x \doteq .382$.

b. Another way to solve the equation $x + \sqrt{x} = 1$ for x is to write $\sqrt{x} = 1 - x$ and square both sides of the equation:

$$\sqrt{x} = 1 - x$$
$$x = (1 - x)^2$$
$$x = 1 - 2x + x^2$$
$$0 = 1 - 3x + x^2.$$

Now the quadratic formula gives values for x:

$$x = \frac{-(-3) + \sqrt{(-3)^2 - 4(1)(1)}}{2(1)} \qquad \text{or} \qquad x = \frac{-(-3) - \sqrt{(-3)^2 - 4(1)(1)}}{2(1)}$$

$$= \frac{3 + \sqrt{5}}{2} \qquad\qquad\qquad\qquad = \frac{3 - \sqrt{5}}{2}$$

$$\doteq 2.618 \qquad\qquad\qquad\qquad\quad \doteq .382.$$

We must check both of these values in the equation $x + \sqrt{x} = 1$. By doing this, you will see that .382 is an approximate solution. (In fact, if you check the calculator approximation for $\dfrac{3 - \sqrt{5}}{2}$ in the expression $\sqrt{x} + x$, the display is 1.) However, $\sqrt{2.618} + 2.618 \neq 1$. The left side of this equation is a number greater than 4. Thus 2.618 is not a solution to $x + \sqrt{x} = 1$.   □

If, in searching for solutions to an equation, we square the two sides of the equation, we will not lose any solutions, but we may get extra values that are solutions to the final equation and not solutions to the original equation. This happens because two numbers can have the same squares without being equal (4 and −4, for example, both have square 16). Thus it is especially important to check values that arise when an equation is squared to be sure they are solutions to the original equation.

Example 3   Solve $6t^{-2} + t^{-1} - 2 = 0$ for t.

This equation is quadratic in t^{-1}. It can be solved by factoring. We start by replacing t^{-2} with $(t^{-1})^2$:

$$6t^{-2} + t^{-1} - 2 = 0$$
$$6(t^{-1})^2 + t^{-1} - 2 = 0$$
$$(3t^{-1} + 2)(2t^{-1} - 1) = 0$$

$$3t^{-1} + 2 = 0 \qquad \text{or} \qquad 2t^{-1} - 1 = 0$$
$$3t^{-1} = -2 \qquad\qquad\qquad 2t^{-1} = 1$$
$$t^{-1} = -\frac{2}{3} \qquad\qquad\qquad t^{-1} = \frac{1}{2}.$$

To find a value for t from a value for t^{-1}, remember that t^{-1} is the reciprocal of t. Thus the equations are

$$\frac{1}{t} = -\frac{2}{3} \qquad \text{or} \qquad \frac{1}{t} = \frac{1}{2}$$
$$t = -\frac{3}{2} \qquad\qquad t = 2.$$

If you had elected to use the quadratic formula to solve $6t^{-2} + t^{-1} - 2 = 0$, you would have obtained values for t^{-1}. Then you would need to take reciprocals to find t.   □

Example 4 The fourth power of a number is equal to 4 times its square. Find the number.

If the number is denoted by x, the problem states that $x^4 = 4x^2$, or $x^4 - 4x^2 = 0$. This equation is quadratic in x^2 and can be solved by factoring:

$$x^4 - 4x^2 = 0$$
$$x^2(x^2 - 4) = 0.$$

Either

$$x^2 = 0 \quad \text{or} \quad x^2 = 4$$
$$x = 0 \qquad\qquad x = 2 \text{ or } -2.$$

Thus there are three numbers with the property that the fourth power of the number equals 4 times its square: 0, 2, and -2.   □

Exercises 9.6

Solve the equations in Exercises 1–14.

1. $y^4 - 5y^2 + 4 = 0$

2. $4x^4 - 5x^2 - 9 = 0$

3. $2x - 5\sqrt{x} - 3 = 0$

4. $6y - 7\sqrt{y} + 2 = 0$

5. $t^{-2} + t^{-1} - 12 = 0$

6. $6t^{-2} + 7t^{-1} - 3 = 0$

7. $2x^4 + 9x^2 + 4 = 0$

8. $36x^4 - 47x^2 + 15 = 0$

9. $3x^{-2} + 11x^{-1} - 4 = 0$

10. $4y^{-2} + 4y^{-1} - 3 = 0$

11. $x^6 + 7x^3 - 8 = 0$

12. $8x^6 - 7x^3 - 1 = 0$

13. $x + 2\sqrt{x} = 2$

14. $4x + 8\sqrt{x} = 1$

15. A number minus its square root equals 1. Find the number(s).

16. Five times the square root of a number minus the number equals 6. Find the number(s).

17. A number minus its cube equals zero. Find the number(s).

18. The fourth power of a number plus its square equals 2. Find the number(s).

9.7 / QUADRATIC INEQUALITIES

If we have the graph of the quadratic equation $y = x^2 - x - 2$, we can use the graph to answer a question like this one:

For what values of x is $x^2 - x - 2 > 0$?

However, we can also solve inqualities involving quadratic expressions without first drawing a graph by using important characteristics of the graph.

Example 1 Solve the inequality $x^2 - x - 2 > 0$.

The graph of $y = x^2 - x - 2$ is a parabola that opens up, since the coefficient of x^2 is a positive number. The inequality asks for those values of x that make $y > 0$; thus we look for the values of x where the graph is above the horizontal axis. Since $x^2 - x - 2$ can be factored, it is easy to see where the parabola crosses the horizontal axis (that is, where $y = 0$):

$$x^2 - x - 2 = 0$$
$$(x + 1)(x - 2) = 0$$
$$x = -1 \quad \text{or} \quad x = 2.$$

Thus the parabola crosses the horizontal axis at $x = -1$ and at $x = 2$. Because it opens up, the parabola is below the axis between $x = -1$ and $x = 2$, and it is above the axis to the left of -1 and to the right of 2. Thus the solutions to $x^2 - x - 2 > 0$ are all $x < -1$ and all $x > 2$.   □

Example 2 Solve the inequality $-4x^2 - 8x + 1 \geq 0$.

The graph of $y = -4x^2 - 8x + 1$ is a parabola that opens down. We can find the two points where the parabola crosses the x-axis by solving $-4x^2 - 8x + 1 = 0$. We use the quadratic formula.

$$x = \frac{-(-8) + \sqrt{(-8)^2 - 4(-4)(1)}}{2(-4)} \quad \text{or} \quad x = \frac{-(-8) - \sqrt{(-8)^2 - 4(-4)(1)}}{2(-4)}$$

$$= \frac{8 + \sqrt{80}}{-8} \qquad\qquad\qquad = \frac{8 - \sqrt{80}}{-8}$$

$$= \frac{8 + 4\sqrt{5}}{-8} \qquad\qquad\qquad = \frac{8 - 4\sqrt{5}}{-8}$$

$$= \frac{2 + \sqrt{5}}{-2} \qquad\qquad\qquad = \frac{2 - \sqrt{5}}{-2}$$

$$\doteq -2.12 \qquad\qquad\qquad\qquad \doteq .12.$$

Since the parabola opens down and crosses the axis at $x = \dfrac{2 + \sqrt{5}}{-2}$ $\doteq -2.12$ and $x = \dfrac{2 - \sqrt{5}}{-2} \doteq .12$, we know the graph is above the axis between $\dfrac{2 + \sqrt{5}}{-2}$ and $\dfrac{2 - \sqrt{5}}{-2}$. Thus the solutions to $-4x^2 - 8x + 1 \geq 0$ include the two values $\dfrac{2 + \sqrt{5}}{-2}$ and $\dfrac{2 - \sqrt{5}}{-2}$ and all x values between $\dfrac{2 + \sqrt{5}}{-2}$ and $\dfrac{2 - \sqrt{5}}{-2}$. We can write these values as $\dfrac{2 + \sqrt{5}}{-2} \leq x \leq \dfrac{2 - \sqrt{5}}{-2}$.

□

Using Sign Charts to Solve Inequalities

In Examples 1 and 2 above, we solved inequalities by using the graph of the quadratic expression. There is a second way to solve Example 1 because the quadratic expression can be factored. This method works in any inequality that can be expressed in factored form. It can also be used to solve inequalities that are not quadratic, provided they can be factored into linear factors.

Example 1 asks for all x that make $x^2 - x - 2 > 0$, or $(x + 1)(x - 2) > 0$. The product $(x + 1)(x - 2)$ will be a positive number if $x + 1$ and $x - 2$ are both positive or if $x + 1$ and $x - 2$ are both negative:

$x + 1$ is 0 if $x = -1$,

$x + 1$ is negative if $x < -1$,

$x + 1$ is positive if $x > -1$,

$x - 2$ is 0 if $x = 2$,

$x - 2$ is negative if $x < 2$,

$x - 2$ is positive if $x > 2$.

These statements can be summarized in what is called a **sign chart**. In this example, there are two important values of x: $x = -1$ and $x = 2$.

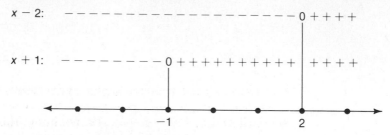

Now it is easy to visualize when $x + 1$ and $x - 2$ are both positive and when they are both negative:

Both positive: $x > 2$,

Both negative: $x < -1$.

Thus the solutions to $(x + 1)(x - 2) > 0$ are all $x > 2$ and all $x < -1$.

Example 3   Solve the inequality $(x - 1)(x + 2)(x - 3) > 0$.

The product $(x - 1)(x + 2)(x - 3)$ will be positive in two cases:

All factors are positive;

Two factors are negative and one is positive.

A sign chart shows both of these cases.

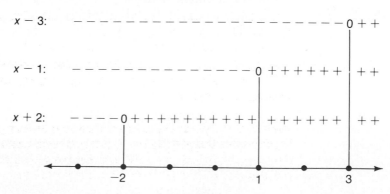

All factors are positive: $x > 3$.

Two factors are negative and one is positive: $-2 < x < 1$.

Thus the solutions to $(x - 1)(x + 2)(x - 3) > 0$ are all x such that $x > 3$ or $-2 < x < 1$. □

Notice that computing the product $(x - 1)(x + 2)(x - 3)$ gives a polynomial of degree 3. The graph of $y = (x - 1)(x + 2)(x - 3)$ crosses the horizontal axis at $x = 1$, $x = -2$, and $x = 3$. The sign chart indicates that y is negative for x to the left of -2, positive between -2 and 1, negative between 1 and 3, and positive to the right of 3. From this analysis we have a good idea of the shape of the graph of $y = (x - 1)(x + 2)(x - 3)$ (Fig. 9.13).

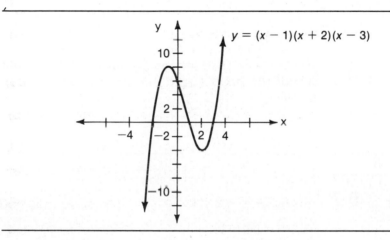

Figure 9.13

Example 4 An object is shot into the air from ground level with an initial speed of 1000 feet per second. How long will it be at least 4600 feet above ground?

Remember that the height y of an object t seconds after launch is given by the equation $y = -16t^2 + 1000t$. The problem asks for the values of t that give a height greater than or equal to 4600 feet; that is, the solutions to

$$-16t^2 + 1000t \geq 4600$$

$$-16t^2 + 1000t - 4600 \geq 0.$$

To use a sign chart we need to be able to factor the quadratic expression; since that is difficult, we can think about the graph of $y = -16t^2 + 1000t - 4600$, a parabola that opens down. The y value will be positive between the two points where the graph crosses the horizontal axis; thus we must find the two points where $-16t^2 + 1000t - 4600 = 0$ or $-2t^2 + 125t - 575 = 0$. The quadratic formula gives

$$x = \frac{-125 + \sqrt{(125)^2 - 4(-2)(-575)}}{2(-2)} \quad \text{or} \quad x = \frac{-125 - \sqrt{(125)^2 - 4(-2)(-575)}}{2(-2)}$$

$$= \frac{-125 + 105}{-4} \qquad\qquad\qquad = \frac{-125 - 105}{-4}$$

$$= 5 \qquad\qquad\qquad\qquad\qquad = 57.5.$$

Thus the object shot at 1000 feet per second will be at height 4600 feet twice, once on the way up (at $t = 5$ seconds) and again on the way down (at $t = 57.5$ seconds). Between these times the object will be at a height greater than 4600 feet, so the object is at least 4600 feet above ground for all the times t with $5 \leq t \leq 57.5$. Thus the length of time is 52.5 seconds.   □

Exercises 9.7

Solve the inequalities in Exercises 1–16.

1. $-x^2 + x + 2 < 0$

2. $-3x^2 + 10x + 8 \geq 0$

3. $x^2 - 3x \geq 0$

4. $6x^2 + x - 2 < 0$

5. $x^2 - 9 > 0$

6. $2x^2 - 8 \leq 0$

7. $x^2 + x + 1 \geq 0$

8. $4x^2 - 4x + 1 \leq 0$

9. $2x^2 - 4x > 1$

10. $-3x^2 - 6x \leq 2$

11. $x(x - 2) \geq x - 2$

12. $(x^2 - 1)(x^2 - 9) \leq 0$

13. $(x + 2)(x - 2)(5x + 4) < 0$

14. $x(x - 3)(2x + 3) \geq 0$

15. $x(3 - x)(x + 2) > 0$

16. $(x + 1)^3 \geq 0$

17. An object is shot into the air from ground level with an initial speed of 800 feet per second. How long will it be at least 6400 feet above the ground?

18. Find the values of x for which the graph of $y = x^2$ lies below the graph of $y = x$.

19. Find the values of x for which the graph of $y = x^3$ lies above the graph of $y = x$.

20. Find the values of x for which the graph of $y = x^4$ lies above the graph of $y = x^2$.

21. Find all the numbers whose squares are less than three times the number.

22. Determine k so that $2x^2 + kx + 2 = 0$ has two distinct real solutions.

23. Determine k so that $x^2 + 2kx + 9 = 0$ has no real solutions.

9.8 / SYSTEMS OF NONLINEAR EQUATIONS

Whenever we have two equations in two variables, we can look for common solutions to the equations. Each equation has a graph; inspecting the two graphs suggests how many solutions the equations have in common as well as their approximate values. A simultaneous solution is a pair of numbers: a value for x and a value for y. To find the exact values of the solutions, we use algebraic techniques (principally substitution) similar to what we used in solving systems of linear equations.

Example 1 Find the simultaneous solutions to the two equations

$$y = x^2 - 3$$
$$x + y = -1.$$

The first equation in the example has a parabola for its graph; the second equation has a line for its graph (Fig. 9.14).

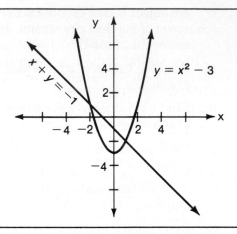

Figure 9.14

We can see that the graphs have two points in common. To find the coordinates of these points, we observe that since $y = x^2 - 3$ in the first equation, we can replace y by $x^2 - 3$ in the second equation to get an equation only in the variable x:

$$x + y = -1$$
$$x + (x^2 - 3) = -1$$
$$x^2 + x - 3 = -1$$
$$x^2 + x - 2 = 0.$$

This quadratic equation in x can be solved either by factoring or by the quadratic formula. Using factoring, we get

$$(x + 2)(x - 1) = 0$$
$$x = -2 \quad \text{or} \quad x = 1.$$

These values are the x-coordinates of the two common points. To get the corresponding y-coordinates, we can use either of the original equations. From the first,

$$\text{if } x = -2, \text{ then } y = x^2 - 3 = (-2)^2 - 3 = 1$$
$$\text{if } x = 1, \quad \text{then } y = x^2 - 3 = 1^2 - 3 = -2.$$

Thus $x = -2$, $y = 1$ is one simultaneous solution and $x = 1$, $y = -2$ is the second.   □

Example 1 demonstrates how to find common solutions to a quadratic equation and a linear equation in two variables. It is also possible to solve simultaneously two quadratic equations in two variables. Remembering that such equations have parabolas for graphs, you should be able to visualize that the number of common solutions for two different parabolas can be 0, 1, or 2. Graphing the equations shows the number of solutions and their approximate values, but finding the exact solutions usually requires solving algebraically.

Example 2 Find the simultaneous solutions to the two equations

$$y = x^2 + x - 2$$
$$y = -x^2 + 3.$$

Inspecting the graphs of the two equations (Fig. 9.15) suggests that the number of simultaneous solutions in this example is two. (Remember that each solution is a pair, a value for x and a value for y.)

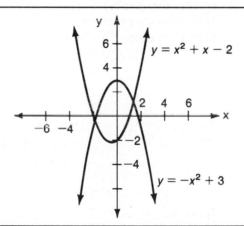

Figure 9.15

Since we seek values for x and y that make $y = x^2 + x - 2$ and also $y = -x^2 + 3$, it must be that

$$x^2 + x - 2 = -x^2 + 3$$
$$2x^2 + x - 5 = 0.$$

Using the quadratic formula

$$x = \frac{-1 + \sqrt{1 - 4(2)(-5)}}{2(2)} \quad \text{or} \quad x = \frac{-1 - \sqrt{1 - 4(2)(-5)}}{2(2)}$$

$$= \frac{-1 + \sqrt{41}}{4} \qquad\qquad = \frac{-1 - \sqrt{41}}{4}$$

$$\doteq 1.35 \qquad\qquad\qquad \doteq -1.85.$$

We can use either of the original equations to compute the y value that corresponds to each of these two x values. The second equation is less complicated:

if $x \doteq 1.35$, then $y = -x^2 + 3 \doteq -(1.35)^2 + 3 \doteq 1.18,$

if $x \doteq -1.85$, then $y = -x^2 + 3 \doteq -(-1.85)^2 + 3 \doteq -.42.$

Thus the simultaneous solutions are $x \doteq 1.35$, $y \doteq 1.18$ and $x \doteq -1.85$, $y \doteq -.42$. Look back at the graphs of the two equations to see that these solutions are reasonable. $\square$

In the example above, the y values are computed in terms of approximations of the x values. Thus both the x and y values are approximations. To get exact y values, we must use exact x values:

if $x = \dfrac{-1 + \sqrt{41}}{4}$, then $y = -x^2 + 3$

$$= -\left(\frac{-1 + \sqrt{41}}{4}\right)^2 + 3$$

$$= -\left(\frac{1 - 2\sqrt{41} + 41}{16}\right) + 3$$

$$= -\left(\frac{42 - 2\sqrt{41}}{16}\right) + 3$$

$$= \frac{-42 + 2\sqrt{41}}{16} + \frac{48}{16}$$

$$= \frac{6 + 2\sqrt{41}}{16}$$

$$= \frac{3 + \sqrt{41}}{8}.$$

$$\text{if } x = \frac{-1 - \sqrt{41}}{4}, \text{ then } y = -x^2 + 3$$

$$= -\left(\frac{-1 - \sqrt{41}}{4}\right)^2 + 3$$

$$= -\left(\frac{1 + 2\sqrt{41} + 41}{16}\right) + 3$$

$$= -\left(\frac{42 + 2\sqrt{41}}{16}\right) + 3$$

$$= \frac{-42 - 2\sqrt{41}}{16} + \frac{48}{16}$$

$$= \frac{6 - 2\sqrt{41}}{16}$$

$$= \frac{3 - \sqrt{41}}{8}.$$

Since exact values are seldom more useful than good approximations, and since computing with radicals is much more time consuming than using a calculator to compute with decimal approximations, we will usually give approximations to the exact solutions of systems of equations in the examples. You may also give approximations in the exercises.

In Example 1, a linear equation in two variables and a quadratic equation are combined to give a quadratic equation in one variable. Similarly, in Example 2, two quadratic equations in two variables are combined to give a quadratic equation in one variable. Since we can always solve a quadratic equation in one variable, we are always able to find the simultaneous solutions to problems of these types if they exist. If we seek simultaneous solutions to equations where one has degree greater than 2, we may not be able to find the solutions algebraically because we may not be able to combine the equations to get an equation that can be readily solved. Linear and quadratic equations in one variable are the only equations for which we will develop techniques for solving in every case. However, sometimes there can be a special pair of equations in two variables that can be solved by the same methods as shown in Examples 1 and 2, even though one equation, or both, has degree greater than 2.

Example 3 Find the simultaneous solutions to the two equations

$$y = x^3 + 2$$
$$y = x + 2.$$

The graphs of the equations in Fig. 9.16 suggest that there are three simultaneous solutions.

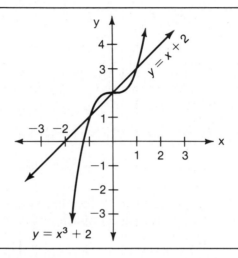

Figure 9.16

To find the solutions, we replace y in the second equation by $x^3 + 2$:

$$y = x + 2$$
$$x^3 + 2 = x + 2$$
$$x^3 - x = 0.$$

The polynomial on the left side factors. This fact permits us to solve the equation:

$$x(x^2 - 1) = 0$$
$$x(x - 1)(x + 1) = 0.$$

Thus $x = 0$ or $x = 1$ or $x = -1$. To find the corresponding y values, we can use the second of the two original equations:

if $x = 0$,   then $y = x + 2 = 0 + 2 = 2,$

if $x = 1$,   then $y = x + 2 = 1 + 2 = 3,$

if $x = -1$, then $y = x + 2 = -1 + 2 = 1$.

The three simultaneous solutions are

$$x = 0, \quad y = 2$$
$$x = 1, \quad y = 3$$
$$x = -1, y = 1.$$

Check these points on the graph in Fig. 9.16.   □

Exercises 9.8

Graph the two equations in Exercises 1 and 2 on the same coordinate system. Then approximate their simultaneous solutions from the graph.

1. $3x - y = 4$
    $y = 2x^2 - 4x - 1$

2. $y = -x^2 + 10$
    $y = x^2 - 2x - 2$

Find the simultaneous solutions to the equations in Exercises 3–15 algebraically.

3. $y = x - 3$
    $y = x^2 - 4x + 1$

4. $y = 2x + 3$
    $y = 2x^2 + 4x - 1$

5. $y = -3x + 1$
    $y = -3x^2 + 6x + 1$

6. $2x + y + 5 = 0$
    $y = -2x^2 - 4x - 1$

7. $y = -x^2 + 5$
    $y = x^2 + 2x - 19$

8. $y = x^2 + 4x + 10$
    $y = 2x^2 + 6x - 5$

9. $y = 6x^2 - 7x$
    $y = x^2 + 7x + 3$

10. $y = -x^2 + 4x$
     $y = x^2 + 2x - 1$

11. $y = -2x^2 + 4x + 9$
     $y = 3x^2 - 6x - 11$

12 $y = 4x + 1$
    $y = x^3 + 1$

13. $y = x^3$
     $y = 4x^2$

14. $y = 2x + 3$
     $xy = 2$

15. $y = \sqrt{x}$
     $x^2 + y^2 = 2$

16. Find two numbers whose product is 48 and whose sum is 14.

17. Find the dimensions of a rectangle with an area of 36 square feet and a perimeter of 26 feet.

18. The sum of two numbers is 5 and the sum of their squares is 53. Find the numbers.

9.9 / CHAPTER 9 PROBLEM COLLECTION

1. Part of the graph of a parabola with a line of symmetry $x = 1$ is given below. Complete the graph.

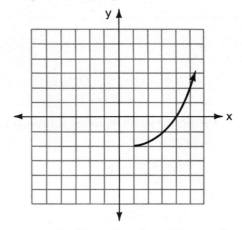

Supply the missing terms in Exercises 2 and 3 to make each expression the square of a binomial.

2. $x^2 + 4x + (?)$ 3. $x^2 - (?)x + 9$

Write the equations in Exercises 4–6 in a form that gives the vertex and line of symmetry of each graph.

4. $y = x^2 + 4x - 7$ 5. $y = -x^2 + 2x + 1$

6. $y = 2x^2 + 2x - 1$

Give the coordinates of the vertex, the equation of the line of symmetry, and draw the graph of each of the equations in Exercises 7–11.

7. $y = 2\left(x + \dfrac{3}{2}\right)^2 - 1$ 8. $y = x^2 - 4x + 5$

9. $y = -x^2 - 8x - 11$ 10. $y = -3x^2 + 6x - 1$

11. $y = (x + 4)(x - 2)$

Simplify the radicals in Exercises 12–16 by computing the square root of perfect square factors.

12. $\sqrt{200}$

13. $\sqrt{\dfrac{72}{49}}$

14. $\dfrac{3 + \sqrt{18}}{6}$

15. $\sqrt{32x^4y^7}$

16. $\sqrt{\dfrac{x^3z^2}{y^4}}$

Simplify each term in the expressions in Exercises 17–19 and then combine like terms. Leave your answers in radical form.

17. $\sqrt{98} - 2\sqrt{50} + 3\sqrt{32}$

18. $\sqrt{125} - 2\sqrt{27} + \sqrt{48}$

19. $\sqrt{x^4y^3} + 2\sqrt{x^2y} - 3\sqrt{y^5}$

Perform the indicated operations in Exercises 20–23 and simplify the resulting expressions. Leave your answers in radical form.

20. $\dfrac{2 + 3\sqrt{7}}{5} + \dfrac{2 - 3\sqrt{7}}{5}$

21. $(3 + \sqrt{5})(3 - \sqrt{5})$

22. $\left(\dfrac{3 - 2\sqrt{2}}{2}\right)\left(\dfrac{3 + 2\sqrt{2}}{2}\right)$

23. $\left(\dfrac{1 - 2\sqrt{6}}{2}\right)^2$

Find the solutions of the equations and inequalities in Exercises 24–44.

24. $x^2 + 3x = 0$

25. $9y^2 - 4 = 0$

26. $x^2 - x - 6 = 0$

27. $x^2 - 2x - 4 = 0$

28. $4x^2 + 4x - 3 = 0$

29. $3x^2 - 6x + 5 = 0$

30. $-6t^2 - t + 1 = 0$

31. $2x^2 - 10x + 11 = 0$

32. $4x^2 - 12x + 1 = 0$

33. $4x^2 + 4x - 1 = 0$

34. $x^4 - 4x^2 - 5 = 0$

35. $y - 4\sqrt{y} + 1 = 0$

36. $4t^{-2} - 4t^{-1} = 3$

37. $x^6 + 9x^3 + 8 = 0$

38. $(4x - 5)(x + 2) < 0$

39. $x^2 - 16 \geq 0$

40. $4x^2 + 7x - 2 > 0$

41. $2x^2 + 5x \geq 3$

42. $4x^2 - 4x - 7 \leq 0$

43. $2x(x - 1) - 11 > (x + 1)^2 + 9$

44. $(x + 2)(x - 2)(x - 5) > 0$

Find the simultaneous solutions to the equations in Exercises 45–48.

45. $y = 4x - 13$
    $y = -2x^2 + 12x - 13$

46. $y = x^2 - 6x + 3$
    $y = 3x^2 - 3x + 1$

47. $y = -x^2 + 2$
    $y = x^2 - 4x$

48. $y = x^3 - x^2$
    $y = x^2 - x$

Write a quadratic equation with integer coefficients and solutions as specified in each of Exercises 49–51.

49. $5, -3$

50. $-\dfrac{2}{3}, \dfrac{4}{5}$

51. $\dfrac{2 + \sqrt{2}}{3}, \dfrac{2 - \sqrt{2}}{3}$

Determine whether the equations in Exercises 52 and 53 have 0, 1, or 2 solutions without actually solving the equations.

52. $4x^2 - 4x + 1 = 0$

53. $2x^2 + 8x + 9 = 0$

54. Determine k so that the vertex of $y = 2x^2 - 4x + k$ is (1,5).

55. Determine k so that the line of symmetry of $y = -3x^2 + kx + 1$ is $x = -\dfrac{1}{2}$.

56. Determine k so that $\dfrac{1}{2}$ is a solution of $2x^2 + kx - 4 = 0$, and then find the other solution.

57. Determine the value(s) of k so that the equation $2x^2 + kx + 1 = 0$ has exactly one real solution.

58. Determine the value(s) of k so that the equation $kx^2 - 2x + 1 = 0$ has two distinct solutions.

59. Determine the value(s) of k so that the equation $3x^2 + 2kx + 3 = 0$ has no real solutions.

60. Find two numbers whose sum is 19 and whose product is 84.

61. The length of a rectangle is 1 foot less than three times the width.

If the area of the rectangle is 184 square feet, find the dimensions of the rectangle.

62. The sum of the squares of three consecutive odd integers is 251. Find the integers.

63. A rectangular pool of dimensions 30 feet by 50 feet is surrounded by a walk of uniform width. If the area of the walk is 801 square feet, find the width of the walk.

64. A page of a book is 2 inches longer than it is wide. The margin at top and bottom is 1 inch, and on each side is $1\frac{1}{2}$ inches. If the area of the printed matter is 18 square inches, find the dimensions of the page.

65. Find the dimensions of a rectangle with a perimeter of 61 feet and an area of 225 square feet.

66. An object is shot into the air from ground level at a speed of 448 feet per second.

(a) When will the object be 2112 feet above the ground?

(b) When will the object reach its maximum height?

(c) What is the maximum height attained by the object?

(d) When will the object strike the ground?

(e) How long will the object be at least 2560 feet above the ground?

Rational Expressions and Fractional Equations

10.1 / EQUIVALENT RATIONAL EXPRESSIONS

Rational expressions have the form of one polynomial divided by another polynomial. Here are some rational expressions:

$$\frac{1}{x+1},$$

$$\frac{x^2 - 2}{3x^3 + 4x^2 - x + 6},$$

$$\frac{2x^4 + x^3 - 3x}{x^2 - 4}.$$

A notation that stands for any rational expression is $\frac{p(x)}{q(x)}$, where $p(x)$ denotes the polynomial in the numerator and $q(x)$ denotes the polynomial in the denominator.

In this chapter we will describe when two rational expressions are equal and how to add, subtract, multiply, and divide rational expressions. The model for this arithmetic is the arithmetic of the rational numbers (or fractions). Remembering when two rational numbers are equal and how to add, subtract, multiply, and divide rational numbers suggests how to perform the arithmetic for rational expressions. There is one important difference: the expression $\frac{1}{x+1}$, for example, has meaning when the variable x is not -1; it has no meaning when $x = -1$. In general, a rational expression has meaning for the values of the variable that make the denominator different from 0 and no meaning for the values of the variable that make the denominator equal to 0. When we write the expression $\frac{1}{x+1}$, we assume that the variable x is not equal to -1. When we write any rational expression, we assume that the variable does not represent a value that makes the denominator 0.

Remember that the two fractions $\frac{a}{b}$ and $\frac{c}{d}$ are equal if $ad = bc$. Thus $\frac{8}{20} = \frac{14}{35}$ because $(8)(35) = (20)(14)$. The definition for equal (or, we also say, equivalent) rational expressions is similar to the definition for equal fractions.

DEFINITION

We say that $\frac{p(x)}{q(\text{x})} = \frac{r(x)}{s(x)}$ if $p(x) \cdot s(x) = q(x) \cdot r(x)$.

Example 1 Determine if $\dfrac{x^2 - x}{x^2} = \dfrac{x^2 + x - 2}{x^2 + 2x}$.

To see if these two rational expressions are equal, we must compare the products $(x^2 - x)(x^2 + 2x)$ and $x^2(x^2 + x - 2)$:

$$(x^2 - x)(x^2 + 2x) = x^4 + x^3 - 2x^2$$
$$x^2(x^2 + x - 2) = x^4 + x^3 - 2x^2.$$

Since the two polynomial products are equal, we conclude that

$$\frac{x^2 - x}{x^2} = \frac{x^2 + x - 2}{x^2 + 2x}.$$

Observe that there are values of x that make the denominators of these rational expressions equal to 0:

$$x^2 = 0 \qquad x^2 + 2x = 0$$
$$x = 0 \qquad x(x + 2) = 0$$
$$x = 0, x = -2.$$

The expression

$$\frac{x^2 - x}{x^2}$$

is a number for every value of x except $x = 0$. The expression

$$\frac{x^2 + x - 2}{x^2 + 2x}$$

is a number for every value of x except $x = 0$ and $x = -2$. Thus the statement

$$\frac{x^2 - x}{x^2} = \frac{x^2 + x - 2}{x^2 + 2x}$$

is meaningful as long as $x \neq 0$ and $x \neq -2$. When we write

$$\frac{x^2 - x}{x^2} = \frac{x^2 + x - 2}{x^2 + 2x}$$

we assume that the condition $x \neq 0$ and $x \neq -2$ is understood. □

One way to generate equivalent rational expressions is to multiply the numerator and denominator of a rational expression by the same polynomial. For example,

$$\frac{5}{4x - 3} = \frac{5(x - 2)}{(4x - 3)(x - 2)}.$$

To check that this is true, compare the products

$$5(4x - 3)(x - 2) \quad \text{and} \quad (4x - 3)(5)(x - 2).$$

These products are the same because they contain identical factors.

Example 2   Write a rational expression with denominator $2x^2 + 9x - 5$ that is equivalent to $\dfrac{x}{2x - 1}$.

The problem is to determine the numerator that makes

$$\frac{x}{2x - 1} = \frac{?}{2x^2 + 9x - 5}.$$

The answer becomes more apparent if we factor $2x^2 + 9x - 5$:

$$\frac{x}{2x - 1} = \frac{?}{(2x - 1)(x + 5)}.$$

Since the new denominator is $x + 5$ times the old denominator, the new numerator will be $x + 5$ times x:

$$\frac{x}{2x - 1} = \frac{x(x + 5)}{(2x - 1)(x + 5)} = \frac{x^2 + 5x}{2x^2 + 9x - 5}. \qquad \square$$

We observed above that if $\frac{p(x)}{q(x)}$ is a rational expression and $t(x)$ is any polynomial other than 0, then

$$\frac{p(x)}{q(x)} = \frac{p(x) \cdot t(x)}{q(x) \cdot t(x)}.$$

When this equation is read from right to left, it states that

$$\frac{p(x) \cdot t(x)}{q(x) \cdot t(x)} = \frac{p(x)}{q(x)}.$$

Thus if a rational expression has the same factor, $t(x)$, in the numerator and denominator, that factor can be removed from the numerator and denominator to give a simpler equivalent rational expression. When we factor the numerator and denominator of a rational expression and remove all the common factors, we say we have **reduced** the expression to lowest terms.

Example 3   Reduce $\dfrac{2x^4 - 16x}{3x^3 + x^2 - 14x}$ by removing common factors from the numerator and denominator.

To see what factors are common to the numerator and denominator, we first factor the polynomials in the numerator and denominator:

$$\frac{2x^4 - 16x}{3x^3 + x^2 - 14x} = \frac{2x(x^3 - 8)}{x(3x^2 + x - 14)}$$

$$= \frac{2x(x - 2)(x^2 + 2x + 4)}{x(3x + 7)(x - 2)}.$$

The numerator and denominator contain the common factors x and $x - 2$. Removing them gives

$$\frac{2x^4 - 16x}{3x^3 + x^2 - 14x} = \frac{2(x^2 + 2x + 4)}{3x + 7}.$$ □

Exercises 10.1

Determine whether each pair of rational expressions in Exercises 1–3 are equal.

1. $\dfrac{3}{4}, \dfrac{15}{20}$

2. $\dfrac{x^2 - x}{x^2 + 2x}, \dfrac{x - 1}{x + 2}$

3. $\dfrac{2x - 1}{2x^2 + 3x - 2}, \dfrac{x - 3}{x^2 - x - 6}$

Find the values of x in Exercises 4–8 for which the rational expressions represent a number.

4. $\dfrac{x^2 + x}{x + 3}$

5. $\dfrac{2x^2 - 1}{x^2 - 4}$

6. $\dfrac{x + 5}{x^2 + x}$

7. $\dfrac{2x - 1}{x^2 + 1}$

8. $\dfrac{x^3 + x}{x^2 - 2x + 1}$

Supply the missing numerator or denominator in Exercises 9–16 so that the pair of fractions will be equal.

9. $\dfrac{25}{36}, \dfrac{50}{?}$

10. $\dfrac{15}{81}, \dfrac{?}{27}$

11. $\dfrac{x^2 y}{z}, \dfrac{?}{z^2}$

12. $\dfrac{x - 2}{x - 3}, \dfrac{?}{3 - x}$

13. $\dfrac{x - y}{x + y}, \dfrac{y - x}{?}$

14. $\dfrac{a + b}{2a - b}, \dfrac{?}{4a^2 - b^2}$

15. $\dfrac{3x + 1}{2x - 1}, \dfrac{3x^2 + x}{?}$

16. $\dfrac{2x^2 - 4x}{x^2 - x - 2}, \dfrac{?}{x + 1}$

Reduce the fractions in Exercises 17–30 to their lowest terms.

17. $\dfrac{540}{126}$

18. $\dfrac{248}{279}$

19. $\dfrac{x^3 y^2 z}{x y^3 z}$

20. $\dfrac{a^4 b c^3}{a^3 b^2 c^2}$

21. $\dfrac{2x^2 - 3x - 2}{x^2 - 4}$

22. $\dfrac{4x^2 + 4x - 3}{2x^2 - x}$

23. $\dfrac{x^2 - xy}{x^2 - 2xy + y^2}$

24. $\dfrac{a^2 + 2ab + b^2}{a^2 - b^2}$

25. $\dfrac{x^4 - y^4}{2x^2 + xy - y^2}$

26. $\dfrac{x^3 - y^3}{x^2 - y^2}$

27. $\dfrac{x^3 + y^3}{x^2 + xy}$

28. $\dfrac{x^3 + x^2 y + xy^2}{x^3 - y^3}$

29. $\dfrac{2x^2 - 7x + 6}{2x^3 + x^2 - 6x}$

30. $\dfrac{x^4 - y^4}{x^3 + xy^2}$

10.2 / MULTIPLYING AND DIVIDING RATIONAL EXPRESSIONS

The operations of multiplication and division for rational expressions are parallel to multiplication and division for rational numbers. The formal definitions can be written this way.

DEFINITIONS

Assume that $\dfrac{p(x)}{q(x)}$ and $\dfrac{r(x)}{s(x)}$ are rational expressions. Then

$$\frac{p(x)}{q(x)} \cdot \frac{r(x)}{s(x)} = \frac{p(x) \cdot r(x)}{q(x) \cdot s(x)} \text{ and}$$

$$\frac{p(x)}{q(x)} \div \frac{r(x)}{s(x)} = \frac{p(x)}{q(x)} \cdot \frac{s(x)}{r(x)}.$$

The statements in the definitions above assume that x does not represent any number that makes a denominator equal to 0.

Example 1 Compute the product $\dfrac{3x}{x^2 - 7x + 12} \cdot \dfrac{4 - x}{x}.$

To compute the product of the two rational expressions, we multiply the numerators and the denominators:

$$\frac{3x}{x^2 - 7x + 12} \cdot \frac{4 - x}{x} = \frac{3x(4 - x)}{(x^2 - 7x + 12)x}$$

It is always possible that the resulting rational expression can be reduced to a simpler form. Thus instead of multiplying out the product of the polynomials in the numerator and the product of the polynomials in the denominator, we factor completely and remove common factors:

$$\frac{3x}{x^2 - 7x + 12} \cdot \frac{4 - x}{x} = \frac{3x(4 - x)}{(x^2 - 7x + 12)x}$$

$$= \frac{3x(4 - x)}{(x - 3)(x - 4)(x)}$$

$$= \frac{3(4 - x)}{(x - 3)(x - 4)}$$

$$= \frac{3(-1)(x - 4)}{(x - 3)(x - 4)}$$

$$= \frac{-3}{x - 3}.$$

Observe that the factor $4 - x$ in the numerator was rewritten as $(-1)(x - 4)$ to correspond to the factor $x - 4$ in the denominator.   □

Example 2   Compute the quotient $\dfrac{4x^2 + 8x - 5}{x^2} \div \dfrac{8x^3 - 1}{4x^2 - 3x}$.

The quotient is computed by multiplying the first rational expression times the reciprocal of the second. Observe that we factor the numerator and denominator of the quotient and reduce it to simpler form:

$$\frac{4x^2 + 8x - 5}{x^2} \div \frac{8x^3 - 1}{4x^2 - 3x} = \frac{4x^2 + 8x - 5}{x^2} \cdot \frac{4x^2 - 3x}{8x^3 - 1}$$

$$= \frac{(4x^2 + 8x - 5)(4x^2 - 3x)}{x^2(8x^3 - 1)}$$

$$= \frac{(2x - 1)(2x + 5)(x)(4x - 3)}{x^2(2x - 1)(4x^2 + 2x + 1)}$$

$$= \frac{(2x + 5)(4x - 3)}{x(4x^2 + 2x + 1)}.$$ □

For rational expressions, as for polynomials, if we need to evaluate a product at some value of the variable, we can either multiply the rational expressions together and then evaluate the product, or we can evaluate the factors in the product and then compute the product of their values. Both give the same value.

Example 3 Find the value of $\dfrac{x^2 - 1}{x} \cdot \dfrac{2x}{4x^2 + 3}$ when $x = -2$.

a. We can first compute the product and then replace x by -2:

$$\frac{x^2 - 1}{x} \cdot \frac{2x}{4x^2 + 3} = \frac{(x^2 - 1)(2x)}{x(4x^2 + 3)}$$

$$= \frac{2(x^2 - 1)}{4x^2 + 3}$$

$$= \frac{2x^2 - 2}{4x^2 + 3}$$

$$= \frac{2(-2)^2 - 2}{4(-2)^2 + 3}$$

$$= \frac{8 - 2}{16 + 3}$$

$$= \frac{6}{19}.$$

b. We can evaluate $\dfrac{x^2 - 1}{x}$ when $x = -2$, evaluate $\dfrac{2x}{4x^2 + 3}$ when $x = -2$, and then compute the product of these numbers:

$$\frac{x^2 - 1}{x} \cdot \frac{2x}{4x^2 + 3} = \frac{(-2)^2 - 1}{(-2)} \cdot \frac{2(-2)}{4(-2)^2 + 3}$$

$$- \frac{4 - 1}{-2} \quad \frac{-4}{16 + 3}$$

$$= \frac{3}{-2} \cdot \frac{-4}{19}$$

$$= \frac{-12}{-38}$$

$$= \frac{6}{19}. \qquad \square$$

The operations of division, addition, and subtraction for rational expressions are similar to multiplication with respect to evaluating the expressions for specific values of the variable. You can either perform the operation first and then evaluate, or you can first evaluate each rational expression and then perform the operation on the values.

Complex Fractions

You will sometimes see an expression like

$$\frac{\dfrac{x}{x^2 + 1}}{\dfrac{5}{x^3}}$$

This is another way of denoting the quotient

$$\frac{x}{x^2 + 1} \div \frac{5}{x^3}.$$

Observe the result of simplifying the quotient:

$$\frac{\dfrac{x}{x^2 + 1}}{\dfrac{5}{x^3}} = \frac{x}{x^2 + 1} \div \frac{5}{x^3}$$

$$= \frac{x}{x^2 + 1} \cdot \frac{x^3}{5}$$

$$= \frac{x \cdot x^3}{5(x^2 + 1)}.$$

The numerator $x \cdot x^3$ is the product of the numerator of the top fraction and the denominator of the bottom fraction; the denominator $5(x^2 + 1)$ is the product of the denominator of the top fraction and the numerator of the bottom fraction:

$$\frac{\dfrac{x}{x^2 + 1}}{\dfrac{5}{x^3}} \qquad \begin{matrix} x \cdot x^3 \\ \overline{5(x^2 + 1)} \end{matrix}$$

Although a complex fraction can always be simplified using the definition for division of rational expressions, it saves time to remember the above pattern.

Example 4 Simplify the complex fraction

$$\frac{\dfrac{x^2 - 4}{x^2 + 4}}{\dfrac{2 - x}{2x}}.$$

We use the pattern above and then reduce the quotient by removing common factors:

$$\frac{\dfrac{x^2 - 4}{x^2 + 4}}{\dfrac{2 - x}{2x}} \qquad \frac{2x(x^2 - 4)}{(2 - x)(x^2 + 4)}$$

$$= \frac{2x(x - 2)(x + 2)}{(2 - x)(x^2 + 4)}$$

$$= \frac{2x(x - 2)(x + 2)}{(-1)(x - 2)(x^2 + 4)}$$

$$= \frac{2x(x + 2)}{-(x^2 + 4)}. \qquad \qquad \square$$

Exercises 10.2 Perform the indicated operations in Exercises 1–28 and express your answers in reduced form.

1. $\dfrac{12}{35} \cdot \dfrac{275}{72}$

2. $-\dfrac{13}{49} \div \dfrac{169}{28}$

3. $\dfrac{\dfrac{20}{17}}{\dfrac{50}{11}}$

4. $x^2\left(\dfrac{1}{x}\right)$

5. $\dfrac{1}{ab}(a^2b^3)$

6. $\dfrac{x^2y}{w^2z} \cdot \dfrac{w^3z^2}{x^3y^2}$

7. $\dfrac{a^4d}{b^3c^2} \div \dfrac{a^2d^2}{b^3c}$

8. $\dfrac{\dfrac{r^2s}{t^3}}{-\dfrac{r^2s^2}{t^2}}$

9. $\dfrac{x^2 - y^2}{2x^2 + xy} \cdot \dfrac{4x^2 + 4xy + y^2}{x^2 - xy - 2y^2}$

10. $\dfrac{x^2 + x}{2x^2 - x} \cdot \dfrac{2x^2 + 5x - 3}{x^2 - x - 2}$

11. $\dfrac{xy - x^2}{xy - y^2} \cdot \dfrac{y}{x}$

12. $\dfrac{6x^2 - x - 1}{x^2 - 1} \cdot \dfrac{x^2 - 4x + 3}{3x^2 - 5x - 2}$

13. $\dfrac{x^3 - 4x}{x^2 + 2x} \cdot \dfrac{x + 2}{x^2 + x - 6}$

14. $\dfrac{4x^2 - 9}{x^2 - x} \div \dfrac{2x^2 + x - 6}{x^2 + 4x}$

15. $\dfrac{x^3 - 1}{x^2 + 4x + 4} \div \dfrac{x^3 + x^2 + x}{x^2 + 2x}$

16. $\dfrac{x^4 - y^4}{2x^2 + xy - y^2} \div \dfrac{x^3 + xy^2}{2x^2 - 3xy + y^2}$

17. $(x - 2)\left(\dfrac{x}{x - 2}\right)$

18. $(y - 1)\left(\dfrac{y}{1 - y}\right)$

19. $\dfrac{a^2 + 4ab + 4b^2}{a^2 + 2ab} \div \dfrac{a^2 - 4b^2}{a^3 - 2a^2b}$

20. $\dfrac{\dfrac{x - y}{x}}{\dfrac{x + y}{x}}$

21. $\dfrac{\dfrac{a^2 - ab}{b}}{\dfrac{ab - b^2}{b^2}}$

22. $\dfrac{\dfrac{x^2 - xy}{x + y}}{\dfrac{y^2 - xy}{x^2 + 2xy + y^2}}$

23. $\dfrac{\dfrac{8 - x^3}{x + 3}}{\dfrac{x^2 - 4}{x^2 + 5x + 6}}$

24. $\dfrac{\dfrac{ac + 2bc - ad - 2bd}{ab}}{\dfrac{a^2 + 2ab}{a^2b^2}}$

25. $\dfrac{x^{-2}y^{-2}}{x^{-1}y^{-1}}$

26. $\dfrac{\dfrac{a^3b^{-2}}{a^{-1}b^2}}{\dfrac{ab^{-1}}{a^{-1}b^2}}$

27. $\dfrac{x^6 - y^6}{x^2 - y^2} \cdot \dfrac{y}{x^3 + x^2y + xy^2} \cdot \dfrac{x}{x^2y - xy^2 + y^3}$

28. $\dfrac{x^2 - 3xy + 2y^2}{2x^2 + xy - y^2} \cdot \dfrac{2x^2 + 3xy - 2y^2}{x^2 - y^2} \cdot \dfrac{x^2 + 4xy + 3y^2}{x^2 - 4y^2}$

10.3 / ADDING AND SUBTRACTING RATIONAL EXPRESSIONS

Addition and subtraction for rational expressions are natural extensions of addition and subtraction for rational numbers. If two rational expressions have the same denominator, they are added or one subtracted from the other by combining the numerators. Study these examples:

$$\frac{x^2}{x+5} + \frac{3x-2}{x+5} = \frac{x^2 + (3x-2)}{x+5} = \frac{x^2 + 3x - 2}{x+5},$$

$$\frac{x^2}{x+5} - \frac{3x-2}{x+5} = \frac{x^2 - (3x-2)}{x+5} = \frac{x^2 - 3x + 2}{x+5}.$$

Observe in the second example that the entire quantity $3x - 2$ is subtracted from x^2.

If two rational expressions have different denominators, we change them to equivalent expressions with the same denominator before adding or subtracting. This common denominator is a multiple of each denominator in the sum or difference.

Example 1   Find the sum $\dfrac{2x}{x^2 - 3} + \dfrac{1}{x}$.

Both rational expressions can be written with denominator $(x^2 - 3)(x)$ and then added:

$$\frac{2x}{x^2 - 3} + \frac{1}{x} = \frac{(2x)(x)}{(x^2 - 3)(x)} + \frac{x^2 - 3}{(x^2 - 3)(x)}$$

$$= \frac{2x^2 + x^2 - 3}{(x^2 - 3)(x)}$$

$$= \frac{3x^2 - 3}{(x^2 - 3)(x)}. \qquad \square$$

One way to get a common denominator for two rational expressions is to take the product of the denominators of the expressions as we did in Example 1. We can write the definitions for the sum and difference of two rational expressions using this common denominator.

DEFINITIONS

Assume that

$$\frac{p(x)}{q(x)} \quad \text{and} \quad \frac{r(x)}{s(x)}$$

are two rational expressions. Then

$$\frac{p(x)}{q(x)} + \frac{r(x)}{s(x)} = \frac{p(x)s(x) + q(x)r(x)}{q(x)s(x)}, \text{ and}$$

$$\frac{p(x)}{q(x)} - \frac{r(x)}{s(x)} = \frac{p(x)s(x) - q(x)r(x)}{q(x)s(x)}.$$

Sometimes it is possible, and more efficient, to use a common denominator that has lower degree than the product of the denominators. If the denominators are polynomials with integer coefficients, we can form their least common multiple from the irreducible factors in the same way we formed the least common multiple of two whole numbers from their prime factors. First we factor the polynomials into irreducible factors and then we form the product of all factors, repeating as many times as necessary to make a multiple of both polynomials.

Example 2   Find the least common multiple of the two polynomials

$$4x^2 - 4x + 1 \quad \text{and} \quad 2x^3 + 5x^2 - 3x.$$

First we factor the polynomials into irreducible factors:

$$4x^2 - 4x + 1 = (2x - 1)(2x - 1) = (2x - 1)^2,$$
$$2x^3 + 5x^2 - 3x = x(2x^2 + 5x - 3) = x(2x - 1)(x + 3).$$

The least common multiple must contain all factors of both polynomials: x, $(2x - 1)^2$, and $x + 3$. Thus the least common multiple is $x(2x - 1)^2 (x + 3)$.   □

Example 3   Find the difference $\dfrac{x}{2x^2 - x - 6} - \dfrac{x - 4}{2x^2 + 5x + 3}$.

We factor the two denominators and form the least common multiple:

$$2x^2 - x - 6 = (2x + 3)(x - 2),$$

$$2x^2 + 5x + 3 = (2x + 3)(x + 1).$$

The least common multiple contains the factors $2x + 3$, $x - 2$, and $x + 1$, each appearing once:

$$\frac{x}{2x^2 - x - 6} - \frac{x - 4}{2x^2 + 5x + 3} = \frac{x}{(2x + 3)(x - 2)} - \frac{x - 4}{(2x + 3)(x + 1)}$$

$$= \frac{x(x + 1)}{(2x + 3)(x - 2)(x + 1)} - \frac{(x - 4)(x - 2)}{(2x + 3)(x + 1)(x - 2)}$$

$$= \frac{x(x + 1) - (x - 4)(x - 2)}{(2x + 3)(x - 2)(x + 1)}$$

$$= \frac{x^2 + x - (x^2 - 6x + 8)}{(2x + 3)(x - 2)(x + 1)}$$

$$= \frac{x^2 + x - x^2 + 6x - 8}{(2x + 3)(x - 2)(x + 1)}$$

$$= \frac{7x - 8}{(2x + 3)(x - 2)(x + 1)}. \qquad \square$$

Example 4   Perform the indicated operations: $a - 1 + \dfrac{1}{a + 1}$.

We use $a + 1$ for a common denominator and change each term to a rational expression with $a + 1$ in the denominator:

$$a - 1 + \frac{1}{a + 1} = (a - 1) + \frac{1}{a + 1}$$

$$= \frac{(a - 1)(a + 1)}{a + 1} + \frac{1}{a + 1}$$

$$= \frac{(a - 1)(a + 1) + 1}{a + 1}$$

$$= \frac{a^2 - 1 + 1}{a + 1}$$

$$= \frac{a^2}{a + 1}.$$

☐

Example 5   Perform the indicated operations:

$$\frac{2x}{x + 4} + \frac{1}{x - 1} - \frac{5}{x^2 + 3x - 4}.$$

Since $x^2 + 3x - 4 = (x + 4)(x - 1)$, the least common multiple of the three denominators is $(x + 4)(x - 1)$:

$$\frac{2x}{x + 4} + \frac{1}{x - 1} - \frac{5}{x^2 + 3x - 4} = \frac{2x(x - 1)}{(x + 4)(x - 1)} + \frac{x + 4}{(x + 4)(x - 1)} - \frac{5}{(x + 4)(x - 1)}$$

$$= \frac{2x(x - 1) + (x + 4) - 5}{(x + 4)(x - 1)}$$

$$= \frac{2x^2 - 2x + x + 4 - 5}{(x + 4)(x - 1)}$$

$$= \frac{2x^2 - x - 1}{(x + 4)(x - 1)}$$

$$= \frac{(2x + 1)(x - 1)}{(x + 4)(x - 1)}$$

$$= \frac{2x + 1}{x + 4}.$$

☐

Complex Fractions

We can write fractions that have sums and differences of rational expressions in their numerators and denominators. Two common ways of simplifying these complex fractions are illustrated in the following example.

Example 6   Simplify

$$\frac{\dfrac{1}{x} - 1}{x - \dfrac{1}{x^2}}.$$

a. The first method is to write the numerator as a single fraction and the denominator as a single fraction before simplifying further:

$$\frac{\dfrac{1}{x} - 1}{x - \dfrac{1}{x^2}} = \frac{\dfrac{1}{x} - \dfrac{x}{x}}{\dfrac{x^3}{x^2} - \dfrac{1}{x^2}}$$

$$= \frac{\dfrac{1-x}{x}}{\dfrac{x^3 - 1}{x^2}} \quad \begin{array}{c} x^2(1-x) \\ x(x^3 - 1) \end{array}$$

$$= \frac{x^2(1-x)}{x(x-1)(x^2 + x + 1)}$$

$$= \frac{x^2(-1)(x-1)}{x(x-1)(x^2 + x + 1)}$$

$$= \frac{-x}{x^2 + x + 1}.$$

b. A second way to proceed is first to multiply the numerator and denominator of the complex fraction by the least common multiple of the denominators that appear in the top and bottom of the complex fraction. The procedure removes fractions from the numerator and denominator and leaves a simple fraction. In this example, the least common multiple of x and x^2 is x^2:

$$\frac{\dfrac{1}{x} - 1}{x - \dfrac{1}{x^2}} = \frac{x^2\left(\dfrac{1}{x} - 1\right)}{x^2\left(x - \dfrac{1}{x^2}\right)}$$

$$= \frac{x - x^2}{x^3 - 1}$$

$$= \frac{x(1-x)}{(x-1)(x^2 + x + 1)}$$

$$= \frac{x(-1)(x-1)}{(x-1)(x^2 + x + 1)}$$

$$= \frac{-x}{x^2 + x + 1}. \qquad \square$$

**Exercises
10.3**

Find the least common multiple of the polynomials in Exercises 1–4.

1. $(x-1)^2(x+2), \ (x-1)(x+4)$

2. $x^2(x-1)(x+1), \ x(x-1)^2(x+2)$

3. $(x-2)(x+3), \ x(x+1)^2$

4. $(x-2)^2(x+1), \ (x-2)^2(x+1)^2$

Perform the indicated operations in Exercises 5–28 and express your answers in reduced form.

5. $\dfrac{5}{12}+\dfrac{7}{18}-\dfrac{11}{6}$

6. $\dfrac{4}{15}-\dfrac{6}{35}+\dfrac{5}{21}$

7. $\dfrac{\dfrac{1}{2}+\dfrac{1}{3}}{\dfrac{1}{2}-\dfrac{1}{3}}$

8. $\dfrac{\dfrac{1}{4}-9}{\dfrac{1}{2}-3}$

9. $\dfrac{1}{x-1}-\dfrac{1}{x^2-1}$

10. $\dfrac{2}{x-5}-\dfrac{2}{x+5}$

11. $x-\dfrac{1}{x+1}$

12. $\dfrac{3x}{x^2-x}-\dfrac{2x}{x^2-2x+1}$

13. $\dfrac{3}{2x^2+3xy}+\dfrac{2}{3y^2+2xy}$

14. $\dfrac{x+1}{x^2-y^2}-\dfrac{1}{2x-2y}$

15. $1+\dfrac{1}{a}+\dfrac{1}{a^2}$

16. $\dfrac{1}{x+1}-\dfrac{1}{(x+1)^2}+\dfrac{1}{(x+1)^3}$

17. $x+\dfrac{3x}{x-2}-\dfrac{6}{x-2}$

18. $\dfrac{1}{x^2-4x+4}+\dfrac{2}{x^2+4x+4}-\dfrac{1}{x^2-4}$

19. $\dfrac{x}{x-1}+\dfrac{2}{x^2+x+1}+\dfrac{2x^2+4}{x^3-1}$

20. $\dfrac{x}{x+2}+\dfrac{2}{x-1}-\dfrac{6}{x^2+x-2}$

21. $\dfrac{1}{x^2-1}-\dfrac{1}{x^3-1}$

22. $\dfrac{\dfrac{1}{x^2}-9}{\dfrac{1}{x}-3}$

23. $\dfrac{1 + \dfrac{1}{x}}{1 - \dfrac{1}{x}}$

24. $\dfrac{1 + \dfrac{1}{x}}{1 - \dfrac{1}{x^2}}$

25. $\dfrac{\dfrac{1}{y^2} - \dfrac{1}{x^2}}{\dfrac{x^2}{y^2} - \dfrac{y^2}{x^2}}$

26. $\dfrac{\dfrac{1}{r} + \dfrac{1}{s}}{\dfrac{1}{r} - \dfrac{1}{s}}$

27. $\dfrac{\dfrac{a+b}{a-b} + \dfrac{a-b}{a+b}}{\dfrac{a+b}{a-b} - \dfrac{a-b}{a+b}}$

28. $\dfrac{\dfrac{1}{b} - \dfrac{1}{a}}{\dfrac{1}{a^2} - \dfrac{1}{b^2}}$

10.4 / GRAPHING RATIONAL EXPRESSIONS

We have seen that a rational expression $\frac{p(x)}{q(x)}$ has a numerical value for every value of x except for those values that make the polynomial $q(x)$ equal 0. One way to display all the numerical values that $\frac{p(x)}{q(x)}$ equals for different values of x is to graph the equation $y = \frac{p(x)}{q(x)}$. Then for each x value the corresponding y value is the value of the rational expression $\frac{p(x)}{q(x)}$.

A simple rational expression is $\frac{x}{x-1}$. We have already graphed $y = \frac{x}{x-1}$, but we want to consider the graph again, especially those parts of the graph near the x value 1 and the parts described by very large and very small values of x (far right and far left on the horizontal axis).

If we attempt to evaluate $y = \frac{x}{x-1}$ when $x = 1$, we get 0 in the denominator and conclude that there is no y value when $x = 1$. However, we can compute a y value for every other value of x. If we choose x values near 1 but somewhat larger than 1, we see that the graph shoots up on the right side of $x = 1$ (Fig. 10.1).

x	2	1.7	1.5	1.3	1.1	1.08	1.05
$y = \dfrac{x}{x-1}$	2	2.4	3	4.3	11	13.5	21

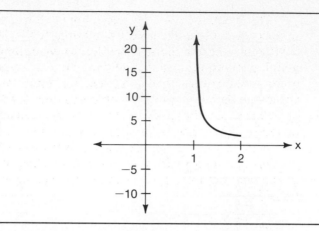

Figure 10.1

On the left side of 1 but close to 1, the y values are negative because the denominator of $\frac{x}{x-1}$ is a negative number while the numerator is positive (Fig. 10.2).

x	0	.3	.5	.7	.9	.92	.95
$y = \dfrac{x}{x-1}$	0	−.4	−1	−2.3	−9	−11.5	−19

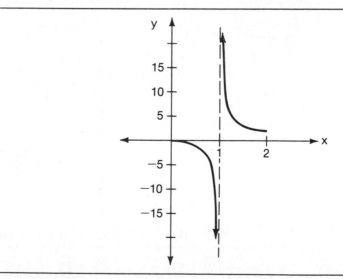

Figure 10.2

The graph of $y = \frac{x}{x-1}$ has no point for $x = 1$. The graph becomes almost vertical for the x values close to 1. We describe this behavior by saying that the graph has a **vertical asymptote** at $x = 1$. This particular graph goes up on one side of the vertical asymptote and down on the other; other graphs with a vertical asymptote may have the two parts of the graph both going up or both going down at the asymptote.

The reason that $y = \frac{x}{x-1}$ has a vertical asymptote at $x = 1$ is that the denominator of this rational expression is 0 when $x = 1$ and the numerator is not 0 when $x = 1$. There are examples where the numerator and denominator of a rational expression are both 0 for some value of x; the graph does not necessarily have a vertical asymptote at that value of x. Consider

$$y = \frac{x^2 - x}{x - 1} = \frac{x(x - 1)}{x - 1}.$$

This equation has no y value when $x = 1$. The denominator is 0, but the numerator is also 0. Plotting points close to $x = 1$ shows that this graph does not have a vertical asymptote at $x = 1$ (Fig. 10.3).

x	0	.5	.7	.9	1.1	1.3	1.5	2
$y = \dfrac{x(x-1)}{x-1}$	0	.5	.7	.9	1.1	1.3	1.5	2

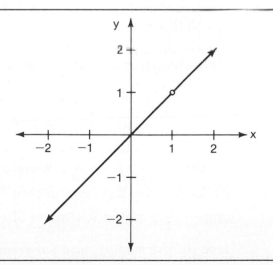

Figure 10.3

In the cases where an x value makes both the numerator and the denominator of a rational expression equal to 0, you will need to consider carefully the points near that x value to see the behavior of the graph.

It is not surprising that the graph of $y = \dfrac{x(x-1)}{x-1}$ is a line with one point missing. Except for $x = 1$, the rational expression $\dfrac{x(x-1)}{x-1} = x$.

Thus, except for $x = 1$, we have graphed the equation $y = x$.

End Behavior of Graphs of Rational Expressions

When we describe the behavior of a graph to the far right and the far left, we sometimes say we have described the **end behavior** of the graph. The division algorithm for polynomials helps us to predict the end behavior of rational expressions.

Example 1 Describe the end behavior of the graph of $y = \dfrac{x}{x-1}$ and draw the complete graph.

a. One way to see how a graph behaves for very large and very small values of x is to plot many points. You have had considerable practice with this method.

b. Since the expression $\frac{x}{x-1}$ means $x \div (x-1)$, we can use the division algorithm to get a quotient and remainder:

$$
\begin{array}{r}
1 \\
x - 1 \overline{)\,x} \\
\underline{x - 1} \\
1 .
\end{array}
$$

Thus $\dfrac{x}{x-1} = 1 + \dfrac{1}{x-1}$. When the rational expression is written in this form, we see that large values of x give y values close to 1, since the term $\dfrac{1}{x-1}$ is almost 0. Observe also that if $y = 1 + \dfrac{1}{x-1}$ and x is a large positive number, then y is greater than 1 because $\dfrac{1}{x-1}$ is a positive number; if x is negative, y is less than 1 because $\dfrac{1}{x-1}$ is a negative number. Thus this graph approaches the line $y = 1$ from above for large x values and approaches the same line from below for small x values (Fig. 10.4). In this case we say the graph has a **horizontal asymptote**, $y = 1$.

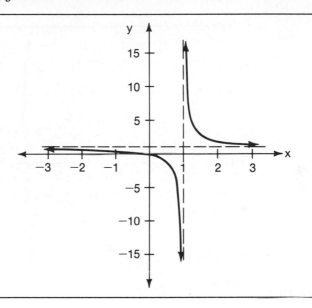

Figure 10.4

□

Example 2 Graph the equation $y = \dfrac{x^2 + 1}{x}$.

The denominator of $\dfrac{x^2 + 1}{x}$ is 0 when $x = 0$. Since the numerator is not 0 when $x = 0$, we expect a vertical asymptote at $x = 0$. We can also predict the end behavior of the graph by rewriting the quotient $\dfrac{x^2 + 1}{x}$:

$$
\begin{array}{r}
x \\
x\ \overline{\smash{)}\ x^2 + 1} \\
\underline{x^2} \\
1\,.
\end{array}
$$

Thus $\dfrac{x^2 + 1}{x} = x + \dfrac{1}{x}$. For large values of x, the expression $\dfrac{1}{x}$ is very small; if $y = \dfrac{x^2 + 1}{x} = x + \dfrac{1}{x}$ and x is a large number, then y equals

x plus a very small number. We expect the graph of $y = \dfrac{x^2 + 1}{x}$ to come close to the graph of $y = x$ to the far right and to the far left. Plotting a few points gives the graph in Fig. 10.5.

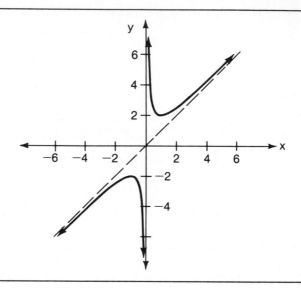

Figure 10.5

Two lines guide this graph: $x = 0$ and $y = x$. Both are said to be asymptotes of the graph; they are not part of the graph of $y = \dfrac{x^2 + 1}{x}$ but we draw them to help sketch the graph of $y = \dfrac{x^2 + 1}{x}$.   □

Example 3   Draw the graph of $y = \dfrac{x^2}{x^2 - 1}$.

Vertical asymptotes: Since $\dfrac{x^2}{x^2 - 1} = \dfrac{x^2}{(x - 1)(x + 1)}$, we see that the denominator of this rational expression is 0 when $x = 1$ and when $x = -1$. The numerator is not 0 when $x = 1$ or when $x = -1$, so the graph has vertical asymptotes at $x = 1$ and $x = -1$. Notice that $y = \dfrac{x^2}{x^2 - 1}$ is positive when $x^2 > 1$ (that is, when $x > 1$ or $x < -1$);

$y = \dfrac{x^2}{x^2-1}$ is negative when $x^2 < 1$ (that is, when x is between -1 and 1). This information about vertical asymptotes describes the graph of $y = \dfrac{x^2}{x^2-1}$ near $x = -1$ and $x = 1$ (Fig. 10.6).

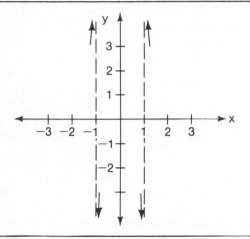

Figure 10.6

Horizontal asymptotes: To understand the end behavior of the graph, we rewrite $\dfrac{x^2}{x^2-1}$ as a polynomial plus a rational expression where this rational expression has the degree of the numerator less than the degree of the denominator:

$$x^2 - 1 \overline{\smash{\big)}\ x^2} \atop {\underline{x^2 - 1}} \atop 1.$$

Thus $y = \dfrac{x^2}{x^2-1} = 1 + \dfrac{1}{x^2-1}$. If x is a large positive number or a small negative value, the corresponding y values will be approximately 1. Thus $y = 1$ is a horizontal asymptote. If x is large and positive or small and negative, $\dfrac{1}{x^2-1}$ will be positive, so y will be greater than 1. Thus the graph approaches $y = 1$ from above at both extremes of the graph. Using this information, we can add the extremes of the graph (Fig. 10.7).

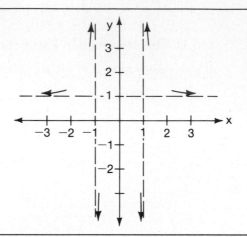

Figure 10.7

Other observations: Notice that $y = 0$ when $x = 0$. Also, negative values of x give the same y values as their opposite positive values. We have already observed that y is negative between -1 and 1 and positive for $x > 1$ and $x < -1$. This information (and perhaps plotting a few points) gives the complete graph, as shown in Fig. 10.8.

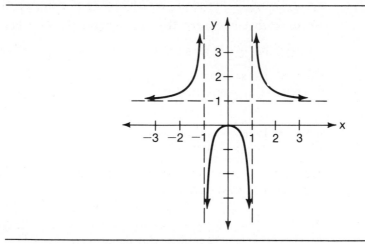

Figure 10.8

□

The three examples in this section have illustrated an important fact about end behavior of graphs of rational equations. The quotient

$q(x)$ obtained by dividing the numerator by the denominator of a rational expression always determines the end behavior of the graph, that is to say, the graph of the rational equation approaches the graph of $y = q(x)$. If $q(x)$ is a polynomial of degree 0 or 1, the graph of $y = q(x)$ is a line and the graph of the rational equation has $y = q(x)$ as an asymptote.

Exercises 10.4

Find the vertical asymptotes, if any, of the rational expressions in Exercises 1–5.

1. $y = \dfrac{x + 3}{x(x + 1)}$

2. $y = \dfrac{(x + 2)^3}{(x + 2)^2(x + 1)}$

3. $y = \dfrac{x(x - 1)^2}{(x - 1)^2(x - 3)}$

4. $y = \dfrac{x(x - 4)}{(x - 4)^2(x + 1)}$

5. $y = \dfrac{x(x - 1)^2}{x + 1}$

Use the division algorithm to analyze the end behavior of the graphs of the equations in Exercises 6–10. If the graph has a nonvertical asymptote, give its equation.

6. $y = \dfrac{2x^2 - 2x + 3}{x^2 - x}$

7. $y = \dfrac{2x + 1}{x^2 + x + 1}$

8. $y = \dfrac{x^2 - x}{2x - 1}$

9. $y = \dfrac{2x^3 - 3x + 1}{x^2 + 4}$

10. $y = \dfrac{2x^5 - 3x^4 + x - 1}{3x^4 - 2x}$

Graph the rational equations in Exercises 11–17. Use dashed lines to represent the asymptotes.

11. $y = \dfrac{x^2 + 1}{x^2}$

12. $y = \dfrac{x^2}{x^2 + x - 2}$

13. $y = \dfrac{2x - 1}{x^2 - x}$

14. $y = \dfrac{1}{x^2 - 4}$

15. $y = \dfrac{x^2 + 1}{x - 1}$

16. $y = \dfrac{x^3 + 1}{x}$

17. $y = \dfrac{x^2 + x}{x + 1}$

10.5 / SOLVING FRACTIONAL EQUATIONS

We have already solved some equations that contain rational expressions. For example, if we have one rational expression equal to another, we can use cross multiplication to get a solution.

Example 1 Solve $\dfrac{5}{x} = \dfrac{4}{3}$.

The equation $\dfrac{5}{x} = \dfrac{4}{3}$ is equivalent to $5 \cdot 3 = x \cdot 4$. Thus we solve $15 = 4x$,

$\dfrac{15}{4} = x$. □

In this section we will develop procedures for solving equations that contain sums and differences of rational expressions. Rather than adding or subtracting the rational expressions, the general method is to multiply both sides of the equation by the least common multiple of the denominators of the rational expressions. This gives an equation with polynomials on both sides, *but* the second equation may not be equivalent to the original equations. It may have some solutions that are not solutions to the original equation. However, any solution to the original equation will appear as a solution to the second equation. The important thing to remember is to check the solutions. Here is an example.

Example 2 Solve the equation $\dfrac{x}{x + 1} + \dfrac{2}{x - 3} = \dfrac{8}{x^2 - 2x - 3}$.

We could compute the sum of the fractions,

$$\dfrac{x}{x + 1} + \dfrac{2}{x - 3},$$

and then use cross multiplication. Another method is to notice that the denominator $x^2 - 2x - 3 = (x + 1)(x - 3)$. Thus the least common denominator of the three denominators, $x + 1$, $x - 3$, and $x^2 - 2x - 3$, is $(x + 1)(x - 3)$. If both sides of the equation are multiplied by $(x + 1)(x - 3)$, the denominators will disappear:

$$\frac{x}{x + 1} + \frac{2}{x - 3} = \frac{8}{x^2 - 2x - 3}$$

$$(x + 1)(x - 3)\left(\frac{x}{x + 1} + \frac{2}{x - 3}\right) = (x + 1)(x - 3)\left(\frac{8}{x^2 - 2x - 3}\right)$$

$$(x + 1)(x - 3)\left(\frac{x}{x + 1}\right) + (x + 1)(x - 3)\left(\frac{2}{x - 3}\right) = 8$$

$$x(x - 3) + 2(x + 1) = 8$$

$$x^2 - 3x + 2x + 2 = 8$$

$$x^2 - x + 2 = 8$$

$$x^2 - x - 6 = 0$$

$$(x - 3)(x + 2) = 0$$

$$x = 3 \quad \text{or} \quad x = -2.$$

When x is replaced by 3 in the original equation, the left side becomes $\frac{3}{3 + 1} + \frac{2}{3 - 3}$. Since the second term has denominator 0, we conclude that 3 is not a solution to the original equation. If we replace x by -2 on both sides, we get

$$\frac{-2}{-2 + 1} + \frac{2}{-2 - 3} = \frac{8}{(-2)^2 - 2(-2) - 3}$$

$$\frac{-2}{-1} + \frac{2}{-5} = \frac{8}{4 + 4 - 3}$$

$$2 - \frac{2}{5} = \frac{8}{5}$$

$$\frac{10}{5} - \frac{2}{5} = \frac{8}{5}.$$

Since this is a true statement, we conclude that -2 is a solution. □

By looking at a simple equation, we can explain what part of the computation in the above example caused 3 to occur as a candidate for a solution when in fact it is not a solution. Consider the equation $x = 1$. It has a single solution 1. However, if both sides of the equation are multiplied by the variable x, we get

$$x^2 = x$$
$$x^2 - x = 0$$
$$x(x - 1) = 0$$
$$x = 0 \quad \text{or} \quad x = 1.$$

The second equation has two solutions, 0 and 1, but only one of these is a solution to the original equation. When we multiply an equation by a polynomial expression that contains the variable, we run the risk of creating an equation that has more solutions than the the original equation. These extra solutions are called **extraneous** solutions. In Example 2, we produced an extraneous solution when we multiplied the original equation by $(x + 1)(x - 3)$. Having done that, both sides of the new equation are 0 when $x = 3$. However, 3 is *not* a solution to the original equation.

Example 3   Solve the equation $\dfrac{x}{x + 1} + \dfrac{2}{2x + 1} = \dfrac{4}{2x^2 + 3x + 1}$.

Since $2x^2 + 3x + 1 = (x + 1)(2x + 1)$, multiplying both sides of the equation by $(x + 1)(2x + 1)$ removes the denominators:

$$\frac{x}{x + 1} + \frac{2}{2x + 1} = \frac{4}{2x^2 + 3x + 1}$$

$$(x + 1)(2x + 1)\left(\frac{x}{x + 1} + \frac{2}{2x + 1}\right) = (x + 1)(2x + 1)\left(\frac{4}{2x^2 + 3x + 1}\right)$$

$$(2x + 1)(x) + (x + 1)(2) = 4$$
$$2x^2 + x + 2x + 2 = 4$$
$$2x^2 + 3x + 2 = 4$$
$$2x^2 + 3x - 2 = 0$$
$$(2x - 1)(x + 2) = 0$$

$$2x - 1 = 0 \quad \text{or} \quad x + 2 = 0$$
$$2x = 1 \qquad x = -2.$$
$$x = \frac{1}{2}$$

We must check both the values $\frac{1}{2}$ and -2 in the original equation:

$$\frac{\frac{1}{2}}{\frac{1}{2} + 1} + \frac{2}{2\left(\frac{1}{2}\right) + 1} = \frac{4}{2\left(\frac{1}{2}\right)^2 + 3\left(\frac{1}{2}\right) + 1}$$

$$\frac{\frac{1}{2}}{\frac{3}{2}} + \frac{2}{1 + 1} = \frac{4}{\frac{2}{4} + \frac{3}{2} + 1}$$

$$\frac{1}{3} + 1 = \frac{4}{3}.$$

Since this statement is true, we conclude that $\frac{1}{2}$ is a solution.

$$\frac{(-2)}{(-2) + 1} + \frac{2}{2(-2) + 1} = \frac{4}{2(-2)^2 + 3(-2) + 1}$$

$$\frac{-2}{-1} + \frac{2}{-3} = \frac{4}{8 - 6 + 1}$$

$$2 - \frac{2}{3} = \frac{4}{3}.$$

Since this statement is true, we conclude that -2 is also a solution. In this example, no extraneous solutions are introduced when we multiply by an expression containing the variable. □

Sometimes in an equation with two variables, we need to solve for one variable in terms of the other. If the equations contain rational expressions, we can use procedures similar to those above.

Example 4   Solve the equation $\dfrac{1}{x} + \dfrac{1}{y} = 2$ for y.

A common denominator for the fractions is xy. Multiplying both sides of the equation by xy removes the denominators:

$$\frac{1}{x} + \frac{1}{y} = 2$$

$$xy\left(\frac{1}{x} + \frac{1}{y}\right) = xy(2)$$

$$y + x = 2xy$$

$$y - 2xy = -x$$

$$y(1 - 2x) = -x$$

$$y = \frac{-x}{1 - 2x}.$$

□

Equations describing problem situations may contain rational expressions. An example is a problem involving distance, rate, and time. Since $d = rt$, if an equation is written in terms of rate ($r = \frac{d}{t}$) or time ($t = \frac{d}{r}$), there are likely to be fractions in the equation.

Example 5   Each weekend Clyde Johnson drives 200 miles from the city to his home town. If he were to increase his speed by 10 mph, he would save 1 hour on the trip. What is his present speed?

We can summarize the information in the problem this way.

	Distance (miles)	Rate (mph)	Time (hours)
At present speed	200	x	$\dfrac{200}{x}$
With increased speed	200	$x + 10$	$\dfrac{200}{x + 10}$

The problem states that the time at the present speed equals the time at increased speed plus 1 hour:

$$\frac{200}{x} = \frac{200}{x + 10} + 1$$

$$x(x + 10)\frac{200}{x} = x(x + 10)\left[\frac{200}{x + 10} + 1\right] \qquad \text{[Multiply both sides by } x(x + 10)\text{]}$$

$$(x + 10)(200) = x(200) + x(x + 10)(1)$$

$$200x + 2000 = 200x + x^2 + 10x$$

$$200x + 2000 = 210x + x^2$$

$$0 = x^2 + 10x - 2000$$

$$0 = (x + 50)(x - 40)$$

$$x = -50 \quad \text{or} \quad x = 40.$$

Since x represents the speed of an automobile, only $x = 40$ can be a solution. In fact, if Clyde is now driving an average of 40 mph and were to increase his speed to 50 mph, he would require 4 hours rather than 5 for his 200-mile trip. □

Exercises 10.5

Solve the equations in Exercises 1–18.

1. $\dfrac{2}{x} = \dfrac{8}{5}$

2. $-\dfrac{10}{3} = \dfrac{5}{x}$

3. $\dfrac{5}{x + 2} = \dfrac{3}{x - 3}$

4. $\dfrac{1}{x} + \dfrac{1}{2} = \dfrac{1}{3}$

5. $\dfrac{y - 1}{y - 4} = \dfrac{y + 1}{y + 3}$

6. $\dfrac{z}{z + 1} = 2$

7. $\dfrac{1}{2}x + \dfrac{1}{3}x - 7 = \dfrac{1}{4}x$

8. $\dfrac{x}{x + 3} - \dfrac{1}{2x - 1} = \dfrac{3x - 5}{2x^2 + 5x - 3}$

9. $\dfrac{x}{x + 4} - \dfrac{2}{x + 3} = \dfrac{3x + 7}{x^2 + 7x + 12}$

10. $\dfrac{x + 3}{x - 2} - \dfrac{5}{x + 2} = \dfrac{15}{x^2 - 4}$

11. $\dfrac{2x + 1}{x - 2} - \dfrac{4}{x - 2} = \dfrac{2}{x^2 - 2x}$

12. $\dfrac{x}{x - 2} + \dfrac{4}{2x + 1} = \dfrac{4}{2x^2 - 3x - 2}$

13. $\dfrac{x}{x-2} - \dfrac{2}{x-1} = \dfrac{x}{x^2 - 3x + 2}$ 14. $2 + \dfrac{3x}{2x-1} = \dfrac{x+1}{x+5}$

15. $3 - \dfrac{x+2}{x+4} = \dfrac{x-1}{x-3}$ 16. $\dfrac{y^2}{(y+1)^2} + \dfrac{y}{y+1} - 6 = 0$

17. $t^{-2} - 3t^{-1} + 2 = 0$ 18. $\dfrac{9}{x^4} - \dfrac{37}{x^2} + 4 = 0$

Solve the equations in Exercises 19–24 for the specified variables.

19. $P = 2l + 2w$; for l.   20. $A = \dfrac{1}{2}bh$; for b.

21. $I = P(1 + r)$; for r.   22. $7x - 2y = 1$; for y.

23. $\dfrac{x}{2} + \dfrac{y}{3} = 1$; for y.   24. $F = g\dfrac{mM}{d^2}$; for m.

25. Bill drives 150 miles in the same time that Jim drives 125 miles. If Bill's speed is 10 mph faster than Jim's, find the rate of each.

26. Jane finds that it takes one hour longer to complete a 360-mile trip by car than by train. If the average speed of the train is 12 mph faster than the car, find the rate of the car and the rate of the train.

10.6 / USING FRACTIONAL EQUATIONS TO SOLVE PROBLEMS

In the last section we described problems involving rate and time with fractional equations. In this section we will analyze other problem situations that can be described with fractional equations.

The first application uses the fact that if finishing a task requires n hours, then after 1 hour, the fraction of the task completed will be $\frac{1}{n}$. For example, if a task requires 2 hours to be completed, then after 1 hour the task will be $\frac{1}{2}$ done. If a task requires 3 hours to be completed, after 1 hour, $\frac{1}{3}$ of the task will be completed. If a task requires 3.5 hours to be completed, in 1 hour $\frac{1}{3.5}$ will be the part of the job finished.

Example 1   Jim can mow the family yard in 2 hours. It takes his younger brother 3 hours. If the two boys mow the yard together, how long will it take them?

We know before we start, since Jim can do the job alone in 2 hours, that the two working together will get it done in less than 2 hours. One way to describe the problem with an equation is to let x represent the number of hours required when working together. Then the fraction of the job that is done in 1 hour when the boys work together is $\frac{1}{x}$. Another way to show the fraction of the task done in 1 hour when the boys work together is to observe that Jim completed $\frac{1}{2}$ of it and his brother completed $\frac{1}{3}$ of it, so together in 1 hour they do $\frac{1}{2} + \frac{1}{3}$ of the total job. Thus $\frac{1}{x} = \frac{1}{2} + \frac{1}{3}$. Both sides describe the fraction of the job done by the two boys working together for 1 hour. We can solve the equation for x as follows:

$$\frac{1}{x} = \frac{1}{2} + \frac{1}{3}$$

$$6 = 3x + 2x \qquad \text{[Multiply both sides by } (2)(3)(x)]$$

$$6 = 5x$$

$$\frac{6}{5} = x.$$

Thus in $\frac{6}{5}$ or $1\frac{1}{5}$ hours, the two boys can complete the lawn mowing.

□

Example 2   If the north drain of the swimming pool is opened, the pool empties in 6 hours. If the south drain only is opened, the pool empties in 5.5 hours. How long does it take the pool to empty if both drains are open?

Let x denote the time required to empty the pool if both drains are open. Then the

portion of pool emptied in 1 hour with both drains open is $\frac{1}{x}$, portion of pool emptied in 1 hour with only the north drain open is $\dfrac{1}{6}$,

portion of pool emptied in 1 hour with only the south drain open is $\dfrac{1}{5.5}$.

Thus

$$\frac{1}{x} = \frac{1}{6} + \frac{1}{5.5}$$

$$6(5.5) = 5.5x + 6x \qquad \text{[Multiply both sides by } (6)(5.5)(x)]$$

$$33 = 11.5x$$

$$\frac{33}{11.5} = x$$

$$2.87 \doteq x.$$

The amount of time required to drain the pool with both drains open is approximately 2.87 hours. □

We learn in physics that an airplane with a speed of 600 miles per hour in still air travels $600 + w$ miles per hour when flying with a wind of w miles per hour. When flying against a wind of w miles per hour, the plane moves at $600 - w$ miles per hour. For example, going with a 15-mph wind, the plane moves at 615 mph; going against a 15-mph wind, the plane moves at 585 mph.

The current in a river has the same effect on the speed of a boat as the wind has on the speed of an airplane. For example, a boat that travels 10 mph in still water will go $10 + 2 = 12$ mph when it goes downstream in a river with a 2-mph current; it will go $10 - 2 = 8$ mph when it goes upstream in the same river.

Example 3   A plane flies 300 mph in still air. It makes a round trip of 1200 miles in a total of $4\frac{1}{4}$ hours, one way with the jet stream and the other way against the jet stream. What is the speed of the jet stream?

Let w denote the speed of the jet stream. The information in the problem can be summarized in this chart:

	Distance (miles)	Rate (mph)	Time (hours)
With jet stream	600	$300 + w$	$\dfrac{600}{300 + w}$
Against jet stream	600	$300 - w$	$\dfrac{600}{300 - w}$

Since the problem states that the total time for the round trip is $4\frac{1}{4}$ hours, the time *with the jet stream* plus the time *against the jet stream* must equal $4\frac{1}{4}$.

$$\frac{600}{300 + w} + \frac{600}{300 - w} = 4.25$$

$$600(300 - w) + 600(300 + w) = 4.25(300 + w)(300 - w)$$

[Multiply both sides by $(300 + w)(300 - w)$]

$$180,000 - 600w + 180,000 + 600w = 4.25(90,000 - w^2)$$

$$360,000 = 382,500 - 4.25w^2$$

$$4.25w^2 = 22,500$$

$$w^2 \doteq 5294.1176$$

$$w \doteq 72.76.$$

The speed of the jet stream is approximately 73 mph.   □

Example 4 A boat that travels 15 mph in still water takes $1\frac{1}{2}$ hours less time to go 60 miles downstream than to return the same distance upstream. What is the rate of the current?

If r denotes the rate of the current, then $15 + r$ is the rate of the boat as it travels downstream and $15 - r$ is the rate as it travels upstream. Since the time required for the trip is distance divided by rate, we can write the time going downstream as $\dfrac{60}{15 + r}$ and the time going upstream as $\dfrac{60}{15 - r}$. All of this information can be summarized this way:

	Distance (miles)	Rate (mph)	Time (hours)
Downstream	60	$15 + r$	$\dfrac{60}{15 + r}$
Upstream	60	$15 - r$	$\dfrac{60}{15 - r}$

The equation that describes the problem is one that states that the time going downstream is $1\frac{1}{2}$ hours less than the time going upstream:

$$\frac{60}{15 + r} = \frac{60}{15 - r} - 1.5$$

$$60(15 - r) = 60(15 + r) - 1.5(15 + r)(15 - r)$$

$$900 - 60r = 900 + 60r - 1.5(225 - r^2)$$

$$900 - 60r = 900 + 60r - 337.5 + 1.5r^2$$

$$0 = 1.5r^2 + 120r - 337.5$$

$$0 = r^2 + 80r - 225$$

$$r = \frac{-80 + \sqrt{80^2 - 4(-225)}}{2} \quad \text{or} \quad r = \frac{-80 - \sqrt{80^2 - 4(-225)}}{2}$$

$$r = \frac{-80 + \sqrt{7300}}{2} \doteq 2.72 \qquad r = \frac{-80 - \sqrt{7300}}{2} \doteq -82.72$$

Since r is the rate of the current, r cannot be negative, so r is approximately 2.72 miles per hour.   □

Exercises 10.6

1. A swimming pool has two pipes to fill the pool with water. The larger pipe can fill the pool in 6 hours and the smaller pipe in 8 hours. How long will it take to fill the pool if both pipes are used?

2. A swimming pool can be filled by the main pipe in 6 hours. If a hose is also used, it takes 5 hours to fill the pool. How long would it take to fill the pool if only the hose were used?

3. A tank can be filled by the inlet pipe in 5 hours and drained by the outlet pipe in 7 hours. If the outlet pipe is accidently left open, how long will it take to fill the tank?

4. John can mow the lawn in 2.5 hours. If it takes Brian 3 hours to mow the lawn, how long will it take them to mow the lawn working together?

5. Tom can paint a house in 12 hours. When his son helps, it only takes them 9 hours. How long would it take his son to paint the house working alone?

6. Joe drove 15 miles to a train station and then took a train 25 miles to work. The rate of the train was 20 mph faster than that of the car. If the total traveling time was 1 hour, find the rate of the car and the rate of the train.

7. A bus traveled 112 miles at one speed and then 96 miles at twice that speed. If the entire trip took 5 hours, what was the original speed?

8. An airplane travels 1375 miles in the same time that a bus travels 121 miles. If the rate of the airplane is 570 mph faster than the bus, find the rates of the airplane and the bus.

9. A plane flies 500 mph in still air. It makes a round trip of 12,320 miles in a total of 25 hours, one way with the jet stream and the other way against the jet stream. Find the speed of the jet stream.

10. A boat can travel at a speed of 24 mph in still water. It makes a round trip of 280 miles in a total of 12 hours, one way with the current and the other way against the current. Find the rate of the current.

11. Sarah lives 20 miles from her place of work. She drives part of the way at 30 mph and walks the remaining distance at 5 mph. If it takes 1 hour and 48 minutes to complete the trip, how far did she drive and how far did she walk?

12. A river is flowing at 6 mph. If it takes as long to go 90 miles downstream as it does to go 60 miles upstream, find the speed of the boat in still water.

13. On a calm day, Joy can pedal her ten-speed bike at 18 mph. When

the wind is blowing, it takes her as long to go 21 miles against the wind as it does to go 33 miles with the wind. Find the rate of the wind.

14.   One number is twice another. Find the numbers if the sum of their reciprocals is 1.

10.7 / VARIATION

In certain situations involving two quantities, an increase in one quantity causes an increase in the other. For example, in the situation of an automobile trip where the average speed is given, we can focus on time traveled and distance traveled. An increase in time produces an increase in distance. An increase in distance means an increase in time. If the average speed is 50 mph and if x represents the number of hours traveled and y the number of miles traveled, then $y = 50x$. This equation shows how the distance and time are related. The situation can be described by saying that *distance varies directly with time*. The phrase means no more than that distance equals a number (called the **constant of variation**) times the time.

> If x and y are variables, then y varies directly with x provided there is a nonzero number k such that $y = kx$.

In this definition x can be replaced with any algebraic expression containing x.

Example 1   The distance required to stop a car varies directly with the square of the speed of the car at the time the brakes are applied. If Brian's Chevrolet requires 190 feet to stop when it is going 50 mph, how much distance will it require when it is going 65 mph?

The first sentence of the example can be summarized in the equation $d = ks^2$ where d denotes distance, s denotes speed at the time the brakes are applied, and k is the constant of variation. The fact that the car stops in 190 feet when it is going 50 mph permits us to find the constant k:

$$190 = k \cdot 50^2$$

$$190 = 2500k$$

$$\frac{190}{2500} = k$$

$$.076 = k.$$

Replacing k by .076 in the equation $d = ks^2$ gives $d = .076s^2$, an equation that permits us to find distance for any speed. In this problem we want to find the stopping distance that corresponds to the speed 65 mph:

$$d = .076s^2$$

$$d = .076(65)^2 = 321.1.$$

Thus Brian's car requires 321.1 feet to stop if the car is going 65 mph when the brakes are applied. □

A second relationship that can occur in a situation involving two quantities is that when one quantity increases, the other decreases. For example, assume that the gear on the pedal of a bicycle has 36 teeth and turns at 60 mph. Consider the number of teeth on the back gear and the speed of the back gear. If the number of teeth is small, then the speed will be fast; if the number of teeth is large, the speed will be slower. Remember that the number of teeth times the speed on one gear must equal the number of teeth times the speed on the other. Thus

$$T \cdot S = (36)(60)$$

$$T = \frac{(36)(60)}{S}$$

$$T = \frac{2160}{S}.$$

Here T is the number of teeth and S is the speed of the back gear. This situation can be described by saying that T *varies inversely with S*. The number 2160 is again called the **constant of variation**.

> If x and y are variables, then y *varies inversely with* x provided there is a nonzero number k such that $y = \dfrac{k}{x}$.

Again, x in the definition can be replaced with any algebraic expression containing x.

Example 2   The intensity of sound (loudness) varies inversely with the square of the distance from the sound. If a rock band measures 150 decibels at 50 feet, what does it measure at 5 feet?

The information in the first sentences gives the equation

$$L = \frac{k}{D^2}$$

where L denotes the number of decibels and D is the distance from the sound. The constant of variation in this problem can be found from the information that $L = 150$ when $D = 50$:

$$L = \frac{k}{D^2}$$

$$150 = \frac{k}{50^2}$$

$$150(50^2) = k \qquad \text{(Multiply both sides by } 50^2\text{)}$$

$$375{,}000 = k.$$

Replacing k by 375,000 in the equation $L = \dfrac{k}{D^2}$ gives a way to find L for any D. For example, say we want L when $D = 5$:

$$L = \frac{375,000}{D^2}$$

$$L = \frac{375,000}{5^2} = 15,000.$$

Thus at 5 feet, the sound intensity is 15,000 decibels. Beware!   □

You should observe that saying y varies inversely with x (that is, $y = \frac{k}{x}$) means the same as saying y varies directly with $\frac{1}{x}$ (that is, $y = k \cdot \frac{1}{x}$). We can use either phrase.

Variation Involving More Than Two Variables

In a situation involving three variables, it is possible that one variable will vary directly with another and inversely with a third. The language has the same meaning as in the cases with two variables.

Example 3   The variable z varies directly with x and inversely with the square of y. If $z = 9$ when $x = 12$ and $y = 2$, find z when $x = 20$ and $y = 4$.

If k is the constant of variation, the first sentence says that

$$z = k \cdot \frac{x}{y^2}.$$

We can find the value of k from the information that $z = 9$ when $x = 12$ and $y = 2$:

$$9 = k \cdot \frac{12}{2^2}$$

$$9 = 3k$$

$$3 = k.$$

Thus the relationship between the variables is given by $z = 3 \cdot \frac{x}{y^2}$. Now we can find z when $x = 20$ and $y = 4$.

$$z = 3 \cdot \frac{x}{y^2}$$

$$= 3 \cdot \frac{20}{4^2}$$

$$= \frac{60}{16}$$

$$= 3.75.$$

□

The procedure for solving variation problems can be summarized this way:

- Write an equation describing the type of variation and using the constant k.
- Use information in the problem to find the value of k.
- Replace k in the equation by its value and use the equation to answer the question in the problem.

Be sure you can find these three steps in the example above.

Example 4 If a string is stretched tight, fastened at both ends, and plucked, the rate of vibration varies directly as the square root of the tension (from plucking) and inversely as the product of its length and diameter. A string of diameter .01 feet and length 3 feet vibrates 28 times per second if it is plucked with a force of 10 pounds. What would be the rate of vibration for the same string if it were plucked with a force of 5 pounds?

We must assign several variables.

V: rate of vibration
t: tension on string
l: length of string
d: diameter of string.

If k is the constant of variation, then from the first sentence of the problem,

$$V = k \cdot \frac{\sqrt{t}}{ld}.$$

The problem states that $V = 28$ when $d = .01$, $l = 3$, and $t = 10$. This information lets us find the value of k:

$$V = k \cdot \frac{\sqrt{t}}{ld}$$

$$28 = k \cdot \frac{\sqrt{10}}{(3)(.01)}$$

$$28 \doteq k(105.41)$$

$$.266 \doteq k.$$

Thus the equation that describes the vibration for any string is

$$V = \frac{.266\sqrt{t}}{ld}.$$

We want to find V when $t = 5$, $l = 3$, and $d = .01$:

$$V = \frac{.266\sqrt{5}}{3(.01)} \doteq 19.8.$$

Thus under a tension of 5 pounds, the string will vibrate approximately 19.8 times per second.   □

Exercises 10.7

In Exercises 1–6, tell how the specified variable varies with the remaining variables and give the constant of variation for each.

1. $C = 2\pi r$; C

2. $A = \pi r^2$; A

3. $d = 16t^2$; d

4. $I = \dfrac{12}{d^2}$; I

5. $V = \dfrac{22}{7} r^2 h$; V

6. $V = \dfrac{22}{7} r^2 h$; h

Write an equation to specify the variation described in Exercises 7–12.

7. The pressure acting at a point in a liquid varies directly with the depth and the density of the liquid.

8. The volume of a cylinder varies directly with the square of the radius and the height of the cylinder.

9. The work done by a force moving an object varies directly with the mass of the object and the distance it is moved.

10. The surface area of a sphere varies directly with the square of the radius of the sphere.

11. z varies directly with the square root of x and inversely with the cube of y.

12. z varies directly with the square of x and inversely with the cube of y.

Write an equation to specify each variation described in Exercises 13–17, and then find the constant of variation.

13. The length of a shadow varies directly with the height of an object. A 4-foot pole casts a 7-foot shadow.

14. The distance required to stop a car varies directly with the square of the speed of the car. It takes 270 feet to stop a car going 60 mph.

15. y varies directly with the cube of x and $y = 4$ when $x = 2$.

16. z varies directly with the square of x and inversely with the cube of y; $z = 4$ when $x = 2$ and $y = 3$.

17. z varies directly with the square root of x and inversely with the cube of y; $z = 6$ when $x = 9$ and $y = 4$.

18. If y varies directly with x and $y = 6$ when $x = 2$, find y when $x = 4$.

19. If y varies inversely with x and $y = 3$ when $x = 2$, find y when $x = 6$.

20. z varies directly with the square root of x and inversely with the cube of y. If $z = 5$ when $x = 4$ and $y = 1$, find z when $x = 9$ and $y = 2$.

21. w varies directly with the product of x and y and inversely with the square of z. If $w = 25$ when $x = 3$, $y = 5$, and $z = 2$, find w when $x = 4$, $y = 3$, and $z = 4$.

22. The rate of vibration of a stretched string varies directly with the square root of the tension and inversely with the product of its length and diameter. The rate of vibration of a string with a length of 60 cm and a diameter of .0025 cm is 30 per second when a tension of 6.5×10^6 dynes is applied. Find the rate of vibration of a string that is 50 cm long and has a diameter of .004 cm if the tension applied is 7.6×10^8 dynes.

23. Newton's law of gravitation states that any two objects attract each other with a force that varies directly with the product of their masses and inversely with the square of the distance between them. If objects of mass 5 gm and 300 gm at a distance of 50 cm attract each other with a force of 4×10^{-8} dynes, find the force of attraction between two objects with masses of 15 gm and 70 gm that are at a distance of 30 cm.

24. Galileo's law states that, neglecting air resistance, the distance a freely falling body falls from rest varies directly with the square of the time of fall. If a body falls 64 feet in 2 seconds, how far does the body fall in 4 seconds? How far does it fall in 8 seconds?

25. The volume a gas occupies varies directly with its temperature and inversely with its pressure. If a gas occupies 2 cubic feet at a temperature of 65° and a pressure of 6.875 pounds per square inch, find the volume if the temperature is 70° and the pressure is 14.7 pounds per square inch.

26. The intensity of illumination from a source of light varies inversely with the square of the distance from the source. If a light has an intensity of 1000 candlepower at 50 feet, what is the intensity at a distance of 200 feet?

27. Hooke's law states that the distance a hanging spring is stretched varies directly with the weight of an attached object. If a weight of 6 pounds stretches the spring a distance of 4.5 inches, how far will a weight of 10 pounds stretch the spring?

10.8 / CHAPTER 10 PROBLEM COLLECTION

Perform the indicated operations in Exercises 1–14 and express your answers in reduced form.

1. $\dfrac{1}{x^2 y}(x^3 y^2)$

2. $\dfrac{x^4 y^2 z}{x^2 y^3 z^2} \cdot \dfrac{x y^2 z}{x^3 y^3 z^2}$

3. $\dfrac{a^2 b^3}{c^3 d^2} \div \dfrac{a b^4}{c^4 d}$

4. $\dfrac{1}{b - a}(a^2 - b^2)$

5. $\dfrac{x^3 - x^2}{2x^2 + 5x - 3} \cdot \dfrac{4x^2 - 4x + 1}{x^3 - x}$

6. $\dfrac{x^2 - x}{2x - x^2} \cdot \dfrac{x^2 - 2x}{x - x^2}$

7. $\dfrac{x^2 + 6x + 9}{3x^2 + 2x - 1} \div \dfrac{x^2 - 9}{3x^2 - 10x + 3}$

8. $\dfrac{\dfrac{x^2 - 4x + 4}{x^3 - 1}}{\dfrac{x^2 - 2x}{x^3 + x^2 + x}}$

9. $\dfrac{3}{x^2 + 2x} - \dfrac{2}{x^2 + 4x + 4}$

10. $\dfrac{2x}{x + 3} - \dfrac{1}{x - 2} + \dfrac{5}{x^2 + x - 6}$

11. $\dfrac{1}{x + 2} + \dfrac{3}{x^2 - 2x} - \dfrac{12}{x^3 - 4x}$

12. $\dfrac{4 - \dfrac{1}{x^2}}{2 - \dfrac{1}{x}}$

13. $\dfrac{1 - x^{-2}}{x^{-1} + x^{-2}}$

14. $\dfrac{x^{-1} - y^{-1}}{x^{-2} - y^{-2}}$

Find the vertical and horizontal asymptotes of the rational functions in Exercises 15–17.

15. $y = \dfrac{x^2 - 2x + 1}{x^2 - x}$

16. $y = \dfrac{x}{x^2 + x - 6}$

17. $y = \dfrac{x^2 + 1}{x + 1}$

Use the division algorithm to analyze the end behavior of the graphs of the equations in Exercises 18 and 19. If the graph has a nonvertical asymptote, give its equation.

18. $y = \dfrac{x^3 + x^2 - 1}{x - 2}$

19. $y = \dfrac{x^2 + 2x - 1}{x^2 - 1}$

Graph the rational equations in Exercises 20–22. Use dashed lines to represent the asymptotes.

20. $y = \dfrac{x - 2}{x - 4}$

21. $y = \dfrac{x}{4x^2 - 9}$

22. $y = \dfrac{1}{x^2 + 1}$

Solve the equations in Exercises 23–29.

23. $\dfrac{x + 1}{x + 2} = \dfrac{x - 3}{x + 4}$

24. $\dfrac{1}{x + 3} + \dfrac{x + 1}{x^2 + x - 6} = 0$

25. $\dfrac{x}{x + 2} - \dfrac{3}{3x - 1} = \dfrac{6x - 9}{3x^2 + 5x - 2}$

26. $\dfrac{x}{x - 1} - \dfrac{4}{x + 1} = \dfrac{8}{x^2 - 1}$

27. $\dfrac{x - 4}{x + 5} + \dfrac{2}{x - 3} = \dfrac{18}{x^2 + 2x - 15}$

28. $\dfrac{1}{x} + \dfrac{1}{y} = 2$; for y

29. $x^2 + xy = y + 1$; for y

30. A swimming pool has two pipes to fill the pool and one pipe to drain it. The pool can be filled in 5 hours by the larger pipe, in 7 hours by the smaller pipe, and drained in 4 hours by the drain pipe. If all three pipes are open, how long will it take to fill the pool?

31. Pam drove 10 miles to a train station and then she traveled by train 25 miles to visit her family. The rate of the train was 10 mph faster than that of the car. If the total travel time was 45 minutes, find the rates of the car and the train.

32. A river is flowing at 4 mph. If it takes as long to go 72 miles downstream as it does to go 54 miles upstream, find the speed of the boat in still water.

33. A plane flies 390 mph in still air. It makes a round trip of 5040 miles in a total of 13 hours, one way with the jet stream and the other way against the jet stream. Find the speed of the jet stream.

34. If y varies directly with x^3 and $y = 12$ when $x = 2$, find y when $x = 3$.

35. If y varies inversely with the cube of x and $y = 5$ when $x = 2$, find y when $x = 3$.

36. z varies directly with the square of y and inversely with the square root of x. If $z = 6$ when $x = 0.81$ and $y = 2$, find z when $x = 4$ and $y = 1.2$.

37. w varies directly with the square root of the product of x and y and inversely with the square of z. If $w = 4$ when $x = 3$, $y = 12$, and $z = 3$, find w when $x = 5$, $y = 5$, and $z = 2$.

38. The electrical resistance of a wire varies directly with its length and inversely with the square of its diameter. If a wire 20 feet long, with a diameter of 0.01 inches has a resistance of 5 ohms, find the resistance in a wire of the same material that has a diameter of 0.02 inches and a length of 50 feet.

39. The area of a circle varies directly with the square of the radius. What happens to the area if the radius is doubled? What happens if the radius is tripled?

40. The volume of a sphere varies directly with the cube of its radius. What happens to the volume if the radius is doubled? What happens if the radius is tripled?

Measurement Geometry and Trigonometry

11.1 / AREA OF GEOMETRIC FIGURES

To describe the area of a geometric figure, we first agree that a square with sides that are each 1 unit in length has an area of 1 square unit. This square is called a **unit square**. Its area could be 1 square inch, 1 square meter, 1 square mile, or simply 1 square unit, depending on the unit that is used to measure the side of the square.

The area of any geometric figure is the number of nonoverlapping unit squares that are required to fill the figure completely. Area need not be a whole number. For example, there are figures that have areas of 3.5 square inches, 2.25 square meters, and $4\frac{2}{3}$ square miles. Further, area is never a negative number.

As we have seen in earlier sections, in a rectangle whose sides have whole number lengths, we can count the number of unit squares needed to fill the rectangle.

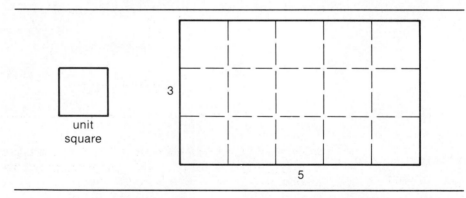

Figure 11.1

The area of the rectangle shown in Fig. 11.1 is 15 square units. The area can be computed by multiplying the length of the base of the rectangle times its height.

The area for any rectangle is, in fact, computed by multiplying the length of the base of the rectangle times the height. We write this fact as the formula $A = bh$ (rectangle).

Area of Parallelograms

Knowing how to compute the area of a rectangle helps us to find shortcuts for computing areas of other figures. For example, it would generally be difficult to estimate the number of unit squares needed to fill a parallelogram. However, a parallelogram can be cut and reassembled to form a rectangle with the same area. The idea is to cut a right triangle off one end of the parallelogram and move it to the other end (Fig. 11.2).

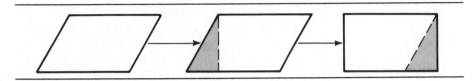

Figure 11.2

The rectangle that results has area equal to the base times the height, so the parallelogram has the same area. It is important to see where these two measurements are made in the parallelogram.

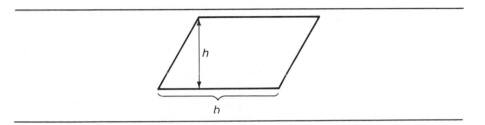

Figure 11.3

The height of the parallelogram is the perpendicular distance between two bases (Fig. 11.3). With this understanding, we can write $A = bh$ (parallelogram).

Area of Triangles

A right triangle is a triangle with one right angle (that is, an angle of 90°). A right triangle can always be viewed as half of a rectangle (Fig. 11.4).

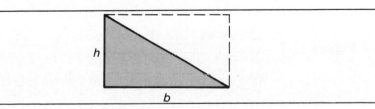

Figure 11.4

Thus the area of a right triangle is $\frac{1}{2}$ times the area of a rectangle, or $\frac{1}{2}bh$. Any triangle can be viewed as half a parallelogram (Fig. 11.5).

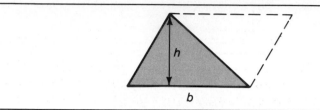

Figure 11.5

Thus the area of the triangle is $\frac{1}{2}$ times the area of the parallelogram, or $\frac{1}{2}bh$. The height is measured on a perpendicular from a vertex of the triangle to the opposite base. Sometimes this perpendicular lies outside the triangle (Fig. 11.6).

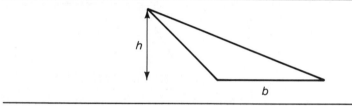

Figure 11.6

The formula for the area of a triangle is the same in every case: $A = \frac{1}{2}bh$ (triangle).

Area of Other Figures

Geometric figures that have their boundaries made up of line segments can be cut into triangles. Consider the three figures in Fig. 11.7.

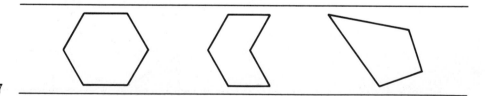

Figure 11.7

One way to cut each of these figures into triangles is shown in Fig. 11.8.

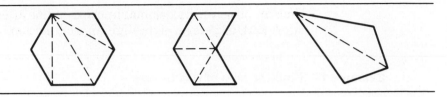

Figure 11.8

Thus it is possible to compute the area of such figures, called **polygons**, by cutting the figure into triangles, finding the areas of the triangles, and taking their sum. This may be a time-consuming task requiring the measuring of bases and heights for several triangles, but it provides one way to find the area of these figures.

Figures that do not have boundaries made of line segments require other techniques for finding area. You can see that finding the number of unit squares that exactly fills each of the figures in Fig. 11.9 is a different kind of problem than finding the areas of polygons.

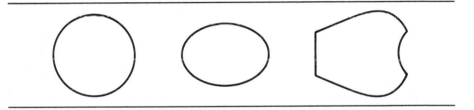

Figure 11.9

Formulas are known for computing the areas of some of these figures. The most common is probably the circle. The formula for the area of a circle is $A = \pi r^2$ (circle). In this formula, r denotes the radius of the circle; π is the number obtained when the circumference (perimeter) of a circle is divided by its diameter (Fig. 11.10).

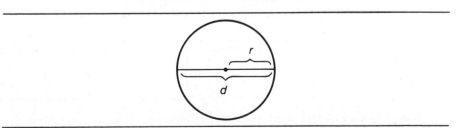

Figure 11.10

The number π is not a rational number. If your calculator has a π key, you can observe the decimal approximation used for π by your calculator: 3.1415927 is an eight-digit approximation.

Example 1   Find the area of this figure.

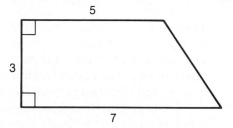

We want to cut the figure into smaller figures whose areas can be readily computed. One way is to cut it into a rectangle and a triangle as shown.

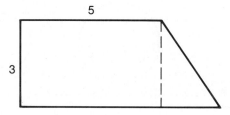

To compute the area of the triangle, we need its base and height. These measurements can be determined by the lengths of the sides in the original sketch.

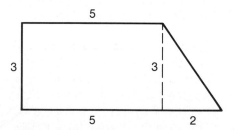

The rectangle has area $5 \cdot 3 = 15$; the triangle has area $\frac{1}{2} \cdot 2 \cdot 3 = 3$. Thus the original figure has area $15 + 3 = 18$.   □

Example 2   Find the area of this figure.

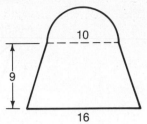

This figure has been obtained by cutting the top off a triangle and replacing it with half a circle (a **semicircle**). To find the area, we can cut the figure into the semicircle, a parallelogram, and a triangle.

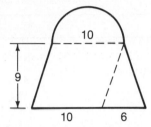

The semicircle has area $\frac{1}{2}(\pi \cdot 5^2) = \frac{25}{2}\pi$. The parallelogram has base 10 and height 9, so it has area $(10)(9) = 90$. The triangle has base 6 and height 9, so it has area $\frac{1}{2}(6)(9) = 27$. Thus the total area is

$$\frac{25}{2}\pi + 90 + 27 = \frac{25}{2}\pi + 117$$

$$\doteq 156.27. \qquad \square$$

Example 3   Find the area of the triangle that has vertices $(-4,5)$, $(-2,1)$, and $(3,1)$.

Regard the line segment joining $(-2,1)$ and $(3,1)$ as the base of the triangle (Fig. 11.11). Using the horizontal axis as a measure, it is apparent that the base has a length of 5. This length can be computed by taking the difference of the two x-coordinates: $3 - (-2) = 5$. The height of the triangle is measured on the perpendicular from the vertex $(-4,5)$ to the line containing the base. Using the vertical axis to measure, you should see that the height is 4 (Fig. 11.12). This can be computed by taking the difference of the two y-coordinates: $5 - 1 = 4$.

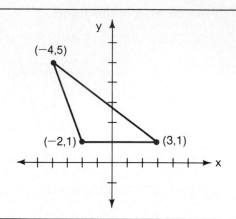

Figure 11.11

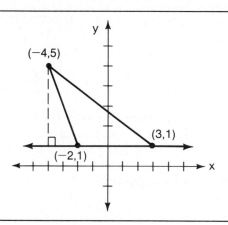

Figure 11.12

Thus the area of the triangle is given by

$$A = \frac{1}{2}bh$$

$$= \frac{1}{2}(5)(4)$$

$$= 10.$$ □

Exercises All parts of the figures below that appear to be circles and semicircles
11.1 are, in fact, circles and semicircles; all angles that appear to be right

angles are right angles; all segments that appear to be parallel are parallel. Find the areas of the figures in Exercises 1–8.

1.
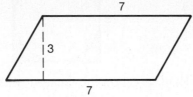

2.
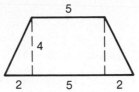

3.
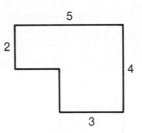

4.
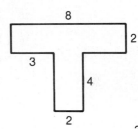

5.
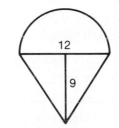

6.
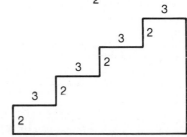

7.
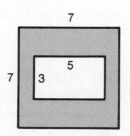

8.
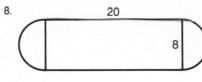

Find the areas of the shaded regions in Exercises 9–13.

9.
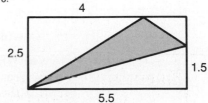

10.
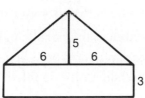

11. 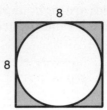   12.   13.

14. Find the width of a rectangle whose area is 90 and whose length is 12.

15. Find the height of a triangle whose area is 16 and whose base is 4.

16. Find the radius of a circle whose area is 9π.

17. Find the height of a parallelogram whose area is 75 and whose base is 12.

18. The base of a triangle is 5 and the height is 6.

   (a) Compute the area of the triangle.

   (b) Compute [and compare with (a)] the area of the triangle if the base and height are doubled.

   (c) Compute [and compare with (a)] the area of the triangle if the base and height are tripled.

19. By what factor is the area of a triangle increased if the base and height are doubled? Tripled?

20. The base of a parallelogram is 4 and the height is 7.

   (a) Compute the area of the parallelogram.

   (b) Compute [and compare with (a)] the area of the parallelogram if the base and height are doubled.

   (c) Compute [and compare with (a)] the area of the parallelogram if the base and height are tripled.

21. By what factor is the area of a parallelogram increased if the base and height are doubled? Tripled?

22. One inch on a map represents 100 feet in the real world.

   (a) What is the area in the real world in square feet if a rectangular garden measures 1.5 inches by 2.6 inches on the map?

(b) A rectangular garden measures 225 feet by 350 feet in the real world. What is the area of the garden on the map in square inches?

23. The area of a rectangle is 180 square feet. How many square inches is that?

24. The area of a rectangle is 32,256 square inches. How many square feet is that?

Find the areas of the polygons determined by the points in Exercises 25–33.

25. (1,1), (2,5), (6,1)

26. (2,1), (2,5), (6,4)

27. (−2,1), (0,4), (3,1)

28. (−3,−4), (−3,3), (2,−1)

29. (−3,0), (0,2), (3,0), (0,−2)

30. (−4,−3), (−4,2), (0,5), (4,2), (4,−3)

31. (−5,−3), (−5,3), (0,7), (5,3), (5,−3), (0,−7)

32. (0,0), (2,3), (7,3), (5,0)

33. (−4,−2), (−2,2), (5,2), (3,−2)

11.2 / APPROXIMATING DISTANCE WITH RULER MEASUREMENT

Finding the areas of geometric figures involves determining lengths of line segments. In Section 11.3 we will develop a method for computing the exact length of a line segment in terms of the coordinates of its endpoints. In this section we will approximate the lengths of line segments using rulers and applying what we already know about ratios.

If two points lie on a line parallel to the x-axis, the graph scale distance between them is the difference of the two x-coordinates. If two points lie on a line parallel to the y-axis, the graph scale distance between them is the difference of their y-coordinates. A more interesting problem is to find the distance between two points that do not lie on a line parallel to one of the axes. Assume, for example, that the scale of the coordinate system is taken so that 1 graph unit is $\frac{1}{4}$ inch on a ruler. Then we can use a ruler to measure the number of inches between two points and convert that number to units of the coordinate system.

Example 1   Approximate the graph scale distance d between points (2,1) and (4,5).

On this graph paper, $\frac{1}{4}$ inch represents 1 graph unit (Fig. 11.13). The ruler distance between (2,1) and (4,5) is approximately $1\frac{1}{8}$ inches. To find the number d of graph units represented by $1\frac{1}{8}$ inches, we can use the proportion

$$\text{inches} : \text{graph units} \qquad \text{inches} : \text{graph units}$$

$$\frac{1}{4} : 1 \qquad = \qquad 1\frac{1}{8} : d$$

$$\frac{1}{4}d = 1\frac{1}{8}$$

$$d = 1\frac{1}{8} \div \frac{1}{4} = 4.5 \text{ units.} \qquad \square$$

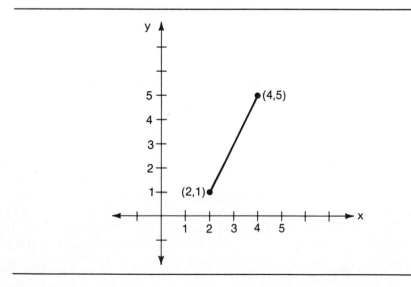

Figure 11.13

We could have obtained a better approximation to the graph scale distance d between two points in Example 2 by using a bigger scale on the axes. For example, if 1 inch represents 1 graph unit (Fig. 11.14), the ruler distance turns out to be $4\frac{15}{32}$ inches, so the graph scale distance d between the points is approximately 4.46875 units.

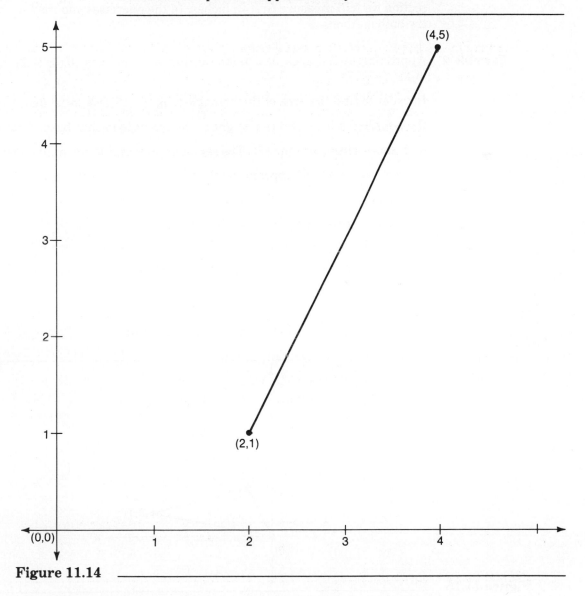

Figure 11.14

Of course we do not need to measure with a ruler and then use a proportion to convert inches to graph units. We could take a second piece of graph paper and use it to measure directly using graph units. However, the resulting approximation is likely to be less accurate than that obtained from a more finely divided ruler. In the examples in this section we will measure first in inches. In the exercises you may wish to try both methods.

Example 2   Approximate the area of a triangle that has vertices (0,0), (1,2), and (5,1).

In order to find the area of this triangle (Fig. 11.15), we must determine the length of a base and the height of the triangle to that base. Again, $\frac{1}{4}$ inch represents 1 graph unit. The segment with endpoints (0,0) and (5,1) has a ruler length of approximately $1\frac{5}{16}$ inches. Thus the approximate scale length of the base b is given by the proportion

inches : graph units   inches : graph units

$$\frac{1}{4} : 1 \qquad = \qquad 1\frac{5}{16} : b$$

$$\frac{1}{4}b = 1\frac{5}{16}$$

$$b = 1\frac{5}{16} \div \frac{1}{4} \doteq 5.25 \text{ units.}$$

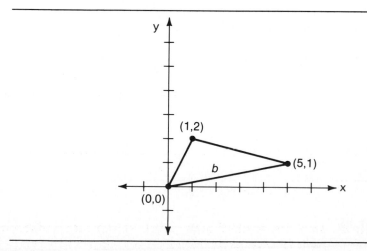

Figure 11.15

One way to sketch a perpendicular line from the vertex (1,2) to the opposite side (Fig. 11.16) is to lay one side of a sheet of paper along the base, letting another side of the paper go through (1,2). The sheet of paper gives a right angle. The height of the triangle measures $\frac{7}{16}$ inches. Thus the approximate number of scale units is given by

inches : graph units   inches : graph units

$$\frac{1}{4} : 1 \qquad = \qquad \frac{7}{16} : h$$

$$\frac{1}{4}h = \frac{7}{16}$$

$$h = \frac{7}{16} \div \frac{1}{4} \doteq 1.75 \text{ units.}$$

The number of square units in the triangle is $\frac{1}{2}bh \doteq \frac{1}{2}(5.25)(1.75) \doteq 4.6.$

□

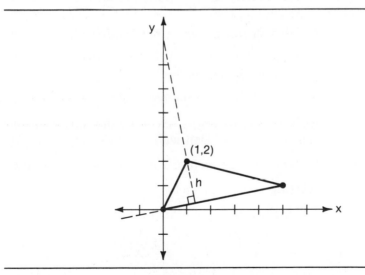

Figure 11.16

Example 3   Dan first walks 5 miles due east and then walks 4 miles at 30° north of east. How far is he from his starting point?

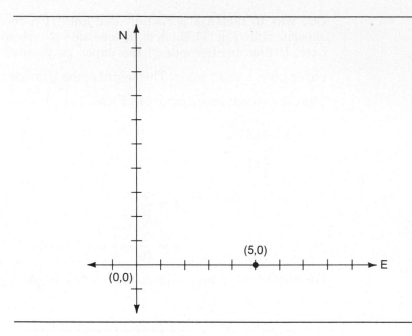

Figure 11.17

On this graph paper $\frac{1}{4}$ inch represents 1 mile (Fig. 11.17). To get a picture of Dan's route, we start at (0,0) and go to (5,0). This represents the 5 miles he walks due east. Now we use a protractor to draw a 30° angle at (5,0) using the horizontal axis as one ray (Fig. 11.18). Since $\frac{1}{4}$ inch represents 1 mile, we locate the point P one inch from (5,0) to represent the 4 miles Dan walks in this direction. The ruler distance from (0,0) to P is $2\frac{5}{32}$ inches. Thus the distance in miles that Dan is from his starting point is given by

inches : miles   inches : miles

$$\frac{1}{4} : 1 \quad = \quad 2\frac{5}{32} : d$$

$$\frac{1}{4}d = 2\frac{5}{32}$$

$$d = 2\frac{5}{32} \div \frac{1}{4} = 8.625 \text{ miles.}$$

□

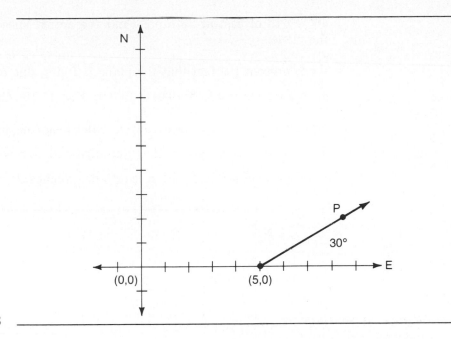

Figure 11.18

Later in this chapter when we compute distance exactly rather than approximating by scale measurement, we will compute the exact distance in Example 3 accurate to 7 decimal places as 8.6971844 miles. Of course, no one measures distance in miles accurate to 7 decimal places, but in the world of computation we pretend such accuracy is possible.

Navigation is an area in which approximating distance by scale measurement is useful. In earlier problems we have used the fact that a plane that travels 500 mph in still air will have an actual speed of (500 + 70) mph if it moves with a 70-mph jet stream; it will have an actual speed of (500 − 70) mph if it moves against a 70-mph jet stream. If the jet stream acts at an angle to the flight path of the plane, then its effect on the speed and direction of the plane can be represented very much like Dan's walking path in Example 3. The next example illustrates what we know from physics about the effect of wind speed and direction on the speed and direction of an airplane.

Example 4   A plane starts due east at 500 mph. A wind is blowing in the direction

40° north of east at 75 mph. Find the actual speed and direction of the plane.

We represent the fact that the plane is flying due east at 500 mph by drawing an arrow from (0,0) to (500,0) (Fig. 11.19). On this graph $\frac{1}{2}$ inch represent 100 mph, so the arrow has ruler length $2\frac{1}{2}$ inches. To represent the direction and speed of the wind, draw an arrow starting at (500,0) at an angle of 40° and with a length of $\frac{3}{8}$ inches (to represent 75 mph).

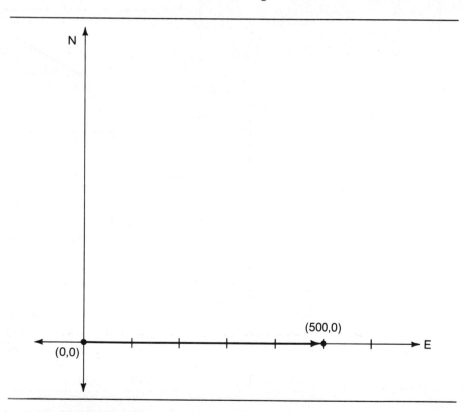

Figure 11.19

The arrow from (0,0) to P in Fig. 11.20 represents the speed and direction of the plane. A protractor measures the angle at the origin as 5°. The ruler length of the arrow is $2\frac{25}{32}$ inches. Since $\frac{1}{2}$ inch represents 100 mph, the speed of the plane is given by

inches : mph   inches : mph

$$\frac{1}{2} : 100 \;\; = \;\; 2\frac{25}{32} : s$$

$$\frac{1}{2}s = 278.125$$

$$s = 556.25 \text{ mph.}$$

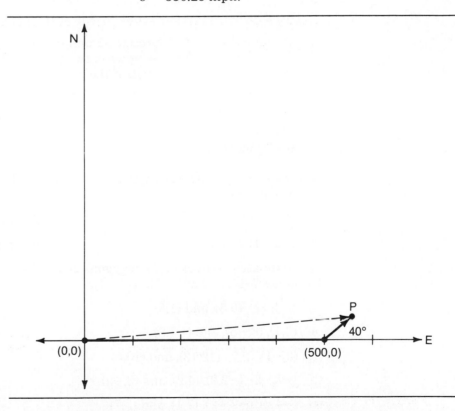

Figure 11.20

Thus the wind causes the plane to fly at 5° north of east at a speed of 556.25 mph. If the plane in this example actually wanted to go due east, it would need to head south of east to compensate for the wind. □

Exercises 11.2

1. The ruler distance between two points on a piece of graph paper is $2\frac{1}{2}$ inches. Determine the distance between these points in terms of graph-paper units if

   (a) $\frac{1}{4}$ inch represents 1 unit on this graph paper

   (b) $\frac{1}{2}$ inch represents 1 unit on this graph paper

   (c) 1 inch represents 1 unit on this graph paper

   (d) 2 inches represent 1 unit on this graph paper

2. Assume that $\frac{1}{2}$ inch represents 1 unit on a piece of graph paper. Determine the distance between two points on this graph paper in terms of graph-paper units if the ruler distance is

   (a) $5\frac{1}{2}$ inches   (b) $2\frac{1}{4}$ inches

   (c) $3\frac{7}{8}$ inches

Approximate the distances between the points in Exercises 3–6 in terms of graph-paper units:

3. $(1,2)$ and $(5,5)$ 4. $(-4,1)$ and $(8,6)$

5. $(-2,-1)$ and $(4,5)$ 6. $(-2,3)$ and $(5,-1)$

Approximate the areas of the polygons determined by the points in Exercises 7–11.

7. $(-3,-1)$, $(6,5)$, and $(1,6)$

8. $(0,1)$, $(8,-5)$, and $(-1,8)$

9. $(0,-1)$, $(2,7)$, $(12,13)$, and $(10,5)$

10. $(-3,-2)$, $(-2,2)$, $(4,2)$, and $(3,-2)$

11. $(-4,4)$, $(-2,-2)$, $(7,1)$, and $(5,7)$

12. Jim first walks 6 miles due east and then walks 2 miles at 70° north of east. How far is he from his starting point?

13. Sarah first walks 4 miles due west, then 3 miles at 50° north of east, and then 1 mile due north. How far is she from her starting point?

14. A plane starts 30° north of east at 400 mph. A wind is blowing due north at 50 mph. Approximate the actual speed and direction of the plane.

15. A plane starts due north at 600 mph. A wind is blowing in the direction 35° north of east at 60 mph. Find the actual speed and direction of the plane.

16. The point B lies on the line determined by $A = (-1,-2)$ and $C = (5,4)$. Estimate the coordinates of B if the length of $\overline{AB}$ is

   (a) $\frac{1}{2}$ the length of $\overline{AC}$ (b) $\frac{1}{3}$ the length of $\overline{AC}$

   (c) $\frac{2}{3}$ the length of $\overline{AC}$ (d) $\frac{3}{2}$ the length of $\overline{AC}$

17. Approximate x if $(x,1)$ is 3 units away from the origin.

11.3 / THE PYTHAGOREAN THEOREM AND THE DISTANCE FORMULA

In the last section we approximated the distance between two points by measuring the line segment with a ruler and then using the scale of the graph to convert the ruler measurement to units of distance. There is a property of right triangles that provides a way of computing distance exactly. This property, which we will study in this section, was known to the ancient Egyptians and Babylonians; because its proof is credited to Pythagoras and his school in the fifth century B.C., the statement is called the Pythagorean theorem. You should remember that in a right triangle, the side opposite the right angle is called the **hypotenuse**, and the other two sides are called **legs** of the triangle.

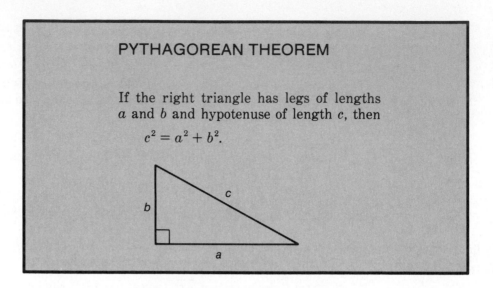

PYTHAGOREAN THEOREM

If the right triangle has legs of lengths a and b and hypotenuse of length c, then

$$c^2 = a^2 + b^2.$$

The Pythagorean theorem is actually a statement about the area of squares. It says that if three squares are formed, one on each side of a right triangle (Fig. 11.21), then the area of the square on the hypotenuse equals the sum of the areas of the squares on the other two sides. In

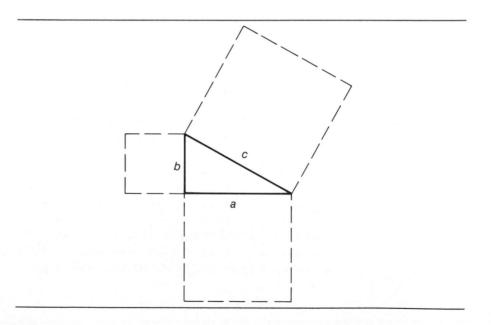

Figure 11.21

fact, the two smaller squares can be cut up and reassembled to cover the square on the hypotenuse exactly.

The principle use we want to make of the Pythagorean theorem is in computing distance. The next two examples illustrate how this is done.

Example 1   Julie walks 3 miles due north and then 4 miles due east. How far is she from her starting point?

Julie's route can be sketched as in Fig. 11.22; c represents the distance (as the crow flies) from the starting point to the finishing point. Because c is the length of the hypotenuse of a right triangle, the Pythagorean theorem says that

$$c^2 = 3^2 + 4^2$$
$$c^2 = 9 + 16$$
$$c^2 = 25.$$

Thus the distance c must be 5 miles.   □

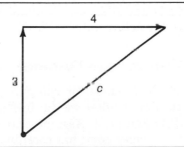

Figure 11.22

Example 2   Find the distance between the two points (2,1) and (4,5).

The line segment joining (2,1) and (4,5) is the hypotenuse of a right triangle as shown in Fig. 11.23. The vertex at the right angle has coordinates (4,1). Finding the lengths of the two legs is easy because they are parallel to the axes. The vertical leg has length $5 - 1$; the horizontal leg has length $4 - 2$. Thus the distance d between (2,1) and (4,5) is given by

$$d^2 = (4 - 2)^2 + (5 - 1)^2$$

$$d^2 = 2^2 + 4^2$$
$$d^2 = 4 + 16$$
$$d^2 = 20$$
$$d = \sqrt{20} \doteq 4.47.$$

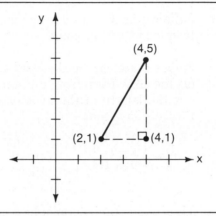

Figure 11.23

In Example 1 of the last section, we approximated the distance between (2,1) and (4,5) as 4.5 units. We see that the approximation was close to the actual value.   □

Finding the Distance between Any Two Points

The procedure for computing the distance between any two points is similar to what we did in Example 2 to compute the distance between (2,1) and (4,5). Assume that the first point has coordinates (x_1, y_1) and the second point has coordinates (x_2, y_2). (In this notation the numbers 1 and 2 are called **subscripts**; their job is to show which point a coordinate belongs to. They are not part of the numerical value of the coordinate.)

No matter where the two points are located, the line segment joining them can be regarded as the hypotenuse of a right triangle (Fig. 11.24). The vertical leg of the triangle has length $y_1 - y_2$ (or maybe $y_2 - y_1$ depending on which of the y-coordinates is larger). The horizontal leg of the triangle has length $x_1 - x_2$ (or maybe $x_2 - x_1$). Thus the distance d between the two points is given by

$$d^2 = (x_1 - x_2)^2 + (y_1 - y_2)^2$$
$$d = \sqrt{(x_1 - x_2)^2 + (y_1 - y_2)^2}.$$

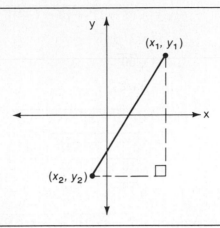

Figure 11.24

This formula permits us to compute the distance between any two points in terms of the coordinates of the points. The process is to compute the difference of the x-coordinates and square, compute the difference of the y-coordinates and square, add these two numbers, and, finally, take the square root of the sum.

Example 3 Find the perimeter of the triangle that has vertices (2,3), (−2,2), and (1,−4).

To find the perimeter we must find the length of each side and then add the lengths (Fig. 11.25). Each length is the distance between two points.

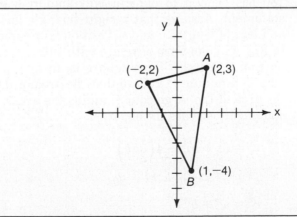

Figure 11.25

$$d_{\overline{AB}} = \sqrt{(2-1)^2 + [3-(-4)]^2}$$
$$= \sqrt{1^2 + 7^2}$$
$$= \sqrt{50}$$
$$= 5\sqrt{2},$$

$$d_{\overline{CA}} = \sqrt{(-2-2)^2 + (2-3)^2}$$
$$= \sqrt{(-4)^2 + (-1)^2}$$
$$= \sqrt{16+1}$$
$$= \sqrt{17},$$

$$d_{\overline{BC}} = \sqrt{[1-(-2)]^2 + (-4-2)^2}$$
$$= \sqrt{3^2 + (-6)^2}$$
$$= \sqrt{9+36}$$
$$= \sqrt{45}$$
$$= 3\sqrt{5}.$$

Thus the perimeter equals $5\sqrt{2} + \sqrt{17} + 3\sqrt{5} \doteq 17.9$.   □

Proof of the Pythagorean Theorem

What we know about areas of squares and triangles, together with the fact that the sum of the measures of the angles in a triangle equals 180°, is enough to explain why the Pythagorean theorem is a true statement. Assume that a right triangle has a hypotenuse of length c and legs of lengths a and b. Position four copies of the triangle as shown in Fig. 11.26 to form a square with sides of length $a+b$. The unshaded portion that looks like a square is, in fact, a square. (We will explain why in a minute.) Observe that the area of the unshaded square equals the area of the large square minus the areas of the four triangles. This statement becomes the equation

$$c^2 = (a+b)^2 - 4\left(\frac{1}{2}ab\right)$$
$$c^2 = a^2 + 2ab + b^2 - 2ab$$
$$c^2 = a^2 + b^2. \qquad\qquad \text{The Pythagorean Theorem!}$$

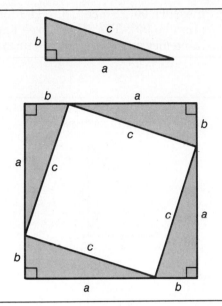

Figure 11.26

To complete the argument, we must explain why the unshaded portion in the drawing in Fig. 11.27 is a square. Since its four sides all have length c, what must be checked is that each angle measures 90°. In particular, we need to explain why the angle marked g is 90°. The measures of the angles f and d together equal 90° because they are the acute angles in a right triangle, and the sum of the angles of the triangle must be 180°. But angles d and e are the same, so angles f and e together

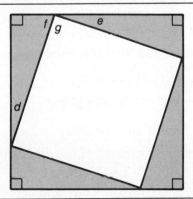

Figure 11.27

measure 90°. Since f, g, and e form a straight angle, this means that g must be 90°. The other three angles in the unshaded portion have the same measure as g; thus the unshaded portion is a square.

Exercises 11.3

Solve for x in Exercises 1–5.

1.

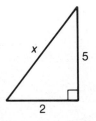

2.

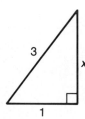

3.

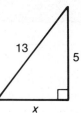

4.

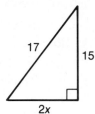

5.

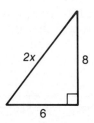

Find the distances between the points in Exercises 6–11.

6. (1,2) and (4,6)

7. (−2,1) and (6,7)

8. (−3, −8) and (5,7)

9. (−5,2) and (7,−3)

10. (0,0) and (1,1)

11. (−2,1) and (2,3)

Find the length of the diagonals of the polygons determined by each set of points in Exercises 12 and 13.

12. (0,0), (0,3), (4,3), and (4,0)

13. (1,2), (2,6), (5,6), and (4,2)

Find the perimeter of the polygons determined by each set of points in Exercises 14 and 15.

14. $(-3,1)$, $(4,6)$, and $(2,-4)$

15. $(-8,1)$, $(1,7)$, $(7,-2)$, and $(-2,-8)$

16. Find the length of the diagonals of a square with sides of lengths

(a) 50 feet   (b) 60 feet

(c) 80 feet

17. A baseball diamond is square in shape. If the base paths are 90 feet in length, find the distance from home base to second base.

18. Pete first walks 12 miles due south and then walks 5 miles due east. How far is he from his starting point?

19. At 10:00 A.M. a boat left Port Clinton heading due west at 35 mph. At 11:00 A.M. another boat left Port Clinton heading due north at 25 mph. What is the distance between the boats at 1:00 P.M.?

20. A plane leaves an airport at 1:00 P.M. heading due east at 350 mph. At 2:00 P.M. a second plane leaves the same airport heading due north at 400 mph. How far apart are the two planes at 3:00 P.M.?

21. A 20-foot ladder just reaches a window of a house. If the foot of the ladder is 3 feet from the base of the house, how far is the window above ground level?

22. A guy wire attached to the top of a 30-foot antenna just reaches the ground at a point 7 feet from the base of the antenna. How long is the guy wire?

23. Determine x so that the distance between $(x,1)$ and $(2,3)$ is 3 units.

24. Write an equation that expresses the fact that the distance between (x,y) and $(1,2)$ is 3 units.

11.4 / APPLICATIONS OF THE DISTANCE FORMULA

We have seen that the distance d between a point with coordinates (x_1,y_1) and a point with coordinates (x_2,y_2) is given by

$$d = \sqrt{(x_1 - x_2)^2 + (y_1 - y_2)^2}\,.$$

This fact is useful in analyzing many geometric problems.

Example 1   Find the points on the y-axis that are 2 units from the point (1,1).

A point on the y-axis has coordinates (0,b) for some number b. We want to find b so that the distance from (0,b) to (1,1) equals 2 (Fig. 11.28).

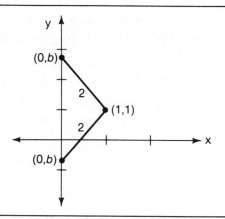

Figure 11.28

Therefore we must have

$$\sqrt{(0-1)^2 + (b-1)^2} = 2$$
$$\sqrt{1 + b^2 - 2b + 1} = 2$$
$$\sqrt{b^2 - 2b + 2} = 2$$
$$b^2 - 2b + 2 = 4 \qquad \text{(Squaring both sides of the equation)}$$
$$b^2 - 2b - 2 = 0.$$

The quadratic formula gives two values for b:

$$b = \frac{2 + \sqrt{4+8}}{2} \qquad \text{or} \qquad b = \frac{2 - \sqrt{4+8}}{2}$$

$$= \frac{2 + \sqrt{12}}{2} \qquad\qquad\qquad = \frac{2 - \sqrt{12}}{2}$$

$$= \frac{2 + 2\sqrt{3}}{2} \qquad\qquad\qquad = \frac{2 - 2\sqrt{3}}{2}$$

$$= 1 + \sqrt{3} \qquad\qquad = 1 - \sqrt{3}$$
$$\doteq 2.73 \qquad\qquad \doteq -.73.$$

There are two points on the y-axis that are 2 units away from $(1,1)$: $(0,1 + \sqrt{3})$ and $(0,1 - \sqrt{3})$. □

Example 2   Write an equation whose graph is a circle of radius 3 with center at $(4,-1)$.

We want to write an equation that describes when a point (x,y) lies on the circle in Fig. 11.29. A point (x,y) will lie on the circle if its distance from $(4,-1)$ is 3 units. Thus we must have

$$\sqrt{(x - 4)^2 + [y - (-1)]^2} = 3$$
$$(x - 4)^2 + [y - (-1)]^2 = 9 \qquad \text{(Square both sides of the equation)}$$
$$(x - 4)^2 + (y + 1)^2 = 9$$

This quadratic equation in x and y has for its graph the circle of radius 3 with center $(4,-1)$ because every point on the graph is 3 units away from $(4,-1)$. □

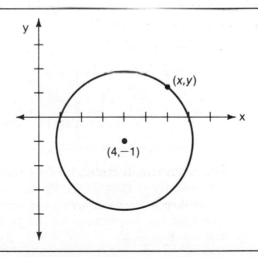

Figure 11.29

If we want to describe a circle with center (h,k) and radius

r, reasoning similar to that in the example above gives the equation $(x - h)^2 + (y - k)^2 = r^2$.

In Section 11.2, we approximated the area of a triangle by measuring the length of the base and the height with a ruler and then converting to scale units. Now we are able to compute distance between points exactly, and we can compute areas of triangles in terms of the coordinates of their vertices.

Example 3   Find the area of the triangle with vertices (0,0), (1,5), and (6,4).

We can start by computing the length of the base $\overline{AC}$ (Fig. 11.30):

$$\begin{aligned} d_{\overline{AC}} &= \sqrt{(6 - 0)^2 + (4 - 0)^2} \\ &= \sqrt{36 + 16} \\ &= \sqrt{52} \\ &= 2\sqrt{13}\,. \end{aligned}$$

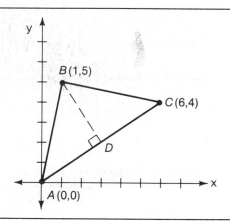

Figure 11.30

A more difficult task is to find the height of the triangle, that is, the distance from B to D. To do this we first find the coordinates of D. There are four steps: write the equation of $\overline{AC}$, write the equation of $\overline{BD}$, find the coordinates of the point D where $\overline{AC}$ and $\overline{BD}$ intersect, and compute $d_{\overline{BD}}$.

(1)   Write the equation of $\overline{AC}$.

Since $\overline{AC}$ has slope $\dfrac{4 - 0}{6 - 0} = \dfrac{2}{3}$ and y-intercept 0, it has the equation

$$y = \frac{2}{3}x + 0$$

$$y = \frac{2}{3}x.$$

(2) Write the equation of $\overline{BD}$.

Since $\overline{BD}$ is perpendicular to $\overline{AC}$, the slope of $\overline{BD}$ is $-\frac{3}{2}$. The equation is

$$y = -\frac{3}{2}x + b$$

where b is the y-intercept. Because $(1,5)$ is one point on the line, we can find b by replacing x by 1 and y by 5 in the equation:

$$5 = -\frac{3}{2}(1) + b$$

$$\frac{13}{2} = b.$$

Thus the equation of $\overline{BD}$ is $y = -\frac{3}{2}x + \frac{13}{2}$.

(3) Find the coordinates of the point D where $\overline{AC}$ and $\overline{BD}$ intersect. We can solve the system of equations

$$y = \frac{2}{3}x$$

$$y = -\frac{3}{2}x + \frac{13}{2}$$

by substituting $\frac{2}{3}x$ for y in the second equation.

$$\frac{2}{3}x = -\frac{3}{2}x + \frac{13}{2}$$

$$4x = -9x + 39 \qquad \text{(Multiply both sides by 6)}$$

$$13x = 39$$

$$x = 3$$

$$y = \frac{2}{3}x = \frac{2}{3}(3) = 2.$$

Thus D has coordinates (3,2).

(4) Compute $d_{\overline{BD}}$.

The height of the triangle is

$$d_{\overline{BD}} = \sqrt{(1-3)^2 + (5-2)^2}$$
$$= \sqrt{4+9}$$
$$= \sqrt{13} .$$

Now we know the length of the base $2\sqrt{13}$ and the height $\sqrt{13}$, so we can compute the area:

$$A = \frac{1}{2}(2\sqrt{13})(\sqrt{13})$$
$$A = 13.$$ □

The Pythagorean theorem says that if a right triangle has legs of lengths a and b and hypotenuse of length c, then $a^2 + b^2 = c^2$. A different statement, called the **converse** of this theorem, is also true. It says that if a triangle has sides of lengths a, b, and c with $a^2 + b^2 = c^2$, then the triangle is a right triangle. We will not prove this statement, but we can use it to show that certain triangles are right triangles.

Example 4   Show that the triangle with vertices (1,1), (2,8), and (5,5) is a right triangle.

a. We have previously solved this type of problem by computing the slopes of $\overline{BC}$ and $\overline{AC}$ (Fig. 11.31) and showing that one is the negative reciprocal of the other.

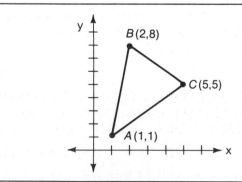

Figure 11.31

b. Another way to solve the problem is to compute the lengths of all three sides and observe that the sum of the squares of the lengths of the two shorter sides equals the square of the length of the longest side:

$$d_{\overline{BC}} = \sqrt{(2-5)^2 + (8-5)^2} = \sqrt{9+9} = \sqrt{18}$$
$$d_{\overline{CA}} = \sqrt{(5-1)^2 + (5-1)^2} = \sqrt{16+16} = \sqrt{32}$$
$$d_{\overline{BA}} = \sqrt{(2-1)^2 + (8-1)^2} = \sqrt{1+49} = \sqrt{50} .$$

Observe that $(\sqrt{18})^2 + (\sqrt{32})^2 = 18 + 32 = 50$. Also $(\sqrt{50})^2 = 50$. Thus the triangle is a right triangle. □

Exercises 11.4

Write an equation whose graph is a circle with center and radius as given in Exercises 1–5.

1. Center at (2,1), radius 2
2. Center at (−3,2), radius 1
3. Center at (−1,−2), radius $\sqrt{5}$
4. Center at (1,−3), radius 3
5. Center at (0,0), radius 5
6. Find all points on the y-axis that are 5 units from (4,3).
7. Find all points on the x-axis that are 13 units from (4,5)

In Exercises 8–11, determine if the given points form a right triangle.

8. (−2,−2), (5,7), and (7,1)
9. (−4,1), (2,5), and (4,2)
10. (0,−2), (4,1), and (7,−3)
11. (0,1), (2,6), and (4,3)

Prove that the points in Exercises 12 and 13 are the vertices of a square. (Remember that a square has four right angles and four sides of equal length.)

12. (0,0), (0,5), (5,5), (5,0)
13. (−6,1), (−1,−4), (4,1), (−1,6)

In Exercises 14 and 15, first write the equation of the line through A and B, then write the equation of the line through C perpendicular to $\overline{AB}$, and, finally, find the point of intersection of the two lines.

14. $A = (-2,-1)$, $B = (5,6)$, and $C = (1,4)$
15. $A = (-3,-1)$, $B = (6,5)$, and $C = (1,6)$

Find the area of each of the polygons determined by the points in Exercises 16–21.

16. $(-2,-1)$, $(5,6)$, and $(1,4)$

17. $(-3,-1)$, $(6,5)$, and $(1,6)$ (Compare your answer to Exercise 7 in Section 11.2)

18. $(0,2)$, $(8,-4)$, and $(-1,9)$

19. $(-2,0)$, $(0,6)$, and $(2,2)$

20. $(-1,3)$, $(3,6)$, $(6,2)$, and $(2,-1)$

21. $(0,-1)$, $(2,7)$, $(12,13)$, and $(10,5)$ (Compare your answer to Exercise 9 in Section 11.2)

22. Determine r so that the circle $(x-3)^2 + (y+3)^2 = r^2$ passes through the point $(0,0)$.

23. Determine k so that the circle $(x-3)^2 + (y-k)^2 = 25$ passes through the point $(-1,6)$.

11.5 / GEOMETRIC ARGUMENTS USING COORDINATES

There are several ways to investigate important concepts of geometry. One way is to establish some definitions and basic assumptions (axioms) and study those properties of geometric figures that are consequences of the definitions and assumptions. This approach is called an **axiomatic approach** to geometry. Another approach is to identify points in geometry with numerical coordinates and then use arithmetic and algebra to investigate geometric ideas. This can be called a **coordinate** or **analytic approach**. It is the approach we will take in this section.

Coordinates of the Midpoint of a Segment

The midpoint of a line segment is the point in the middle of the segment (Fig. 11.32). If the segment could be folded, the midpoint could be found by putting one endpoint on top of the other endpoint. If the coordinates of the two endpoints are known, then the coordinates of the midpoint can be computed in terms of the coordinates of the endpoints.

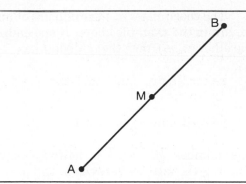

Figure 11.32

Example 1   Find the coordinates of the midpoint of the segment that has endpoints (1,2) and (6,4).

Saying that M is midway between A and B means that its x-coordinate is halfway between 1 and 6 and its y-coordinate is halfway between 2 and 4. Thus the

x-coordinate is $\frac{1}{2}(1 + 6) = 3.5$; and the

y-coordinate is $\frac{1}{2}(2 + 4) = 3$.

You can verify these coordinates from the graph in Fig. 11.33.

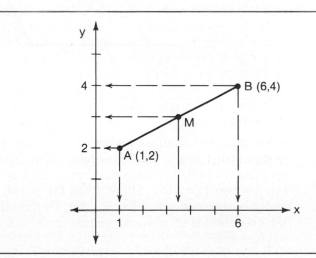

Figure 11.33

The coordinates of the midpoint of any segment can be computed as we did in the example above. If one endpoint has coordinates (x_1, y_1) and the other (x_2, y_2), then the midpoint has

x-coordinate $\frac{1}{2}(x_1 + x_2)$, and

y-coordinate $\frac{1}{2}(y_1 + y_2)$.

The number $\frac{1}{2}(x_1 + x_2)$ is halfway between x_1 and x_2. The number $\frac{1}{2}(y_1 + y_2)$ is halfway between y_1 and y_2.

Geometric Properties Involving Midpoints

Any parallelogram can be placed in a coordinate system with one vertex at the origin and one side along the x-axis, as shown in Fig. 11.34.

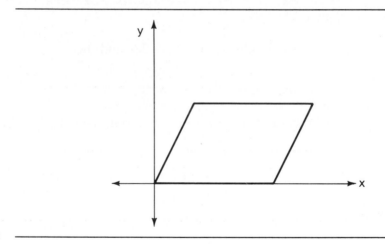

Figure 11.34

What numbers shall we assign as coordinates to the four vertices? The vertex at the origin has coordinates $(0,0)$. The vertex on the x-axis has coordinates $(a,0)$ for some number a. Say we call the coordinates of a third vertex (c,d) (Fig. 11.35). Then the fourth vertex has y-coordinate d and x-coordinate $a + c$ (Fig. 11.36). This method of using variables for the coordinates of vertices permits us to talk about all parallelograms, not just one.

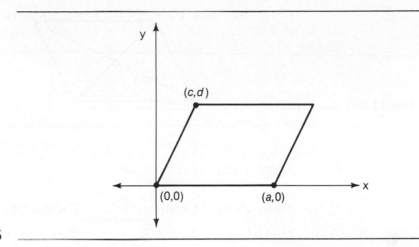

Figure 11.35

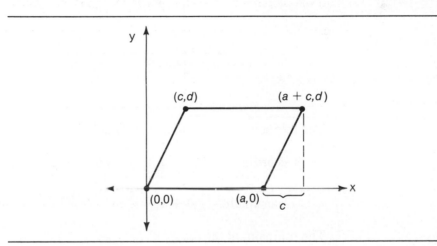

Figure 11.36

Example 2   Show that in any parallelogram, the diagonals bisect each other.

The diagonals of a parallelogram are the line segments joining opposite vertices (Fig. 11.37). We want to show that the diagonals cut each other in half, that is to say, the midpoint of one is the same point as the midpoint of the other. We can show this is true by assigning coordinates to the vertices and computing the coordinates of the midpoints of the diagonals (Fig. 11.38).

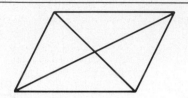

Figure 11.37

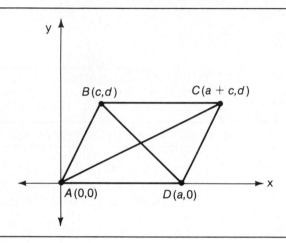

Figure 11.38

The midpoint of the diagonal $\overline{AC}$ has

  x-coordinate: $\dfrac{1}{2}(0 + a + c) = \dfrac{1}{2}(a + c),$

  y-coordinate: $\dfrac{1}{2}(0 + d) = \dfrac{1}{2}d.$

The midpoint of the diagonal $\overline{BD}$ has

  x-coordinate: $\dfrac{1}{2}(a + c),$

  y-coordinate: $\dfrac{1}{2}(0 + d) = \dfrac{1}{2}d.$

Thus the diagonals have the same midpoint, so they cut each other in half.   □

There is a geometric property involving midpoints that is somewhat more surprising than the result in Example 2. It says that if we take any four-sided polygon (or **quadrilateral**) and connect the midpoints of the four sides, we will draw a parallelogram (Fig. 11.39).

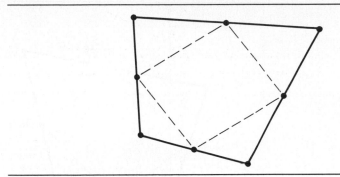

Figure 11.39

To show that this statement is true using coordinates, we can position one vertex of the quadrilateral at the origin and put one side along the x-axis (Fig. 11.40). This helps us to label coordinates for two vertices, but if we are describing any quadrilateral, the remaining two vertices must have coordinates that can be any numbers.

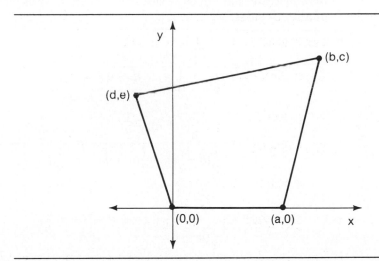

Figure 11.40

Example 3 Show that if the midpoints of any quadrilateral are connected, the result is a parallelogram.

In Fig. 11.41, we have labeled the coordinates of the midpoints and joined these points.

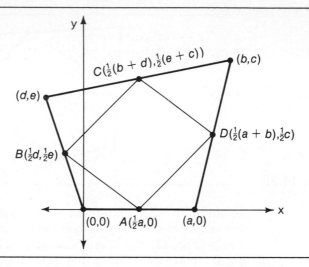

Figure 11.41

To show that the quadrilateral $ABCD$ is a parallelogram, we must demonstrate that opposite sides are parallel. Thus we compute the slopes of all four sides:

$$\text{slope } \overline{AB} = \frac{\frac{1}{2}e}{\frac{1}{2}d - \frac{1}{2}a} = \frac{e}{d - a},$$

$$\text{slope } \overline{DC} = \frac{\frac{1}{2}(e + c) - \frac{1}{2}c}{\frac{1}{2}(b + d) - \frac{1}{2}(a + b)} = \frac{\frac{1}{2}e}{\frac{1}{2}d - \frac{1}{2}a} = \frac{e}{d - a},$$

$$\text{slope } \overline{AD} = \frac{\frac{1}{2}c}{\frac{1}{2}(a + b) - \frac{1}{2}a} = \frac{\frac{1}{2}c}{\frac{1}{2}b} = \frac{c}{b},$$

$$\text{slope } \overline{BC} = \frac{\frac{1}{2}(e + c) - \frac{1}{2}e}{\frac{1}{2}(b + d) - \frac{1}{2}d} = \frac{\frac{1}{2}c}{\frac{1}{2}b} = \frac{c}{b}.$$

We have shown that opposite sides have the same slope and, hence, are parallel.   □

To use coordinates in making an argument about triangles, we can position one vertex at the origin and one side along the x-axis, as shown in Fig. 11.42.

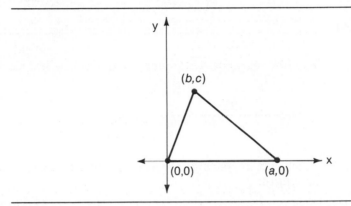

Figure 11.42

Example 4 Show that in any triangle, the line segment joining the midpoints of two sides is parallel to the third side and has a length that is half the length of the third side.

The situation might look like that shown in Fig. 11.43.

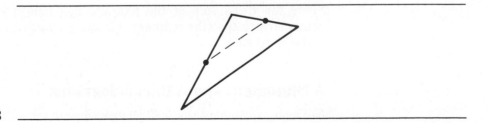

Figure 11.43

In order to use coordinates in the argument, we can put the *third* side on the x-axis with one of its endpoints at the origin, assign the coordinates of the vertices, and compute the coordinates of the midpoints (Fig. 11.44).

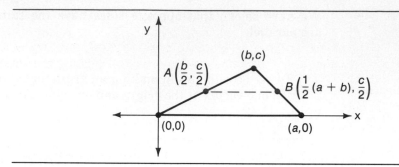

Figure 11.44

The line segment $\overline{AB}$ joining the midpoints has slope

$$\frac{\dfrac{c}{2} - \dfrac{c}{2}}{\dfrac{1}{2}(a + b) - \dfrac{b}{2}} = \frac{0}{\dfrac{a}{2}} = 0.$$

Thus $\overline{AB}$ is parallel to the horizontal axis. We can also compute its length:

$$d_{\overline{AB}} = \sqrt{\left(\frac{1}{2}(a + b) - \frac{b}{2}\right)^2 + \left(\frac{c}{2} - \frac{c}{2}\right)^2}$$

$$= \sqrt{\left(\frac{a}{2}\right)^2}$$

$$= \frac{a}{2}.$$

Since the third side of the triangle has length a, the computation demonstrates that the segment $\overline{AB}$ has a length that is half the length of the third side. □

A Statement about Parallelograms

Not every statement that can be proved with a coordinate argument is a statement about midpoints, of course. The next example provides an answer to the question of when a parallelogram is a rectangle. One answer is that a parallelogram with right angles is a rectangle. Another condition that guarantees that a parallelogram is a rectangle is given in this example.

Example 5   A parallelogram that has diagonals of equal length is a rectangle.

We start in Fig. 11.45 with a parallelogram positioned so that one vertex is at the origin and one side is along the x-axis. We would like to

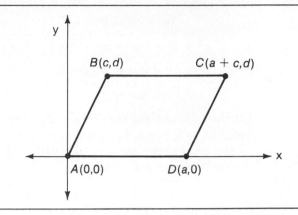

Figure 11.45

argue that if the diagonals of the parallelogram have the same length, then the point B lies on the y-axis. That will mean that the parallelogram is a rectangle. B lies on the y-axis if its x-coordinate c is 0, so we want to investigate what knowing that the diagonals are the same length says about the coordinate c (Fig. 11.46):

Length of $\overline{AC}$: $\sqrt{(a+c)^2 + d^2}$

Length of $\overline{BD}$: $\sqrt{(c-a)^2 + d^2}$.

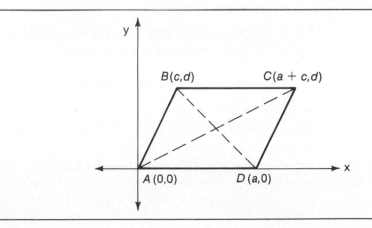

Figure 11.46

The statement that the lengths of the diagonals are equal gives this equation:

$$\sqrt{(a+c)^2 + d^2} = \sqrt{(c-a)^2 + d^2}$$

$$(a+c)^2 + d^2 = (c-a)^2 + d^2 \qquad \text{(Square both sides)}$$

$$(a+c)^2 = (c-a)^2$$

$$a^2 + 2ac + c^2 = c^2 - 2ac + a^2$$

$$2ac = -2ac$$

$$4ac = 0.$$

If the product $4ac$ is 0, then either a or c must be 0. But a is the length of $\overline{AD}$, so a is not 0. Thus c is 0. This means that in spite of the way we have drawn the picture, the vertex B is on the vertical axis and the parallelogram is a rectangle. □

**Exercises
11.5**

Find the coordinates of the midpoint of the line segment determined by each pair of points in Exercises 1–6.

1. $(1,2)$, $(4,7)$ 2. $(-3,2)$, $(1,5)$

3. $(-4,-2)$, $(4,6)$ 4. $(-6,-2)$, $(-2,5)$

5. $(-8,-6)$, $(-3,-1)$ 6. $(-5,1)$, $(5,-1)$

7. If $(3,4)$ is the midpoint of the line segment joining $(1,2)$ and (a,b), find a and b.

Let $A = (2,1)$ and $B = (6,5)$. Find the coordinates of a point C on $\overline{AB}$ that satisfies the condition given in Exercises 8–11.

8. The length of $\overline{AC}$ is $\frac{1}{3}$ the length of $\overline{AB}$.

9. The length of $\overline{AC}$ is $\frac{2}{3}$ the length of $\overline{AB}$.

10. The length of $\overline{AC}$ is $\frac{1}{4}$ the length of $\overline{AB}$.

11. The length of $\overline{AC}$ is $\frac{3}{2}$ the length of $\overline{AB}$.

12. Determine D so that $A = (1,2)$, $B = (3,5)$, $C = (7,7)$, and D are vertices of a parallelogram. (There are three answers.)

13. Let $A = (3,0)$, $B = (-1,2)$, and $C = (5,4)$.

    (a) Prove that the triangle determined by A, B, and C is isosceles but not equilateral.

    (b) Prove that the line through A and the midpoint of $\overline{BC}$ is perpendicular to $\overline{BC}$.

14. Let $A = (0,0)$, $B = (2,6)$, and $C = (5,0)$.

    (a) Write an equation for each of the three lines determined by a vertex and the midpoint of the opposite side of the triangle ABC.

    (b) Prove that the three lines in (a) intersect in one point.

15. Prove that the line segments joining the midpoints of the opposite sides of any quadrilateral bisect each other.

16. Prove that the diagonals of a rectangle are equal in length.

17. Prove that the diagonals of a square are perpendicular.

18. Prove that if the diagonals of a rectangle are perpendicular, then the rectangle is a square.

19. Prove that the midpoint of the hypotenuse of a right triangle is the same distance from each of the three vertices.

20. Let $A = (-2,-1)$ and $B = (6,5)$.

    (a) Write an equation for the line that is the perpendicular bisector of $\overline{AB}$.

    (b) Find the coordinates of a point C on the perpendicular bisector of $\overline{AB}$ so that ABC is an equilateral triangle.

11.6 / TRIG RATIOS IN RIGHT TRIANGLES

The word **trigonometry** has two parts: *trigon* comes from the Greek word for triangle and *metry* means to measure. For convenience, we frequently use the word **trig** rather than *trigonometry*. In this section we will investigate measurements of a right triangle. We have already seen that the Pythagorean theorem gives us a way to compute the length of one side of a right triangle if the lengths of the other sides are known. There are also relationships between the lengths of the sides and the size of the angles in a right triangle.

In labeling a right triangle (Fig. 11.47), we will use capital letters to designate the three angles and corresponding lower case letters to indicate the lengths of the sides opposite the angles. C is a right angle and has measure 90°; A and B both have measure less than 90° (and thus are said to be **acute** angles). Since there are three sides of the triangle, there are six ratios that can be written using the lengths of the sides, namely, $a : b$, $a : c$, $b : a$, $b : c$, $c : a$, $c : b$.

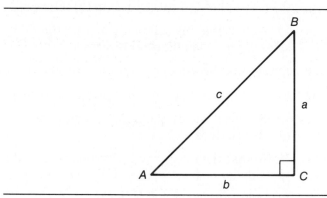

Figure 11.47

If the same angle A occurs in two different right triangles, the lengths of the sides of the triangles may be different, but the ratios of corresponding lengths will be the same (Fig. 11.48). To see this, observe that angle A and angle C are in both triangles, so angle B must be the same in both triangles, because the measures of the three angles add up to 180°. That is to say, if two triangles have two angles that are the same, all three angles must be the same. This means the two triangles are similar, so the ratios of lengths of corresponding sides are equal. Thus the six ratios of the lengths of the sides depend on the measure of angle A and not on what right triangle A is in.

It is common to call the side of length a the side **opposite** A and the side of length b the side **adjacent** to A. The six ratios can be described using this vocabulary as well as the letters a, b, and c. Trigonometry is a very old area of mathematics, with some records of it dating back to

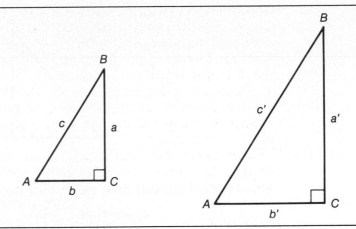

Figure 11.48

the second century B.C. The historic names of the ratios, although perhaps strange sounding to a modern ear, are still used. In this discussion and for computation, we write the ratios as fractions:

$$\text{sine } A = \frac{a}{c} = \frac{\text{opposite side}}{\text{hypotenuse}}$$

$$\text{cosine } A = \frac{b}{c} = \frac{\text{adjacent side}}{\text{hypotenuse}}$$

$$\text{tangent } A = \frac{a}{b} = \frac{\text{opposite side}}{\text{adjacent side}}$$

$$\text{cotangent } A = \frac{b}{a} = \frac{\text{adjacent side}}{\text{opposite side}}$$

$$\text{secant } A = \frac{c}{b} = \frac{\text{hypotenuse}}{\text{adjacent side}}$$

$$\text{cosecant } A = \frac{c}{a} = \frac{\text{hypotenuse}}{\text{opposite side}} \cdot$$

The trigonometric ratios for angle B are described in an analogous way using the side opposite B and the side adjacent to B.

Example 1   A right triangle has sides of lengths as shown in Fig. 11.49. Find the six trigonometric ratios for angle A and the six trigonometric ratios for angle B.

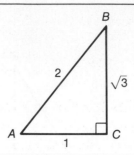

Figure 11.49

Using the definitions above, we first list the ratios for angle A:

$$\text{sine } A = \frac{\text{opposite side}}{\text{hypotenuse}} = \frac{\sqrt{3}}{2} \doteq .87$$

$$\text{cosine } A = \frac{\text{adjacent side}}{\text{hypotenuse}} = \frac{1}{2} = .5$$

$$\text{tangent } A = \frac{\text{opposite side}}{\text{adjacent side}} = \frac{\sqrt{3}}{1} \doteq 1.73$$

$$\text{cotangent } A = \frac{\text{adjacent side}}{\text{opposite side}} = \frac{1}{\sqrt{3}} \doteq .58$$

$$\text{secant } A = \frac{\text{hypotenuse}}{\text{adjacent side}} = \frac{2}{1} = 2$$

$$\text{cosecant } A = \frac{\text{hypotenuse}}{\text{opposite side}} = \frac{2}{\sqrt{3}} \doteq 1.15.$$

Observe that the side opposite B has length 1 and the side adjacent to B has length $\sqrt{3}$. Thus the trigonometric ratios for angle B have these values:

$$\text{sine } B = \frac{\text{opposite side}}{\text{hypotenuse}} = \frac{1}{2} = .5$$

$$\text{cosine } B = \frac{\text{adjacent side}}{\text{hypotenuse}} = \frac{\sqrt{3}}{2} \doteq .87$$

$$\text{tangent } B = \frac{\text{opposite side}}{\text{adjacent side}} = \frac{1}{\sqrt{3}} \doteq .58$$

$$\text{cotangent } B = \frac{\text{adjacent side}}{\text{opposite side}} = \frac{\sqrt{3}}{1} \doteq 1.73$$

$$\text{secant } B \quad = \frac{\text{hypotenuse}}{\text{adjacent side}} = \frac{2}{\sqrt{3}} \doteq 1.15$$

$$\text{cosecant } B \quad = \frac{\text{hypotenuse}}{\text{opposite side}} = \frac{2}{1} = 2.$$

□

Ratios for Complementary Angles

In the example above, the side opposite B is the side adjacent to A; the side adjacent to B is the side opposite A. For this reason, the trigonometric ratios for A and B compare this way:

sine A = cosine B

cosine A = sine B

tangent A = cotangent B

cotangent A = tangent B

secant A = cosecant B

cosecant A = secant B.

The sum of the measures of the angles A and B must be 90° (because the sum of the measures of all three angles is 180°). Such angles are said to be **complementary**. This word has contributed the *co* part of cosine, cotangent, and cosecant.

Using a Calculator to Find Trig Ratios

If the measure of angle A is known, all trig ratios for A can be found directly from the calculator. In Example 1, angle A is 60°. You can see this by copying the triangle across the segment $\overline{BC}$ to form a new triangle ABD (Fig. 11.50). In the triangle ABD, all sides have length 2. Thus the triangle is equilateral and the three angles must have the same measure, so each is 60°. We can compare the calculator values for sine 60°, cosine 60°, and tangent 60° with the values we computed as ratios in Example 1.

	Keying sequence	Display
sine 60°	60 (sin)	.86602541
cosine 60°	60 (cos)	.5
tangent 60°	60 (tan)	1.7320508

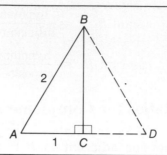

Figure 11.50

Notice that the $\boxed{=}$ key is not needed in computing the trig ratios.

Because angle A is 60° and angle C is 90° in Example 1, angle B is 30°. You should compare the calculator values for sine 30°, cosine 30°, and tangent 30° with the values computed using ratios in the example. Even though the calculator does not contain keys for cotangent, secant, and cosecant, we will see in Section 11.7 how these ratios can also be found using the calculator if the measure of an angle is known.

Finding All Trig Ratios from One Trig Ratio

Although six trig ratios can be formed from the lengths of three sides of a right triangle, if one is known, the other five can be computed from it using the Pythagorean theorem.

Example 2   If A is an acute angle and cosecant $A = \dfrac{5}{3}$, find the other trig ratios for A.

Since cosecant $A = \dfrac{\text{hypotenuse}}{\text{opposite side}} = \dfrac{5}{3}$, we can draw a triangle containing A that looks like Fig. 11.51. In order to find all the trig ratios, we need to find the length b of the third side.

By the Pythagorean theorem,

$$b^2 + 3^2 = 5^2$$
$$b^2 + 9 = 25$$

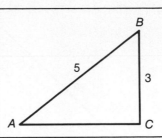

Figure 11.51

$$b^2 = 16$$
$$b = 4.$$

Now we can write all ratios. (If you have not yet memorized the meaning of each, you should stop and memorize them now.)

sine A	$= \dfrac{3}{5}$	cotangent A	$= \dfrac{4}{3}$
cosine A	$= \dfrac{4}{5}$	secant A	$= \dfrac{5}{4}$
tangent A	$= \dfrac{3}{4}$	cosecant A	$= \dfrac{5}{3}.$

□

Exercises 11.6   Write the six trigonometric ratios for angles A and B in Exercises 1–4.

1.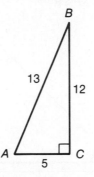

2.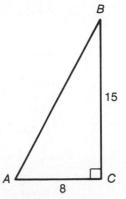

3.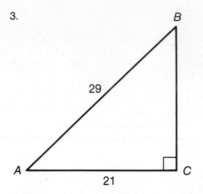

4.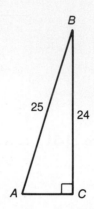

5. Use the triangle below to find the sine, cosine, and tangent of 30°; then compare with the calculator values.

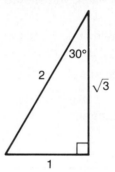

6. Use the triangle below to find the sine, cosine, and tangent of 45°; then compare with the calculator values.

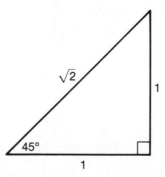

Find all six trigonometric ratios of the angle *A* using the information given in Exercises 7–12.

7. sine $A = \dfrac{4}{5}$

8. cosine $A = \dfrac{3}{4}$

9. tangent $A = \dfrac{6}{5}$

10. cotangent $A = \dfrac{7}{4}$

11. secant $A = \dfrac{8}{5}$

12. cosecant $A = \dfrac{7}{3}$

Use your calculator to find the sine, cosine, and tangent of the angles given in Exercises 13–20.

13. $22°$

14. $37.5°$

15. $2°$

16. $1°$

17. $88°$

18. $89°$

19. $0°$

20. $90°$

Explain why the statements in Exercises 21–24 are true by using the relationships among the lengths of the sides of a right triangle.

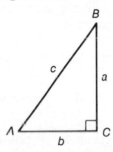

21. $0 < \text{sine } A < 1$

22. $0 < \text{cosine } A < 1$

23. secant $A > 1$

24. cosecant $A > 1$

25.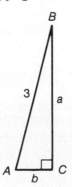

Find a and b if sine $A = \dfrac{4}{5}$.

26.

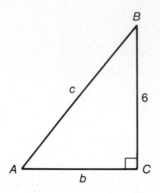

Find b and c if
tangent $A = \dfrac{5}{4}$.

11.7 / USING TRIG RATIOS FOR ACUTE ANGLES

An acute angle A in a right triangle has six trigonometric ratios. Rather than write the names of these ratios in full, we adopt the following abbreviations:

$$\text{sine } A \quad = \sin A = \frac{a}{c}$$

$$\text{cosine } A \quad = \cos A = \frac{b}{c}$$

$$\text{tangent } A \quad = \tan A = \frac{a}{b}$$

$$\text{cotangent } A = \cot A = \frac{b}{a}$$

$$\text{secant } A \quad = \sec A = \frac{c}{b}$$

$$\text{cosecant } A \quad = \csc A = \frac{c}{a}.$$

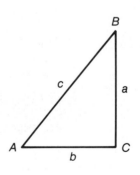

There are many relationships among the six ratios. Notice that cot A (the ratio $\frac{b}{a}$) is the reciprocal of tan A (the ratio $\frac{a}{b}$), sec A is the reciprocal of cos A, and csc A is the reciprocal of sin A:

$$\cot A = \frac{1}{\tan A}$$

$$\sec A = \frac{1}{\cos A}$$

$$\csc A = \frac{1}{\sin A}.$$

Using the Calculator to Find Trig Ratios

The relationships involving reciprocals are important for computation on a calculator. The calculator has values for $\sin A$, $\cos A$, and $\tan A$ for any angle A. To get values for $\cot A$, $\sec A$, and $\csc A$ we use the facts that

$$\csc A = \frac{1}{\sin A}$$

$$\sec A = \frac{1}{\cos A}$$

$$\cot A = \frac{1}{\tan A}.$$

Example 1 Find the value of all trig ratios for an angle A of $25°$.

The three ratios $\sin A$, $\cos A$, and $\tan A$ are keyed directly using $\boxed{\sin}$, $\boxed{\cos}$, and $\boxed{\tan}$ keys. The ratios $\csc A$, $\sec A$, and $\cot A$ must be keyed as reciprocals of the other three. The following chart summarizes the keying sequences and the values.

	Keying sequence	Display
sin 25°	25 $\boxed{\sin}$	.42261826
csc 25°	25 $\boxed{\sin}$ $\boxed{1/x}$	2.3662016
cos 25°	25 $\boxed{\cos}$	.90630779
sec 25°	25 $\boxed{\cos}$ $\boxed{1/x}$	1.1033779

	Keying sequence	Display
tan 25°	25 (tan)	.46630766
cot 25°	25 (tan) (1/x)	2.1445069

□

Any two right triangles that have one acute angle of 30° are the same shape. If two right triangles have the side opposite the 30° angle a particular length, say 5 cm, they are the same shape and the same size. In fact, if one acute angle and one side are known in a right triangle, the other sides and the other angles can be computed.

Example 2   If angle A in a right triangle is 30° and $a = 5$, find B, C, b, and c.

C is the right angle in the triangle, so C has measure 90°, and the measure of B and A together equals 90° (Fig. 11.52). Thus B is a 60° angle. To find c, remember that $\sin A = \dfrac{5}{c}$. Thus

$$\sin 30° = \frac{5}{c}$$

$$c(\sin 30°) = 5$$

$$c = \frac{5}{\sin 30°}.$$

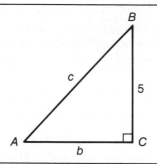

Figure 11.52

To evaluate $\dfrac{5}{\sin 30°}$, we key (5) (÷) 30 (sin) (=). The display is 10. Since $a = 5$ and $c = 10$, we can use the Pythagorean theorem to find b:

$$a^2 + b^2 = c^2$$
$$5^2 + b^2 = 10^2$$
$$b^2 = 75$$
$$b \doteq 8.66.$$
□

In Example 2, we used information about two parts of a right triangle to compute the value of all six parts. This process is sometimes called **solving the triangle**.

Computations like those in Example 2 can be used to measure indirectly distances that would be difficult to measure directly. For example, if we wished to measure the distance across a river, we could stand at a point C across from an object B on the other side. To compute the distance from C to B, we walk a convenient distance, say 100 feet, along the shore at a right angle to $\overline{BC}$, marking a third point A. Using a transit or other instrument, we measure the angle at A. Now we can compute the distance from C to B without getting wet feet.

Example 3   Find the distance across the river in Fig. 11.53, assuming that $d_{\overline{AC}} = 100$ feet and that A is a 38° angle.

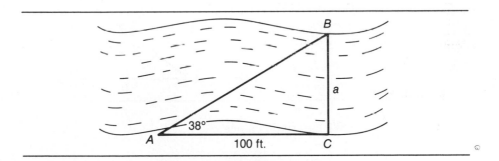

Figure 11.53

We choose a trig ratio that uses the distances a and b, since a is the number we want to find and $b = 100$ is the number we know. There are two choices: $\tan A = \frac{a}{b}$ and $\cot A = \frac{b}{a}$. The first choice gives this equation:

$$\tan 38° = \frac{a}{100}$$

$$100(\tan 38°) = a$$
$$78.13 \doteq a.$$

The river is approximately 78 feet wide.   □

Finding an Angle When a Trig Ratio is Known

We have seen that if we know the measure of an angle, we can find all the trig ratios of the angle using a calculator. In addition, if we know one trig ratio for an acute angle, the calculator can compute the measure of the angle. The new key that is used in this type of computation is $\boxed{\text{INV}}$, the inverse key. To find the size of angle A if we know, for example, that $\sin A = .5$, we use this keying sequence:

Display
.5 $\boxed{\text{INV}}$ $\boxed{\sin}$ 30

Angle A is 30° if $\sin A = .5$.

Example 4   In the right triangle shown in Fig. 11.54, $a = 15$ and $c = 25$. Find b, A, and B (that is to say, solve the triangle).

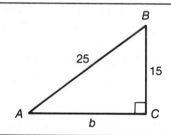

Figure 11.54

We can find A since

$$\sin A = \frac{15}{25} = .6.$$

To find the angle whose sine is .6, we key

Display
.6 $\boxed{\text{INV}}$ $\boxed{\sin}$ 36.869898

Thus A is approximately 36.87°. Angle B has measure approximately

$90 - 36.87 = 53.13°$. Side length b either can be computed using the Pythagorean theorem, or we can observe

$$\cos A = \frac{b}{25}$$

$$\cos 36.87 = \frac{b}{25}$$

$$25 \cos 36.87 = b$$

$$20 \doteq b. \qquad\qquad\qquad \square$$

Example 5   A 30-foot ladder is leaning against a house. If the base of the ladder is 22 feet from the house, what angle does the ladder make with the ground?

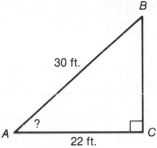

Because the lengths b and c are known, we can compute the value of either $\cos A$ or $\sec A$: $\cos A = \frac{22}{30}$. To recover A from $\cos A$, we key

$$\underline{22} \;\boxed{\div}\; \underline{30} \;\boxed{=}\; \boxed{\text{INV}} \;\boxed{\text{cos}}$$

Display
42.833428

Thus angle A is approximately $42.83°$. $\qquad\qquad \square$

Example 6   If $\sec A = \frac{30}{22}$, find the measure of A.

The equation $\sec A = \frac{30}{22}$ is another equation describing angle A in Example 5. Since the calculator has no $\boxed{\text{sec}}$ key, we first find $\cos A = \frac{1}{\sec A}$ and then recover A from $\cos A$. This is the keying sequence:

Display

30 $\div$ 22 $=$ 1/x INV cos   42.833428   □

Exercises 11.7

Use the meanings of the trig ratios to prove the statements in Exercises 1–3.

1. $\tan A = \dfrac{1}{\cot A}$ 2. $\cos A = \dfrac{1}{\sec A}$

3. $\sin A = \dfrac{1}{\csc A}$

4. Complete the following table:

A	$\sin A$	$\csc A$	$\cos A$	$\sec A$	$\tan A$	$\cot A$
30°						
45°						
60°						
36.4°						
1°						
0°						
89°						
90°						

Exercises 5–14 refer to Fig. 11.55. Solve the right triangles described in Exercises 5–14, that is, find the missing parts.

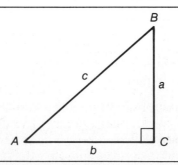

Figure 11.55

5. $A = 46°, a = 5$

6. $A = 62.5°, c = 7$

7. $B = 34.6°, a = 6$

8. $B = 12.8°, c = 9$

9. $a = 4, c = 7$

10. $a = 6, b = 9$

11. $\tan A = 2.1635, a = 8$

12. $\cos A = 0.7562, c = 12$

13. $\sin A = 0.2351, b = 11$

14. $\cot A = 1.7652, c = 10$

Find the measure of angle A in Exercises 15–20.

15. $\sin A = 0.5821$

16. $\cos A = 0.3425$

17. $\sec A = 1.8651$

18. $\tan A = 0.4565$

19. $\cot A = 3.5678$

20. $\csc A = 2.7575$

Find the other five trig ratios from the trig ratio given in Exercises 21–24.

21. $\cos A = 0.6565$

22. $\tan A = 1.2134$

23. $\sec A = 1.9595$

24. $\csc A = 2.1515$

25. A 20-foot ladder just reaches a windowsill with the base of the ladder 3 feet from the house. How high is the window and what angle does the ladder make with the ground?

26. A 25-foot ladder just reaches the top of a wall when the ladder makes an angle of 80° with the ground. How tall is the wall?

27. A surveyor sights a tree directly across a river and then measures 80 feet along the bank. At the new location, she finds that the angle between the river bank and her line of sight to the tree is 42°. Find the approximate width of the river.

11.8 / APPLICATIONS OF TRIG RATIOS FOR ACUTE ANGLES

Trig ratios are commonly used to compute distances that are difficult to measure directly. The general idea in this kind of problem is to make the angle measurements and distance measurements that can be made readily, then choose a trig ratio that expresses the desired measurement in terms of the ones that are known.

Example 1   A wire is stretched from the top of a pole to the ground 50 feet from the base of the pole. If the angle between the wire and the ground is 50°, what is the height of the pole?

In this problem it is easy to measure the distance from the base of the pole to the place where the wire is anchored and to measure the angle the wire makes with the ground. The ratio of the height of the pole h to the distance 50 is the tangent of 50°.

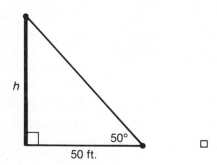

$$\frac{h}{50} = \tan 50°$$

$$h = 50(\tan 50°)$$

$$h = 50(1.1917536)$$

$$h \doteq 59.59 \text{ feet.}$$

Thus the pole is almost 60 feet tall.

In describing problems that involve actual measurement, we sometimes use the phrases *angle of elevation* and *angle of depression*. If a point P is higher than a point O, the angle between the line of sight $\overline{OP}$ and the horizontal $\overline{OH}$ is called the angle of elevation of P at O. If P is lower than O, the angle between the line of sight $\overline{OP}$ and the horizontal $\overline{OH}$ is called the **angle of depression** of P at O.

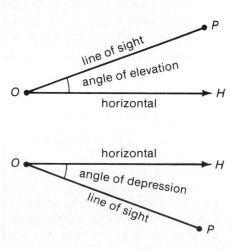

Example 2   Find the angle of elevation of the sun if a 50-foot pole casts a 22-foot shadow.

Here we want the angle made with the horizontal by a line joining A to the sun (Fig. 11.56). Such a line represents a ray of the sun and goes

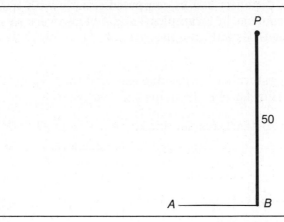

Figure 11.56

through point P. We can write the tangent of A in terms of the distances that are known (Fig. 11.57): $\tan A = \dfrac{50}{22}$. To recover the angle A, we must key

					Display
50 ÷	22 =	INV	tan		66.250506

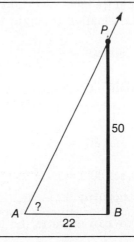

Figure 11.57

Thus the angle of elevation is approximately 66°.   □

In section 11.2 we approximated distances by making a drawing, measuring with a ruler, and converting the ruler distance to scale units. These distances can be computed exactly with trig ratios. Observe that the problems in Examples 3 and 4 below were solved using approximation methods in Examples 3 and 4 of Section 11.2.

Example 3   Dan first walks 5 miles due east and then walks 4 miles at 30° north of east. How far is he from his starting point?

The problem asks for the distance from O to P (Fig. 11.58). In this

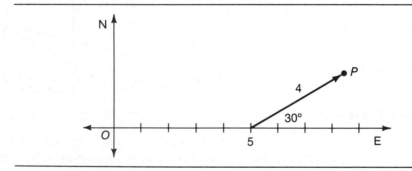

Figure 11.58

problem we have two right triangles to work with (Fig. 11.59). First we use the smaller triangle to find the distances h and k:

$$\sin 30° = \frac{k}{4}$$

$$4(\sin 30°) = k$$

$$2 = k$$

$$\cos 30° = \frac{h}{4}$$

$$4(\cos 30°) = h$$

$$3.4641016 \doteq h.$$

Now we can move to the larger right triangle to find $d_{\overline{OP}}$ (Fig 11.60).

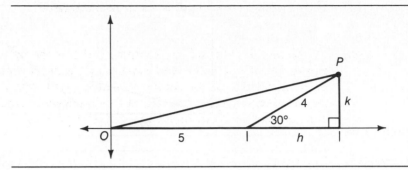

Figure 11.59

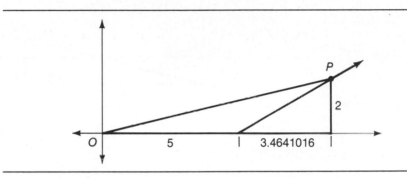

Figure 11.60

Since $\overline{OP}$ is the hypotenuse of a right triangle in which we know the lengths of the legs, we can use the Pythagorean theorem:

$$d^2 \doteq 2^2 + (8.4641016)^2$$

$$d^2 \doteq 75.641016$$

$$d \doteq 8.6971844.$$

Our approximation in Section 11.2 for this distance was 8.625 miles, a very good approximation indeed. □

Example 4   A plane starts due east at 500 mph. A wind is blowing in the direction 40° north of east at 75 mph. Find the actual speed and direction of the plane.

Fig. 11.61 indicates that the plane flies 500 mph due east with a 75 mph wind in the direction 40° north of east. We want to compute the distance from A to P and the angle at A. Again, we have two right triangles (Fig. 11.62):

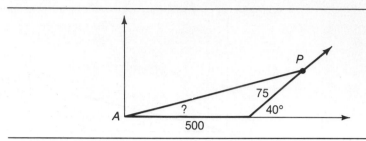

Figure 11.61

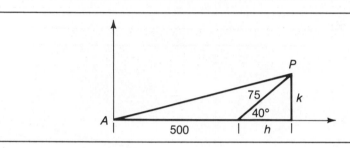

Figure 11.62

$$\sin 40° = \frac{k}{75}$$

$$75(\sin 40°) = k$$

$$48.209071 \doteq k$$

$$\cos 40° = \frac{h}{75}$$

$$75(\cos 40°) = h$$

$$57.453333 \doteq h.$$

Since $\overline{AP}$ is the hypotenuse of a right triangle with legs of lengths 48.209071 and 557.453333, the length of $\overline{AP}$ can be computed from the Pythagorean theorem (Fig. 11.63):

$$d^2 \doteq (48.209071)^2 + (557.453333)^2$$

$$d^2 \doteq 313078.33$$

$$d \doteq 559.53403.$$

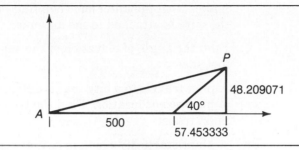

Figure 11.63

To find the angle at A, we observe that

$$\sin A = \frac{48.209071}{d} = \frac{48.209071}{559.53403}$$

$$= .08615932.$$

To recover angle A from the sine of A, we key

Display
.08615932 (INV) (sin)   4.9426935

Thus the angle A is approximately 4.9°. In Section 11.2, our approximation was 5°.   □

Exercises 11.8

1. Find the height of a vertical wall if the angle of elevation of the top of the wall from a point 75 feet from the base of the wall is 56°.

2. Find the height of a tree if the angle of elevation of the top of the tree from a point 100 feet from the base of the tree is 44°.

3. The angle of elevation of the top of a lighthouse from a boat is 22°. If the lighthouse is 120 feet tall, how far from the lighthouse is the boat?

4. The angle of depression from the top of a 250-foot-tall building to a car is 38°. How far is the car from the base of the building?

5. A guy wire 50 feet long runs from an antenna to a point on level ground 10 feet from the base of the antenna. Find the angle the guy wire makes with the horizontal and also the angle the wire

makes with the antenna. Then find how far from above the ground the wire is attached to the antenna.

6. Find the angle of elevation of the sun if a 100-foot pole casts a 20-foot shadow.

7. A pilot travels to a city that is 150 miles east and 200 miles north of his present location. Find the bearing and the distance the plane must travel.

8. A plane starts 30° north of east at 400 mph. A wind is blowing due north at 50 mph. Find the speed and direction of the plane. (Compare with your answer to Problem 14 in Section 11.2.)

9. Jim first walks 6 miles due east and then walks 2 miles at 70° north of east. How far is he from his starting point? (Compare with your answer to Problem 12 in Section 11.2.)

10. Sarah first walks 4 miles due west, then 3 miles at 50° north of east, and then 1 mile due north. How far is she from her starting point? (Compare with your answer to Problem 13 in Section 11.2.)

11. A plane starts 20° north of west at 350 mph. A wind is blowing in the direction 80° north of west at 60 mph. Find the speed and direction of the plane.

12. A plane leaves the ground at an angle of elevation of 40° traveling 600 mph. How long will it take for the plane to be 2 miles above the ground?

13. An airplane flying at 8000 feet passes directly over an observation station. One minute later, the angle of elevation of the plane from the station is 48°. What is the speed of the plane?

14. A ship leaves port at 11:00 A.M. and sails in the direction 32° north of east at 24 mph. Another ship leaves the same port at 1:00 P.M. and sails in the direction 42° south of east at 32 mph. How far apart are the ships at 3:00 P.M.?

15. An airplane flying at 5000 feet is 10 miles from an airstrip. What angle of descent should the plane maintain to reach the front edge of the airstrip?

16. The string on a kite is taut and makes an angle of 62° with the

horizontal. If 200 feet of string are out and the string is held 5 feet above the ground, how high is the kite above the ground?

17. A flagpole is located on top of a building. From a point on level ground 100 feet from the building, the angle of elevation of the top of the flagpole is 52° and the angle of elevation of the top of the building is 46°. How tall is the building and how tall is the flagpole?

18. From the top of a tower 500 feet from a building, the angle of elevation of the top of the building is 79° and the angle of depression of the bottom of the building is 63°. How tall is the building?

11.9 / TRIG RATIOS FOR ANGLES THAT ARE NOT ACUTE

We have defined trig ratios for acute angles in right triangles. An acute angle has measure between 0° and 90°. It is also possible to describe angles that have measure greater than 90° and angles that have negative measure. In this section we will describe these angles and say what the trig ratios mean for them.

Angles that Are Not Acute

Any angle can be positioned with its vertex at the origin of a coordinate system and one side along the positive x-axis. We can think of an angle as being formed by a specific rotation of the second side of the angle, turning from the first side that is on the x-axis. A curved arrow helps to show the rotation.

The angle in (a) of Fig. 11.64 is acute; its measure is approximately 40°. The angle in (b) is a right angle; it has measure 90°. The angle in (c) has measure more than 90°; in fact, its measure is approximately 135°. (An angle with measure between 90° and 180° is sometimes called **obtuse**.) The angle in (d) is called a **straight** angle; you can see that it is made up of two right angles and has measure 180°.

Using the idea of rotation, we can describe angles that have measure

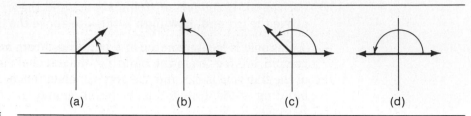

Figure 11.64

more than 180° and angles that have measure less than 0°. By convention, if the rotation is counterclockwise from the x-axis, we say that the measure of the angle is positive, whereas if the rotation is clockwise, we say the measure is a negative number. More examples are shown in Fig. 11.65.

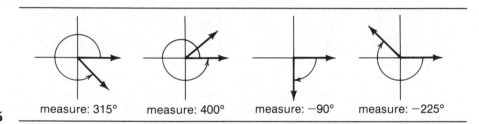

Figure 11.65

Trig Ratios for Angles that Are Not Acute

We cannot look at right triangles to see what trig ratios mean for angles that are not acute; these angles do not fit into right triangles. However, look again at an acute angle, as in Fig. 11.66, positioned at

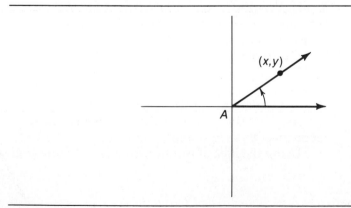

Figure 11.66

the origin and choose any point (x,y) different from the origin on its second side (commonly called the **terminal** side).

To write the trig ratios for the acute angle A, we can form a right triangle containing A (Fig. 11.67). If r denotes the distance from (x,y) to the origin, the trig ratios for A can be given in terms of x, y, and r.

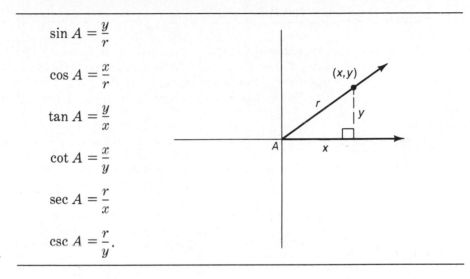

$$\sin A = \frac{y}{r}$$

$$\cos A = \frac{x}{r}$$

$$\tan A = \frac{y}{x}$$

$$\cot A = \frac{x}{y}$$

$$\sec A = \frac{r}{x}$$

$$\csc A = \frac{r}{y}.$$

Figure 11.67

These ratios can be formed not only for acute angles; they can be formed for angles of any measure. If B is any angle, choose a point (x,y) on its terminal side and let r denote the distance from (x,y) to the origin (Fig. 11.68):

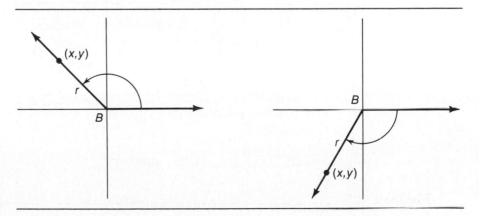

Figure 11.68

$$\sin B = \frac{y}{r}$$

$$\cos B = \frac{x}{r}$$

$$\tan B = \frac{y}{x}$$

$$\cot B = \frac{x}{y}$$

$$\sec B = \frac{r}{x}$$

$$\csc B = \frac{r}{y}.$$

The distance r is always a positive number, but x and y can be positive or negative depending on where the terminal side of the angle is positioned. If we visualize the axes as dividing the coordinate plane into four quadrants, we can say in terms of quadrants which of the trig ratios are positive and which are negative for angles that have their terminal side in the quadrant.

A point in quadrant II, for example, has negative x-coordinate and positive y-coordinate. Thus if an angle B has its terminal side in quadrant II,

$$\sin B = \frac{y}{r} \text{ is a positive number,}$$

$$\cos B = \frac{x}{r} \text{ is a negative number,}$$

$\tan B = \dfrac{y}{x}$ is a negative number.

We can make a chart to summarize these observations:

II		I	
sin +	cot −	sin +	cot +
cos −	sec −	cos +	sec +
tan −	csc +	tan +	csc +

III		IV	
sin −	cot +	sin −	cot −
cos −	sec −	cos +	sec +
tan +	csc −	tan −	csc −

Example 1   Find the trig ratios for an angle C of 135° and an angle D of −45°.

a. To choose a point on the terminal side of the angle C, we observe that the side bisects the right angle of the second quadrant (Fig. 11.69).

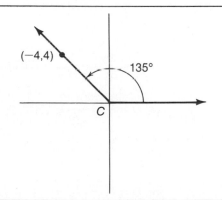

Figure 11.69

Thus we can choose the point (−4,4), for example, on the terminal side. The distance from (−4,4) to (0,0) is

$$r = \sqrt{(-4)^2 + 4^2}$$
$$= \sqrt{32}$$
$$= 4\sqrt{2}.$$

Thus

$$\sin C = \frac{y}{r} = \frac{4}{4\sqrt{2}} = \frac{1}{\sqrt{2}} \doteq .70710678$$

$$\cos C = \frac{x}{r} = \frac{-4}{4\sqrt{2}} = -\frac{1}{\sqrt{2}} \doteq -.70710678$$

$$\tan C = \frac{y}{x} = \frac{4}{-4} = -1$$

$$\cot C = \frac{x}{y} = \frac{-4}{4} = -1$$

$$\sec C = \frac{r}{x} = \frac{4\sqrt{2}}{-4} = -\frac{\sqrt{2}}{1} \doteq -1.4142136$$

$$\csc C = \frac{r}{y} = \frac{4\sqrt{2}}{4} = \frac{\sqrt{2}}{1} \doteq 1.4142136.$$

To find the trig ratios for an angle of $-45°$, we can choose any point on the terminal side, say $(1,-1)$ (Fig. 11.70). The distance from $(1,-1)$ to $(0,0)$ is

$$r = \sqrt{1^2 + (-1)^2}$$
$$= \sqrt{2}.$$

Thus

$$\sin D = \frac{y}{r} = \frac{-1}{\sqrt{2}} \doteq -.70710678$$

$$\cos D = \frac{x}{r} = \frac{1}{\sqrt{2}} \doteq .70710678$$

$$\tan D = \frac{y}{x} = \frac{-1}{1} = -1$$

$$\cot D = \frac{x}{y} = \frac{1}{-1} = -1$$

$$\sec D = \frac{r}{x} = \frac{\sqrt{2}}{1} \doteq 1.4142136$$

$$\csc D = \frac{r}{y} = \frac{\sqrt{2}}{-1} \doteq -1.4142136.$$

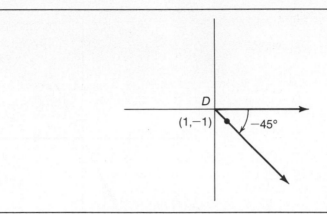

Figure 11.70

b. Another way to find the trig ratios for 135° and −45° is to use the calculator. The values of sin, cos, and tan can be keyed for any angle; the values of cot, sec, and csc must be keyed as the reciprocals of the other three, as we did for acute angles. Here are some examples:

		Display
sin 135°:	135 (sin)	.70710678
cot 135°:	135 (tan) (1/x)	−1
cos −45°:	45 (+/−) (cos)	.70710678
csc −45°:	45 (+/−) (sin) (1/x)	−1.4142136 □

Example 2   Find the trig ratios for an angle A whose terminal side lies in the third quadrant on the line $y = 2x$.

In order to draw the angle, we first graph $y = 2x$. It is shown as the dashed line in Fig. 11.71. We can compute the trig ratios for A in terms of any point on $y = 2x$ in the third quadrant, for example, $(-1,-2)$. The distance from $(-1,-2)$ to $(0,0)$ is

$$r = \sqrt{(-1)^2 + (-2)^2} = \sqrt{5}.$$

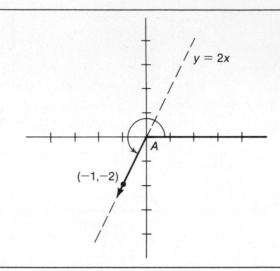

Figure 11.71

Thus

$$\sin A = \frac{y}{r} = \frac{-2}{\sqrt{5}} = -.89442719$$

$$\cos A = \frac{x}{r} = \frac{-1}{\sqrt{5}} = -.4472136$$

$$\tan A = \frac{y}{x} = \frac{-2}{-1} = 2$$

$$\cot A = \frac{x}{y} = \frac{-1}{-2} = .5$$

$$\sec A = \frac{r}{x} = \frac{\sqrt{5}}{-1} \doteq -2.236068$$

$$\csc A = \frac{r}{y} = \frac{\sqrt{5}}{-2} \doteq -1.118034.$$ □

Graphing the Trig Functions

If we restrict our attention to one of the six trig ratios, we are able to say for any angle what the trig ratio is. Thus we are able to draw a graph that shows the relationship between angles and the trig ratio of the angles. You may recall that in Section 6.1 we drew the graphs of $y = \sin x$, $y = \cos x$, and $y = \tan x$ by using the appropriate calculator keys even before we knew what the meanings of the trig ratios were. The graphs looked like Figs. 11.72, 11.73, and 11.74.

$y = \sin x$ Remember that the sine of an angle is the ratio $\frac{y}{r}$. Since r is never 0, there is a value of the sine for every angle.

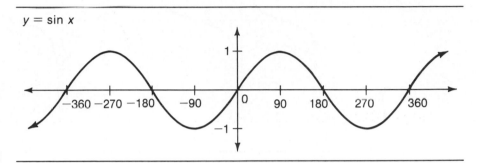

Figure 11.72

$y = \cos x$ Since the cosine of an angle is $\frac{x}{r}$ and r is never 0, there is a value of the cosine for every angle.

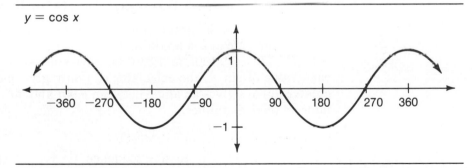

Figure 11.73

$y = \tan x$ The tangent of an angle is the ratio $\frac{y}{x}$. If x is 0, this fraction has no value. Thus there are some angles that do not have a tangent value. The x-coordinate of a point is 0 if the point lies on the y-axis, so any angle with a terminal side on the y-axis will not have a value for tangent. In fact, the graph of $y = \tan x$ has vertical asymptotes at these angles.

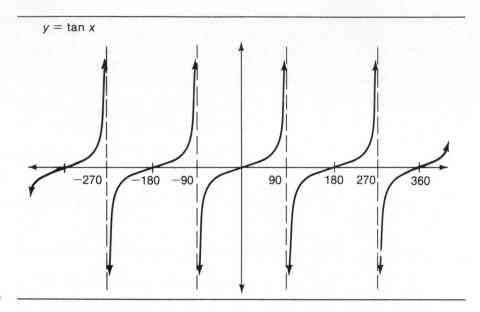

$y = \tan x$

Figure 11.74

The graphs of the other three trig functions will be drawn as part of the exercises.

Finding the Angle When a Trig Ratio Is Known

You can see from the graph of $y = \sin x$ that there are many answers to the question, "What is angle A if $\sin A = 0$?" For example, angles of $-360°$, $-180°$, $0°$, $180°$, and $360°$ all have sine equal to 0. However, for each value of $\sin A$, the calculator can only give one angle A; the calculator always gives an angle between $-90°$ and $90°$. Check your calculator on these examples:

Value of sin A	Keying sequence to get angle A	Display
-1	⬚1 +/− INV sin	-90
$-.8$	.8 +/− INV sin	-53.130102
$-.4$	.4 +/− INV sin	-23.578179
0	⬚0 INV sin	0

Value of sin A	Keying sequence to get angle A	Display
.4	.4 [INV] [sin]	23.578179
.8	.8 [INV] [sin]	53.130102
1	[1] [INV] [sin]	90

Look back at the graph of $y = \sin x$ to see that every possible value of $\sin x$ (all numbers between -1 and 1) occurs for angles between $-90°$ and $90°$. This is the reason the calculator can give an angle between $-90°$ and $90°$ when you enter the sine of an angle and press [INV] [sin].

Example 3 Find the angle A between $90°$ and $270°$ that has $\sin A = -.8$.

We can see in the chart above that the angle between $-90°$ and $90°$ that has sine equal to $-.8$ is approximately $-53.13°$. The graph of $y = \sin x$ in Fig. 11.75 shows that there is one angle A between $90°$ and $270°$ that has $\sin A = -.8$.

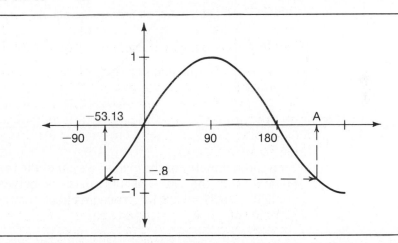

Figure 11.75

From the symmetry of the graph, we conclude that the distance between 180 and A is the same as the distance between -53.13 and 0. Thus $A \doteq 180 + 53.13 = 233.13°$. You can check on your calculator that $\sin 233.13° \doteq -.8$. □

The cosine is like the sine: there are many angles that have the same cosine value. For each value of the cosine, the calculator always gives an angle between 0° and 180° (rather than between −90° and 90°, as in the case of sine). If another angle is needed with the same cosine value, it can be found by inspecting the graph of $y = \cos x$.

Example 4   Find an angle A between 180° and 360° that has $\cos A = .8$.

If we key .8 [INV] [cos] , the calculator gives an angle of approximately 36.87°. To find the corresponding angle between 180° and 360°, we consider the part of the graph of $y = \cos x$ shown in Fig. 11.76.

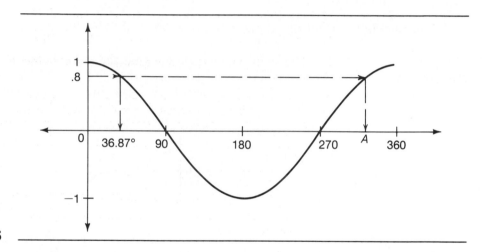

Figure 11.76

From the symmetry of the graph, we conclude that the distance between 360 and A is the same as the distance between 0 and 36.87. Thus $A \doteq 360 − 36.87 = 323.13°$. You can check that $\cos 323.13° \doteq .8$.   □

If the keys [INV] [tan] are used to find angle A for some value of $\tan A$, the calculator always gives angle A between −90° and 90°. Other angles with the same tangent can be determined from the graph of $y = \tan A$ in a manner similar to what we have done for sine and cosine.

**Exercises
11.9**

1. Determine the quadrant in which the following angles lies:

    (a) $-200°$ (b) $750°$

    (c) $-400°$ (d) $560°$

Draw each angle with its vertex at the origin of a coordinate system and one side along the positive x-axis using the measure as given in Exercises 2–11.

2. $75°$ 3. $140°$

4. $220°$ 5. $300°$

6. $470°$ 7. $-40°$

8. $-110°$ 9. $-200°$

10. $-320°$ 11. $-400°$

12. Use your calculator to complete the following table:

A	$\sin A$	$\csc A$	$\cos A$	$\sec A$	$\tan A$	$\cot A$
$43.5°$						
$132.8°$						
$232.6°$						
$302°$						
$-76.5°$						
$-290°$						

Find all six trig ratios for the angles in Exercises 13–18.

13. The point $(2,-1)$ lies on the terminal side of the angle.

14. The point $(-2,-3)$ lies on the terminal side of the angle.

15. The terminal side lies on the line $y = -3x$ in the second quadrant.

16. The terminal side lies on the line $y = \frac{1}{2}x$ in the third quadrant.

17. The terminal side lies on the line $y = \frac{2}{3}x$ in the first quadrant.

18. The terminal side lies on the line $y = -\frac{3}{4}x$ in the fourth quadrant.

Solve for x in Exercises 19–22.

19. $\sin x = 0.3456$, x between $90°$ and $180°$

20. $\cos x = -0.6742$, x between $180°$ and $270°$

21. $\sec x = 2.3142$, x between $-90°$ and $0°$

22. $\cot x = 3.1672$, x between $-180°$ and $-90°$

23. Solve for y and r:

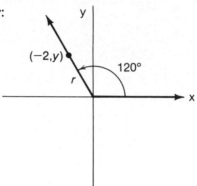

24. Solve for x and r:

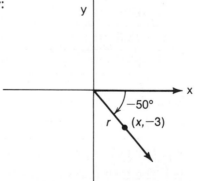

Determine the quadrant in which the angle that satisfies the conditions in Exercises 25–28 lies.

25. $\cos A < 0$, $\sin A > 0$ 26. $\tan A > 0$, $\cos A < 0$

27. $\sec A > 0$, $\csc A < 0$ 28. $\cot A < 0$, $\sec A < 0$

Draw the graph of the functions in Exercises 29–31.

29. $y = \sec x$, $-360° \le x \le 360°$ 30. $y = \csc x$, $-360° \le x \le 360°$

31. $y = \cot x$, $-360° \le x \le 360°$

11.10 / CHAPTER 11 PROBLEM COLLECTION

All parts of the figures below that appear to be circles and semicircles are, in fact, circles and semicircles; all angles that appear to be right angles are right angles; all segments that appear to be parallel are parallel. Find the areas of the figures in Exercises 1–3.

1.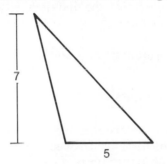

2.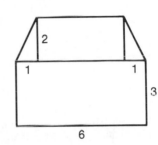

3.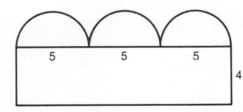

Find the areas of the shaded regions in Exercises 4 and 5.

4.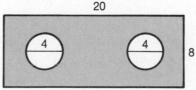

5.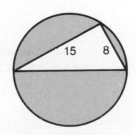

Find the areas of the polygons determined by the points in Exercises 6–9.

6. $(-6,3)$, $(-3,1)$, and $(2,2)$

7. $(-2,2)$, $(2,4)$, and $(4,0)$

8. $(-5,8)$, $(-3,0)$, $(5,2)$, and $(3,10)$

9. $(-2,2)$, $(0,8)$, $(8,6)$, and $(6,0)$

10. Assume that 1 inch on a map represents 80 feet in the real world.

(a) A rectangular lot measures 2.6 inches by 3.5 inches on the map. What is its actual area in square feet?

(b) A rectangular lot measures 200 feet by 110 feet. What is the area of the lot on the map in square inches?

11. The ruler distance between two points on a piece of graph paper is $1\frac{7}{32}$ inches. Determine the distance between these points in terms of graph-paper units if

(a) $\frac{1}{4}$ inch represents 1 unit on this graph paper

(b) $\frac{1}{2}$ inch represents 1 unit on this graph paper

12. Assume that 1 inch represents 4 units on a piece of graph paper. Determine the distance between two points on this graph paper in terms of graph-paper units if the ruler distance is

(a) $3\frac{1}{2}$ inches

(b) $4\frac{7}{16}$ inches

Find the distances between the points in Exercises 13 and 14.

13. $(-3,2)$ and $(5,-1)$

14. $(-4,-2)$ and $(7,-5)$

Solve for x in Exercises 15–20.

15.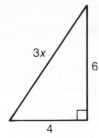

16.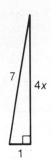

17. $\sin x = 0.4376$, x between $0°$ and $90°$

18. $\tan x = 1.0303$, x between $180°$ and $270°$

19. $\cot x = -0.8943$, x between $90°$ and $180°$

20. $\sec x = -2.4253$, x between $90°$ and $180°$

21. Determine y so that the distance between $(2,y)$ and $(4,5)$ is 2 units.

22. Find all points on the y-axis that are 6 units from $(2,-1)$.

23. Find all points on the x-axis that are 7 units from $(-3,-2)$.

Determine whether the given points in Exercises 24 and 25 form a right triangle.

24. $(1,1)$, $(2,7)$, and $(4,2)$

25. $(-3,-1)$, $(3,3)$, and $(7,-3)$

Write an equation whose graph is a circle with center and radius as given in Exercises 26 and 27.

26. Center at $(-2,3)$, radius 4

27. Center at $(3,-4)$, radius $\sqrt{7}$

28. Determine h so that the circle $(x - h)^2 + (y - 3)^2 = 4$ passes through the point $(1, 3 + \sqrt{3})$.

Find the coordinates of the midpoint of the line segment determined by the points in Exercises 29 and 30.

29. $(-3,4)$ and $(5,-3)$

30. (1,4) and (6,2)

31. If $(-2,1)$ is the midpoint of the line segment joining $(-5,-2)$ and (a,b), find a and b.

32. Prove that the line determined by (2,5) and the midpoint of the line segment joining $B = (1,2)$ and $C = (5,4)$ is perpendicular to $\overline{BC}$.

33. Let $A = (0,0)$, $B = (5,1)$, and $C = (2,4)$.

   (a) Write an equation for each of the perpendicular bisectors of the sides of the triangle ABC.

   (b) Show the three lines in (a) have a common point of intersection.

Write the six trig ratios for the angle A in Exercises 34–38.

34.

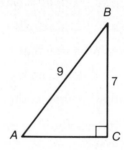

35. $\sin A = 0.8976$ 36. $\sec A = 2.3508$

37. The point $(3,-2)$ lies on the terminal side of angle A.

38. The terminal side of angle A lies on the line $y = \frac{3}{5}x$ and is in the third quadrant.

Solve the right triangles in Exercises 39–43.

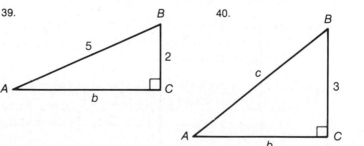

39.

40.

$\sin A = 0.6712$

41.

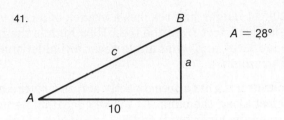

$A = 28°$

42.

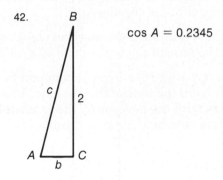

$\cos A = 0.2345$

43.

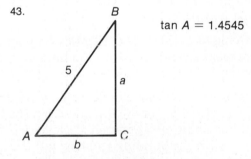

$\tan A = 1.4545$

44. A plane starts 34° south of east at 600 mph. A wind is blowing 22° north of east at 80 mph. Find the actual speed and direction of the plane.

45. Bill first walks 6 miles at 30° south of west, then 8 miles at 45° north of west, and then 2 miles due north. How far is he from his starting point?

46. At 10:30 A.M. a boat leaves port heading 45° south of west at 32 mph. At 11:00 A.M. another boat leaves the same port heading 45° south of east at 26 mph. What is the distance between the boats at 2:00 P.M.?

47. A 40-foot ladder just reaches a branch of a tree with the base of the ladder 5 feet from the tree. How high is the branch above ground level, and what angle does the ladder make with the horizontal?

48. A surveyor sights a tree directly across a river and then measures 120 feet along the bank. At the new location he finds that the angle between the river bank and his line of sight to the tree is 56°. Find the approximate width of the river.

49. Find the height of a tower if the angle of elevation from a point 130 feet from the base of the tower on level ground is 82°.

50. A guy wire 80 feet long runs from an antenna to a point on level ground 18 feet from the base of the antenna. Find the angle the guy wire makes with the horizontal, the angle the wire makes with the antenna, and how far above ground the wire is attached to the antenna.

51. A plane flies to a city that is 220 miles west and 430 miles south of the plane's present location. Find the bearing and the distance the plane must travel.

52. A plane flying at 10,000 feet is 15 miles from an airstrip. What angle of descent should the plane maintain to reach the front edge of the airstrip?

53. A flagpole is 400 feet from a building on level ground. The angle of depression of the top of the flagpole from the top of the building is 23°, and the angle of depression of the base of the flagpole is 26.5°. How tall is the building and how tall is the flagpole?

Answers to odd-numbered exercises

CHAPTER 1   *Exercises 1.1*

1. 34   3. 70   5. 1.6284404   7. 6.043956   9. 55

11. $\boxed{1}\,\boxed{6}\,\boxed{\cdot}\,\boxed{4}\,\boxed{\div}\,\boxed{4}\,\boxed{\cdot}\,\boxed{2}\,\boxed{=}$; 3.9047619

13. $\boxed{4}\,\boxed{3}\,\boxed{\cdot}\,\boxed{2}\,\boxed{+}\,\boxed{3}\,\boxed{5}\,\boxed{\cdot}\,\boxed{7}\,\boxed{+}\,\boxed{1}\,\boxed{6}\,\boxed{\cdot}\,\boxed{8}\,\boxed{=}$
$\boxed{\div}\,\boxed{3}\,\boxed{\cdot}\,\boxed{2}\,\boxed{=}$; 29.90625

15. $(3 \times 5) + (4 \times 6)$ 17. $\dfrac{3 \times 5}{6}$, or $(3 \times 5) \div 6$ 19. $(3 \div 6) \div 5$

21. $5 \times 4 + 3$, or $(5 \times 4) + 3$ 23. $3 \times 8 + 7$, or $(3 \times 8) + 7$

25. $\dfrac{8 + 7}{11}$, or $(8 + 7) \div 11$ 27. $\dfrac{12}{6 + 8}$, or $12 \div (6 + 8)$

29. *(a)*

First number	Second number
8	19
11	25
14	31
13	29

(b) Multiply the first number by 2 and then add 3 to the result.

(c) Subtract 3 from the second number and then divide the result by 2.

31. *(a)*

Width	Length
15	27
20	37
25	47
24	45

(b) Multiply the width by 2 and then subtract 3 from the result.

(c) Add 3 to the length and then divide the result by 2.

Exercises 1.2

1.

+	0	2	3	5	8	12	16
8	8	10	11	13	16	20	24

3.

×	0	3	5	7	10	14	20
5	0	15	25	35	50	70	100

5. $4 + 4 + 4$ 7. $2 + 2 + 2 + 2 + 2 + 2 + 2 + 2$

9. $(2 + 2 + 2) + (2 + 2 + 2) + (2 + 2 + 2) + (2 + 2 + 2)$

11.

Gallons used	1	2	5	10	12	18	25	n
Miles traveled	22	44	110	220	264	396	550	$22n$

13. *(a)* $106   *(b)* $1060   *(c)* $3710   *(d)* $4531.50
    (e) $(N + .06N)$

15. *(a)* $6.20   *(b)* $10.00   *(c)* $24.25   *(d)* $95.50
    (e) $(.95M + .50)$

17. *(a)* 4400 miles   *(b)* 4800 miles   *(c)* 6000 miles
    (d) 7200 miles   *(e)* $400b$ miles

19. 14.833333 cents per ounce

21. 6.5 miles per gallon

23. *(a)* 5.8747×10^{12} miles, or 5,874,652,224,000 miles
    (b) 2.5261×10^{13} miles, or 25,261,004,563,200 miles

25. *(a)*

Number of dimes	Number of quarters	Value ($)
0	40	10.00
10	30	8.50
20	20	7.00
30	10	5.50
40	0	4.00
35	5	4.75

(b) Subtract the number of quarters from 40.

(c) Multiply the number of dimes by .10 and the number of quarters by .25 and then add the results.

27. *(a)*

First number	Second number
2	7
4	31
8	127
12	287
6	71

(b) Square the first number, then multiply the square by 2, and then subtract 1 from the result.

(c) Add 1 to the second number, then divide the sum by 2, and finally find a number whose square is this result.

Exercises 1.3

1. Base -2, exponent 3   3. Base 4, exponent 5

5. Base -3, exponent -2 7. 55

9. 265.86807   11. 468.75

13. $\boxed{(}$ $\underline{3.2}$ $\boxed{+}$ $\underline{4.5}$ $\boxed{)}$ $\boxed{x^2}$; 59.29

15. $\boxed{(}$ $\boxed{2}$ $\boxed{\times}$ $\boxed{3}$ $\boxed{)}$ $\boxed{x^2}$ $\boxed{+}$ $\underline{12}$ $\boxed{=}$; 48

17. $\boxed{3}$ $\boxed{\times}$ $\boxed{2}$ $\boxed{y^x}$ $\boxed{4}$ $\boxed{+}$ $\boxed{4}$ $\boxed{y^x}$ $\boxed{3}$ $\boxed{=}$; 112

19. $4^2 + 6^2$ 21. $(4 + 6)^2$ 23. $3^3 - 4^2$

25. $2^2 + 3^2 + 5^2$ 27. $3^2 + 2^2$ 29. $(3^2)^2$

31. 2^6 33. 3^2 35. $2^2 + 3^4$

37. 100 square inches   39. 216 cubic inches

Exercises 1.4

1. 439.11   3. 210   5. 108

7. $\underline{12}$ $\boxed{\times}$ $\boxed{(}$ $\underline{13}$ $\boxed{+}$ $\underline{15}$ $\boxed{)}$ $\boxed{=}$ and $\underline{12}$ $\boxed{\times}$ $\underline{13}$ $\boxed{+}$ $\underline{12}$ $\boxed{\times}$ $\underline{15}$ $\boxed{=}$

9. $8(3 + 4)$ 11. $6(5 - 3)$ 13. $6(4 + 5)$

15. $5[7 + 3(8 - 4)]$ 17. $3a + 2a = (3 + 2)a = 5a$

19. $y^3 - 0.35y^3 = (1 - 0.35)y^3 = 0.65y^3$

21. $1.3c + 0.3b$ 23. $0.45b^4 + b^2$

25. $0.1x + 900$ 27. $8x^2 + x$

29. $\boxed{7}\,\boxed{\times}\,\boxed{4}\,\boxed{+}\,\boxed{7}\,\boxed{\times}\,\boxed{5}\,\boxed{+}\,\boxed{7}\,\boxed{\times}\,\boxed{6}\,\boxed{+}\,\boxed{7}\,\boxed{\times}\,\boxed{7}$ $\boxed{+}\,\boxed{7}\,\boxed{\times}\,\boxed{8}\,\boxed{=}$ (20 key strokes) and $\boxed{7}\,\boxed{\times}\,\boxed{(}\,\boxed{4}\,\boxed{+}\,\boxed{5}\,\boxed{+}\,\boxed{6}\,\boxed{+}\,\boxed{7}\,\boxed{+}\,\boxed{8}\,\boxed{)}\,\boxed{=}$ (14 key strokes)

31. $40(.99) = 40(1 - .01) = 40 - .40 = \$39.60$

33. (a) $8(6 + 5) = \$88$ and $8(6) + 8(5) = \$88$
    (b) $8.5(6 + 5) = \$93.50$ and $8.5(6) + 8.5(5) = \$93.50$

35. $52(6 + 7 + 8) = 1092$ miles and $52(6) + 52(7) + 52(8) = 1092$ miles

Exercises 1.5

1. $2^2 \cdot 3^2$ 3. $2 \cdot 3$ 5. $2^2 \cdot 3^3 \cdot 5^2 \cdot 7^3 \cdot 11^2$

7. $2^3 \cdot 3^3 \cdot 7^3 \cdot 11^2$ 9. 1, 2, 4, 8   11. 1, 2, 5, 10

13. 1, 2, 3, 4, 6, 8, 12, 24   15. $1, p, p^2, p^3$

17. $2^3 \cdot 3^3 \cdot 5^2$ 19. Prime   21. Prime

23. 29·31   25. 5^6

27. (a) $2 \cdot 3 \cdot 11$ (b) $2^2 \cdot 3$ (c) $2^3 \cdot 3 \cdot 5 \cdot 7^2$ (d) 1

Exercises 1.6

1. 1, 3   3. 1, 2, 3, 6   5. GCF = 1, LCM = $11 \cdot 13 = 143$

7. GCF = $2 \cdot 3 = 6$, LCM = $2^2 \cdot 3^2 \cdot 5 \cdot 7 = 1260$

9. GCF = $3^2 \cdot 5 = 45$, LCM = $3^2 \cdot 5^2 = 225$

11. GCF = 1, LCM = $2^2 \cdot 3^2 \cdot 5 \cdot 7 \cdot 11 = 13,860$

13. GCF = $5 \cdot 7 = 35$, LCM = $2 \cdot 3 \cdot 5 \cdot 7 \cdot 11 = 2310$

15. In 24 minutes, since the LCM of 8 and 12 is 24.

17. 12:36 P.M., since the LCM of 12 and 18 is 36.

19. Every 60 seconds, or 1 minute, since the LCM of the time between blinks, namely, 5 seconds, 4 seconds, and 3 seconds is 60.

21.

Number of stations train moves forward	Stations train stops at	Total number of stations stopped at
1	1, 2, 3, 4, 5, 6, 7, 8, 9, 10, 11, 12	12
2	2, 4, 6, 8, 10, 12	6
3	3, 6, 9, 12	4
4	4, 8, 12	3
5	5, 10, 3, 8, 1, 6, 11, 4, 9, 2, 7, 12	12
6	6, 12	2
7	7, 2, 9, 4, 11, 6, 1, 8, 3, 10, 5, 12	12
8	8, 4, 12	3
9	9, 6, 3, 12	4
10	10, 8, 6, 4, 2, 12	6
11	11, 10, 9, 8, 7, 6, 5, 4, 3, 2, 1, 12	12
12	12	1

Exercises 1.7

1. -7 3. -4

5. $-(2 - 3) = -(-1) = 1$, or $-(2 - 3) = -2 + 3 = 1$

7. $3x^2$ 9. 11

11. -32 13. -18

15. 8.3647552

17. $\underline{64}$ $\boxed{-}$ $\underline{32}$ $\boxed{+/-}$ $\boxed{+}$ $\underline{72}$ $\boxed{+/-}$ $\boxed{-}$ $\underline{49}$ $\boxed{=}$; -25

19. $\underline{17.8}$ $\boxed{+/-}$ $\boxed{\times}$ $\underline{13.4}$ $\boxed{+/-}$ $\boxed{\div}$ $\underline{10.5}$ $\boxed{+/-}$ $\boxed{=}$; -22.71619

21. $-12 + (-2)x$

23. $-2 + (-3)^2$

25. $<$

27. $<$

29. $=$

31. 0, 1, 2, 3, 4, 5, 6

33. 3, 4, 5, 6, 7

35. $-4, -3, -2, -1, 0, 1, 2, \ldots$

37. 0

39. 38

41. $\dfrac{1}{5} = 0.2$

43. $-9x + 2$

45. $-3x^2 + 10x$

47. $(a + b - c) + (c - b - a) = a + b - c + c - b - a = a + b - b - a = a - a = 0$ or $-(a + b - c) = -a - b + c = c - b - a$

49. Net gain of 3

51. $32°\text{F}$

53. 204 pages, $305 - N$ pages

Exercises 1.8

1. 168.56

3. -4483.08

5. -11.37037

7. 26

9. 14,227.824

11. $\underline{22.7}$ $\boxed{-}$ $\underline{49.1}$ $\boxed{+}$ $\underline{86.8}$ $\boxed{=}$ $\boxed{\div}$ $\boxed{(}$ $\underline{13.5}$ $\boxed{+}$ $\underline{29.7}$ $\boxed{)}$ $\boxed{=}$; 1.3981481

13. $\boxed{2}$ $\boxed{\times}$ $\boxed{3}$ $\boxed{x^2}$ $\boxed{+}$ $\boxed{4}$ $\boxed{y^x}$ $\boxed{3}$ $\boxed{=}$; 82

15. $\dfrac{18}{-6}(4)$

17. $5 + (-3)^2$

19. $2 \cdot 3^4 + (-5)^2$

21. $4x + 6$

23. $-7(x + y)$

25. $2 \cdot 4^3 + 3 \cdot 5^4$

27.

Price per pound	7.20	8.32	14.24	10.72	16.16	x	$16y$
Price per ounce	.45	.52	.89	.67	1.01	$x/16$	y

29. *(a)* $30.96   *(b)* 16 gallons

31. *(a)* 137.6 miles   *(b)* 12.5 gallons

33. 1, 2, 5, 10, 25, 50

35. $29 \cdot 31$

37. $2 \cdot 3 \cdot 7 \cdot 17 \cdot 19$

39. $\text{GCF} = 5$, $\text{LCM} = 2^2 \cdot 3^3 \cdot 5 \cdot 7 \cdot 11 \cdot 13 = 540{,}540$

41. $GCF = 2 \cdot 7 = 14$, $LCM = 2^3 \cdot 3^2 \cdot 7^2 \cdot 11 \cdot 13 = 504{,}504$

43. $\underline{13}$ $\boxed{\times}$ $\boxed{(}$ $\underline{18}$ $\boxed{+}$ $\underline{19}$ $\boxed{)}$ $\boxed{=}$ and $\underline{13}$ $\boxed{\times}$ $\underline{18}$ $\boxed{+}$ $\underline{13}$ $\boxed{\times}$ $\underline{19}$ $\boxed{=}$

45. $5.2z^2$ 47. $6x^2 - 5x - 6$

49. $x^3 - 8$

51. $(a - b + c) + (-c + b - a) = a - b + c - c + b - a =$
$a - b + b - a = a - a = 0$ or $-(a - b + c) = -a + b - c =$
$-c + b - a$

53. 1, 2, 3 55. $-2, -1, 0$

57. $\dfrac{3}{2}$, or 1.5 59. $\dfrac{-3}{2}$, or -1.5

61. In 336 seconds, since the LCM of 42 and 48 is 336.

63. 132.25 square inches

65. *(a)*

Number of nickels	Number of quarters	Value of coins ($)
10	30	8.00
20	20	6.00
30	10	4.00
40	0	2.00
25	15	5.00
x	$40 - x$	$.05x + .25(40 - x)$

(b) Subtract the number of nickels from 40.

(c) Multiply the number of nickels by .05 and the quantity 40 minus the number of nickels by .25, then add the results.

67.

First number	Second number	Sum	Product
-2	16	14	-32
-3	11	8	-33
3	11	14	33
5	-5	0	-25
-5	-5	-10	25

CHAPTER 2   *Exercises 2.1*

1. $\dfrac{9}{10}, \dfrac{6}{5}, \dfrac{5}{2}, 3$

3. $-1.8, -0.7, 0.8, 1.5$

5.

```
   -2   -1    0    1    2
```

7.

```
   -2   -1    0    1    2
```

9. .4

11. .3

13. 12, 24, 36, . . .

15. 3, 6, 9, . . .

17. $\dfrac{22}{5}$

19. $-\dfrac{29}{23}$

21. .007

23. $-.23$

25. 1.75

27. $\dfrac{7}{10}$

29. $-\dfrac{2}{100}$, or $-\dfrac{1}{50}$

31. $\dfrac{75}{10^3}$

33. Not possible

35. .55555556

37. .37500000, or .375

39. $\dfrac{3}{5}$ (cut each pizza into 5 equal pieces and give each person 3

pieces); $\dfrac{6}{10}$ (cut each pizza into 10 equal pieces and give each

person 6 pieces); $\dfrac{9}{15}$ (cut each pizza into 15 equal pieces and give

each person 9 pieces).

41. $\dfrac{1}{60}$ of a minute, $\dfrac{1}{3600}$ of an hour, $\dfrac{1}{86,400}$ of a day

43. $\dfrac{4}{32}$ alcohol, $\dfrac{28}{32}$ water

45. $\dfrac{18}{74}$ alcohol, $\dfrac{56}{74}$ water

Exercises 2.2

1. $=$ 3. $>$ 5. $<$ 7. $=$

9. $<$ 11. $<$ 13. $>$ 15. $=$

17. $<$ 19. -2 21. 10   23. $.3, .33, .\overline{3}$

25. $1.6, 1.66, 1.666, 1.\overline{6}$ 27. $-\frac{8}{7}, -1.1427, 0, .8571, \frac{6}{7}$

29. $0, 1, 2, 3, 4, 5, 6$ 31. $-10, -9, \ldots, -1, 0, 1, 2$

33. $0, 1, \ldots, 7$ 35. 3 bars for 41 cents

37. 4.7 ounces for 79 cents   39. the 60-gallon solution

Exercises 2.3

1. $8\frac{7}{13}, 8.5384615$ 3. $-\frac{71}{4}, -17.75$

5. $\frac{172}{10}, 17\frac{2}{10}$; or $\frac{86}{5}, 17\frac{1}{5}$ 7. $-\frac{17567}{1000}, -17\frac{567}{1000}$

9. $\frac{29}{36}$ 11. $\frac{7}{4}$

13. $\frac{101}{12}$ 15. $\frac{5}{12}$

17. $-\frac{29}{78}$ 19. $\frac{9}{6}$, or $-\frac{3}{2}$

21. 0   23. 0

25. Both $\frac{5}{12} + \frac{7}{18}$ and $\frac{29}{36}$ are $\doteq .80555556$.

Both $\frac{8}{15} - \frac{4}{35}$ and $\frac{44}{105}$ are $\doteq .41904762$.

Both $1\frac{1}{3} - 2\frac{5}{6}$ and $-\frac{9}{6}$ are $\doteq -1.5$.

27.

	Recreation	Picnicking	Parking	Remaining part
Shadeywood	$\frac{3}{8}$	$\frac{1}{4}$	$\frac{3}{16}$	$\frac{3}{16}$
10th Avenue	$\frac{1}{3}$	$\frac{2}{9}$	$\frac{1}{18}$	$\frac{7}{18}$
Linden	$\frac{2}{5}$	$\frac{1}{3}$	$\frac{1}{6}$	$\frac{3}{30}$, or $\frac{1}{10}$

29. $1\frac{2}{5}$ degrees above normal, $\frac{14}{15}$ degrees below normal, $4\frac{7}{10}$ degrees above normal

31. $\frac{119}{10}$, or $11\frac{9}{10}$ feet

33. $-3\frac{1}{2} + 2\frac{1}{5}$

35. $\frac{17}{6}x + 5y$

37. $-\frac{1}{2}z^3 + 4$

39. $\frac{2}{3}a^2 - \frac{1}{6}b$

41. (a) $\frac{1}{3} + \frac{1}{9} = \frac{4}{9}$

    (b) $\frac{1}{3} + \frac{1}{9} + \frac{1}{27} = \frac{13}{27}$

    (c) $\frac{1}{3} + \frac{1}{9} + \frac{1}{27} + \frac{1}{81} = \frac{40}{81}$

    (d) $\frac{1}{3} + \frac{1}{9} + \frac{1}{27} + \frac{1}{81} + \frac{1}{243} = \frac{121}{243}$

        $\frac{1}{3} + \frac{1}{9} + \frac{1}{27} + \frac{1}{81} + \frac{1}{243} + \frac{1}{729} = \frac{364}{729}$

        $\frac{1}{3} + \frac{1}{9} + \frac{1}{27} + \frac{1}{81} + \frac{1}{243} + \frac{1}{729} + \frac{1}{2187} = \frac{1093}{2187}$

    (e) $\frac{1}{3} + \frac{1}{9} + \cdots + \frac{1}{3^{101}}$

    (f) $\dfrac{\frac{3^{101} - 1}{2}}{3^{101}}$

43. (a) .44444444
    (b) .48148148
    (c) .49382716
    (d) .49794239, .49931413, .49977138
    (f) $\doteq .5$

Exercises 2.4

1. $\frac{1}{6}$

3. $-\frac{21}{11}$

5. $-\frac{143}{3}$

7. $\frac{45}{4}$

9. $\frac{1}{7}$

11. $-\frac{33}{26}$

13. Both $\dfrac{20}{63} \cdot \dfrac{21}{40}$ and $\dfrac{1}{6} \doteq .16666667$.

Both $\dfrac{247}{187} \div \dfrac{39}{34}$ and $\dfrac{38}{33} \doteq 1.1515152$.

Both $\left(16\dfrac{1}{4}\right)\left(-2\dfrac{14}{15}\right)$ and $-\dfrac{143}{3} \doteq -47.666667$.

Both $\left(3\dfrac{3}{7}\right) \div \left(1\dfrac{11}{14}\right)$ and $\dfrac{48}{25} = 1.92$.

15. -3.4135 17. $.045679$

19. 4 21. 3

23. $-\dfrac{315}{28}$ 25. $\dfrac{30}{252}$

27. $\dfrac{11}{3}x - \dfrac{31}{12}$ 29. $\dfrac{1}{2}a^2 - \dfrac{1}{2}b^2$

31. $\dfrac{1}{2} + \dfrac{1}{4} - \dfrac{1}{8}$ 33. $2 \cdot \dfrac{1}{-5} \cdot 7 \cdot \dfrac{1}{3}$, or $\dfrac{2}{-5} \cdot \dfrac{7}{3}$

35. *(a)* $59.40   *(b)* 5.5 hours

37. *(a)* $140   *(b)* $230   *(c)* $356.80   *(d)* $40y$

39. *(a)* 11 minutes   *(b)* 15 minutes   *(c)* 23 minutes

    (d) $\left(\dfrac{x - .45}{.12} + 3\right)$ minutes

41. $24

43.

Wholesale price ($)	10	15	20.50	25	40	x
Retail price ($)	14	21	28.70	35	56	$x + \dfrac{2}{5}x$, or $\dfrac{7}{5}x$

45.

First number	Second number	Sum
4	3	7
10	6	16
-16	-7	-23
22	12	34
-24	-11	-35
18	10	28
x	$\dfrac{1}{2}x + 1$	$x + \dfrac{1}{2}x + 1$, or $\dfrac{3}{2}x + 1$

47.

Smallest piece	Largest piece	Other piece	Total length
4	16	6	26
10	40	15	65
16	64	24	104
22	88	33	143
18	72	27	117
x	$4x$	$\frac{3}{2}x$	$x + 4x + \frac{3}{2}x$, or $\frac{13}{2}x$

Exercises 2.5

1. $\dfrac{67}{10}$

3. $-\dfrac{374}{10}$, or $-\dfrac{187}{5}$

5. $\dfrac{234{,}567}{10{,}000}$

7. $\dfrac{48}{2695}$

9. $-\dfrac{78}{5}$

11. $\dfrac{13}{14}$

13. $\dfrac{11}{7}$

15. 2.056565656

17. 1.2436363636

19. 2.05

21. -2.6

23. $\dfrac{5}{24}$

25. $\dfrac{71}{15}$

27. $-\dfrac{38}{15}$

29. $-0.5x + 15.1$

31. $-2.9x^2 - 8.84x$

33. $\dfrac{38}{99}$

35. $\dfrac{72}{999}$, or $\dfrac{8}{111}$

37. 1

39. $\dfrac{6}{5}$, or 1.2

41. 2.5, 3, 3.5

43. $-.5$, 0, .5, 1, 1.5

Exercises 2.6

1.

Percent form (%)	Decimal form	Fraction form
12.6	.126	$\dfrac{126}{1000}$
132	1.32	$\dfrac{132}{100}$
.5	.005	$\dfrac{5}{1000}$
$\dfrac{1}{2}$	.005	$\dfrac{5}{1000}$
38	.38	$\dfrac{38}{100}$
167	1.67	$\dfrac{167}{100}$
.4	.004	$\dfrac{4}{1000}$
.05	.0005	$\dfrac{5}{10000}$
72	.72	$\dfrac{72}{100}$
246	2.46	$\dfrac{246}{100}$
3.5	.035	$\dfrac{35}{1000}$
37.5	.375	$\dfrac{3}{8}$

3. 75

5. 112%

7. $275

9. $3.75, $.15$N$

11. 38.709677%, or 38.7%

13. $460

15. $611.90

17. $196

19. $119

21.

Wholesale price ($)	Markup ($)	Retail price ($)
10	4	14
20	8	28
30	12	42
40	16	56
50	20	70
55	22	77
x	$.4x$	$x + .4x$, or $1.4x$
90	36	126

23.   37.5% salt, 62.5% water

25.   15.2 liters alcohol, 24.8 liters water

27. $28.80   29. $14,500

Exercises 2.7

1.   1.035353535   3.   2.3542424242

5. $q = 6$, $r = 24$; $6(72) + 24 = 456$

7. $q = 0$, $r = 75$; $0(83) + 75 = 75$

9.   Quotient 5, remainder 72   11. $46\dfrac{7}{18}$

13. $-172\dfrac{53}{78}$ 15. $\dfrac{215}{12}$

17. $-\dfrac{728}{99}$ 19.   .125

21. $.08\overline{3}$ 23. $.0\overline{45}$

25. $.\overline{123}$ 27. $.12\overline{34}$

29. $.\overline{052631578947368421}$

Exercises 2.8

1. $787 = 14(56) + 3$ 3.   GCF = 12, LCM = 563,940

5.   GCF = 1147, LCM = 14,911

7.   GCF = 1, LCM = 12,612,600 = (4312)(2925)

9. $\dfrac{49,539}{1,856,400}$

11. $\dfrac{111,679}{2,807,434}$, or $\dfrac{111679}{(938)(2993)}$

13. $\dfrac{495}{637}$

15. $\dfrac{1186}{1575}$

17. $\left(\dfrac{1853}{72} - 25\right) 72$

19. GCF = 1

Exercises 2.9

1. $-\dfrac{1}{10}$, or -0.1

3. $\dfrac{11}{36}$, or $.3055556$

5. $\dfrac{2}{3}$, or $.66666667$

7. $\dfrac{1}{6}$, or $.16666667$

9. -2.1111829

11. 0

13. -247.5

15. 7, 8, 9, 10, 11

17. $-11, -10, \ldots, -1, 0, 1$

19. $\dfrac{6}{5} - \dfrac{3}{8}$

21. $\left(\dfrac{235}{34} - 6\right) 34$

23. $50 - .123(50)$

25. $-\dfrac{17}{60}, -\dfrac{11}{45}, 0, \dfrac{11}{45}, \dfrac{17}{60}$

27. 5 pounds 4 ounces for $2.09

29. *(a)* $52.50   *(b)* 7.2 hours, or 7 hours 12 minutes

31. $\dfrac{29}{24}$

33. $-3\dfrac{11}{12}$, or $-\dfrac{47}{12}$

35. $\dfrac{1}{3}, \dfrac{2}{3}, 1, \dfrac{4}{3}, \dfrac{5}{3}, 2, \dfrac{7}{3}, \dfrac{8}{3}$

37. 4

39.

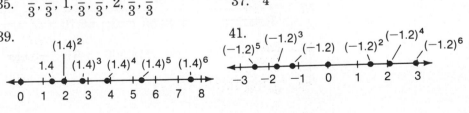

43. $\dfrac{122}{99}$

45. $\dfrac{594}{900}$, or $\dfrac{33}{50}$

47.

Decimal form	Percent form (%)	Fraction form
0.67	67	$\dfrac{67}{100}$
1.23	123	$\dfrac{123}{100}$
0.452	45.2	$\dfrac{452}{1000}$
0.37	37	$\dfrac{37}{100}$
1.12	112	$\dfrac{112}{100}$
0.756	75.6	$\dfrac{756}{1000}$
0.13	13	$\dfrac{13}{100}$
1.42	142	$\dfrac{142}{100}$
0.017	1.7	$\dfrac{17}{1000}$

49. 99.2

51. 137.5%

53. $38.\overline{18}$

55. 31.2 gallons alcohol, 28.8 gallons water

57.

Wholesale ($)	30	40	46	50	x
Retail ($)	43.50	58	66.70	72.50	$x + .45x$, or $1.45x$

59. (a) $14,784   (b) $14,800   61. $-0.68x^2 + 0.29y$

63. $\dfrac{x^2}{6} - \dfrac{x}{6}$

65. (a) GCF $= 84$, LCM $= 75{,}516$ (b) $\dfrac{31}{29}$

(c) $-\dfrac{84}{75{,}516}$, or $\dfrac{-1}{899}$

67. $16,620.40

69. $\frac{3}{2}$, or 1.5

71. -4

73.

Number of nickels	Number of dimes	Number of quarters	Value of coins ($)
4	7	8	2.90
6	9	12	4.20
10	13	20	6.80
x	$x + 3$	$2x$	$.05x + .10(x + 3) + .25(2x)$, or $.65x + .3$

75. 2:09 P.M., since the LCM of 54 and 60 is 540 seconds, or 9 minutes.

CHAPTER 3　　*Exercises 3.1*

1. $<$

3. $=$

5. 5

7. 5

9. 10

11. 7

13. 1024

15. -19683

17. -108

19. 5,764,801

21. -59

23. 80

25. -342.60832

27. $\boxed{3}\boxed{+/-}\boxed{2}\boxed{y^x}\boxed{5}\boxed{+}\boxed{4}\boxed{\times}\boxed{3}\boxed{y^x}\boxed{6}\boxed{=}$; 2820

29. $4(5)^2 + 3^4$

31. $-3(2^4) - 2^2(3^2)$

33.

35. $(-1)^n = \begin{cases} 1 \text{ if } n \text{ is even} \\ -1 \text{ if } n \text{ is odd} \end{cases}$

Exercises 3.2

1. $=$

3. $<$

5. x^{15}

7. x^9

9. $x^{12}y^8$

11. $-6x^6y^5$

13. y^{12}

15. $8x^{11}y^{12}$

17. x^8

19. $-x^{10}y^5$ 21. $(x + 2)^5$ 23. 4

25. 3 27. 3 29. 3

31. -216 33. 1 35. -32

37. $x^4 - y^4$ 39. $xy^3 - xy^4$

41. $[3(4^3) - 5(6^2)] \div [7^4 - 2(5^3)]$, or $\dfrac{3(4^3) - 5(6^2)}{7^4 - 2(5^3)}$

Exercises 3.3

1. $\dfrac{1}{8}$, or .125 3. -8

5. $-\dfrac{1}{8}$, or $-.125$ 7. 8

9. -1 11. 64

13. -5 15. 1

17. 0.38888889, or $\dfrac{7}{18}$ 19. 14

21. -4 23. -4

25. 5 27. 0, 3^{-4}, 3^{-3}, 3^{-2}, 3^{-1}, 3^0

29. $\dfrac{1}{a^6}$ 31. $\dfrac{1}{a^2 b^3}$

33. $\dfrac{2^5}{r^4}$ 35. $-\dfrac{a^7}{b^2}$

37. $\dfrac{2^6 x^7}{3^3 y^4}$ 39. $2y^2 + \dfrac{y}{4}$

41. $(5^{-3})^{-4}$ 43. $x + \dfrac{1}{x^2}$

45. $(-1)^n = \begin{cases} 1 \text{ if } n \text{ is even} \\ -1 \text{ if } n \text{ is odd} \end{cases}$

Exercises 3.4

1. 3 3. $\dfrac{16}{9}$, or 1.7777778 5. $\dfrac{4^3}{5^3}$, or 0.512

7. -4 9. x^6 11. a^8

13. $\dfrac{1}{a^4}$ 15. $\dfrac{1}{a^6 b^8}$ 17. $\dfrac{y^6}{x^9}$

19. x^2 21. $\dfrac{a^6}{b^9}$ 23. a^4

25. $(x+3)^3$ 27. $\dfrac{x^4 z^8}{y^{12}}$ 29. $\dfrac{1}{b^9}$

31. $\dfrac{-1}{a^6 b^{11} c^2}$ 33. $\dfrac{y^{15}}{x^5}$ 35. 1

Exercises 3.5

1. 3,670,000   3. 374,000,000   5. 0.00007
7. 5.67×10^6 9. -4.32×10^{-5} 11. 6
13. -6 15. 6   17. $<$
19. $=$ 21. $<$ 23. 2.6308×10^{18}
25. 3.5265×10^{-19} 27. 9×10^{10} 29. 3.6512×10^7
31. 1.4×10^{-3}
33. *(a)* 1.6095×10^{10} miles   *(b)* 1.1266×10^{11} miles
    (c) 5.8586×10^{12} miles   *(d)* 2.9293×10^{13} miles
    (e) $(1.6095 \times 10^{10})x$ miles   *(f)* $(5.8586 \times 10^{12})y$ miles
35. 3.3278×10^5

Exercises 3.6

1. 1.234568   3. 1.2346
5. 1.23   7. 1
9. 1.3446, 1.3447, 1.3448, 1.3449, 1.3450, 1.3451, 1.3452, 1.3453
11. 2.5665, 2.5666, 2.5667, 2.5668, 2.5669, 2.5670, 2.5671, 2.5672, 2.5673, 2.5674
13. $3^2 - 3$ 15. $3^2 + 3^4$
17. $1.414; -1.414$ 19. $2.031; -2.031$
21. $0.79; -0.79$

Exercises 3.7

1. *(a)* $20   *(b)* $27.50   *(c)* $43.75

3. $1703.80   5. $1440.49

7. *(a)* $5372.48   *(b)* $7161.30   *(c)* $7557.80

9. *(a)* $6431.48   *(b)* $8490.79   *(c)* $8542.64

11. *(a)* $3718.75   *(b)* $4739.28   *(c)* $6417.38

13. *(a)* 5.49%   *(b)* 6.15%

15. *(a)* $5854.31   *(b)* $5776.75

17. *(a)* 16 years   *(b)* 62 quarters, or $15\frac{1}{2}$ years

Exercises 3.8

1. 3.5%   3. 4.5%

5.

Time	Number of bacteria
9:00 A.M.	2000
10:00 A.M.	2400
11:00 A.M.	2880
1:00 P.M.	4147
8:00 P.M.	14,860

7.

Year	Value of car ($)
1975	6000
1976	4800
1977	3840
1978	3072
1981	1572.86
1985	644.25

9.

Year	Number of kilowatt hours
1980	400,000
1981	450,000
1982	506,250
1992	1,643,956.3
2000	4,218,037.5

11. *(a)* $1.20   *(b)* $5.86   *(c)* $(1.09)(1.104)^t$ *(d)* 1986

13. *(a)* $5100   *(b)* $2262.90   *(c)* $(6000)(0.85)^t$

15. *(a)* 2.5%   *(b)* 210,125   *(c)* 268,978
    (d) $(200,000)(1.025)^t$

17. *(a)* 5%   *(b)* 90,250   *(c)* 54,036
    (d) $(100,000)(0.95)^t$

19. *(a)* 5%   *(b)*
    (c) Between 13 and
         14 days

Time elapsed	Number of grams remaining
0 days	10
1 day	9.5
2 days	9.03
5 days	7.74
10 days	5.99
t days	$10(0.95)^t$

21. $100(0.965)^t$; 19 minutes

23. 6.696701%

Exercises 3.9

1. 1.414   3. 3   5. 1.414
7. 0.039   9. 2.683   11. 5
13. 2200.810   15. 48.735   17. $<$

19. $>$ 21. $=$ 23. $\dfrac{8}{15}$

25. $\dfrac{3}{4}$ 27. 4

29. 1.072, or -1.072 31. 1.257, or -1.257
33. -0.109, or -1.891 35. 0.147, or -2.147
37. 2744   39. 58.095

41.

Time	Number of grams
8:00 A.M.	60
9:00 A.M.	50.45
10:00 A.M.	42.43
11:00 A.M.	35.68

Time	Number of grams
12 noon	30
1:00 P.M.	25.23
2:00 P.M.	21.21
3:00 P.M.	17.84
4:00 P.M.	15

XX

43.

Date	Savings account ($)
Jan. 1, 1970	2000
Jan. 1, 1971	2244.92
Jan. 1, 1972	2519.84
Jan. 1, 1973	2828.42
Jan. 1, 1974	3174.80
Jan. 1, 1975	3563.59

Date	Savings account ($)
Jan. 1, 1976	3999.99
Jan. 1, 1977	4489.84
Jan. 1, 1978	5039.67
Jan. 1, 1979	5656.83
Jan. 1, 1980	6349.58
Jan. 1, 1981	7127.16

45.

Time	Number of grams
8:00 A.M.	100
9:00 A.M.	79.37
10:00 A.M.	63
11:00 A.M.	50
12 noon	39.68
1:00 P.M.	31.50
2:00 P.M.	25
3:00 P.M.	19.84
4:00 P.M.	15.75

47.

Year	Population
1968	300,000
1969	283,162
1970	267,270
1971	252,269
1972	238,110
1973	224,746
1974	212,132
1975	200,226
1976	188,988
1977	178,381
1978	168,369
1979	158,920
1980	150,000

49. *(a)* 14.86984%/hour   *(b)* 7.17735%/hour   *(c)* 4.72941%/day

Exercises 3.10

1. 162

3. $-\dfrac{1}{9}$, or $-.\overline{1}$

5. 9

7. 25

9. 5

11. 9.4629×10^9

13. 11

15. 1.7782794

17. 7.3003721

19. 3

21. $\frac{1}{7}$, or 0.14285714

23. $\frac{2}{3}$, 0.66666667

25. >

27.

29. 36,700,000

31. 2.67×10^6

33. 1

35. $\frac{4}{3}$

37. 3

39. -5

41. $\frac{9}{4}$

43. $8x^5y^5$

45. a^{22}

47. x

49. $-x^5$

51. 1

53. $2x^4$

55. $-x^2y + 2\dfrac{y^2}{x^2}$

57. *(a)* $1618.06   *(b)* $3199.74

59. *(a)* $1687.50   *(b)* $1931.22   *(c)* $1987.49

61. *(a)* $3648.99   *(b)* $3614.75

63. 2.4% per day

65.

Time	Number of grams
$t = 0$	60
1 day	55.8
2 days	51.89
5 days	41.74
10 days	29.04
t days	$60(0.93)^t$

67.

Year	Gallons used
1980	1,000,000
1981	943,874.16
1982	890,898.42
1983	840,896
1984	793,700
1985	749,152.92
$1980 + t$	$10^6(0.94387416)^t$

69. *(a)* $7500(0.82)^t$ *(b)* $3683.22

71. *(a)* 259,815   *(b)* $(250,000)(2^{t/18})$, or $(250,000)(1.0392592)^t$
    (c) 283,330

73. 1.710   75. -0.167 77. 2.618   79. 100,000

CHAPTER 4   In this chapter, those answers that give values read from graphs are approximations to actual solutions. Students may get answers that differ slightly from these.

Exercises 4.1

1. $A(1.5, 1.5)$, $B(3, 4)$, $C(-4, 2)$, $D(-3, -3)$, $E(2, -4)$

3. 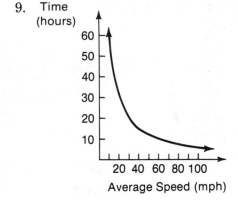   5.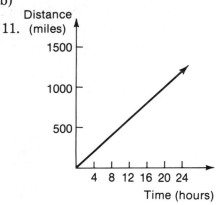

7. *(a)*   0, 1, 2, 3   *(b)* $0, \frac{1}{3}, \frac{2}{3}, 1, \frac{4}{3}, \frac{5}{3}, 2, \frac{7}{3}, \frac{8}{3}, 3$

    (c)   Length is 1 in (a) and $\frac{1}{3}$ in (b)

9.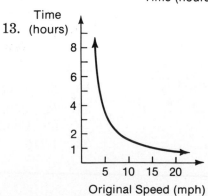

11.

13.

15.

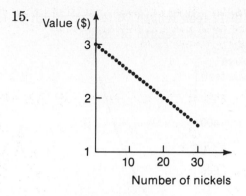

Exercises 4.2

1. *(a)* **12 feet, 22 feet**   *(b)* **60 square feet, 220 square feet**
   (c) **34 feet, 64 feet**

3. *(a)* **275, 900, 750**   *(b)* **Width 0 and 60 for area 0,**
                                   Width 10 and 50 for area 500,
                                   Width 15 and 45 for area 675.

5. *(a)* **12 feet, 6 feet, 4 feet, 3 feet, and 2 feet**
   (b)

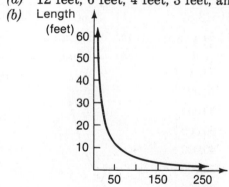

7. *(a)*

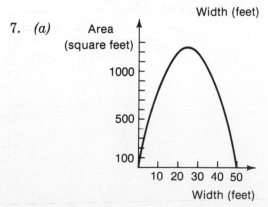

(b) Width 12 feet, length 76 feet or width 38 feet, length 24 feet

(c) Width 25 feet, length 50 feet

9. *(a)*

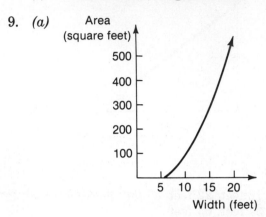

(b) 18 feet by 25 feet, $14\frac{1}{4}$ feet by $17\frac{1}{2}$ feet

Exercises 4.3

1. *(a)* $1648.87   *(b)* $1651.64

3. *(a)* 33.64 grams   *(b)* 10 grams

5. *(a)* Amount   *(b)* 4.9 years   *(c)* 17.5 years

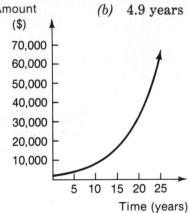

7. *(a)* Number of   *(b)*   6 hours, or 3:00 P.M.   *(c)*   1500
           bacteria

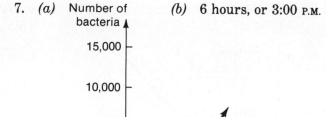

t = 0 (9 AM)   Time (hours)

9. *(a)* Amount of isotope   *(b)*   6.5 years   *(c)*   13 years later
           (grams)

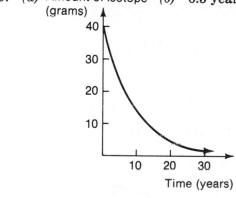

Time (years)

Exercises 4.4

1. 22.4   3. $51.20, $79.36
5. *(a)* 15, 27.5, −17.5   *(b)*   120, 110, −60
7. Amount accumulated   *(a)* $10,400   *(b)* $7000
          ($)

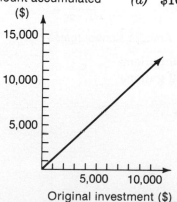

Original investment ($)

9. Sale price ($)

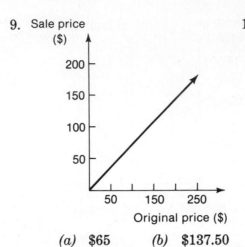

Original price ($)

(a) $65   (b) $137.50

11. Value ($)

Number of $20 bills

(a) 2 twenties and
    58 fifties

(b) No solution

Exercises 4.5

1. 20%

3. (a) $39
   (b) $1.95

5. Percent Alcohol

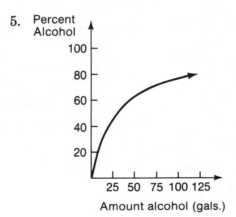

Amount alcohol (gals.)

(a) 95 gallons
(b) 12 gallons

7. Percent Salt

Amount of 8% salt solution (gals)

(a) 30 gallons of 8% solution,
    10 gallons of 28% solution

(b) 5 gallons of 8% solution,
    35 gallons of 28% solution

9. Value per pound ($)

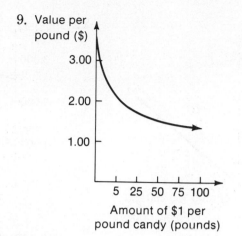

Amount of $1 per pound candy (pounds)

(a) 42.5 pounds

(b) 105 pounds

Exercises 4.6

1. $A\left(\dfrac{7}{2}, \dfrac{3}{2}\right)$, $B(-2, 3)$, $C(-4, -1)$, $D\left(\dfrac{5}{2}, -3\right)$

3. (a) 40 feet, 16 feet   (b) 120 feet, 132 feet

5. 30.4

7. (a) 864   (b) 1269

9. (a) 56.63 grams   (b) 42.43 grams   (c) 15 grams

11. 28%

13. (a) $960, $660   (b) $6000, $4800

15. (a) $1, $2.20   (b) Between 0 and 2 minutes; between 4 and 5 minutes

17. Amount ($)

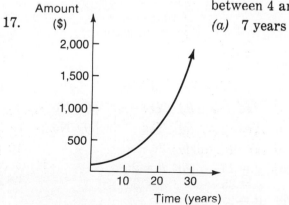

Time (years)

(a) 7 years   (b) 21 years

19. Sale price

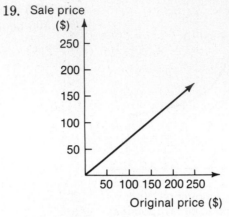

(a) $38
(b) $125

21. Interest

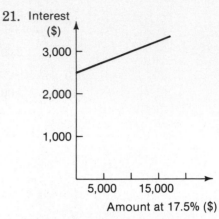

$12,200 at 17.5% and $7800 at 12.5%

23. Distance

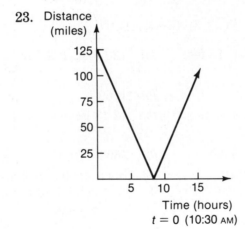

$8\frac{1}{3}$ hours, or 6:50 P.M.

CHAPTER 5   *Exercises 5.1*

1. 40   3. 2   5. −8

7. *(a)* Yes   *(b)* Yes   *(c)* No

9. *(a)* Yes   *(b)* Yes   *(c)* No

11. *(a)* 18   *(b)* $30 - x$

13. *(a)* $d = 17$ *(b)* $d = 25$
         $q = 35$ $q = 23$
     (c) $d = 2x + 1$
         $q = 59 - 3x$, or $60 - [x + (2x + 1)]$

15. $1.92(1.123)^x$; $3.43

17.   *(a)* $.05x + .10(x - 3) + .25(x + 2)$
       (b) $.05x + .10(x - 3) + .25(x + 2) = 2.60$

19.   *(a)* $w + .40w$ *(b)* $w + .40w = 27.75$

21.   *(a)* $2.50x + 3.75(2000 - x)$
       (b) $2.50x + 3.75(2000 - x) = 5812.50$

23.   *(a)* $61t - 48t$ *(b)* $61t - 48t = 100$

25.   *(a)* $\ell \cdot w = 80$ *(b)* $\ell \cdot w = A$

Exercises 5.2

1.   Linear, one   3.   Not linear   5.   Not linear

7.   Yes   9.   No   11.   12

13.   62   15.   7   17.   1

19. -6 21. -3 23. $-\dfrac{3}{4}$

25.   0   27.   18   29. $\dfrac{13}{6}$

31. -4 33. -10 35. -2

37. $-\dfrac{3}{2}$ 39. -2 41.   No solution

43. $-\dfrac{5}{4}$ 45.   1   47. $48 = \dfrac{2}{3}x$; 72

49. $x + .22x = 244$; 200

Exercises 5.3

1.   1250   3. -14

5.   250   7.   12,500

9.   *(a)*

Retail price ($)	Sales tax ($)	Consumer cost ($)
45	$.05(45) = 2.25$	$45 + .05(45) = 47.25$
57	$.05(57) = 2.85$	$57 + .05(57) = 59.85$
x	$.05x$	$x + .05x$, or $1.05x$

       (b) $x + .05x = 86.10$; $x = \$82$

11. $.25x = 72$ 13. $.50 = .20x$
     $x = 288$ $x = \$2.50$

15. $.30x = 20000$ 17. $x - .32x = 43$
     $x = \$66,666.67$ $x = 63.235294$

19. (a) $\$15,877.50$ (b) $x + .095x = 15,603.75; x = \$14,250$

21. $.09x + .065(20000 - x) = 1637.75;$ $\$13,510$ @ 9% and $\$6490$ @ 6.5%

23. $.66(x + 15) = x + .4(15); x = 11.470588$ ounces

25. $.35x + .65(30 - x) = .47(30);$ 18 gallons of 35% and 12 gallons of 65%

Exercises 5.4

1. -25 3. 37.6 5. 21

7. 13 9. 30

11. $.05(3x + 2) + .10x + .25[65 - (3x + 2) - x] = 6.85;$
     38 nickels, 12 dimes, 15 quarters

13. $x + 3x + (x + 5) = 75;$ 14, 19, and 42 feet.

15. $55t = 75(t - 2),$ or $\dfrac{d}{55} = \dfrac{d}{75} + 2; d = 412.5$ miles

17. $10t + 8t = 63; t = 3.5$ hours

19. $20r = 3(r + 140); r \doteq 24.7$ mph (boat), $r + 140 \doteq 164.7$ mph (plane)

21. $2l + 2(l - 3) = 123; w = 29.25, l = 32.25$

23. $s + (4s + 6) + (s + 24) = 180; 25°, 49°, 106°$

25. $2(3x + 6) + x = 180; 24°, 78°, 78°$

Exercises 5.5

1. 10 3. 80 5. $\dfrac{70}{3},$ or $23\dfrac{1}{3}$

7. -18 9. 250

11. (a) 2790 (b) 2880 13. 7.5 inches

15. 22.5 feet, 25.5 feet   17. 64.75 feet

19. 240 rpm   21. 54

23. $\$16,000; \$28,000$

25. (a) 396,000 inches, or 33,000 feet, or 6.25 miles
     (b) 0.00005333 miles, or 0.2816 feet, or 3.3792 inches

27. *(a)* 2.7963 miles   *(b)* 3.2185388 kilometers

29. *(a)* $73\frac{1}{3}$ feet/second   *(b)* $\dfrac{88x}{60}$ feet/second

Exercises 5.6

1. *(a)* Yes
   (b) Yes
   (c) No
   (d) Yes

3. *(a)* Yes
   (b) No
   (c) Yes
   (d) Yes

5. *(a)* $2, 0, -\dfrac{3}{2}$
   (b) $4, 0, -2$

7. *(a)* $0, \dfrac{8}{3}, 2$

   (b) $0, -\dfrac{3}{2}, 3$

9. $l = 1, w = 100;$
   $l = 10, w = 10;$
   $l = 20, w = 5$

11. $x = 0, y = 6;$
    $x = 2, y = 3;$
    $x = 4, y = 0$

13. $\dfrac{2}{3}y$, or $\dfrac{y}{1.5}$

15. $\dfrac{7}{8}x$

17. $\dfrac{1 - 2x}{3}$

19. $\dfrac{y - 6}{3}$

21. $\dfrac{2 - 4y}{3}$

23. $\dfrac{5 - 6x}{2}$

25. $\dfrac{x - 3}{6}$

27. $\dfrac{y - 2}{.17}$

29. $\dfrac{d}{r}$

31. $\dfrac{s - 2}{s}$

33. $m = \dfrac{k}{0.6214} \doteq 1.6092694k$

35. $\dfrac{C}{2\pi}$

Exercises 5.7

1. *(a)* 2   *(b)* $x > 2$ *(c)* $x < 2$

3. $x < \dfrac{4}{3}$ 5. $x < \dfrac{6}{5}$ 7. $x \le -2$

9. $x > 6$ 11. $x \ge -\dfrac{4}{7}$ 13. $x < 1250$

15. $x < 30$ 17. $x \le -\dfrac{16}{3}$ 19. No solution

21. *(a)* $n < -80$ *(b)* $n \ge 60$

23. $1.50s + 2.75\left(\dfrac{s}{2}\right) > 7350;\ s > 2556$

25. $.10x + .25(60 - x) \ge 900;\ x \le 40$

Exercises 5.8

1. 1457.606

3. −12

5. Linear, one

7. Not linear

9. Yes

11. Yes

13. *(a)* $\left[\dfrac{.32x + .58(30 - x)}{30}\right](100)$

 (b) $.32x + .58(30 - x) = .43(30)$

 (c) $.32x + .58(30 - x) < .43(30)$

15. $2l + 2w = 45$

17. *(a)*

Amount at 7.5% ($)	Amount at 5.6% ($)	Total interest ($)
4000	$19000 - 4000 = 15000$	$.075(4000) + .056(19000 - 4000) = 1140$
6000	$19000 - 6000 = 13000$	$.075(6000) + .056(19000 - 6000) = 1178$
x	$19000 - x$	$.075x + .056(19000 - x)$

 (b) $.075x + .056(19000 - x) = 1219.80;\ x = \$8200$

19. 42.75

21. 12.6

23. 24

25. −4

27. 16

29. No solution

31. $\dfrac{4}{3}$

33. −28

35. $\dfrac{3x - 4}{2}$

37. $\dfrac{21 - 14x}{15}$

39. $x \le \dfrac{1}{3}$

41. $x < -\dfrac{2}{3}$

43. $x \le \dfrac{24}{7}$

45. $m = \dfrac{15}{22}f$

47. 342 rpm

49. 432 rpm

51. \$29,250; \$47,250

53. 1068

55. $5x + 10(x + 3) + 20(2x - 1) = 1055$; 19 5-dollar bills, 22 10-dollar bills, 37 20-dollar bills

57. $2x + 3.5(2x) + 2.5(4422 - 3x) = 12930$; 1250 @ \$2, 2500 @ \$3.50, 672 @ \$2.50

59. $60t - 55t = 48$; 9.6 hours

61. $2w + 2(2w - 3) = 99$; 17.5 feet × 32 feet

63. $.12x + .24(30 - x) = .20(30)$; 10 gallons at 12%, 20 gallons at 24%

65. $.22x + .10(30 - x) = .15(30)$; 12.5 gallons

67. $\dfrac{x}{77 - x} = \dfrac{3}{4}$; 33, 44

69. $.10x + .60(40) \le .20(x + 40)$; $x \ge 160$ gallons

CHAPTER 6　　In this chapter, those answers that give values read from graphs are approximations to actual solutions. Students may get answers that differ slightly from these.

Exercises 6.1

1. (a) $-1, 0, 1, 2, 3, 4, 5$　　(b) $-3, -2, -1, 0, 1$
   (c) None　　(d) $-2, -1, 0, 1, 2, 3, 4, 5, 6$

3. (a) .55555556　　(b) $-.47619048$
   (c) 3.8729833　　(d) Error
   (e) .70710678　　(f) .08715574
   (g) -4.7046301　　(h) 1
   (i) 2.0794415　　(j) Error

5. (a)

x	y	x	y
-4	16	.5	.25
-3.5	12.25	1	1
-3	9	1.5	2.25
-2.5	6.25	2	4
-2	4	2.5	6.25
-1.5	2.25	3	9
-1	1	3.5	12.25
$-.5$	.25	4	16
0	0		

(b)

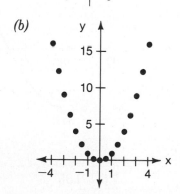

(c) For $x > 4$ the values of y are greater than 16 and increasing rapidly; for $x < -4$ the values of y are greater than 16 and increasing rapidly.

(d)

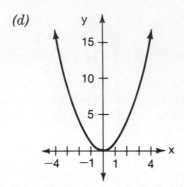

7. *(a)*

x	y	x	y	x	y
−180	−1	−50	.64	70	.34
−170	−.98	−40	.77	80	.17
−160	−.94	−30	.87	90	0
−150	−.87	−20	.94	100	−.17
−140	−.77	−10	.98	110	−.34
−130	−.64	0	1	120	−.50
−120	−.50	10	.98	130	−.64
−110	−.34	20	.94	140	−.77
−100	−.17	30	.87	150	−.84
−90	0	40	.77	160	−.94
−80	.17	50	.64	170	−.98
−70	.34	60	.50	180	−1
−60	.50				

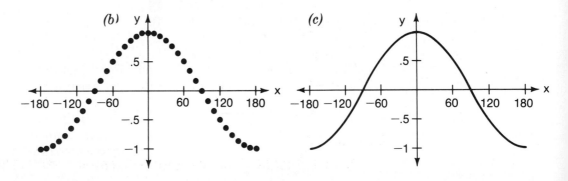

(b)

(c)

9. *(a)*

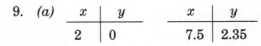

x	y	x	y
2	0	7.5	2.35

x	y	x	y
2.5	.71	8	2.45
3	1	8.5	2.55
3.5	1.22	9	2.65
4	1.41	9.5	2.74
4.5	1.58	10	2.83
5	1.73	10.5	2.92
5.5	1.87	11	3
6	2	11.5	3.08
6.5	2.12	12	3.16
7	2.24		

(b)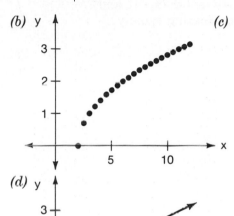

(c) When the x values are close to 2 but greater than 2, the y values are positive and close to 0. When the x values are close to 2 but less than 2, you get an error message. When the x values are greater than 12, the y values are greater than 3.16 and increasing slowly.

(d)

Exercises 6.2

1. 0, 8.448, −8.448   3. 66.55104, −66.55104, −238.61669

5. (a)

x	y	x	y
−3	−15	.5	−1.875
−2.5	−5.625	1	−3

x	y	x	y
−2	0	1.5	−2.625
−1.5	2.625	2	0
−1	3	2.5	5.625
−0.5	1.875	3	15
0	0		

(b) For $x > 3$ the y values are greater than 15 and increasing rapidly; for $x < -3$ the y values are less than −15 and decreasing rapidly.

(c)

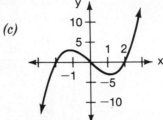

7.   2.46

9.   Raise each point in the graph of $y = x^3 - 4x$ by 3 units.

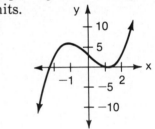

11. *(a)*   2.12, −2.12   *(b)*   No solution

13. *(a)* $-2 < x < 0, 0 < x < 2$ *(b)* $x < -2.11, x > 2.11$

15. *(a)*

x	y	x	y
−2.5	11.81	.25	3.69
−2.25	4.32	.5	2.81
−2	0	.75	1.50
−1.75	−1.93	1	0
−1.5	−2.19	1.25	−1.37
−1.25	−1.37	1.5	−2.19
−1	0	1.75	−1.93
−.75	1.50	2	0
−.5	2.81	2.25	4.32
−.25	3.69	2.5	11.81
0	4		

(b) For $x > 2.5$ the y values are greater than 11.81 and increasing rapidly; for $x < -2.5$ the y values are greater than 11.81 and increasing rapidly.

(c)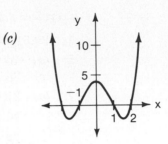

17. (a)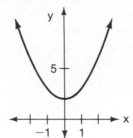

(b)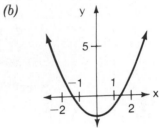

(c)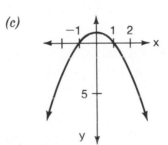

(d)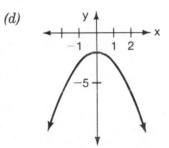

Exercises 6.3

1. 0, 1.1797753, .02009799

3. No value, 1.5124717, 1.9706891

5. -4 7. 0, 2

9. (a) $x = 0$ (b) No value
   (c) $x < -1, x > 0$ (d) $-1 < x < 0$

11. (a) No values   (b) 0.5, 1.5
    (c) All $x \neq 1$ (d) No values

13. (a) $-.75, 2.75$ (b) No values
    (c) $-2 < x < -.75, x > 2.75$

15. (a) $-1.7, 1.7$ (b) $-2 < x < -1.7, 1.7 < x < 2$

Exercises 6.4

1. 20.515563

3. .11579073

5. (a) 10

(b) 18

(c) 38

(d) 1386

7. (a) 21, .25

(b) 1.75, −1.25

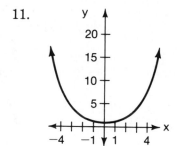

9.

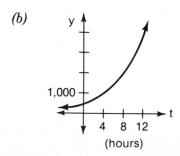

11.

13. (a) $y = (500)2^{t/4}$

(b)

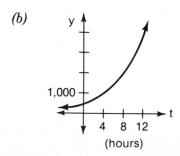

The portion of the graph for which $t \geq 0$ is the graph of the problem.

Exercises 6.5

1. *(a)*

x	-2	-1	0	1	2	3
y	-1	1	3	5	7	9

(b)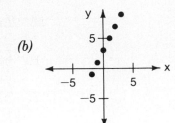

(c)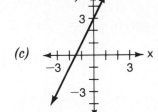

(d) 0, 2, 4, 6

(e) $-\dfrac{5}{2}, -\dfrac{1}{4}, \dfrac{5}{2}$

3.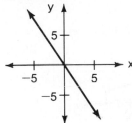

5.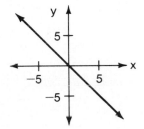

7.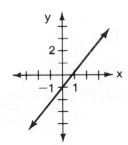

9.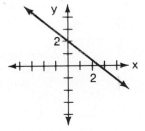

11. Slope $\dfrac{5}{3}$, y-intercept: 0   13. Slope $\dfrac{1}{4}$, y-intercept: -2

15. $y = -x + 3$

17. $y = -2x - 3$, $y = -2x$, $y = -2x + 4$

19. 2, 3   21. 5, 3

23. 3, 2

25. L_1: slope $\frac{1}{2}$, y-intercept 1, $y = \frac{1}{2}x + 1$

L_2: slope $\frac{2}{3}$, y-intercept -2, $y = \frac{2}{3}x - 2$

L_3: slope $-\frac{1}{2}$, y-intercept 1, $y = -\frac{1}{2}x + 1$

L_4: slope $-\frac{2}{7}$, y-intercept 2, $y = -\frac{2}{7}x + 2$

27.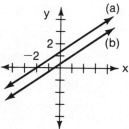

29. *(a)* $y = .22x + .54(40 - x)$ or $y = -.32x + 21.6$

(b) 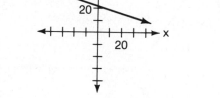   *(c)* Slope: $-.32$, y-intercept: 21.6

Exercises 6.6

1. Slope: $\frac{1}{3}$, y-intercept: -4

3.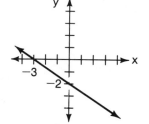

5. Slope: $-\frac{1}{2}$, y-intercept: $\frac{1}{2}$

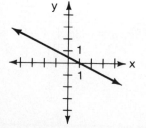

7. Slope: $-\frac{4}{3}$, y-intercept: $-\frac{2}{3}$

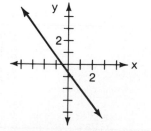

9. Slope: -1, y-intercept: 5

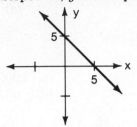

11. x-intercept: 4, y-intercept: -2

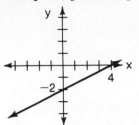

13. x-intercept: 2, y-intercept: 3

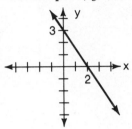

15. x-intercept: 4, y-intercept: 4

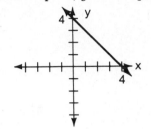

17. x-intercept: $-\dfrac{3}{2}$, no y-intercept

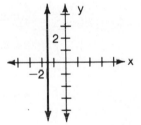

19.

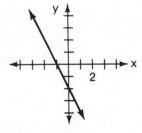

21.

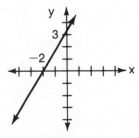

23.

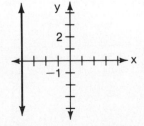

25.

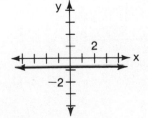

27. *(a)* $y = -2$ *(b)* $y = -1$
    (c) $y = 2$ *(d)* $y = 4$

Exercises 6.7

1. *(a)* .25881905 *(b)* .90308999
   (c) 22.498671 *(d)* .00794328

3. $-1, -.75, -.5, -.25, 0, .25, .5, .75, 1, 1.25, 1.5, 1.75, 2, 2.25, 2.5,$
   $2.75, 3$

5. No value, $-.46956522$

7.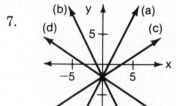

9. Slope: $-\dfrac{5}{3}$, y-intercept: $\dfrac{4}{3}$

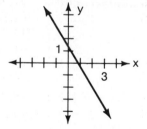

11. No slope, no y-intercept

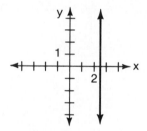

13. Slope: $\dfrac{3}{2}$, y-intercept: 0

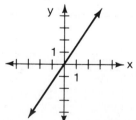

15. x-intercept: 3, y-intercept: 5

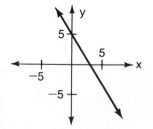

17.   No x-intercept, y-intercept: $\dfrac{3}{2}$

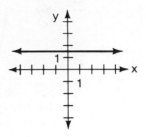

19.   *(a)*

x	y		x	y
−90	No value		100	−5.76
−80	5.76		110	−2.92
−70	2.92		120	−2
−60	2		130	−1.56
−50	1.56		140	−1.31
−40	1.31		150	−1.15
−30	1.15		160	−1.06
−20	1.06		170	−1.02
−10	1.02		180	−1
0	1		190	−1.02
10	1.02		200	−1.06
20	1.06		210	−1.15
30	1.15		220	−1.31
40	1.31		230	−1.56
50	1.56		240	−2
60	2		250	−2.92
70	2.92		260	5.76
80	5.76		270	No value
90	No value			

(b)

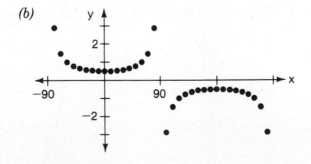

(c)

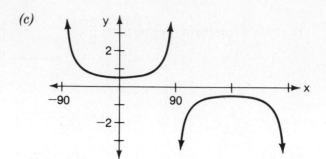

21.

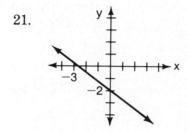

23.

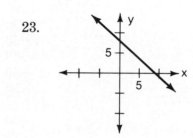

25.

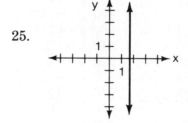

27.

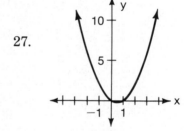

29.

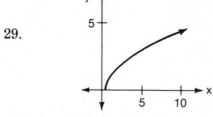

31.

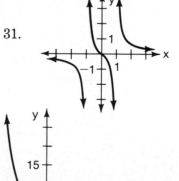

33.

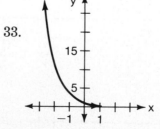

35. L_1: slope $\frac{3}{2}$, y-intercept 0, $y = \frac{3}{2}x$

      L_2: slope $-\frac{4}{3}$, y-intercept 3, $y = -\frac{4}{3}x + 3$

37. $y = 3x - 2$ 39. $y = -1$

41.   (a) $y = \dfrac{x}{50} + \dfrac{300 - x}{80}$, or $y = \dfrac{3}{400}x + \dfrac{15}{4}$

      (b)   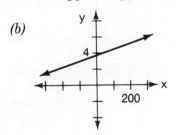   The portion of the graph for which
                                         $0 \le x \le 300$ represents the graph of
                                         the problem.

                                   (c)   31.25 miles

CHAPTER 7   *Exercises 7.1*

1.   (a)   5, 5   (b)   1

3.   (a)   Right, up   (b) $\dfrac{3}{4}$

5. $\dfrac{5}{4}$ 7. $-\dfrac{3}{5}$

9.   Parallel y-axis; undefined   11. $\dfrac{10}{3}$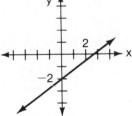

13. -3 15.

17.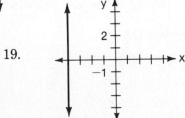

19.

21.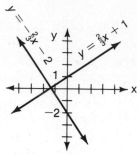

23. $y = \dfrac{-2}{3}x + 3$

25. $\overline{AB}$ and $\overline{CD}$ have slope $\dfrac{2}{7}$ and are parallel;

$\overline{AD}$ and $\overline{CB}$ have slope $\dfrac{7}{3}$ and are parallel.

27. No   29. Yes

Exercises 7.2

1. x-intercept $= 2$, y-intercept $= \dfrac{4}{3}$

3. Slope $= -\dfrac{3}{4}$, y-intercept $= 3$

5. -2 7. 4

9. $\dfrac{1}{2}$ 11. 0

13. $y = -\dfrac{3}{5}x - \dfrac{1}{5}$ 15. $y = \dfrac{4}{3}x + 4$

17. $x = 2$ 19. $y = 2x + \dfrac{3}{2}$

21. $y = \dfrac{1}{2}x - \dfrac{5}{2}$ 23. $y = \dfrac{-3}{4}x + \dfrac{13}{4}$

25. $y = \dfrac{5}{3}x + 2$ 27. $y = 2$

29. Taking the points pairwise, the slope is always 2.

Exercises 7.3

1. $x = -2$, $y = 1$ 3. Lines are identical;
infinitely many solutions.

5. $x = -1$, $y = 2$

7. Lines are identical;
   infinitely many solutions.

9. $x = 3$, $y = 9$; $x = -1$, $y = 1$

11. $x = 0$, $y = 0$; $x = 2$, $y = 8$; $x = -2$, $y = -8$

13. $x + y = 20{,}000$; $.11x + .06y = 1800$;
    \$12,000 at 11% and \$8,000 at 6%

15. $x + y = 100$; $1.20x + 1.80y = 1.35(100)$;
    75 pounds at \$1.20 and 25 pounds at \$1.80

Exercises 7.4

1. $x = 1$, $y = -\dfrac{1}{2}$

3. $x = -3$, $y = \dfrac{17}{3}$

5. $x = 3$, $y = 1$

7. $x = -1$, $y = -2$

9. Same line; infinitely many solutions

11. Same line; infinitely many solutions

13. $x + y = 40$; $.24x + .52y = .3625(40)$;
    22.5 gallons of 24% and 17.5 gallons of 52%

15. $q = 3d - 1$; $.10d + .25q = 26.95$;
    32 dimes and 95 quarters

17. $l = 2w - 5$; $2l + 2w = 146$;
    length = 47 feet, width = 26 feet

19. $x + y = 20$; $y = x + 3$;
    8.5 feet and 11.5 feet

21. $3(x + y) = 45$; $5(x - y) = 45$;
    12 miles per hour = boat's rate and 3 miles per hour = river's
    rate

23. $-1 = 2m + b$; $5 = -m + b$;
    $m = -2$ and $b = 3$

Exercises 7.5

1. $(1, 1, 1)$

3. $(13, 17, 20)$

5. $(-1, 2, 1)$

7. Infinitely many solutions
   $\left(\dfrac{4}{7}, \dfrac{1}{7}, 0\right)$, $(1, -1, -1)$, $\left(\dfrac{1}{7}, \dfrac{9}{7}, 1\right)$

9. $(2.5, 0, 2)$

11. $(6, -2, -3)$

13. $0.5n + .10d + .25q = 8.85$;
$d = 2n + 1$;
$n + d + q = 60$;
12 nickels, 25 dimes, 23 quarters

15. $x + y + z = 2000$;
$y = x + 250$;
$2.50x + 3.75y + 3z = 6328.75$;
565 students, 815 adults, 620 senior citizens

17. $x + y + z = 40$;
$z = 3x$;
$y = z - 2$;
6 feet, 16 feet, 18 feet

19. $x + y + z = 60$;
$\dfrac{x + y}{z} = 2$;
$x - y = 4$;
22, 18, 20

Exercises 7.6

1.
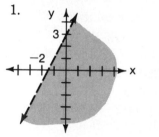

3.
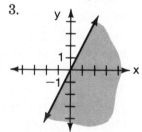

5.
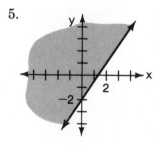

7.
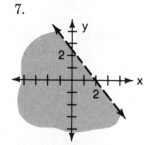

9.
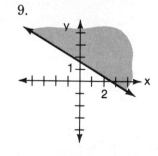

11.
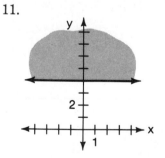

13.
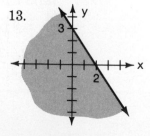

15.
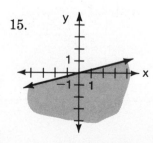

17. $y > -x + 4$ 19. $y \le 2x - 3$ 21. $y > -3$

23. $2.50S + 3.75A > 1500$ 25. $.105x + .167y \ge .13(x + y)$

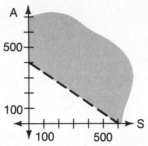

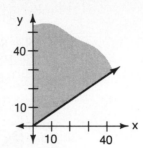

27. $.25H + .15F - 600 > 400$

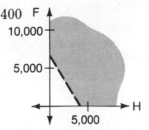

Exercises 7.7

1.
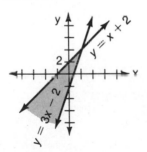

3.
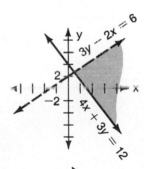

5.
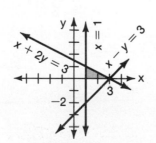

7.
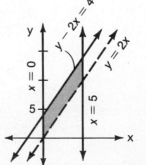

9. $y > 2x + 1$ and $y < -3x + 2$ 11. $y < 2x + 1$ and $y < -2x + 6$

13. $y > \dfrac{1}{2}x$ and $x > 0$ and $y < -\dfrac{1}{2}x + 4$

15. $5S + 8P > 2000$ and
    $S + P \le 10{,}000$ and
    $S \ge 0$ and $P \ge 0$

17. $\dfrac{5}{4}P + \dfrac{3}{4}F \le 800$ and $\dfrac{P}{2} + \dfrac{3}{4}F$
    ≤ 500 and $\dfrac{P}{4} + \dfrac{F}{2} \le 300$ and
    $P \ge 0$ and $F \ge 0$

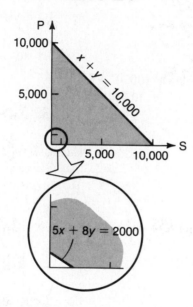

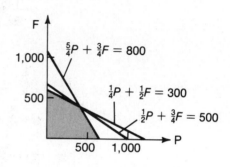

19. 4 suits and 2 gowns

Exercises 7.8

1.

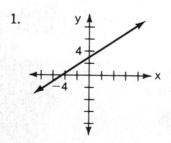

3.

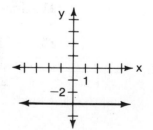

5. Slope $= \dfrac{2}{3}$, y-intercept $= \dfrac{4}{3}$

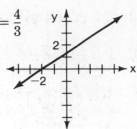

7.

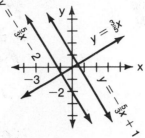

9. 4

11. $-\dfrac{3}{5}$

13. No slope, vertical line

15. $y = \dfrac{2}{3}x - 1$

17. $y = -\dfrac{3}{2}x - 2$

19. $y = \dfrac{-2}{3}x - \dfrac{1}{3}$

21. $y = 1$

23. Slope $\overline{AB}=$ slope $\overline{CD} = \dfrac{5}{2}$ and slope $\overline{AD}=$ slope $\overline{BC} = \dfrac{3}{2}$

25. $x = \dfrac{3}{2}$, $y = -\dfrac{1}{2}$

27. $x = 2$, $y = -1$

29. $x = \dfrac{1}{2}$, $y = \dfrac{1}{3}$

31. Lines are identical; infinitely many solutions

33. $(2, -1, 1)$

35. No solution

37. 37 and 25

39. $13,700 at 6.7% and $8300 at 5.2%

41. 22 feet by 67 feet

43. 21 nickels, 42 dimes, 44 quarters

45.

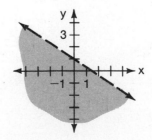

47.

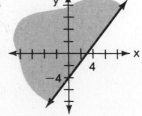

49.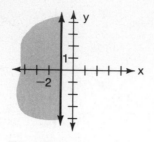

51. $y > \frac{1}{2}x + 2$

53. $\dfrac{W}{5} + \dfrac{B}{30} \leq 3$, where
$W = $ distance walking and
$B = $ distance by bus

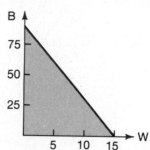

55.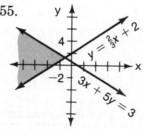

57.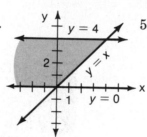

59.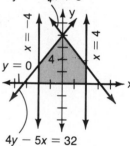

61. $y < \dfrac{8}{5}x + 1$ and $y < -\dfrac{1}{2}x + 4$ and $y > \dfrac{5}{11}x - \dfrac{19}{11}$

63. 4 suits and 6 dresses earns \$440

CHAPTER 8 *Exercises 8.1*

1. $3x^3 - 5x^2 + 7x - 1$

3. $3xy - 1$

5. $5x - y$

7. $y^2 - x^2$

9. $6x^2 - x - 2$

11. $4x^2 - 12x + 9$

13. $y^2 - 9$

15. $6x^4 + x^2 - 2$

17. $4x$

19. $a^2 + b^2 + c^2 + 2ab + 2ac + 2bc$

21. $a^2 - b^2 - c^2 - 2bc$

23. $x^3 - x$

25. $x^4 + x^2 + 1$

27. $x^3 - 1$

29. $x^3 + 8y^3$

31. $x^3 + 3x^2 + 3x + 1$

33. $x^4 - 1$

35. $x^5 + 1$

37. $2x^4 + x^3 + 4x^2 - x + 2$

39. $x^5 - y^5$

41. 44

43. $P = 2x + 2(x + 2) = 4x + 4,$
    $A = x(x + 2) = x^2 + 2x$

45. $A = \left(\dfrac{x}{4}\right)^2 = \dfrac{x^2}{16}$

Exercises 8.2

1. $x^3 - 8$

3. $x^2 - 16$

5. $x^3 + 8$

7. $4x^2 - 9$

9. $x^3 - 8y^3$

11. $9y^2 - 4$

13. $x^3 + 8y^3$

15. $1 - x^2$

17. $x^6 - 27y^3$

19. $1 - x^4$

21. $x^6 + y^6$

23. $4 - x^2y^2$

25. $x^6 + 1$

27. $2y^2 - 11y + 15$

29.

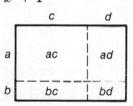

Exercises 8.3

1. $(4x - 3)(2x - 3)$

3. $(4x - 1)(x + 3)$

5. $-8(x - 3y)$

7. $(x + 5)(x - 3)$

9. $(3x - 1)(2x + 1)$

11. $x(3x - 1)(x + 4)$

13. $(5x - 2)(x - 3)$

15. $(x - 3)^2$

17. $x(x - 2y)^2$

19. $x^2y(3x - y)(2x + 5y)$

21. $(3x + 1)^2$

23. $(5y^2 + 4)(2y^2 - 3)$

25. $3xy(x + 2)^2$

27. The only possibility is $2(x + a)$ but $2a = 3$ does not have an integer solution for a.

29. The only possibility is $(y - z)(y - z)$ but $(y - z)(y - z) = y^2 - 2yz + z^2$.

Exercises 8.4

1. $(2x - 3)(2x + 3)$
3. $(2x - 1)(4x^2 + 2x + 1)$
5. $(x + 5)(x - 5)$
7. $(1 - x)(1 + x + x^2)$
9. $y(2y - 1)(4y^2 + 2y + 1)$
11. $(2x + 3z)(y + 2z)$
13. $x^2(x + 1)(x - 1)$
15. $(5r - 2s)(25r^2 + 10rs + 4s^2)$
17. $(x^2 + y)(x^2 - y)$
19. $3(r + 2)(r^2 - 2r + 4)$
21. $(5a - 3b)(x^2 + 2y)$
23. $(xy - 2)(xy + 2)(x^2y^2 + 4)$
25. $x(x - y)(x + y)(x^2 + y^2)$
27. $(x - 1)(x + 1)(y - 2z)$
29. $(x - 2)(x + 2)(x - 3)(x + 3)$

Exercises 8.5

1. $(6x^3 + 5x^2 - 2x - 1) \div (3x + 1) = 2x^2 + x - 1$
$(6x^3 + 5x^2 - 2x - 1) \div (2x^2 + x - 1) = 3x + 1$

3. $(x^4 - 1) \div (x^2 - 1) = x^2 + 1$
$(x^4 - 1) \div (x^2 + 1) = x^2 - 1$
5. $x^2 - x + 1$

7. $x^3 - x$
9. $1, 0$
11. $q = 2x^3 - x + 1, r = 2x - 1$
13. $q = 2x + 1, r = 0$
15. $q = x^2 - x + 1, r = 0$
17. $q = x^3 + 4x^2 - x + 4, r = 5$
19. $q = 2, r = 3$
21. $q = 0, r = x^2 - x$

23. $q = x^3 + 2x^2 + 4x + 8, r = 0$
25. $q = \dfrac{5}{2}x^2 - \dfrac{11}{4}x + \dfrac{19}{8}$,
$r = \dfrac{-27}{8}$

27. $q(x) = 2x^3 + x + 1$,
$r(x) = 3x - 1$
29. $q(x) = x^4 + x^3 + x^2 + x + 1$,
$r(x) = 0$

31. $q(x) = \dfrac{3}{2}x^2 + \dfrac{9}{4}x + \dfrac{13}{8}$, $r(x) = \dfrac{29}{8}x - \dfrac{21}{8}$

33. *(a)* 0   *(b)* 0   *(c)* −6.25   *(d)* −5

Exercises 8.6

1. *(a)* −3   *(b)* 0   *(c)* 0   *(d)* 7.3125
3. 29   5. 0
7. $(x-5)(x-5)(x+1)(x+1) = (x-5)^2(x+1)^2$
9. −3, −8   11. −2, 1
13. Yes   15. No
17. Yes   19. Yes
21. 7, −3   23. −5, 2
25. $(x-2)(x-3) = x^2 - 5x + 6$
27. $(x-1)^2(x+2)$, or $(x-1)(x+2)^2$
29. $\dfrac{15}{14}$

Exercises 8.7

1. 1, −1, 2, −2, 4, −4   3. $0, -2, \dfrac{5}{3}$
5. $(5x-2)(x^2-x+1)$ 7. No rational roots
9. −2   11. $(x-2)(x+3)(2x+1)$
13. $(x-1)(x-1)(x-2)(x-2) = (x-1)^2(x-2)^2$
15. $(2x+1)(3x-2) = 6x^2 - x - 2$
17. $(5x-3)(3x+2)(x+1)^2$, or $(5x-3)(3x+2)^2(x+1)$, or $(5x-3)^2(3x+2)(x+1)$
19. 1

Exercises 8.8

1. $-x^4 - x^2 + 5x - 6$ 3. $2x^2y + 2xy + 4$
5. $15x^2 + 14xy - 8y^2$ 7. $12x^2 + x - 6$
9. $a^2 + b^2 + c^2 - 2ab - 2ac + 2bc$ 11. $6y^4 + 5y^2 + 1$
13. $4 - a^2b^2$ 15. $x^2y - xy^2$
17. $9x^2 - 6x + 1$ 19. $3x(x^2 + 1)$
21. $(x+4)(x-3)$ 23. $(5x-4)(3x-2)$
25. Irreducible   27. $(2a+3b)(3c-2b)$

29. $x(x + 2)(x^2 - 2x + 4)$

31. $(a^2 - 2b^2)(4a^2 + 3b^2)$

33. $a^2b(a + 2b)(2a - 3b)$

35. $2(r - 3)(r^2 + 3r + 9)$

37. $(3x^2 + 4y^2)(x + y)(x - y)$

39. $x^3(x + 1)(x^2 + 1)$

41. $(2x - 1)(3x^3 - x^2 + 5x + 1)$

43. $q = x^4 + x^3 + x^2 + x + 1$, $r = 0$

45. $q = x^3 + x^2 - x + 1$, $r = 4$

47. $q = x^3 + 4x^2 - 5x - 7$, $r = -2x + 3$

49. *(a)* 2   *(b)* 0   *(c)* 0   *(d)* −700   *(e)* 0

51. Yes

53. Yes

55. Yes

57. No

59. $0, -\dfrac{5}{3}, 7$

61. $-2, \dfrac{1}{2}$

63. $(x + 2)^2(x + 1)(2x - 1)$

65. $0, 2, -5$

67. $3, -1, \dfrac{2}{3}$

69. $x(x + 3)(x - 2) = x^3 + x^2 - 6x$

71. $A = x(2x - 1)$, $P = 2x + 2(2x - 1) = 6x - 2$
    $= 2x^2 - x$

CHAPTER 9   *Exercises 9.1*

1. $-1, -1$

3. $5, 5$

5.

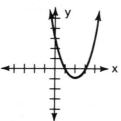

7.

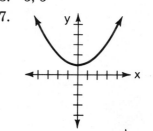

9. $(3,2)$, $x = 3$

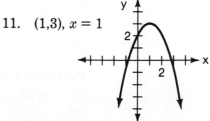

11. $(1,3)$, $x = 1$

13. $(-2, -1)$, $x = -2$ 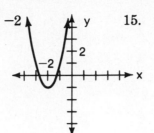   15.   4

17.   9   19. $y = (x - 1)^2 - 2$

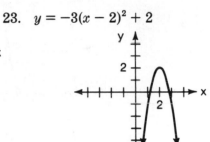

21. $y = -(x - 3)^2 + 9$ 23. $y = -3(x - 2)^2 + 2$

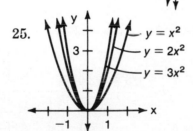

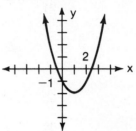

25.

27.   Vertex and line of symmetry same for all a. Bigger a-value gives "skinny" parabola, smaller a-value gives a flatter curve.

29.   7

Exercises 9.2

1.   0, 2   3. $-\dfrac{1}{3}, \dfrac{3}{2}$

5. $-\dfrac{3}{2}, \dfrac{3}{2}$

7. $-4, 3$

9. $\dfrac{2}{3}, -\dfrac{4}{3}$

11. $-2, \dfrac{3}{4}$

13. $8, -2$

15. $7, -3$

17. $-\dfrac{4}{3}, 3$

19. $-\dfrac{4}{3}, 2$

21. 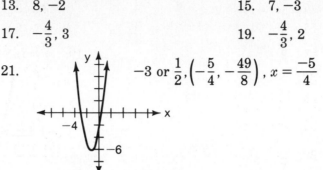 -3 or $\dfrac{1}{2}, \left(-\dfrac{5}{4}, -\dfrac{49}{8}\right), x = \dfrac{-5}{4}$

23. 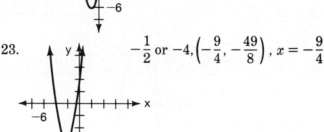 $-\dfrac{1}{2}$ or $-4, \left(-\dfrac{9}{4}, -\dfrac{49}{8}\right), x = -\dfrac{9}{4}$

25. $-3 + \sqrt{7} \doteq -0.354, -3 - \sqrt{7} \doteq -5.646$

27. No solution

29. $-\dfrac{1}{3} + \sqrt{3} \doteq 1.399, -\dfrac{1}{3} - \sqrt{3} \doteq -2.065$

31. $-12, -7$ or $7, 12$

33. 4.5 feet by 10 feet

35. (a) $x^2 - x - 12 = 0$

(b) $2x^2 - 5x - 3 = 0$

(c) $6x^2 - 13x - 5 = 0$

Exercises 9.3

1. $3\sqrt{2}$

3. $15\sqrt{2}$

5. 0.9

7. $\dfrac{5\sqrt{2}}{3}$

9. $\dfrac{-2 + \sqrt{7}}{3}$

11. $ab\sqrt{a}$

13. $2a^3b^3\sqrt{3ab}$

15. $\dfrac{a^2b}{c^2}\sqrt{b}$

17. $(a + b)^2$

19. $3\sqrt{15}$

21. $8\sqrt{6}$

23. $13\sqrt{3}$

25. $8\sqrt{11} - 5\sqrt{7}$

27. $(2x - 3y + xy)\sqrt{xy}$

29. $2 - \sqrt{6}$

31. $\dfrac{-23}{4}$

33. $5 + 2\sqrt{6}$

35. $-4 - 3\sqrt{15}$

37. $x - y$

39. $\left(\dfrac{1 + \sqrt{5}}{2}\right)^2 - \left(\dfrac{1 + \sqrt{5}}{2}\right) - 1 = \dfrac{1 + 2\sqrt{5} + 5}{4} - \dfrac{2 + 2\sqrt{5}}{4} - \dfrac{4}{4} = 0$

41. $\boxed{3}\ \boxed{-}\ \boxed{5}\ \boxed{\sqrt{x}}\ \boxed{=}\ \boxed{\div}\ \boxed{7}\ \boxed{=}\ \boxed{\text{STO}}$
$\underline{49}\ \boxed{\times}\ \boxed{\text{RCL}}\ \boxed{x^2}\ \boxed{-}\ \underline{42}\ \boxed{\times}\ \boxed{\text{RCL}}\ \boxed{+}\ \boxed{4}\ \boxed{=}$ (Display: -2.6 -09)

43. $2x^2 + 2x - 3 = 0$

Exercises 9.4

1. $-5, \dfrac{1}{2}$

3. $\sqrt{3}, -\sqrt{3}$

5. $0, \dfrac{2}{3}$

7. No real solutions

9. $\dfrac{1 + \sqrt{14}}{2}, \dfrac{1 - \sqrt{14}}{2}$

11. $\dfrac{4 + 3\sqrt{3}}{2}, \dfrac{4 - 3\sqrt{3}}{2}$

13. 0

15. 2

17. $(2, -23), x = 2$

19. $(3, 1), x = 3$

21. 0.62, −1.62

23. −0.41, 2.41

25. $c > 1$

Exercises 9.5

1. 14, 15; −14, −15

3. −10, −9, −8; 8, 9, 10

5. −15, 13

7. 12 feet by 15 feet

9. 16 inches by 32 inches

11. 10 feet by 8.5 feet

13. 12 inches by 25 inches

15. 5.600189%

17. $-\dfrac{1}{2}$

19. (a) 12.5 seconds or 37.5 seconds (b) 25 seconds
(c) 10,000 feet (d) 50 seconds

Exercises 9.6

1. 2, −2, 1, −1

3. 9

5. $-\dfrac{1}{4}, \dfrac{1}{3}$

7. No solutions

9. $3, -\dfrac{1}{4}$

11. −2, 1

13. $4 - 2\sqrt{3}$

15. $\dfrac{3 + \sqrt{5}}{2}$

17. 0, 1, −1

Exercises 9.7

1. $x < -1$ or $x > 2$

3. $x \geq 3$ or $x \leq 0$

5. $x < -3$ or $x > 3$

7. All real numbers

9. $x < \dfrac{2 - \sqrt{6}}{2}$ or $x > \dfrac{2 + \sqrt{6}}{2}$

11. $x \leq 1$ or $x \geq 2$

13. $x < -2$ or $-\dfrac{4}{5} < x < 2$

15. $x < -2$ or $0 < x < 3$

17. 30 seconds, from $t = 10$ until $t = 40$

19. $-1 < x < 0$ or $x > 1$

21. $0 < x < 3$

23. $-3 < k < 3$

Exercises 9.8

1. $x = 3, y = 5; x = \dfrac{1}{2}, y = -\dfrac{5}{2}$ 3. $x = 4, y = 1; x = 1, y = -2$

5. $x = 0, y = 1; x = 3, y = -8$

7. $x = -4, y = -11; x = 3, y = -4$

9. $x = -\dfrac{1}{5}, y = \dfrac{41}{25}; x = 3, y = 33$

11. $x = 1 + \sqrt{5}, y = 1; x = 1 - \sqrt{5}, y = 1$

13. $x = 0, y = 0; x = 4, y = 64$ 15. $x = 1, y = 1$

17. 9 feet by 4 feet

Exercises 9.9

1.

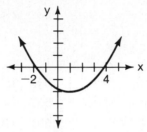

3. 6

5. $y = -(x - 1)^2 + 2$

7. $\left(-\dfrac{3}{2}, -1\right), x = -\dfrac{3}{2}$

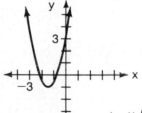

9. $(-4, 5), x = -4$

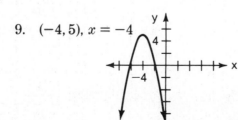

11. $(-1, -9), x = -1$

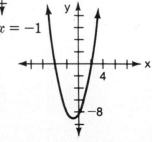

13. $\dfrac{6}{7}\sqrt{2}$

15. $4x^2y^3\sqrt{2y}$

17. $9\sqrt{2}$

19. $(x^2y + 2x - 3y^2)\sqrt{y}$

21. 4

23. $\dfrac{25 - 4\sqrt{6}}{4}$

25. $\dfrac{2}{3}$ or $-\dfrac{2}{3}$

27. $1 + \sqrt{5}$ or $1 - \sqrt{5}$

29. No real solutions

31. $\dfrac{5 + \sqrt{3}}{2}, \dfrac{5 - \sqrt{3}}{2}$

33. $\dfrac{-1 + \sqrt{2}}{2}, \dfrac{-1 - \sqrt{2}}{2}$

35. $7 + 4\sqrt{3}, 7 - 4\sqrt{3}$

37. $-2, -1$

39. $x \le -4$ or $x \ge 4$

41. $x \ge \dfrac{1}{2}$ or $x \le -3$

43. $x < -3$ or $x > 7$

45. $x = 0, y = -13; x = 4, y = 3$

47. $x = 1 + \sqrt{2}, y = -1 - 2\sqrt{2}; x = 1 - \sqrt{2}, y = -1 + 2\sqrt{2}$

49. $x^2 - 2x - 15 = 0$

51. $9x^2 - 12x + 2 = 0$

53. 0

55. $k = -3$

57. $2\sqrt{2}, -2\sqrt{2}$

59. $-3 < k < 3$

61. 8 feet by 23 feet

63. $4\dfrac{1}{2}$ feet

65. 18 feet by 12.5 feet

CHAPTER 10   *Exercises 10.1*

1. Yes

3. Yes

5. $x \ne 2, -2$

7. All x

9. 72

11. x^2yz

13. $-(x + y)$

15. $2x^2 - x$

17. $\dfrac{30}{7}$

19. $\dfrac{x^2}{y}$

21. $\dfrac{2x + 1}{x + 2}$

23. $\dfrac{x}{x - y}$

25. $\dfrac{(x - y)(x^2 + y^2)}{2x - y}$

27. $\dfrac{x^2 - xy + y^2}{x}$

29. $\dfrac{x - 2}{x(x + 2)}$

Exercises 10.2

1. $\dfrac{55}{42}$

3. $\dfrac{22}{85}$

5. ab^2

7. $\dfrac{a^2}{cd}$

9. $\dfrac{(x - y)(2x + y)}{x(x - 2y)}$

11. -1

13. $\dfrac{x + 2}{x + 3}$

15. $\dfrac{x - 1}{x + 2}$

17. x

19. a

21. a

23. $-(x^2 + 2x + 4)$

25. $\dfrac{1}{xy}$

27. 1

Exercises 10.3

1. $(x - 1)^2(x + 2)(x + 4)$

3. $x(x + 1)^2(x - 2)(x + 3)$

5. $-\dfrac{37}{36}$

7. 5

9. $\dfrac{x}{x^2 - 1}$

11. $\dfrac{x^2 + x - 1}{x + 1}$

13. $\dfrac{1}{xy}$

15. $\dfrac{a^2 + a + 1}{a^2}$

17. $x + 3$

19. $\dfrac{x + 2}{x - 1}$

21. $\dfrac{x^2}{(x - 1)(x + 1)(x^2 + x + 1)}$

23. $\dfrac{x + 1}{x - 1}$

25. $\dfrac{1}{x^2 + y^2}$

27. $\dfrac{a^2 + b^2}{2ab}$

Exercises 10.4

1. $x = 0,\ x = -1$

3. $x = 3$

5. $x = -1$

7. The graph approaches the horizontal asymptote $y = 0$ from above for large positive values of x and from below for small negative values of x.

9. The graph approaches the asymptote $y = 2x$ from below for large positive values of x and from above for small negative values of x since $\dfrac{2x^3 - 3x + 1}{x^2 + 4} = 2x + \dfrac{-11x + 1}{x^2 + 4}$.

11.

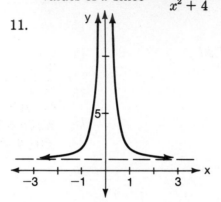

13.

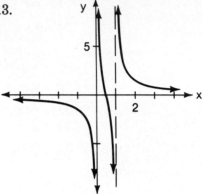

15.

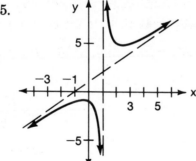

17.

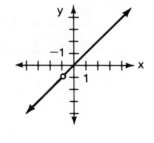

Exercises 10.5

1. $\dfrac{5}{4}$

3. $\dfrac{21}{2}$

5. $-\dfrac{1}{5}$

7. 12

9. 5

11. $-\dfrac{1}{2}$

13. No solution

15. $\dfrac{-1+\sqrt{105}}{2}, \dfrac{-1-\sqrt{105}}{2}$

17. $1, \dfrac{1}{2}$

19. $\dfrac{P-2w}{2}$

21. $\dfrac{I-P}{P}$

23. $\dfrac{6-3x}{2}$

25. 50 mph (Jim), 60 mph (Bill)

Exercises 10.6

1. $\dfrac{24}{7}$ hours

3. $\dfrac{35}{2}$ hours

5. 36 hours

7. 32 mph

9. 60 mph

11. Walk: 6.8 miles,
     Drive: 13.2 miles

13. 4 mph

Exercises 10.7

1. C varies directly with r, 2π is the constant of variation.
3. d varies directly with t^2, 16 is the constant of variation.

5. V varies directly with r^2 and h, $\dfrac{22}{7}$ is the constant of variation.

7. $P = kdD$, $d =$ depth and $D =$ density

9. $W = kmd$

11. $z = \dfrac{k\sqrt{x}}{y^3}$

13. $L = kh$, $k = \dfrac{7}{4}$

15. $y = kx^3$, $k - .5$

17. $z = \dfrac{k\sqrt{x}}{y^3}$, $k = 128$

19. 1

21. 5

23. 7.7778×10^{-8} dynes

25. 1.007326 cubic feet

27. 7.5 inches

Exercises 10.8

1. xy

3. $\dfrac{ac}{bd}$

5. $\dfrac{x(2x-1)}{(x+3)(x+1)}$

7. $\dfrac{x+3}{x+1}$

9. $\dfrac{x+6}{x(x+2)^2}$

11. $\dfrac{x+3}{x(x+2)}$

13. $x-1$

15. $x=0,\ y=1$

17. $x=-1$

19. The graph approaches the horizontal asymptote $y=1$ from above for large positive values of x and from below for small negative values of x, since $\dfrac{x^2+2x-1}{x^2-1}=1+\dfrac{2x}{x^2-1}$.

21.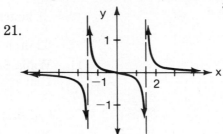

23. $-\dfrac{5}{3}$

25. 3

27. 4, 1

29. $-(x+1)$

31. Car: 40 mph,
Train: 50 mph

33. 30 mph

35. $\dfrac{40}{27}$

37. $\dfrac{15}{2}$

39. Multiplied by 4, by 9

CHAPTER 11   *Exercises 11.1*

1. 21	3. 16	5. 110.55	7. 66
9. 34	11. 37.70	13. 28.54	15. 8
17. 6.25	19. 4, 9	21. 4, 9	23. 25,920
25. 10	27. 7.5	29. 12	31. 100
33. 28			

Exercises 11.2

1. *(a)* 10   *(b)* 5   *(c)* 2.5   *(d)* 1.25

3. 5   5. 8.5   7. 19.5   9. 68.3

11.　60

13.　3.875 miles

15.　637.5 mph at 85° north of east

17.　−2.75 or 2.75

Exercises 11.3

1.　$\sqrt{29} \doteq 5.39$

3.　12

5.　5

7.　10

9.　13

11.　$\sqrt{20} \doteq 4.47$

13.　$\sqrt{20} \doteq 4.47$, $\sqrt{32} \doteq 5.66$

15.　$4\sqrt{117} \doteq 43.27$

17.　$\sqrt{16,200} \doteq 127.28$ feet

19.　$\sqrt{13,525} \doteq 116.3$ miles

21.　$\sqrt{391} \doteq 19.77$ feet

23.　$2 - \sqrt{5} \doteq -0.24$, $2 + \sqrt{5} \doteq 4.24$

Exercises 11.4

1.　$(x - 2)^2 + (y - 1)^2 = 4$

3.　$(x + 1)^2 + (y + 2)^2 = 5$

5.　$x^2 + y^2 = 25$

7.　$(-8, 0)$, $(16, 0)$

9.　Yes, right angle at (2,5)

11.　No

13.　All sides are of length $\sqrt{50}$. Adjacent sides have slopes 1 and −1 which are negative reciprocals of each other, and therefore are perpendicular.

15.　$y = \dfrac{2}{3}x + 1$, $y = -\dfrac{3}{2}x + \dfrac{15}{2}$, $(3, 3)$

17.　19.5

19.　10

21.　68

23.　9, 3

Exercises 11.5

1.　$\left(\dfrac{5}{2}, \dfrac{9}{2}\right)$

3.　$(0, 2)$

5.　$\left(-\dfrac{11}{2}, -\dfrac{7}{2}\right)$

7.　$a = 5$, $b = 6$

9.　$\left(\dfrac{14}{3}, \dfrac{11}{3}\right)$ or $\left(-\dfrac{2}{3}, -\dfrac{5}{3}\right)$

11.　(8, 7) or (−4, −5)

13.　*(a)*　Length of $\overline{AB}$ = length of $\overline{AC} = \sqrt{20}$, length of $\overline{BC} = \sqrt{40} \neq \sqrt{20}$

　　　(b)　The midpoint of $\overline{BC}$ is $D = (2, 3)$. The slope of $\overline{AD}$ is −3,

which is the negative reciprocal of the slope $\frac{1}{3}$ of $\overline{BC}$.

15.

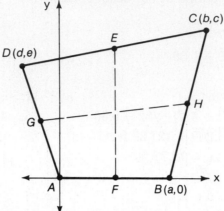

Midpoint of $\overline{DC}$ is $E = \left(\dfrac{b+d}{2}, \dfrac{c+e}{2}\right)$

Midpoint of $\overline{AB}$ is $F = \left(\dfrac{a}{2}, 0\right)$

Midpoint of $\overline{AD}$ is $G = \left(\dfrac{d}{2}, \dfrac{e}{2}\right)$

Midpoint of $\overline{BC}$ is $H = \left(\dfrac{a+b}{2}, \dfrac{c}{2}\right)$

Midpoint of $\overline{EF}$ = midpoint of $\overline{GH} = \left(\dfrac{a+b+d}{4}, \dfrac{c+e}{4}\right)$, so $\overline{EF}$ and $\overline{GH}$ bisect each other

17.

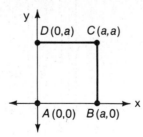

Slope of $\overline{AC} = 1$, which is the negative reciprocal of the slope -1 of $\overline{BD}$, so $\overline{AC}$ is perpendicular to $\overline{BD}$.

19. Midpoint of $\overline{CB}$ is $D = \left(\dfrac{a}{2}, \dfrac{b}{2}\right)$. The distances from D to C, D to B, and D to A are all equal to $\sqrt{\dfrac{a^2}{4} + \dfrac{b^2}{4}}$.

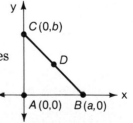

Exercises 11.6

1.   sine A $= \dfrac{12}{13}$ sine B $= \dfrac{5}{13}$

      cosine A $= \dfrac{5}{13}$ cosine B $= \dfrac{12}{13}$

      tangent A $= \dfrac{12}{5}$ tangent B $= \dfrac{5}{12}$

      cotangent $A = \dfrac{5}{12}$ cotangent $B = \dfrac{12}{5}$

      secant A $= \dfrac{13}{5}$ secant B $= \dfrac{13}{12}$

      cosecant A $= \dfrac{13}{12}$ cosecant B $= \dfrac{13}{5}$

3.   sine A $= \dfrac{20}{29}$ sine B $= \dfrac{21}{29}$

      cosine A $= \dfrac{21}{29}$ cosine B $= \dfrac{20}{29}$

      tangent A $= \dfrac{20}{21}$ tangent B $= \dfrac{21}{20}$

      cotangent $A = \dfrac{21}{20}$ cotangent $B = \dfrac{20}{21}$

      secant A $= \dfrac{29}{21}$ secant B $= \dfrac{29}{20}$

      cosecant A $= \dfrac{29}{20}$ cosecant B $= \dfrac{29}{21}$

5.   sine $30° = \dfrac{1}{2} = 0.5$ 7.   cosine A $= \dfrac{3}{5}$

      cosine $30° = \dfrac{\sqrt{3}}{2} \doteq 0.86602541$ tangent A $= \dfrac{4}{3}$

         tangent $30° = \dfrac{1}{\sqrt{3}} \doteq 0.57735027$ cotangent $A = \dfrac{3}{4}$

                                            secant A $= \dfrac{5}{3}$

                                            cosecant A $= \dfrac{5}{4}$

9. $\text{sine } A = \dfrac{6}{\sqrt{61}}$

   $\text{cosine } A = \dfrac{5}{\sqrt{61}}$

   $\text{cotangent } A = \dfrac{5}{6}$

   $\text{secant } A = \dfrac{\sqrt{61}}{5}$

   $\text{cosecant } A = \dfrac{\sqrt{61}}{6}$

11. $\text{sine } A = \dfrac{\sqrt{39}}{8}$

   $\text{cosine } A = \dfrac{5}{8}$

   $\text{tangent } A = \dfrac{\sqrt{39}}{5}$

   $\text{cotangent } A = \dfrac{5}{\sqrt{39}}$

   $\text{cosecant } A = \dfrac{8}{\sqrt{39}}$

13. $\text{sine } 22° = 0.3746$
$\text{cosine } 22° = 0.9272$
$\text{tangent } 22° = 0.4040$

15. $\text{sine } 2° = 0.0349$
$\text{cosine } 2° = 0.9994$
$\text{tangent } 2° = 0.0349$

17. $\text{sine } 88° = 0.9994$
$\text{cosine } 88° = 0.0349$
$\text{tangent } 88° = 28.6363$

19. $\text{sine } 0° = 0$
$\text{cosine } 0° = 1$
$\text{tangent } 0° = 0$

21. $0 < a < c$ so that $0 < \dfrac{a}{c} < 1$ or $0 < \text{sine } A < 1 \left(\text{since sine } A = \dfrac{a}{c}\right)$.

23. $0 < b < c$ so that $0 < \dfrac{b}{c} < 1$ and $\dfrac{c}{b} > 1$. Thus secant $A = \dfrac{c}{b} > 1$.

25. $a = \dfrac{12}{5},\ b = \dfrac{9}{5}$

Exercises 11.7

1. $\tan A = \dfrac{a}{b} = \dfrac{1}{\dfrac{b}{a}} = \dfrac{1}{\cot A}$

3. $\sin A = \dfrac{a}{c} = \dfrac{1}{\dfrac{c}{a}} = \dfrac{1}{\csc A}$

5. $c = 6.95,\ C = 90°,\ b = 4.83,\ B = 44°$

7. $A = 55.4°,\ C = 90°,\ c = 7.29,\ b = 4.14$

9. $A = 34.8°,\ B = 55.2°,\ b = 5.74,\ C = 90°$

11. $A = 65.2°,\ B = 24.8°,\ b = 3.70,\ C = 90°,\ c = 8.81$

13. $A = 13.6°,\ a = 2.66,\ B = 76.4°,\ C = 90°,\ c = 11.32$

15. $35.6°$ 17. $57.6°$ 19. $15.7°$

21. $\sin A = 0.7543$
     $\tan A = 1.1490$
     $\sec A = 1.5232$
     $\csc A = 1.3257$
     $\cot A = 0.8703$

23. $\cos A = 0.5103$
     $\sin A = 0.8600$
     $\tan A = 1.6851$
     $\cot A = 0.5934$
     $\csc A = 1.1628$

25. 19.77 feet, 81.4°

27. 72.03 feet

Exercises 11.8

1. 111.19 feet

3. 297.01 feet

5. 78.5°, 11.5°, 48.99 feet

7. 250 miles, 53.1° north of east

9. 6.94 miles

11. 27.8° north of west, 383.54 mph

13. 81.85 miles/hour or 120.05 feet/second

15. 5.4°

17. 103.55 feet (building), 24.44 feet (flagpole)

Exercises 11.9

1. *(a)* II
    (c) IV

    (b) I
    (d) III

3.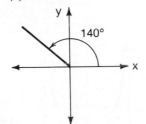

5.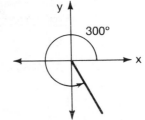

7.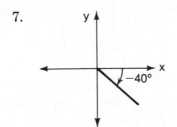

9.

11.

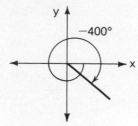

13. $\sin A = -0.4472$
    $\cos A = 0.8944$
    $\tan A = -0.5$
    $\cot A = -2$
    $\sec A = 1.1180$
    $\csc A = -2.2361$

15. $\sin A = 0.9487$
    $\cos A = -0.3162$
    $\tan A = -3$
    $\cot A = -0.3333$
    $\sec A = -3.1623$
    $\csc A = 1.0541$

17. $\sin A = 0.5547$
    $\cos A = 0.8321$
    $\tan A = 0.6667$
    $\cot A = 1.5$
    $\sec A = 1.2019$
    $\csc A = 1.8028$

19. $159.8°$

21. $-64.4°$

23. $r = 4$, $y = 2\sqrt{3} \doteq 3.46$

25. II

27. IV

29.

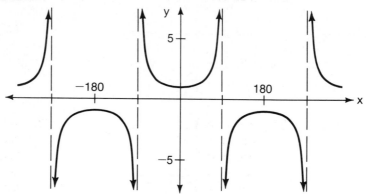

31.

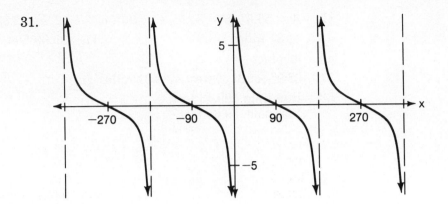

Exercises 11.10

1. $\dfrac{35}{2}$

3. 89.45

5. 166.98

7. 10

9. 52

11. *(a)* 4.875   *(b)* 2.4375

13. $\sqrt{73} \doteq 8.54$

15. $\dfrac{1}{3}\sqrt{52} \doteq 2.40$ 17. 25.95°

19. 131.8°

21. 5

23. $(-3 + 3\sqrt{5}, 0), (-3 - 3\sqrt{5}, 0)$

25. Yes, right angle at $(3, 3)$

27. $(x - 3)^2 + (y + 4)^2 = 7$

29. $\left(1, \dfrac{1}{2}\right)$

31. $a = 1, b = 4$

33. *(a)* $y = -5x + 13, \; y = x - 1, \; y = -\dfrac{1}{2}x + \dfrac{5}{2}$

   (b) They intersect in the point $\left(\dfrac{7}{3}, \dfrac{4}{3}\right)$.

35. $\cos A = 0.4408$
    $\tan A = 2.0362$
    $\csc A = 1.1141$
    $\sec A = 2.2685$
    $\cot A = 0.4911$

37. $\sin A = -0.5547$
    $\csc A = -1.8028$
    $\tan A = -0.6667$
    $\cot A = -1.5$
    $\cos A = 0.8321$
    $\sec A = 1.2019$

39. $A = 23.6°, \; B = 66.4°, \; C = 90°, \; b = 4.58$

41. $C = 90°, \; c = 11.33, \; B = 62°, \; a = 5.33$

43. $A = 55.5°$, $B = 34.5°$, $C = 90°$, $a = 4.12$, $b = 2.83$

45. 11.81 miles   47. 39.69 feet, 82.8°

49. 925 feet

51. 62.9° south of west, 483.01 miles

53. Building: 199.43 feet,
     Flagpole: 29.64 feet

Index